Bovine Health: An Issue of Veterinary Clinics

Bovine Health: An Issue of Veterinary Clinics

Edited by Katie Stevens

SYRAWOOD
PUBLISHING HOUSE
New York

Published by Syrawood Publishing House,
750 Third Avenue, 9th Floor,
New York, NY 10017, USA
www.syrawoodpublishinghouse.com

Bovine Health: An Issue of Veterinary Clinics
Edited by Katie Stevens

International Standard Book Number: 978-1-64740-074-3 (Hardback)

Cataloging-in-Publication Data

Bovine health : an issue of veterinary clinics / edited by Katie Stevens.
p. cm.
Includes bibliographical references and index.
ISBN 978-1-64740-074-3
1. Cattle--Health. 2. Veterinary hospitals. 3. Animal health. 4. Veterinary medicine. I. Stevens, Katie.
SF961 .B68 2022
636.208 96--dc23

TABLE OF CONTENTS

PREFACE

Bovine animals are the most commonly raised animals as livestock for meat, milk, dung and other products, and are often used as draft animals and riding animals. The biological subfamily Bovinae includes cattle, bison, African buffalo, the water buffalo, the yak, etc. These bovine animals need special care and treatment as they suffer from various diseases. One of the most common and serious disease found in bovine animals is bovine spongiform encephalopathy (BSE), also known as mad cow disease. Other diseases like blackleg, bluetongue, foot rot, may also be found in bovine animals. Bovine health is not only characterised as a veterinarian issue, but also a public health issue as many diseases can be transmitted to humans through milk or meat. Veterinary medicine is the sub-field of medicine which involves the prevention, diagnosis and treatment of injury, disease and disorder in animals. This book provides comprehensive insights into the field of bovine health. It also presents researches and studies performed by experts across the globe. It is a resource guide for experts as well as students.

This book unites the global concepts and researches in an organized manner for a comprehensive understanding of the subject. It is a ripe text for all researchers, students, scientists or anyone else who is interested in acquiring a better knowledge of this dynamic field.

I extend my sincere thanks to the contributors for such eloquent research chapters. Finally, I thank my family for being a source of support and help.

Editor

1

Evidence of shared bovine viral diarrhea infections between red deer and extensively raised cattle in south-central Spain

Víctor Rodríguez-Prieto[1*], Deborah Kukielka[1], Belén Rivera-Arroyo[1], Beatriz Martínez-López[1,2], Ana Isabel de las Heras[1], José Manuel Sánchez-Vizcaíno[1] and Joaquín Vicente[2]

Abstract

Background: Bovine viral diarrhea virus (BVDV) is a pestivirus that affects cattle production worldwide and that can infect other ungulates such as cervids and even wild boar (*Sus scrofa*). It is believed that domestic livestock can become infected through contact with wild animals, though it is known that infection can spread among wild animals in the absence of contact with livestock. Little is known about the sharing of BVDV infection between wild and domestic animals in the same habitat, which is important for designing eradication campaigns and preventing outbreaks, especially on hunting estates with high animal densities.

Results: We assessed the sharing of BVDV infections among hunted red deer, wild boar and cattle in south-central Spain. Sampled red deer (*Cervus elaphus*; $n = 267$) and wild boar ($n = 52$) were located on 19 hunting estates, and cattle ($n = 180$) were located on 18 nearby farms. We used ELISA kits for the serological screening, Taqman RT-PCR assay for the virus determination, and subsequent phylogenetic analysis for 17 RT-PCR positive sample amplicons. Fifty-two red deer (19.5 %) and 82 cattle (45.6 %) samples tested positive by ELISA. A high apparent prevalence (22.47 %) was obtained for red deer, while only five cattle farms tested positive by RT-PCR. Conversely, no wild boar tested positive by both ELISA or RT-PCR. Eleven red deer (4.1 %) tested positive by both ELISA and RT-PCR; these animals may have been sampled during the last phase of viremia, or they may represent previously exposed individuals infected by a different BVDV strain. The amplicons shared 92.7–100 % identity and fell within the BVDV subgroup 1b, although nine of these (from four red deer and five cattle pools) formed a separate branch. This suggests that there might be a common BVDV infecting both cattle and red deer. Higher red deer abundance was significantly associated with greater risk that extensively raised cattle would test positive for BVDV by ELISA.

Conclusions: Our findings suggest that BVDV is circulating between cattle and red deer populations in proximity, but further work is required to determine whether they share the same strain(s). These results suggest the potential of BVDV to serve as a surveillance marker in these shared habitats. High seroprevalence of BVDV in red deer from our study area suggests that although BVDV infection is common, animals usually survive the infection. Further research is needed to verify and investigate the role of red deer as a BVDV reservoir.

Keywords: Bovine viral diarrhea, Pestivirus, Taqman RT-PCR, ELISA, GLMM, Wild ungulates, Wildlife reservoirs, Spain

* Correspondence: vrprieto@ucm.es
[1]VISAVET, Veterinary School, Complutense University of Madrid, Puerta de Hierro s/n, 28040, Madrid, Spain
Full list of author information is available at the end of the article

Background

Pestivirus, a genus of RNA viruses belonging to the *Flaviviridae* family, comprises four major members: bovine viral diarrhea virus (BVDV) genotypes one and two, border disease virus (BDV) and classical swine fever virus (CSFV) [1]. Pestiviruses are characterized by affecting a wide variety of ungulates. Specifically, BVDV mainly affects cattle, causing a significant economic impact on livestock production. This disease is considered endemic in many European regions, reaching values up to 60–85 % seropositive cattle [2]. Many European countries have implemented eradication campaigns focused on removing persistently infected (PI) calves. Although PI animals usually make up only 0.5–2 % of cattle populations [2], this proportion is enough to perpetuate BVDV circulation over time, due to the high viral excretion occurring in these individuals. Recently, other modes of transmission have been proposed, such as contact between cattle and infected wild hosts [3].

Transmission of pathogens between wild and domestic hosts is an area of growing concern all over the world [4]. Many current research programs are trying to evaluate the role of wild hosts in the persistence and transmission of infections shared with domestic animals. As more information is gained, more species are designated as reservoirs of infectious diseases, widening classical host ranges. Pestiviruses, originally categorized based on their domestic host, have been shown to infect a wide range of hosts [5], suggesting interspecies transmission. In the case of BVD, serological surveys in free-ranging and captive populations have detected infection in more than 40 species [3]. This list includes a wide variety of ruminants, especially cervids, although other species such as wild boar (*Sus scrofa*) or rabbit (*Oryctolagus cuniculus*) have been proposed as carriers [6]. Thus, the interaction between domestic livestock and wild animals has been proposed as a source of BVDV infection [7, 8]. Interaction may be direct or indirect via common pasturing and shared watering areas [9]. Understanding this "sharing" of BVDV between wild and domestic animals living in proximity is key in order to design effective eradication strategies in domestic populations.

Complicating our understanding of whether and how such BVDV sharing occurs, several studies suggest that contact with livestock is not always required for BVDV infection of wild hosts. Frölich [10] found that BVDV seroprevalence rates in free-ranging deer were not associated with cattle density, suggesting that the virus can be maintained in deer populations without contact with cattle. Elazhary et al. [11] also observed high BVDV seroprevalence in free-ranging Canadian caribou (*Rangifer tarandus*) that showed no documented contact with domestic cattle for 25 years. These findings suggest that BVDV can be maintained in wild populations without transmission from cattle. BVDV spread can even be enhanced when animal densities are high, so any management measure that increases wild ruminant densities and/or aggregation may favor viral circulation within a population. Such management practices, often intended to increase hunting bags, include fencing, supplementation of feeding and watering and animal translocation [12, 13]. When not properly controlled, this human-derived management leads not only to wild species overabundance, but also to create an optimal epidemiological scenario for pathogen transmission.

In Europe, wild boar and deer constitute the major big game species. To satisfy recreational hunting demand, these species are frequently bred under more intensive conditions, but under sometimes inadequate health conditions [4]. In addition, hunting estates usually lie adjacent to farms with extensively raised livestock. In these agroforestry systems wild ungulates can easily interact with extensively raised livestock, leading to the transmission of pathogens from domestic to wild populations or vice versa.

Breeding of red deer (*Cervus elaphus*) for hunting in Europe raises the question of whether this species may serve as a significant BVDV reservoir that may help drive infections in cattle populations. Vertical BVDV transmission has been experimentally demonstrated from pregnant white-tailed deer (*Odocoileus virginianus*) to fawns, which became PI as a result [14–16]. The pregnant deer in these studies had become infected by direct inoculation [14], or through contact with PI calves [15] or PI fawns [16]. These studies suggest that red deer may play a key role in BVDV epidemiology, but further work is needed to determine the details of that role and examine whether infected red deer can transmit the virus horizontally into other animal populations sharing the same habitat. This would allow to identify areas at high risk and increase opportunities to control disease transmission.

The aims of the present study were to (i) detect BVDV circulating in extensively raised cattle and red deer cohabiting, (ii) examine the epidemiological role of wild ungulates and cattle living in proximity concerning BVD, and (iii) assess the potential for using BVDV as a health marker in habitats where wildlife (i.e. red deer and wild boar) co-exist with domestic cattle. The insights gained from this work may advance research on the interactions at the wildlife-cattle interface and aid in the design of husbandry and health policies for more cost-effectively disease control.

Results

We assessed the sharing of BVDV infections among hunted red deer, wild boar and cattle in south-central Spain. Sampled red deer ($n = 267$) and wild boar ($n = 52$)

were located on 19 hunting estates, and the cattle (n = 180) were located on 18 nearby farms. Anti-BVDV antibodies were detected in 52 deer (19.5 %; 95 % CI = 14.7–24.2) and 82 cattle (45.6 %; 95 % CI = 38.3–52.8), but not in any wild boar. Table 1 provides data on overall seroprevalence and apparent prevalence in red deer as a function of estate type (fenced or unfenced), hunting season, age, and sex.

These results were based on the preliminary screening using the indirect enzyme-linked immunosorbent assay (ELISA) and the confirmation of results by the blocking ELISA. There was complete concordance regarding the positive results between both assays. As for the doubtful samples of the indirect test (n = 40), two out of the 33 from red deer were positive and 19 remained doubtful when retested by the blocking ELISA (i.e. not included in the seropositive data). The rest of doubtful samples from red deer and cattle remained negative when retested.

An approach based on the reverse transcriptase-polymerase chain reaction (RT-PCR) was used to determine the apparent prevalence of BVDV in serum samples from red deer, wild boar and cattle, as well as in tissue samples from red deer. First, samples were screened by a one-step real-time RT-PCR (rt RT-PCR). The rt RT-PCR procedure was adapted from the one described by Baxi et al. [17]. Amplicons identified as BVDV in this assay (by comparison between the melting curve of samples and controls) were then confirmed to be positive using a Taqman RT-PCR. The threshold cycle (Ct) of positive samples ranged from 33.9 to 37.1 (median, 36.4), and these values were normally higher than the ones from the positive controls. In the end, 60 red deer (22.5 %; 95 % CI = 17.5–27.5) tested positive to BVDV1 by the Taqman RT-PCR (Table 1). Fourty-three negative samples from red deer showed Ct values higher than the cutoff (37.27–39.98; median, 38.36), while the rest showed no Ct. Four out of the 17 pools of beef cattle sera as well as two samples from bullfighting bulls were positive for BVDV1 RNA detection (Fig. 1c). Results from the diagnostic tests for cattle are shown in Table 2.

Amplicons (68 bp) from the Taqman RT-PCR assay were sequenced for 12 red deer and five cattle. Alignment with related sequences in GenBank indicated that all samples belonged to the BVDV1b subtype. The 92.7–

Table 1 Diagnostic test results for BVDV detection in red deer

Variable	Category	Total samples	No. PCR-positive (%)	No. ELISA-positive (%)
Species	Red deer	267	60 (22.47)	52 (19.47)
Region	MT	154	40 (25.97)	27 (17.53)
	SM	113	20 (17.70)	25 (22.12)
Fenced estate	Yes	135	37 (27.41)	28 (20.74)
	No	132	23 (17.42)	24 (18.18)
Hunting season	2003–04	17	4 (23.53)	9 (52.94)
	2004–05	20	2 (10.00)	8 (40.00)
	2005–06	36	0 (0)	4 (11.11)
	2006–07	29	6 (20.69)	6 (20.69)
	2007–08	46	5 (10.87)	10 (21.74)
	2008–09	30	2 (6.67)	3 (10.00)
	2009–10	89	41 (46.07)	12 (13.48)
Age	Fawn	6	1 (16.67)	1 (16.67)
	Yearling	13	2 (15.38)	1 (7.69)
	Sub-adult	26	5 (19.23)	9 (34.62)
	Adult	222	52 (23.42)	41 (18.47)
Sex	Male	133	37 (27.82)	28 (21.05)
	Female	130	21 (16.15)	23 (17.69)
	(ND)	4	2 (50.00)	1 (25.00)
PCR	Negative	207	——	41 (19.81)
	Positive	60	——	11 (18.33)
ELISA	Negative	196	46 (23.47)	——
	Doubtful	19	3 (15.79)	——
	Positive	52	11 (21.15)	——

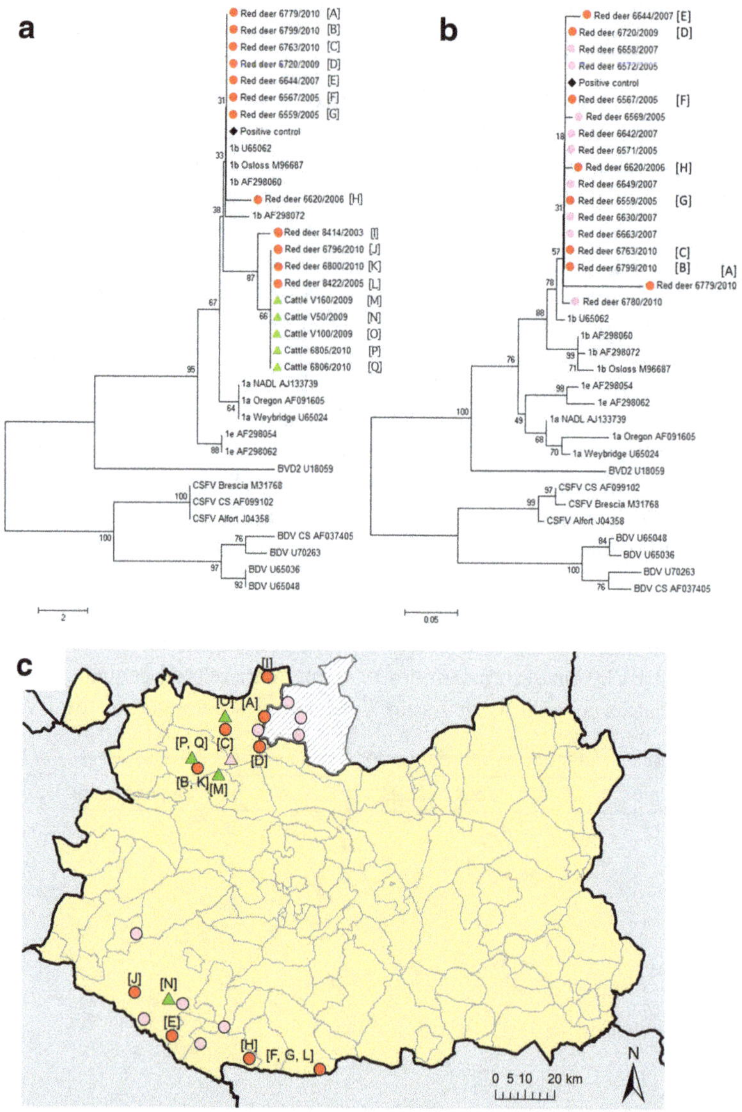

Fig. 1 Phylogenetic trees of BVDV 5'-UTR sequences identified in red deer and cattle in the study area and pestivirus reference sequences [38]. **a** Phylogenetic tree based on sequences amplified from red deer and cattle by Taqman RT-PCR. **b** Phylogenetic tree based on sequences amplified from red deer by nested RT-PCR. Trees were constructed using the neighbor-joining algorithm [45]. Bootstrap percentage values were calculated from 1000 resamplings; values over 50 % are displayed. **c** Location of samples positive by RT-PCR that were taken from red deer and cattle. A black diamond denotes the place for the bovine field sample used as a positive control, while the grey circles indicate red deer samples, and the grey triangles samples from cattle. In the phylogenetic trees, the species, identification number and year of sampling are shown under the corresponding symbol. Letters in brackets (e.g. [A]) correspond to the sampling locations shown in the map in panel C

100 % sequence identity observed among the amplicons is consistent with the fact that the RT-PCR amplified a highly conserved region in the 5' untranslated region (5' UTR). Despite the short length of the sequenced region, samples were subjected to phylogenetic analysis in order to gain preliminary insights into viral molecular patterns. However, results should be interpreted with caution. All sequenced amplicons from red deer and cattle fell within subgroup 1b (Fig. 1a), which was confirmed by performing nested RT-PCR as described by Zhong et al. [18] on a set of red deer samples (Fig. 1b). However, nine of the 17 sequenced amplicons from Taqman RT-PCR clustered in a branch separate from the rest of the amplicons; these nine samples came from red deer ($n = 4$) and cattle ($n = 5$) (Fig. 1a). This grouping was not apparently related to sampling year (Fig. 1a) or location (Fig. 1c).

Of the 18 farms included in the study, 11 (61.1 %) recorded red deer in proximity (i.e. within 2 km). The hunting bag averaged 0.24 deer per year per km^2 (maximum, 1.8). Abundance index of red deer showed a significant positive association with the risk that cattle would test positive for BVDV by ELISA ($F = 6.11$, d.f. =

Table 2 Diagnostic test results for BVDV detection in cattle

Variable	Category	No. samples tested by ELISA	No. ELISA-positive (%)	No. samples tested by RT-PCR[a]	No. RT-PCR-positive (%)
Production type	Bullfighting	10	0 (0)	10	2 (20)
	Beef cattle	170	82 (48.24)	17	4 (23.53)
Region	MT	140	77 (55.00)	10[b]; 13[c]	2[b] (20); 3[c] (23.08)
	SM	40	5 (12.50)	4[b]	1[b] (25)
Hunting season	2008–09	80	26 (32.5)	8[c]	2[c] (25)
	2009–10	100	56 (56)	10[b]; 9[c]	2[b] (20); 2[c] (22.22)

[a]Samples from the same cattle farm were pooled to perform RT-PCR because sample volume was insufficient to allow individual animal testing
[b]:Bullfighting
[c]:Beef cattle

178, $p = 0.01$, $\beta = 28.82$, 95 % CI = 6.09–51.55) (Fig. 2), although no significant relationship was found between cattle risk and red deer presence in an additional model. Female red deer showed significantly higher risk of testing positive for BVDV by ELISA or RT-PCR than did males ($F = 5.45$, d.f. = 258, $p = 0.02$, $\beta = -0{-}77$, 95 % CI − 1.43 to − 0.13; males served as the reference).

Discussion

While the role of cattle in BVDV maintenance is well established, the epidemiology of wild ruminants is not so clear and more difficult to assess. In this study, we determined seroprevalence and apparent prevalence of BVDV in populations of extensively raised cattle, red deer and wild boar living close to one another in south-central Spain. Our results suggest that BVDV is actively circulating in this area in both domestic and wild ruminants. In particular, BVDV infection in red deer seems endemic throughout the area, affecting more than 22 % animals regardless of age. Preliminary phylogenetic analysis based on 68-bp viral sequences suggests that the BVDV strains circulating among red deer and cattle may be related, though more extensive sequencing studies are required to verify this. The abundance of red deer living in proximity to extensively raised cattle significantly affects the likelihood that those cattle will test positive for BVDV. These findings suggest that wildlife-livestock interactions may contribute to the epidemiology of BVD in south-central Spain. This further supports the potential usefulness of BVDV as a health marker to monitor interactions between cattle and wildlife such as red deer.

Seroprevalence of BVDV in our study was 45.6 % in cattle and 19.5 % in red deer. This is high relative to the

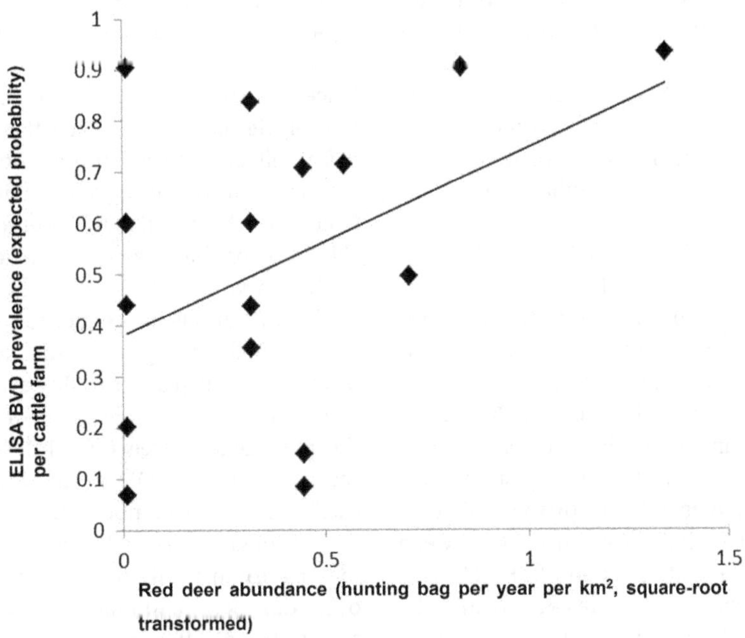

Fig. 2 Relationship between the abundance of red deer in areas adjacent to extensive cattle farms and BVDV ELISA seroprevalence in cattle in these farms. Risk was estimated as an expected probability using a GLMM and red deer abundance data were square-root transformed

prevalence of anti-pestivirus antibodies, including against BVDV, reported in deer populations in many other parts of Europe [8, 19, 20]. However, several other studies have detected seroprevalences up to 60 % in free-ranging wild ruminant populations [11], including deer species [21]. These high prevalences may indicate endemic BVDV circulation. Previous results in the literature are based mostly on ELISA, which can show false positives due to cross-reactions against other pestiviruses including BDV [22]. Therefore we confirmed our positive ELISA results in the present study using a blocking kit, which is highly recommended for its use in wildlife [8], and there was complete concordance of results.

Consistent with high BVDV seroprevalence in deer, these animals also showed 22.5 % apparent prevalence, which we consider to be reliable, given that it is based on a Taqman probe and subsequent sequencing of the PCR products. Taking both values together (seroprevalence and apparent prevalence), endemicity of BVDV in the study area is suggested. Ct values in positive animals were relatively high (# 33.9), suggesting that viral load (or the amount of RNA) found in the samples was low. Such high values may reflect non-specific amplification reactions, but this seems unlikely as the assay has been carefully optimized to reliably detect low number of viral RNA copies [17].

We did not find significant age variation in prevalence among red deer. Other studies showed that infected deer may survive until adolescence [21, 23–26]. Nonetheless, our sampling size for fawns was limited, so future research should address this age variability in detail. Apart from that, high prevalences were maintained from yearlings to adults, evidencing that infection occurs at early ages. However, prevalence in red deer showed a significant bias toward females. This sex difference may reflect, in part, differences in herd structure and behavior: female deer tend to reside at relatively high densities in larger herds, which may enhance viral circulation [27–29].

A substantial proportion of red deer [$n = 41$ (15.4 %)] was found to be positive only by the ELISA. These infections may have occurred long before the study, since anti-BVDV antibody titers can persist for years, at least in cattle [3]. In addition, a small number of red deer [$n = 11$ (4.1 %)] tested positive by both ELISA and RT-PCR; these animals may have been sampled during the last phase of viremia, when antibodies production was just starting. This finding would be likely in the case that viremia lasts longer in red deer than in cattle (which lasts between four and 15 days after the acute infection [30, 31]). Another explanation for why animals tested positive by both ELISA and RT-PCR may be that they were PI fawns that had been infected only a few weeks before. This possibility seems less likely, however, since all 11 animals were older than 6 months, and colostral antibodies in deer last as long as in cattle [32]. A third potential explanation may be related to previously BVDV-exposed individuals recently infected by a different strain.

A substantial proportion of red deer (i.e. $n = 46$, 17.2 %) tested positive only by RT-PCR, showing no detectable antibody titers. This may reflect infections still in their first week, when antibodies were not circulating yet, or it may reflect persistent infection. The first possibility is not very likely given the low probability of capturing such a large proportion of infections within the first week. The second one would imply that PI deer could survive to adulthood, considering that 41 out of these 46 animals were older than 2 years. These results highlight the need of more studies on immunity and pathogenesis of BVDV infections in deer, including persistence of viremia.

The combination of these diagnosis results and high Ct values suggest that the infections detected in the present study reflect viral circulation for at least 8 years rather than a recent outbreak of acute infections. Extended viremias might be responsible for high apparent prevalences in cattle, related to the BVDV long-term persistence in bovine white blood cells [33]. Nevertheless, these findings should be studied and verified in red deer and extended by further serological and molecular analyses, especially viral isolation, in order to clarify BVD epidemiology in the region. Preferably these future studies should be longitudinal and should examine viral infection in greater detail. The insights may help prevent BVDV spread via red deer.

Our failure to detect BVDV in wild boar by either ELISA or RT-PCR is consistent with BVDV seroprevalences of 0.6–4.5 % in the Czech Republic [34], Croatia [35] or Germany [36]. Given that our wild boar samples were collected near hunting estates where BVDV was circulating in red deer, we conclude that wild boar is unlikely to act as a BVDV reservoir. However, attention should be paid to the cross reactions against CSFV [37].

Though virus isolation has not been performed in this study, and amplicon sequences were short, the phylogenetic analysis suggested that all the samples studied belonged to subgroup 1b. However, several samples from cattle and red deer were grouped together in a branch of the phylogenetic tree (Fig. 1a). This is consistent with the idea that a BVDV strain was circulating within the cattle and red deer populations, but this must be verified through more extensive sequencing. Viruses belonging to this subgroup 1b are widespread in cattle over Europe, including Spain [38, 39]. However, it seems that a different BVDV type could be circulating only among red deer. In the case of roe deer, studies in Germany [23], Norway [40] and Switzerland [20] indicate that the infection in this wild host can be caused by wild BVD-like strains. Further studies could focus on obtaining longer sequences and more accessible information on pestiviral strains, not only considering the 5' UTR, but also other genes, such as the N-terminal autoprotease [41]. These studies are needed in order to fully understand the patterns of BVDV circulation and to see whether there is a different pathogen circulating among different wild species and to assess their potential impact in domestic livestock.

Of the 18 farms in our study, 11 (61.1 %) recorded the presence of red deer within a radius of 2 km. Indeed, in south-central Spain, deer populations often reside in and near farms with extensively raised cattle [42]. We observed that the abundance index of red deer showed a significant positive association with the likelihood that cattle would test positive for BVDV. While many factors are likely to influence the number of deer hunted per year and the hunting bag data are therefore likely to provide only a crude indicator of red deer abundance, the positive relationship between deer hunting bag and cattle BVD seroprevalence suggests that the presence of deer may affect the likelihood of BVD infection in cattle. This fact supports the molecular findings suggesting that BVDV transmission occurs at the wildlife-livestock interface. If PI individuals were common among red deer, as our immunological results suggest, it could highlight the role of red deer as reservoir for BVDV. Indeed, red deer may be a true reservoir in our study area, given that they are endemic to the area, and the risk that red deer would test positive for BVDV did not correlate with abundance of cattle or other domestic ruminants in the hunting estates.

Nonetheless, in this correlational study, we cannot infer whether red deer are responsible for transmitting BVDV to cattle. However, in the light of our results, these free-living wild hosts could act as a hidden reservoir of BVDV for domestic animals, as others have previously pointed out [7, 8, 43, 44]. A role for red deer as BVDV reservoir seems more likely in certain epidemiological scenarios, like that in our study area, where both wild and domestic ruminants share a habitat under specific management practices, which is ideal for the virus circulation and maintenance.

Conclusions

In this study we used molecular microbiology, immunology and field epidemiology methods to assess BVDV as a potential marker for monitoring the contact between wildlife and cattle in south-central Spain. Our results suggest that BVDV is infecting red deer populations, and some strains could be circulating among the wildlife. The possibility that the same BVDV strain infected the cattle and red deer in our study area highlights the relevance of red deer as a potential source of infection to cattle. Interactions between extensively raised cattle and wildlife may contribute to BVDV spill back, with implications for BVD surveillance and control programs, as well as wildlife conservation plans. More research is needed to understand how wildlife management, farming practices, habitat, and species-specific behavior and interactions influence BVDV infection in wildlife and cattle.

Methods

Study area and animal sampling

The study was carried out in the provinces of Toledo (TO) and Ciudad Real (CR) in the Castile-La Mancha region (CLM) (38° 32' 45" N to 39° 32' 45" N in latitude; 3° 54' 34" W to 4° 52' 18" W in longitude) in south-central Spain (Fig. 3). The study area is flanked by two mountain chains: Montes de Toledo (MT), located between TO and northern CR; and Sierra Morena (SM), in the southern part of CR (Fig. 3), which comprises interspersed hunting estates and farms.

Hunted red deer ($n = 267$) and wild boar ($n = 52$) were sampled from 2003 to 2010 in 19 hunting estates from CLM, comprising 9 estates in MT and ten in SM. We considered hunting an adequate method for surveying wild ungulate populations given that animals were shot randomly during hunting drives. Hunting took place from October to February. Red deer and wild boar were necropsied in detail and sex was recorded for all animals except one red deer and three wild boar. Based on tooth eruption patterns [45], red deer were classified as fawns if younger than 1 year ($n = 6$), as yearlings if between 1 and 2 years ($n = 13$), as sub-adults if between 2 and 3 years ($n = 26$), and as adults if over 3 years ($n = 222$). Wild boar were classified as piglets if younger than 6 months ($n = 1$), as weaners if between 7 and 12 months ($n = 11$), as juveniles if between 12 and 24 months ($n = 15$), and as adults if over 2 years ($n = 22$).

Blood samples were taken from the heart or thoracic cavity during field necropsies, and sera were stored at −20 °C for further analysis. In several red deer, tissue samples were also included, i.e. spleen ($n = 6$) and retropharyngeal lymph nodes ($n = 10$), and conserved at − 20 °C until the homogenization in sterile PBS in a 1:10 proportion.

Serum samples from 18 extensive cattle farms were obtained as part of regular health measures to monitor bovine tuberculosis. Ten samples from each farm were obtained and stored at − 20 °C until analysis. Since the volume of collected serum proved insufficient to allow virological analysis of individual animals, samples from all animals within one farm were pooled for that part of the study (see below).

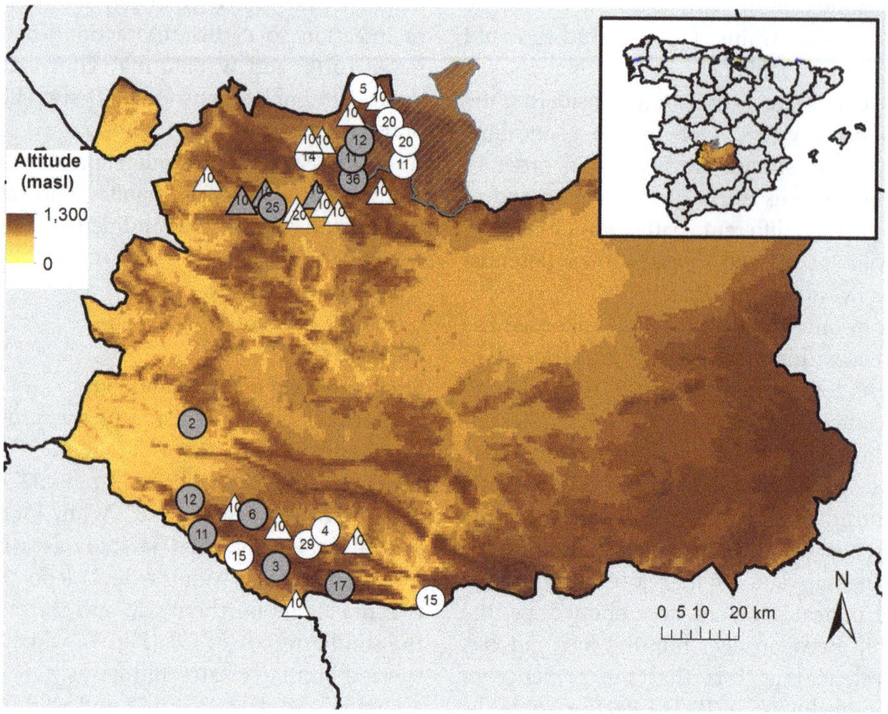

Fig. 3 Location of sampling sites in Castile-La Mancha. Most samples were collected in two of the highest areas of the region: Montes de Toledo, entailing the northern part of Ciudad Real and the Toledo municipality of Los Yébenes (grey hatched), and Sierra Morena in southern Ciudad Real. Circles denote hunting estates, and triangles denote cattle farms. The numbers inside each figure indicate the number of sampled red deer and cattle per hunting estate or farm, respectively. Figures in white indicate hunting estates lacking fence enclosures; black-outlined figures in grey, hunting estates with fence enclosures. The inset in the upper right shows the location of the study area in south-central Spain

Serological analysis

Serological analyses to detect the presence of antibodies against BVDV were performed in all animals, including red deer ($n = 267$), wild boar ($n = 52$) and cattle ($n = 180$), using an indirect ELISA (IDEXX, Liebefeld-Bern, Switzerland). This kit detects antibodies targeted against BVDV-1 and – 2 strains. A blocking ELISA (INGENASA, Madrid, Spain), which detects specific antibodies against the p80 antigen of the BVDV, was also performed in order to clarify doubtful results (seven cattle and 33 red deer), as well as confirm samples that were positive in the indirect ELISA (82 cattle and 50 red deer) . Both kits were used according to the manufacturers' instructions. Manufacturer-specified sensitivity and specificity were 96.3 and 95 % for the indirect ELISA, and 99.5 and 92 % for the blocking ELISA.

Virological analysis

Prior to the virus analyses, total RNA was extracted from serum samples using the Nucleospin® RNA virus kit (Macherey-Nagel, Düren, Germany) and from tissue homogenates using the Nucleospin® RNA II kit (Macherey-Nagel). We followed the manufacturer's recommendations to perform the extractions. Extraction products were stored at – 20 °C until further analysis.

Samples were first screened for the presence of BVDV using an rt RT-PCR assay, which was developed by adapting the RT-PCR described by Baxi et al. [17]. Our rt RT-PCR amplified a 68-bp product (106 bp including primer-derived sequences) from the 5' UTR, which is conserved among all pestiviruses, and then detected the amplicon using SYBR Green. This technique was used for a preliminary screening of samples. The optimized rt RT-PCR reaction (25 μl) contained 2 μl of RNA template, 12.5 μl of SYBR Green Master mix (Quantitative RT-PCR Brilliant II SYBR Green Master Mix), 0.0625 μl of StrataScript RT/RNase Block Enzyme Mixture (both from Stratagene, Palo Alto, CA, USA) and 400 nM of each primer. Reactions were performed in a Stratagene MxPro 3000 thermocycler (Stratagene, Palo Alto, CA, USA) using the primers and probe in Table 3. Thermal cycling conditions were as follows: 30 min at 48 °C for the RT, followed by 10 min at 95 °C for denaturalization; then 40 amplification cycles of 95 °C for 30 s, 60 °C for 60 s, and 72 °C for 30 s; and a final extension at 72 °C for 7 min. Fluorescence was measured at the extension step of each cycle. Dissociation curves were generated by increasing the temperature from 65 °C to 99 °C over 35 cycles, with 1 °C increments every 30 s.

Table 3 Primer and probe sequences for RT-PCR-based detection of BVDV

ID	Sequence (5' - 3')	Size (bp)	Amplicon size (bp)	Region amplified	Target viruses	Source
Pesti-FW	5'-CTAGCCATGCCCTTAGTAG-3'	19	106	5'-UTR	Common for Pestivirus	[17]
Pesti-RS	5'-CGTCGAACCAGTGACGACT-3'	19				
FAM-BVDV1	5'-FAM-TAGCAACAGTGGTGAGTTCGTTGGATGGCT-BHQ-3'	30			BVDV-1	

After samples were screened using rt RT-PCR, they were confirmed to be BVDV1-positive using a Taqman-based RT-PCR assay [17]. Samples whose dissociation curves were similar to the ones produced by the positive controls were especially considered for confirmation, although other samples were also included in order to check the good performance of the assays. The Taqman-based RT-PCR assay has been shown to correlate perfectly with the results of virus isolation [17]. We tested for BVDV1, since it accounts for 90 % of the BVDV isolates in Europe [8]. The final Taqman RT-PCR reaction (20 μl) contained 2 μl of RNA template, 10 μl of Kapa Probe Fast PCR Master Mix (Kapa Biosystems, Woburn, MA, USA), RT Kapa SYBR Fast One-Step qRT-PCR (Kapa Biosystems), 200 nM of each primer and the BVDV1-specific probe (Table 3). Reactions were performed in a Stratagene MxPro 3000 thermocycler. Thermal cycling conditions were as follows: 15 min at 42 °C for RT, 5 min at 95 °C for denaturalization; then 40 amplification cycles of 95 °C for 3 s and 60 °C for 30 s. The detection limit of the Taqman assay was determined to be 100 tissue culture infectious dose 50 ($TCID_{50}$), corresponding to a Ct value of 37.72 ± 0.58 [17]. We considered a sample positive when its Ct in the Taqman-based RT-PCR assay was lower than 37.1.

Controls were included from the RNA extraction through the RT-PCR to rule out contamination. As positive controls, we used two bovine field-isolated strains belonging to the BVDV1b as positive controls. As negative controls, we used sera from a bottlenose dolphin (*Tursiops truncatus*). No-template controls were included in every step and were placed in between the samples and the controls, to monitor the good management during the performance of the techniques. The controls had to perform as expected in order for the RT-PCRs to be accepted as valid.

Taqman RT-PCR products from 17 samples were confirmed by electrophoresis, using 10 μl of the RT-PCR reaction in a 3 % agarose gel (with an expected amplicon size of 106 bp), and subsequently gel-purified and sequenced.

Phylogenetic analysis

A phylogenetic tree was calculated using the sequences of the abovementioned 17 positive samples (12 from red deer and five from cattle). Since these amplicons were not long enough to create a definitive tree, we also sequenced 18 positive samples from red deer using a nested RT-PCR [18]. This assay produced an amplicon of 195 bp (159 bp excluding the primer-derived sequences) from 5' UTR.

Both sets of sequences were aligned with DNA sequences from the GenBank database using Clustal X, version 1.83. Several pestivirus reference and field sequences for BVDV1 (subgroups a, b, and e), BVDV2, CSFV and BDV were used in the alignment files [39]. Phylogenetic analyses were performed using the Kimura model of nucleotide substitution, and phylogenetic trees were constructed using the neighbor-joining method in MEGA5 software [46]. The statistical significance of the tree topology was evaluated by bootstrap re-sampling of the sequences 1000 times. The cutoff value was 50 %. The results and sequences are shown in Fig. 1.

Statistical analysis

For each farm, we collected data about location (x and y coordinates of the farm centroid) and abundance of wildlife species in nearby hunting estates (i.e. within a radius of 2 km). This radius was chosen based on the spatial distribution of red deer in extensive farms in the study area (the authors, unpublished). We calculated an index of relative abundance for red deer per study farm by calculating the overall annual hunting bag (deer hunted per year per km^2) from the nearby hunting estates during the study period. Details on converting hunting bags into an abundance index have been presented elsewhere [47]. We used a GIS layer including the perimeters of all hunting estates, which are the smallest management units. Data from hunting estates were processed using ArcGIS 9.2 (ESRI, Redlands, CA, USA).

We used a generalized linear mixed model (GLMM) to explore a possible relationship between red deer abundance near a farm (square-root transformed, explanatory variable) and the risk that cattle on that farm would test positive for BVDV by ELISA (binomial response variable). Farm and sampling season were included as random factors. Similar models were created to identify factors affecting the risk that individual red deer would test positive for BVDV by ELISA or RT-PCR. Factors included sex, age, deer abundance index and livestock density in the estate. We used hunting estate and season as random factors.

We used binomial error and a logit link in all models. SPSS 19 (Surrey, UK) was used to fit the models. All reported p values were two-sided, and p values < 0.05 were considered statistically significant.

Abbreviations

5' UTR: 5' untranslated region; bp: Base pair; BDV: Border disease virus; BVDV: Bovine viral diarrhea virus; CI: Confidence interval; CLM: Castile-La Mancha; CR: Ciudad real; CSFV: Classical swine fever virus; Ct: Threshold cycle; ELISA: Enzyme-linked immunosorbent assay; GIS: Geographic information system; GLMM: Generalized linear mixed model; MT: Montes de Toledo; PI: Persistently infected; RNA: Ribonucleic acid; rt RT-PCR: One-step real-time reverse transcriptase-polymerase chain reaction; SM: Sierra Morena; $TCID_{50}$: Tissue culture infectious dose 50; TO: Toledo.

Competing interests

The authors declare that there are no competing interests.

Authors' contributions

VRP processed the samples, performed the analyses and drafted the manuscript. DK and AIH supervised the analyses and gave substantial advice about methodology and interpretation of results. BRA and BML provided invaluable advice on technical issues. JMSV and JV supervised the inter-institutional collaboration and made important intellectual contributions to the manuscript. All authors read, edit, review and approved the final manuscript.

Acknowledgements

This work could not have been performed without the personal and technical help of Rocío Sánchez. The present work has benefited from financial aid from the following research grants from the Spanish Ministry of Economy and Competitiveness: MINECO and AGL2013-48523-C3-1-R. This is also a contribution to EU FP7 EMIDA ERA-NET project APHAEA on wildlife disease surveillance. Víctor Rodríguez-Prieto held an FPU pre-doctoral scholarship, funded by the Spanish Ministry of Education and Science. We thank the Agriculture Department of the Junta de Comunidades de Castilla-La Mancha (JCCM) and its Delegation in Ciudad Real for providing samples from health campaigns. We would like to highlight the valuable input given by the reviewers. Finally, we would like to thank Armando Chapin for his writing assistance.

Author details

[1]VISAVET, Veterinary School, Complutense University of Madrid, Puerta de Hierro s/n, 28040, Madrid, Spain. [2]SaBio (Sanidad y Biotecnología) research group, National Wildlife Research Institute IREC (CSIC - Universidad de Castilla-La Mancha), Ronda de Toledo s/n, 13071, Ciudad Real, Spain.

References

1. Flaviviridae. ICTVdb Index of Viruses. International Committee on Taxonomy of Viruses. 2002. http://ictvdb.bio-mirror.cn/Ictv/fs_flavi.htm. Accessed 4 May 2015.
2. Houe H. Epidemiological features and economical importance of bovine virus diarrhoea virus (BVDV) infections. Vet Microbiol. 1999;64(2–3):89–107.
3. Van Campen H, Frölich K, Hofmann M. Pestivirus infections. In: Williams ES, Barker IK, editors. Infectious diseases of wild mammals. Thirdth ed. Ames: Iowa State University Press; 2008. p. 232–44.
4. Gortázar C, Ferroglio E, Höfle U, Frölich K, Vicente J. Diseases shared between wildlife and livestock: a European perspective. Eur J Wildl Res. 2007;53(4):241–56.
5. Giangaspero M, Harasawa R. Species characterization in the genus pestivirus according to palindromic nucleotide substitutions in the 5'-untranslated region. J Virol Methods. 2011;174(1–2):166–72.
6. Wizigmann G. Bovine virusdiarrhoe/mucosal disease und border diaseas. Tierärztliche Umschau. 1984;39:174–8.
7. Meyling A, Houe H, Jensen AM. Epidemiology of bovine virus diarrhoea virus. Rev Sci Tech. 1990;9(1):75–93.
8. Frölich K, Marco I, Moennig V: Pestivirus infections. In: Infectious diseases of Wild Mammals and Birds in Europe. edn. Edited by Gavier-Widén D. DJP, Meredith A. Oxford: Wiley-Blackwell Publishing Ltd.; 2012: 146–167
9. Richomme C, Gauthier D, Fromont E. Contact rates and exposure to inter-species disease transmission in mountain ungulates. Epidemiol Infect. 2006; 134(1):21–30.
10. Frölich K. Bovine virus diarrhea and mucosal disease in free-ranging and captive deer (*Cervidae*) in Germany. J Wildl Dis. 1995;31(2):247–50.
11. Elazhary MA, Frechette JL, Silim A, Roy RS. Serological evidence of some bovine viruses in the caribou (*Rangifer tarandus caribou*) in Quebec. J Wildl Dis. 1981;17(4):609–12.
12. Gortázar C, Acevedo P, Ruiz-Fons F, Vicente J. Disease risks and overabundance of game species. Eur J Wildl Res. 2006;52(2):81–7.
13. Martin C, Pastoret P-P, Brochier B, Humblet M-F, Saegerman C. A survey of the transmission of infectious diseases/infections between wild and domestic ungulates in Europe. Vet Res. 2011;42:70.
14. Passler T, Walz PH, Ditchkoff SS, Givens MD, Maxwell HS, Brock KV. Experimental persistent infection with bovine viral diarrhea virus in white-tailed deer. Vet Microbiol. 2007;122(3–4):350–6.
15. Passler T, Walz PH, Ditchkoff SS, Brock KV, Deyoung RW, Foley AM, et al. Cohabitation of pregnant white-tailed deer and cattle persistently infected with Bovine viral diarrhea virus results in persistently infected fawns. Vet Microbiol. 2009;134(3–4):362–7.
16. Passler T, Ditchkoff SS, Givens MD, Brock KV, DeYoung RW, Walz PH. Transmission of bovine viral diarrhea virus among white-tailed deer (*Odocoileus virginianus*). Vet Res. 2010;41(2):20.
17. Baxi M, McRae D, Baxi S, Greiser-Wilke I, Vilcek S, Amoako K, et al. A one-step multiplex real-time RT-PCR for detection and typing of bovine viral diarrhea viruses. Vet Microbiol. 2006;116(1–3):37–44.
18. Zhong F, Li N, Huang X, Guo Y, Chen H, Wang X, et al. Genetic typing and epidemiologic observation of bovine viral diarrhea virus in Western China. Virus Genes. 2010;42(2):204–7.
19. Köppel C, Knopf L, Thür B, Vogt H, Meli M, Lutz H, et al. Bovine virus diarrhea and the vector-borne diseases anaplasmosis and bluetongue: a sero-surveillance in free-ranging red deer (*cervus elaphus*) in selected areas of Switzerland. Eur J Wildl Res. 2007;53(3):226–30.
20. Casaubon J, Vogt HR, Stalder H, Hug C, Ryser-Degiorgis MP. Bovine viral diarrhea virus in free-ranging wild ruminants in Switzerland: low prevalence of infection despite regular interactions with domestic livestock. BMC Vet Res. 2012;8:204.
21. Van Campen H, Ridpath J, Williams E, Cavender J, Edwards J, Smith S, et al. Isolation of bovine viral diarrhea virus from a free-ranging mule deer in Wyoming. J Wildl Dis. 2001;37(2):306–11.
22. Plant JW, Littlejohns IE, Gardiner AC, Vantsis JT, Huck RA. Immunological relationship between border disease, mucosal disease and swine fever. Vet Rec. 1973;92(17):455.
23. Frölich K, Hofmann M. Isolation of bovine viral diarrhea virus-like pestiviruses from roe deer (Capreolus capreolus). J Wildl Dis. 1995;31(2):243–6.
24. Duncan C, Van Campen H, Soto S, LeVan IK, Baeten LA, Miller MW. Persistent Bovine viral diarrhea virus infection in wild cervids of Colorado. J Vet Diagn Invest. 2008;20(5):650–3.
25. Chase CC, Braun LJ, Leslie-Steen P, Graham T, Miskimins D, Ridpath JF. Bovine viral diarrhea virus multiorgan infection in two white-tailed deer in southeastern South Dakota. J Wildl Dis. 2008;44(3):753–9.
26. Passler T, Walz PH, Ditchkoff SS, Walz HL, Givens MD, Brock KV. Evaluation of hunter-harvested white-tailed deer for evidence of bovine viral diarrhea virus infection in Alabama. J Vet Diagn Invest. 2008;20(1):79–82.
27. Olde Riekerink RG, Dominici A, Barkema HW, de Smit AJ. Seroprevalence of pestivirus in four species of alpine wild ungulates in the High Valley of Susa. Italy Vet Microbiol. 2005;108(3–4):297–303.
28. Martin C, Letellier C, Caij B, Gauthier D, Jean N, Shaffii A, et al. Epidemiology of pestivirus infection in wild ungulates of the French South alps. Vet Microbiol. 2011;147(3–4):320–8.
29. Pruvot M, Kutz S, van der Meer F, Musiani M, Barkema HW, Orsel K. Pathogens at the livestock-wildlife interface in Western Alberta: does transmission route matter? Vet Res. 2014;45:18.
30. Duffell SJ, Harkness JW. Bovine virus diarrhoea-mucosal disease infection in cattle. Vet Rec. 1985;117(10):240–5.
31. Brownlie J, Clarke MC, Howard CJ, Pocock DH. Pathogenesis and epidemiology of bovine virus diarrhoea virus infection of cattle. Ann Rech Vet. 1987;18(2):157–66.

32. Palfi V, Houe H, Philipsen J. Studies on the decline of bovine virus diarrhoea virus (BVDV) maternal antibodies and detectability of BVDV in persistently infected calves. Acta Vet Scand. 1993;34(1):105–7.
33. Collins ME, Heaney J, Thomas CJ, Brownlie J. Infectivity of pestivirus following persistence of acute infection. Vet Microbiol. 2009;138:289–96.
34. Sedlak K, Bartova E, Machova J. Antibodies to selected viral disease agents in wild boars from the Czech Republic. J Wildl Dis. 2008;44(3):777–80.
35. Zupancic Z, Jukic B, Lojkic M, Cac Z, Jemersic L, Staresina V. Prevalence of antibodies to classical swine fever, Aujeszky's disease, porcine reproductive and respiratory syndrome, and bovine viral diarrhoea viruses in wild boars in Croatia. J Vet Med B Infect Dis Vet Public Health. 2002;49(5):253–6.
36. Dahle J, Patzelt T, Schagemann G, Liess B. Antibody prevalence of hog cholera, bovine viral diarrhoea and Aujeszky's disease virus in wild boars in northern Germany. Dtsch Tierarztl Wochenschr. 1993;100(8):330–3.
37. Albina E, Mesplede A, Chenut G, Le Potier MF, Bourbao G, Le Gal S, et al. A serological survey on classical swine fever (CSF), Aujeszky's disease (AD) and porcine reproductive and respiratory syndrome (PRRS) virus infections in French wild boars from 1991 to 1998. Vet Microbiol. 2000;77(1–2):43–57.
38. Arias P, Orlich M, Prieto M, Cedillo Rosales S, Thiel HJ, Alvarez M, et al. Genetic heterogeneity of bovine viral diarrhoea viruses from Spain. Vet Microbiol. 2003;96(4):327–36.
39. Hurtado A, Garcia-Perez AL, Aduriz G, Juste RA. Genetic diversity of ruminant pestiviruses from Spain. Virus Res. 2003;92(1):67–73.
40. Lillehaug A, Vikoren T, Larsen IL, Akerstedt J, Tharaldsen J, Handeland K. Antibodies to ruminant alpha-herpesviruses and pestiviruses in Norwegian cervids. J Wildl Dis. 2003;39(4):779–86.
41. Toplak I, Sandvik T, Barlic-Maganja D, Grom J, Paton DJ. Genetic typing of bovine viral diarrhoea virus: most Slovenian isolates are of genotypes 1d and 1f. Vet Microbiol. 2004;99(3–4):175–85.
42. Rodríguez-Prieto V, Martínez-López B, Barasona JA, Acevedo P, Romero B, Rodriguez-Campos S, et al. A Bayesian approach to study the risk variables for tuberculosis occurrence in domestic and wild ungulates in South Central Spain. BMC Vet Res. 2012;8:148.
43. Uttenthal A, Grondahl C, Hoyer MJ, Houe H, van Maanen C, Rasmussen TB, et al. Persistent BVDV infection in mousedeer infects calves. Do we know the reservoirs for BVDV? Prev Vet Med. 2005;72(1–2):87–91. discussion 215–219.
44. Raizman EA, Pogranichniy R, Levy M, Negron M, Langohr I, Van Alstine W. Experimental infection of white-tailed deer fawns (*Odocoileus virginianus*) with bovine viral diarrhea virus type-1 isolated from free-ranging white-tailed deer. J Wildl Dis. 2009;45(3):653–60.
45. Sáenz de Buruaga M, Lucio AJ, Purroy FJ: Reconocimiento de sexo y edad en especies cinegéticas. Vitoria-Gasteiz: Edilesa (Ediciones Leonesas S.A.); 1991.
46. Tamura K, Dudley J, Nei M, Kumar S. MEGA4: molecular evolutionary genetics analysis (MEGA) software version 4.0. Mol Biol Evol. 2007;24(8): 1596–9.
47. Martínez-Lopez B, Barasona JA, Gortazar C, Rodríguez-Prieto V, Sánchez-Vizcaíno JM, Vicente J. Farm-level risk factors for the occurrence, new infection or persistence of tuberculosis in cattle herds from South-Central Spain. Prev Vet Med. 2014;116(3):268–78.

2

Pestivirus infection in cattle dairy farms: E2 glycoprotein ELISA reveals the presence of bovine viral diarrhea virus type 2 in northwestern Italy

Chiara Nogarol[1], Nicola Decaro[2], Luigi Bertolotti[1], Barbara Colitti[1], Bryan Iotti[1], Stefano Petrini[3], Maria Stella Lucente[2], Gabriella Elia[2], Giovanni Perona[1], Margherita Profiti[1], Canio Buonavoglia[2] and Sergio Rosati[1,4*]

Abstract

Background: Bovine viral diarrhea virus (BVDV) types 1 and 2 are members of the *Pestivirus* genus of the Flaviviridae family. This genus also includes the HoBi-like virus, tentatively classified as BVDV type 3. BVDV-1 is widely distributed in Italy despite the extensive use of BVDV-1-based vaccines, while BVDV-2 and HoBi-like *Pestivirus* have been detected occasionally. Monitoring the occurrence of sporadic or atypical pestiviruses is a useful approach to evaluate the need for additional vaccine strains that can be used in BVDV control programs.

Results: In this study we developed a multiwell antibody ELISA based on the recombinant E2 protein of the three bovine pestiviruses. We evaluated the assay's applicability for surveillance purposes using pooled milk samples, each prepared from a maximum of 35 lactating cows and collected from 176 dairy herds. As expected, the majority of the pooled samples reacted to a greater extent against the BVDV-1 E2 antigen. All three milk pools from a single farm reacted to the BVDV-2 antigen, however. Further analysis using spot tests, antigen detection, and sequence analysis of the 5′-UTR region confirmed the presence of five persistently infected calves carrying a BVDV-2a strain.

Conclusions: This study highlights for the first time that sporadic circulation of BVDV-2 can be predicted by immunoenzymatic methods in the absence of specific vaccination.

Keywords: Bvdv-1, Bvdv-2, HoBi-like *Pestivirus*, E2 glycoprotein, Bulk milk ELISA

Background

Bovine viral diarrhea virus (BVDV) types 1 and 2 belong to the *Pestivirus* genus of the Flaviviridae family, along with border disease virus (BDV) and classical swine disease virus [1]. Other atypical pestiviruses include HoBi-like virus, tentatively classified as BVDV type 3 [2, 3], and wild ungulate pestiviruses. The severity of clinical signs is strictly associated with the viral strain and the type of infection. In most cattle infected transiently with BVDV, disease signs are mild, characterized by low-grade fever, diarrhea, and coughing [1, 4]. Infection is sustained by persistently infected (PI) carriers that acquire the infection with a non cytopathic (NCP) strain early in the fetal stage and remain immunotollerant virus shedders for life. The lifespan of PI animals is usually short due to early culling or the development of mucosal disease (MD), a fatal outcome arising from mutation of the NCP strain to its cytopathic (CP) counterpart.

The distribution of BVD virus types and subtypes is continuously evaluated in some European countries. A well-established procedure of sequencing the 5'UTR viral fragment from PI animals has been widely used, allowing characterization of the circulating viral strains [5]. BVDV-1 is detected with higher frequency than BVDV-2, but BVDV-2 outbreaks have been frequently

* Correspondence: sergio.rosati@unito.it
[1]Department of Veterinary Science, University of Turin, Turin, Italy
[4]Largo Paolo Braccini, 2, 10095 Grugliasco, Italy
Full list of author information is available at the end of the article

associated with severe acute disease [6–8]. Up to 17 BVDV-1 subtypes have been detected in Italy over the course of the last 18 years despite the regular use of BVDV-1 vaccines for nearly two decades. The hypothesis is that some of the newly described subtypes may escape control through herd immunity [9]. In contrast, BVDV-2 is only partially neutralized by BVDV-1-induced immunity and has seldom been identified [7, 10].

Circulation of unusual or atypical pestiviruses has been linked to iatrogenic transmission, including the preparation of vaccines contaminated by NCP strain of BVDV type 2 [11]. The origin of the infection has not been clearly established in some recent reports, for example, the single outbreak of HoBi-like *Pestivirus* in southern Italy [3]. Circulation of BVDV strains that may either not or only partially be covered by vaccine-induced herd immunity raises major concerns [12], since the impact of infection is expected to be similar to that of a naïve population. In this light, monitoring of possible introduction of BVDV-2 and HoBi-like *Pestivirus* could provide a valuable tool to assess the effectiveness of BVDV-1-based vaccination programs and prevent the spread of viral strains. Unfortunately, genetic classification of BVDV strains in the course of PI identification is not routinely carried out by diagnostic laboratories, and analysis of BVDV isolates is often obtained years after the identification of PI animals, limiting investigation into risk factors for the introduction of such strains.

To our knowledge, the only serological diagnostic tool that can differentiate BVDV infections is the cross-seroneutralization test [13]. The E2 glycoprotein is the most divergent antigen between BVDV type 1, 2, and HoBi-like *pestivirus* and it is able to elicit neutralizing antibodies [14, 15]. Accordingly, we developed a multiwell antibody ELISA based on the E2 protein of the aforementioned viruses and evaluated, using pooled milk samples collected from dairy herds, the test's applicability for surveillance purposes. A background of BVDV-1 infection and/or immunization clearly emerged, with greater reactivity against the BVDV-2 antigen noted in a single dairy herd. Further investigation using spot tests (sample of animals of a certain age [16]), individual milk samples, and antigen/RNA detection allowed the identification of several PI animals harboring the BVDV-2a subtype.

Results

Sheep immunized with the inactivated whole virus preparation of BVDV type 1, 2, and HoBi-like *pestivirus* developed a strong immunoresponse against E2 viral glycoprotein (average of homologous virus-neutralization [VN] titers ranged from 445 to 2702). The homologous VN titer was at least three times greater than the heterologous VN titer (Table 1). The three control sera were normalized by diluting the two most reactive sera into

Table 1 Cross-neutralization titer induced in sheep by inactivated preparation of BVDV-1, BVDV-2, and HoBi-like *Pestivirus*, based on 5 independent replicates. Range of titers is reported within brackets

	SN titers		
Serum	Against BVDV-1	Against BVDV-2	Against HoBi-like
#8135 BVDV-1 immunized	**1782.89 [1024–2048]**	48.50 [32–64]	97.01 [64–128]
#9750 BVDV-2 immunized	1.74 [0–2]	**445.72 [256–512]**	27.86 [16–32]
#8148 HoBi-like immunized	97.01 [64–128]	776.05 [512–1024]	**2702.35 [2048–4096]**

negative bovine serum to attain a homologous VN titer equal to the less reactive serum. Checkerboard titration was performed to evaluate antigen expression levels. The results showed a large difference in ELISA reactivity, leading to the assumption that expression levels differed widely, with BVDV type 1 E2 being 100 times more reactive than the HoBi-like one. BVDV-1 antigen was diluted to 1:900, BVDV-2 to 1:150, and Hobi-like to 1:8. Any attempt to improve the signal of the latter antigen, including the use of a baculovirus expression system, has been unsuccessful so far (data not shown).

In order to better evaluate the test performances, a set of 51 negative sera from 8 different farms was used to estimate the cut-off of acceptance (OD cut-off 0.5). In the other hand, sera collected from animals (bovine or rabbits) transiently infected or immunized with a defined BVDV strain were tested and found to be more reactive versus homologous BVDV antigen, although some were raised in response to different clades within the same viral species (Fig. 1). This sample set allowed us to set a discrimination cut-off at 1.42, representing the ratio between reactivity against the homologous and the heterologous antigens.

Sera from immunized sheep always reacted to a greater extent against the E2 protein corresponding to the homologous immunogen (Additional file 1: Figure S1) when used as positive controls in 17 test runs (i.e., ELISA plates). Based on the 17 ELISA runs, each control showed significantly higher reactivity against the homologous antigen as compared to the highest heterologous antigens. For each positive control serum, reactivity against the homologous antigen was compared to the highest OD obtained when the serum was tested against the heterologous antigens. The Wilcoxon signed rank test showed significant differences ($p < 0.001$ in all three cases), with the median ratio higher than 1.42 in all three (BVDV1 6.55, BVDV2 5.76, BVDV3 2.25).

The method was then applied to a total of 436 milk pools collected from 179 dairy farms: the number of pools per farm ranged between 1 and 9. On the basis of the proposed method of classification, 33 farms showed only negative

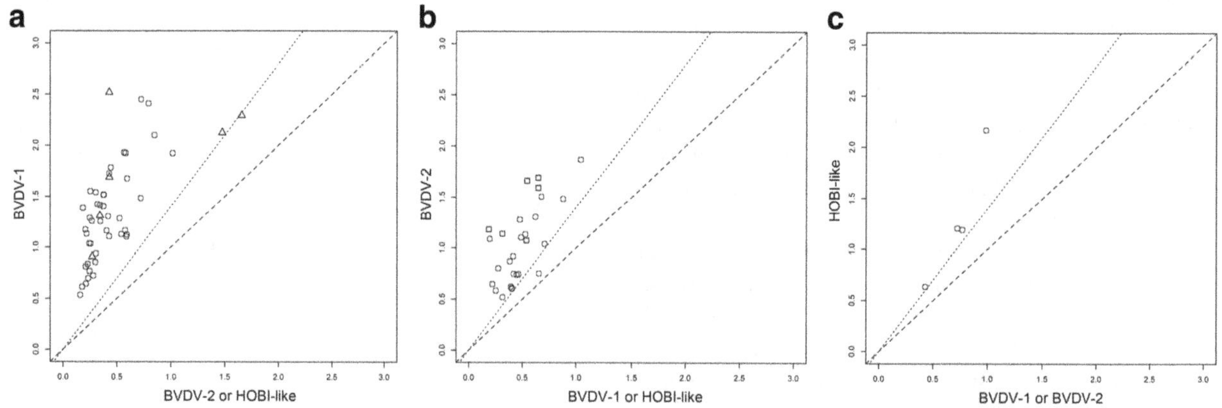

Fig. 1 Dispersion plot of ELISA reactivity using sera from animals transiently infected or immunized with a BVDV-1 (panel **a**), BVDV-2 (panel **b**), and HOBI-like strain (panel **c**). Circles: bovine sera; triangles: rabbit sera; squares: bovine sera from experimentally infected animals. Reactivity against homologous antigen (Y axis) and the most reactive heterologous antigen (X axis) are shown. Dashed lines represent equal reactivity; dotted lines represent cut-off discrimination

pools ($n = 125$). Further investigation revealed that no vaccination programs were in place on any of the farms during sample collection. A total of 111 pools, from 63 farms, showed the greatest reactivity against the BVDV type 1 antigen. A few milk pools from 3 farms reacted against BVDV-2 antigen. In particular, 3 pools from a single farm showed clear reactivity on E2 ELISA (Fig. 2a). Questioning farmers about their animal management practices revealed that a double-strain live vaccine, based on BVDV-1 and 2 strains, was used on the remaining 2 farms. Fourteen pools showed reactivity against HoBi-like *pestivirus* just above the level of discrimination (data not shown). The BVDV type could not be ascertained in 177 pools.

The farm with clear BVDV-2 reactivity was further investigated. Testing of individual young animals that had never been vaccinated confirmed that most were reactive against the type 2 antigen (Fig. 2b). Five PI animals were identified in this group. Sequences of the 5′-UTR region obtained from two of the animals were 100% similar; based on the similarity found between the newly obtained sequence and the sequences available in GenBank database and the tree topology, the analyses confirmed that the animals carried the BVDV-2a strain (GenBank accession number KU878975, Fig. 3). No PI animals were found among lactating cows.

Discussion

Circulation of BVDV-2 in Italy was occasionally documented between 1995 and 2004 [17, 18]. Its genetic characterization accounts for less than 3% of the total

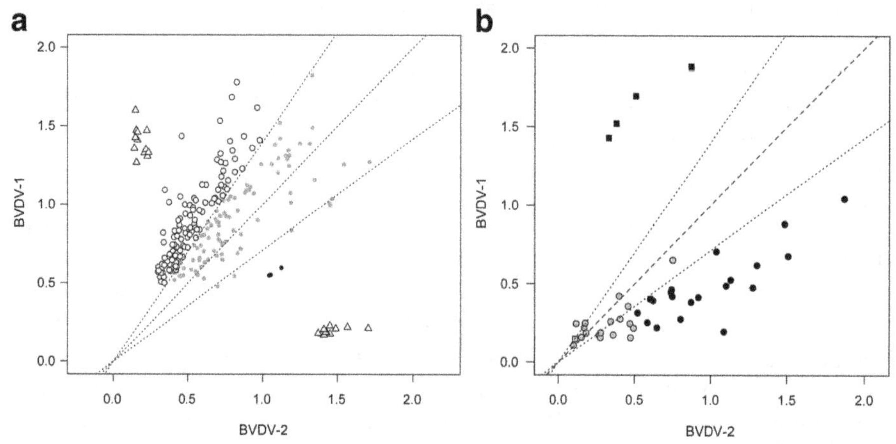

Fig. 2 Dispersion plot of ELISA reactivity versus BVDV-1 recombinant E2 (Y axis) and BVDV-2 recombinant E2 (X axis). Diagonal and envelope lines represent equal reactivity and cut-off discrimination, respectively. Panel **a**: Reactivity of pooled milk samples. Strain specific control sera are shown as white triangles. White dots denote pool milk samples reactive against BVDV-1 antigen. Grey dots denote indeterminate samples. Black dots denote pools from a single dairy farm reactive against BVDV-2 antigen. Panel **b**: Spot test on young animals. Grey dots denote negative or indeterminate samples. Black dots denote individual serum samples reactive against BVDV-2 antigen. Black squares denote individual serum samples from a BVDV-1 positive farm

pestivirus sequences retrieved from the literature. Due to its high sensitivity, RT-PCR has become a cost-effective diagnostic tool for PI identification that is applicable to pooled blood/milk samples. Sequence analyses of positive samples represent an additional cost, which is why they are rarely performed, thus limiting strain characterization at the time of diagnosis. The only genetic information available in many countries comes from periodical screening of stored PCR positive samples. In all cases, atypical or sporadic strains have been detected following PI identification. To our knowledge, this is the first immunoenzymatic assay that can differentiate BVDV species based on seroreactivity against the most variable E2 glycoprotein. The ELISA format was successfully employed on both serum samples and individual or pooled milk samples, demonstrating its use as a cost effective tool for rapid screening of dairy herds. Since pestiviruses display considerable heterogeneity within one species, we first analyzed a panel of sera raised against defined BVDV-1 and BVDV-2 strains and then evaluated the ability of each of the E2 antigens to specifically recognize homologous infection (at the

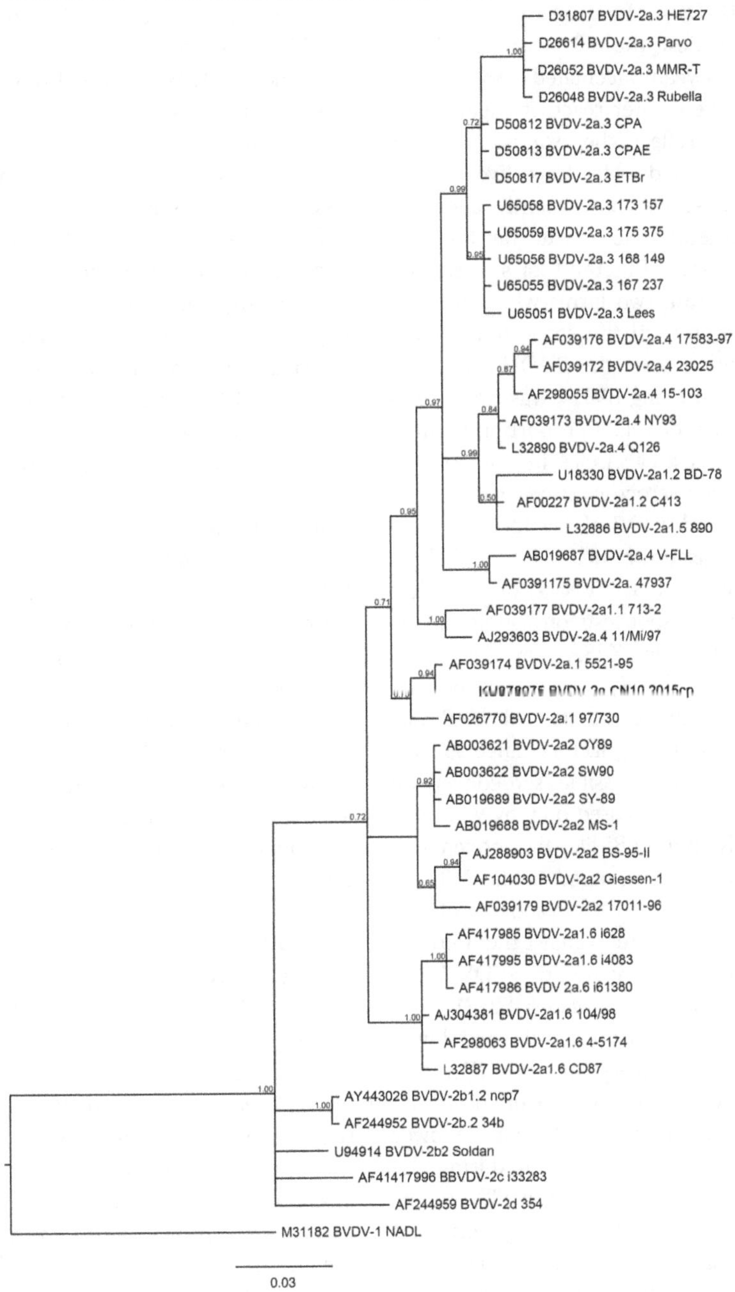

Fig. 3 Bayesian phylogenetic tree. New taxa obtained in this study are given in bold. Posterior probability of each node is given above the branches

species level). The results clearly suggest that discrimination is possible even for sera raised against BVDV species of different subtypes.

It should be noted that the proposed method has been roughly validated for sensitivity, specificity, and discrimination power. A recent study reported that the frequency of detection of BVDV-2 and HoBi-like viruses among 371 UTR sequences was 2.7% and 0%, respectively [19]. The identification of a single BVDV-2 positive herd in the present study is in line with a sporadic level of circulation. On the other hand, a large part of the milk pools reacted against the BVDV-1 antigen, supporting the high frequency of detection of BVDV-1. Because the E2 protein elicits neutralizing antibodies, the E2-based ELISA is unable to discriminate between vaccinated and infected animals. In this context, the greater reactivity against the type 1 antigen may also reflect the extensive use of BVDV-1 vaccination. Since a double-strain live vaccine based on BVDV-1 and 2 strains has recently been approved for the Italian market, its use may further limit the future usefulness of the proposed method for surveillance purposes. The milk pools from two farms where double-strain vaccination had been recently implemented were, in fact, slightly more reactive against the BVDV-2 antigen. Any attempt to identify circulating BVDV was unsuccessful. However, we cannot exclude that a proportion of indeterminate results (shown as grey circles in Fig. 2) may reflect the outcome of mixed seroconversion against both a BVDV-1-based vaccine and a BVDV-2 wild-type strain or against a double-strain based vaccine.

In this scenario, circulation of BVDV-1, 2, or HoBi-like *Pestivirus* may be predicted by spot tests on young, unvaccinated animals, while bulk milk ELISA may serve as a marker for herd immunity following vaccination. The serological identification of the circulation of HoBi-like *Pestivirus* is still free of confounding factors, since specific vaccines are not available, and possible seroconversion against Hobi like antigens can be caused only by wild type viral strains. Unfortunately, the HoBi-like E2 antigen was expressed at very low levels as compared to its BVDV-1 and BVDV-2 counterparts. The lower reactivity of the strain specific HoBi-like control serum against the homologous antigen (median OD 0.75), compared to BVDV-1 and BVDV-2 control sera (median OD 1.5 against E2 BVDV-1 and E2 BVDV-2 antigens, respectively), as well as its lower discriminating power (supplementary material S1), was probably due to a lower level of expression of the transfected cells. Furthermore, we cannot exclude that the lower dilution factor (1:6) of the HoBi-like antigen during the plate coating step might also have influenced ELISA performance. For this reason, we considered as undetermined the 14 pools weakly reacting against HoBi-like antigen; this result may reflect unspecific reactivity against cellular proteins rather than against the E2 antigen (data not shown).

Sequence analysis of BVDV-2 strains isolated from PI animals belonging to the same herd indicated that the strains were all subtype 2a. This subtype was identified in northern Italy more than 20 years ago [7]. It is difficult at this stage to speculate how low levels of circulation in the absence of specific vaccination could have persisted for such a long time. The partial sequence clustered together with strains identified in the United States [20], falling into the BVDV-2a.1 clade. Unfortunately, a clear picture of the distribution and the genetic heterogeneity of BVDV-2 is not available. For this reason, we can merely hypothesize that a recent introduction through biological material and/or animal trade could be responsible for the presence of the BVDV strain identified. Extensive animal trading with European countries (mainly France and Germany) in the last 3 years was recorded for this herd, further supporting the role of this risk factor in the introduction of unusual *Pestiviruses*. It is noteworthy that the herd has been repeatedly vaccinated with the BVDV-1-based inactivated vaccine, so that antibody reactivity of lactating cows against the homologous E2 glycoprotein could be expected. Though vaccination may represent a bias when interpreting the data, outbreaks of infection with different BVDV species have occurred, given the strong immune response the infected animals developed, especially under the infective pressure of the PI animals.

Conclusions

This study provides evidence for how serological surveillance of BVDV can provide a valuable tool for the prompt identification of atypical *Pestivirus*. Serological screening to differentiate BVDV infections can be used for monitoring the introduction and circulation of different *Pestivirus* species. Moreover, serological investigation is a very easy way to follow and control vaccination protocols, especially if different BVDV species are used as the vaccine strain. This approach can improve the surveillance system and help identify possible risk factors associated with atypical BVDV infections.

Methods

Viruses and cells

The BVDV-1 NADL (courtesy of Dr. Ferrari, Istituto Zooprofilattico Sperimentale di Lombardia ed. Emilia Romagna, Brescia, Italy), BVDV-2232/02 [17], and HoBi-like *Pestivirus* Italy-83/10cp [21] cytopathogenic strains were used for immunization procedures (as inactivated preparations) and for the virus neutralization (VN) assays to assess the VN titer in immunized animals. All viruses were propagated in pestivirus-free Madin-Darby bovine kidney cell lines (MDBK; ATCC CCL-22), cultured in Dulbecco's modified essential medium (DMEM) (Sigma-Aldrich, Schnelldorf, Germany) containing 10%

fetal bovine serum (FBS, Gibco, Thermo Scientific, Rodano, Italy, tested negative for pestivirus or anti-BVDV antibody), 2 mM of L-glutamine, 100 IU/ml of penicillin (Sigma-Aldrich), 100 mg/ml of streptomycin (Sigma-Aldrich), and 2.5 mg/ml of amphotericin B. Cells were incubated at 37 °C with 5% CO_2. Human embryo kidney cell lines (HEK293T; ATCC, CRL-1573) were cultured following the same procedure and used for transfection experiments.

For immunization, viral suspensions were inactivated with 0.05% β-propiolactone and emulsified initially with Tween-80 (4.1% *v*/v) and then with a mixture (50% v/v) of Montanide ISA 563 (Seppic Inc., Paris, France), Marcol 52 (Esso Italiana S.r.l., Rome, Italy), and Montane 80 (Seppic Inc.) in a 30:63:7 ratio, as previously described [22].

Strain specific sera

To obtain strain-specific control sera in the ELISA procedure, three female, *Biellese* breed sheep were selected from the *Pestivirus*-free flock belonging to the Veterinary Teaching Hospital at the Department of Veterinary Science, University of Torino. The animals were immunized with the inactivated preparations described above. Each sheep received two inoculations of 2 ml administered 28 days apart. All the animals were housed at the Veterinary Teaching Hospital under standard husbandry conditions. The blood was drawn on day 42 after anesthesia followed by euthanasia. Experiments were carried out at the Veterinary Teaching Hospital laboratories and in compliance with the relevant national legislation on experimental animals and animal welfare (Italian Ministry of Health DL 116/92), upon authorization by the competent authority (Italian Ministry of Health Aut Req 26,062,013). Virus neutralization (VN) titer against homologous and heterologous immunogens was determined for each serum by cross-seroneutralization (5 technical replicates for each serum). Sera were collected and freeze-dried in 5-ml aliquots for long-term storage.

BVDV negative samples were collected from 8 BVDV-free farms for setting the acceptance cut-off. All the samples showed negative results when tested with a commercial IDEXX Ab ELISA test. In addition, sera collected from animals transiently infected with a defined BVDV strain were retrieved from previous studies. The sample set included 44 sera from 3 BVDV-1b positive farms, 30 sera from a BVDV-2a positive farm, 6 sera from a BVDV2c experimental infection (CODA-CERVA, Belgium), and 4 sera from a BVDV-3 positive farm [2].

To take into account the considerable heterogeneity even within one species, such as BVDV 1, sera from rabbits immunized with different members of the clade (BVDV1b, d, g, h, and e) were kindly provided by Gian Mario Demia (Istituto Zooprofilattico Sperimentale di Umbria e Marche, Perugia, Italy). Strain-specific sera were used to identify the best parameters for interpretation of the results.

Gene fragment, cloning and eukaryotic expression

Amino acid sequences, corresponding to the ectodomain of the E2 protein of BVDV-1 (strain NADL), BVDV-2 (strain 1373), and HoBi-like *Pestivirus* (Italy-83/10cp) were retrieved from the GenBank database (accession number CAB91847, AAD38683 and AFL65619.1). Synthetic genes (Eurofins MWG, Ebersberg Germany) were constructed carrying HindIII and XhoI restriction sites to facilitate unidirectional cloning. Genes were subcloned into pSecTag2/Hygro plasmids (Invitrogen, Waltham, MA, USA), allowing the extracellular sorting and secretion in the medium of transiently transfected mammalian cells. These plasmids were used to transform chemically competent *Escherichia coli* (strain BL21) by heat shock. Ampicillin-resistant colonies were subjected to PCR for rapid screening and sequencing to confirm the authenticity and in-frame insertion of each fragment. Endotoxin-free plasmid purification from 500 ml of LB culture was carried out using NucleoBond® Xtra Midi/Midi Plus (Macherey-Nagel, Düren, Germany). Sub-confluent human embryonic kidney (HEK293T) cells, cultured in 75-cm^2 flasks, were transfected with 9 μg of plasmid and 21 μl of LTX transfection reagent (Thermo Fisher Scientific, Waltham, MA, USA) in 6 ml of DMEM [23]. After 6 h at 37 °C with 5% CO_2, the transfection medium was replaced with 6 ml of protein-free medium (ExCell293, Sigma-Aldrich) and the flasks were incubated as described above for an additional 42 h. The medium was then collected, centrifuged at 5000 g for 10 min to remove cell debris, and stored at −80 °C until use. The amino acid sequences of the three E2 proteins used in the present study are shown in Fig. 4.

Farms, milk and serum samples

During a previous study on infectious bovine rhinotracheitis, milk samples were collected from 179 dairy farms located in the Province of Cuneo, Piedmont, in collaboration with the Regional Farmers' Association (Associazione Regionale Allevatori, ARA). Written consent of the milk collection for research purposes for was obtained from all farmers. During that study, individual milk samples were collected from each farm, forming a total of 436 milk pool samples from about 35 lactating animals. The same samples were used for the present work. At the time of sample collection, the BVDV status and vaccination programs for the dairy farms were unknown.

Following milk testing, we performed additional testing on one of the farms that had displayed greater reactivity against the E2 antigen of BVDV-2. Individual blood and milk samples were collected from young unvaccinated animals and lactating cows, respectively,

```
BVDV-1     QGHLDCKPEFSYAIAKDERIGQLGAEGLTTTWKEYSPGMKLEDTMVIAWCEDGKLMYLQR
BVDV-2     QGFPECKEGFQYAISKDRRMGLLGPESLTTTW--HLPTKKIVDSMVHVWCEGKDLKILKM
HoBi-like  HGDLSCKPEFQYAISETDEIDLTGPTGLTTRWYAYSERLHITDSSVDLTCVDGNFLVHRR
           :*  .**  *.***::  .:.  *. .*** *  :    :: *: *   * . .:   :

BVDV-1     CTRETRYLAILHTRALPTSVVFKKLFDGRKQEDVVEMNDNFEFGLCPCDAKPIVRGKFNT
BVDV-2     CTKEERYLVAVHERALSTSAEFMQISDGTIGPDVIDMPDDFEFGLCPCDSKPVIKGKFNA
HoBi-like  CVRKRRYLATIHERALSTSVKFTLVAN-PQDIEDVQMGDDFEFGLCPCDATPVIRGKFNT
           *.:: ***. :* ***.**. *  : :     : ::* *:*********:.*:::****:

BVDV-1     TLLNGPAFQMVCPIGWTGTVSCTSFNMDTLATTVVRTYRRSKPFPHRQGCITQKNLGEDL
BVDV-2     SLLNGPAFQMVCPQGWSGTIECTLANQDTLDTTVIRTYRRTTPFQRRKWCTYEKIIGEDV
HoBi-like  TLLNGPAFQLVCPYGWTGTVECTVVSDSTLKTQVVKRFARYKPFPHRKHCMDQMVVGEDL
           :********:*** **:**:.**  . .** * *:: : * .** :*: *  :  :***:

BVDV-1     HNCILGGNWTCVPGDQLLYKGGSIESCKWCGYQFKESEGLPHYPIGKCKLENETGYRLVD
BVDV-2     YECILGGNWTCITGDHSRLKDGPIKKCKWCGYDFVNSEGLPHYPIGKCMLINESGYRYVD
HoBi-like  YKCLYGGNWTCIPGDRVLYQGGKVKDCKWCGFIFEEPSDLPHFPLGKCRLTNETGYRYVD
           ::*: ******:.**:   :.* ::.*****: * :...***:*:*** * **:*** **

BVDV-1     STSCNREGVAIVPQGTLKCKIGKTTVQVIAMDTKLGPMPCRPYEIISSEGPVEKTACTFN
BVDV-2     DTSCDRGGVAIVPSGTVKCRIGNVTVQVIATNNDLGPMPCSPAEVIASEGPVEKTACTFN
HoBi-like  DTTCNRDGVAITEQGTLRCKIGKVEVRVLALNKDLGPMPCKPSHVTQSEGPVSKTACTFN
           .*:*:* ****. .**::*:**:. *:*:* :..****** * .:  *****.*******

BVDV-1     YTKTLKNKYFEPRDSYFQQYMLKGEYQYWFDLEVTDHHRDY
BVDV-2     YSRTLPNKYYEPRDRYFQQYMLKGEWQYWFDLDSVDHHKDY
HoBi-like  WTETLENKYFEPRDNYFQQYMLKGKYQYWFDLEATDHHQDY
           ::.** ***:**** *********::******: .***:**
```

Fig. 4 Amino acid alignment of BVDV-1, BVDV-2 and HoBi-like E2 glycoprotein ectodomains expressed in recombinant form in the present study. Conserved cysteine residues are given in bold. Asterisks denote identical amino acids, colons (:)conserved substitutions, and dots (.)semi-conserved substitutions

and further analyzed by spot test (5–10 individual tests per group) and for PI identification [16].

ELISA procedure

All control sera were normalized to a dilution corresponding to a VN titer of 430. Due to the different levels of expression of the three E2 proteins, the optimal dilution of each antigen was carried out using control sera by checkerboard titration. Antigens were diluted in 0.1 M carbonate/bicarbonate buffer, pH 9.6, and used to coat Nunc Maxisorp plate wells overnight at 4 °C.

After blocking non-specific sites, the milk or serum samples were assayed undiluted or diluted 1/20 in PBS 1.25% casein, respectively. Samples were incubated for 1 h (blood serum and controls) or 2 h (milk) at room temperature. After washing, peroxidase labeled protein G, diluted to 10 ng/ml in PBS 1.25% casein, was added to the wells and the plates were incubated for 45 min at room temperature. After the final wash, the reaction was developed with 3,3′,5,5′-tetramethylbenzidine (TMB) and stopped with 0.2 M H_2SO_4. Reactivity of the monospecific control sera against the three antigens was evaluated in 10 independent tests and classified depending on the immunogen (supplementary material S1).

Negative sera ($n = 28$) were from animals coming from a not vaccinated farm where all the animals resulted negative to official diagnostic tests. All the samples were tested against the three antigens, and the cut-off for considering unknown sera as positive (i.e., valuable for viral typing) was identified as the mean reactivity plus five standard deviations (mean reactivity 0.145, standard deviation 0.058). Samples showing reactivity ≥ 0.5 (corresponding to a VN titer between 8 and 16) were investigated for further typing.

Sera collected from animals transiently infected or immunized with a defined BVDV strain were then tested against the three antigens. The sample set included 44 sera from 3 BVDV-1b positive farms, 30 sera from a BVDV-2a positive farm, 6 sera from animals experimentally infected with BVDV2c strain, and 4 sera from a BVDV-3 positive farm. We applied a BVDV typing method to analyze the data. The ratio between the homologous antigen and the most reactive heterologous antigen was calculated for each sample. The distribution of ratios was evaluated (Shapiro-Wilk test for normality assumption) and the mean minus 1 standard deviation was considered as the discrimination cut-off for antigen reactivity. This approach indicated that 1.42 is the most restrictive ratio for seropositivity discrimination among antigens. In other words, a sample is considered positive for a given antigen if the optical density (OD) against that antigen is greater than 142% as compared to the ODs obtained against the other antigen with the highest reactivity.

PI identification

PI animals were identified using an antigen ELISA (IDEXX BVDV Ag/Serum Plus, IDEXX, Westbrook, ME,

USA). Two of the 5 PI animals (CN10.2015.813 and CN10.2015.821) were selected for viral strain characterization and blood serum was collected from both. The first animal died after developing MD during the study, and samples from the spleen and intestinal epithelium were collected. Tissue homogenates and blood serum from the animal were used to isolate the viral strain on MDBK culture. Viral RNA was extracted from homogenized tissues and infected cell culture supernatant using an EZ1 Virus kit v2.0 (Qiagen, Hilden, Germany). For each sample, the 5'UTR region was amplified as previously described [5]. Amplicons were purified using Nucleospin Gel and PCR Clean-up kit (Machery Nagel) and directly sequenced in both directions (BMR Genomics, Padua, Italy). Consensus sequences were aligned to homologous reference sequences available in the online databases (Clustal W algorithm embedded in Geneious software ver. 9.0.2), referring to the BVDV-2 classification proposed previously [24]. Phylogenetic analysis was conducted by selecting the best evolutionary model with jModelTest version 2.1.7 [25] and applying the Bayesian heuristic search approach included in MrBayes software [26].

Abbreviations

BDV: Border disease virus; BVDV: Bovine viral diarrhea virus; CP: Cytopathic (strain); DMEM: Dulbecco's modified essential medium; ELISA: Enzyme-linked immunosorbent assay; FBS: Fetal bovine serum; MD: Mucosal disease; NCP: Non cytopathic (strain); OD: Optical density; PBS: Phosphate-buffered saline; PI: Persistently infected (animal); TMB: 3,3',5,5'-tetramethylbenzidine; VN: Virus neutralization (titer)

Acknowledgements

The authors wish to thank the veterinary practitioners and farmers for their help with sample collection.

Funding

This study was funded by the Italian Ministry of Instruction, University and Research (PRIN 2010LLXR94).

Authors' contributions

CN, CB and SR conceived the study; BC and MP conducted the lab work on milk, supported by MSL and GE; LB, NC and BI analyzed the data, supported by SP. All authors were involved in drafting the manuscript; all authors read and approved the final manuscript.

Ethics approval and consent to participate

The sheep used in the study were obtained by the Veterinary Teaching Hospital at the Department of Veterinary Science. Experiments were carried out in compliance with the relevant national legislation on experimental animals and animal welfare, upon authorization by the competent authority (Italian Ministry of Health DL 116/92 Aut Req 26,062,013). Milk and blood samples were collected during routine activities performed by farmers and veterinary practitioners on the farms. Written consent was obtained from all farmers for research purposes.

Consent for publication

Not applicable

Competing interests

The authors declare that they have no competing interests.

Author details

[1]Department of Veterinary Science, University of Turin, Turin, Italy. [2]Department of Veterinary Medicine, University of Bari, Bari, Italy. [3]Istituto Zooprofilattico Sperimentale dell'Umbria e delle Marche, Perugia, Italy. [4]Largo Paolo Braccini, 2, 10095 Grugliasco, Italy.

References

1. Schweizer M, Peterhans E. Pestiviruses. Annu. Rev. Anim. Biosci. 2014;2:141–63.
2. Decaro N, Lucente MS, Losurdo M, Larocca V, Elia G, Occhiogrosso L, et al. HoBi-like Pestivirus and its impact on cattle productivity. Transbound Emerg Dis. 2016;63:469–73.
3. Decaro N, Lucente MS, Mari V, Cirone F, Cordioli P, Camero M, et al. Atypical pestivirus and severe respiratory disease in calves, Europe. Emerg Infect Dis. 2011;17:1549–52.
4. Rebhun WC, French TW, Perdrizet JA, Dubovi EJ, Dill SG, Karcher LF. Thrombocytopenia associated with acute bovine virus diarrhea infection in cattle. J Vet Intern Med. 1989;3:42–6.
5. Vilcek S, Herring AJ, Herring JA, Nettleton PF, Lowings JP, Paton DJ. Pestiviruses isolated from pigs, cattle and sheep can be allocated into at least three genogroups using polymerase chain reaction and restriction endonuclease analysis. Arch Virol. 1994;136:309–23.
6. Aduriz G, Atxaerandio R, Cortabarria N. First detection of bovine viral diarrhoea virus type 2 in cattle in Spain. Vet Rec Open. 2015;2:e000110.
7. Giangaspero M, Harasawa R, Zecconi A, Luzzago C. Genotypic characteristics of bovine viral diarrhea virus 2 strains isolated in northern Italy. J Vet Med Sci Jpn Soc Vet Sci. 2001;63:1045–9.
8. Polak MP, Kuta A, Rybałtowski W, Rola J, Larska M, Żmudziński JF. First report of bovine viral diarrhoea virus-2 infection in cattle in Poland. Vet J. 2014;202:643–5.
9. Giammarioli M, Ceglie L, Rossi E, Bazzucchi M, Casciari C, Petrini S, et al. Increased genetic diversity of BVDV-1: recent findings and implications thereof. Virus Genes. 2015;50:147–51.
10. Pratelli A, Martella V, Cirone F, Buonavoglia D, Elia G, Tempesta M, et al. Genomic characterization of pestiviruses isolated from lambs and kids in southern Italy. J Virol Methods. 2001;94:81–5.
11. Barkema HW, Bartels CJ, van Wuijckhuise L, Hesselink JW, Holzhauer M, Weber MF, et al. Outbreak of bovine virus diarrhea on Dutch dairy farms induced by a bovine herpesvirus 1 marker vaccine contaminated with bovine virus diarrhea virus type 2. Tijdschr Diergeneeskd. 2001; 126:158–65.
12. Bauermann FV, Harmon A, Flores EF, Falkenberg SM, Reecy JM, Ridpath JF. In vitro neutralization of HoBi-like viruses by antibodies in serum of cattle immunized with inactivated or modified live vaccines of bovine viral diarrhea viruses 1 and 2. Vet Microbiol. 2013;166:242–5.
13. Ridpath JF, Neill JD, Frey M, Landgraf JG. Phylogenetic, antigenic and clinical characterization of type 2 BVDV from North America. Vet Microbiol. 2000;77:145–55.
14. Donofrio G, Bottarelli E, Sandro C, Flammini CF. Expression of bovine viral diarrhea virus glycoprotein E2 as a soluble secreted form in a mammalian cell line. Clin Vaccine Immunol CVI. 2006;13:698–701.
15. Harpin S, Hurley DJ, Mbikay M, Talbot B, Elazhary Y. Vaccination of cattle with a DNA plasmid encoding the bovine viral diarrhoea virus major glycoprotein E2. J Gen Virol. 1999;80(Pt 12):3137–44.
16. Lindberg AL, Alenius S. Principles for eradication of bovine viral diarrhoea virus (BVDV) infections in cattle populations. Vet Microbiol. 1999;64:197–222.
17. Decaro N, Camero M, Elia G, Martella V, Pratelli A, Gargano P, et al. Malattia delle mucose da BVDV tipo 2: descrizione di un focolaio in Puglia. Large Anim Rev. 2004;Anno 10, 2:29–34.

18. Luzzago C, Lauzi S, Ebranati E, Giammarioli M, Moreno A, Cannella V, et al. Extended genetic diversity of bovine viral diarrhea virus and frequency of genotypes and subtypes in cattle in Italy between 1995 and 2013. Biomed Res Int. 2014;2014:1–8.
19. Luzzago C, Lauzi S, Ebranati E, Giammarioli M, Moreno A, Cannella V, et al. Extended genetic diversity of bovine viral diarrhea virus and frequency of genotypes and subtypes in cattle in Italy between 1995 and 2013. Biomed Res Int. 2014;2014:1–8.
20. Gripshover EM, Givens MD, Ridpath JF, Brock KV, Whitley EM, Sartin EA. Variation in E(rns) viral glycoprotein associated with failure of immunohistochemistry and commercial antigen capture ELISA to detect a field strain of bovine viral diarrhea virus. Vet Microbiol. 2007;125:11–21.
21. Decaro N, Mari V, Pinto P, Lucente MS, Sciarretta R, Cirone F, et al. Hobi-like pestivirus: both biotypes isolated from a diseased animal. J. Gen. Virol. 2012;93:1976–83.
22. Decaro N, Mari V, Lucente MS, Sciarretta R, Moreno A, Armenise C, et al. Experimental infection of cattle, sheep and pigs with "Hobi"-like pestivirus. Vet Microbiol. 2012;155:165–71.
23. Bertolotti L, Muratore E, Nogarol C, Caruso C, Lucchese L, Profiti M, et al. Development and validation of an indirect ELISA as a confirmatory test for surveillance of infectious bovine rhinotracheitis in vaccinated herds. BMC Vet Res. 2015;11:300.
24. Giangaspero M, Harasawa R, Weber L, Belloli A. Genoepidemiological evaluation of bovine viral diarrhea virus 2 species based on secondary structures in the 5′ untranslated region. J Vet Med Sci. 2008;70:571–80.
25. Darriba D, Taboada GL, Doallo R, Posada D. jModelTest 2: more models, new heuristics and parallel computing. Nat Methods. 2012;9:772.
26. Ronquist F, Teslenko M, van der Mark P, Ayres DL, Darling A, Höhna S, et al. MrBayes 3.2: efficient Bayesian phylogenetic inference and model choice across a large model space. Syst Biol. 2012;61:539–42.

3

Bovine lactoferrin free of lipopolysaccharide can induce a proinflammatory response of macrophages

Nada Zemankova[1,2], Katarina Chlebova[1,2], Jan Matiasovic[1], Jana Prodelalova[1], Jan Gebauer[1,2] and Martin Faldyna[1*]

Abstract

Background: Lactoferrin (LF) is an 80 kDa glycoprotein which is known for its effects against bacteria, viruses and other pathogens. It also has a high potential in nutrition therapy and welfare of people and a variety of animals, including piglets. The ability to bind lipopolysaccharide (LPS) is one of the described anti-inflammatory mechanisms of LF. Previous studies suggested that cells can be stimulated even by LPS-free LF. Therefore, the aim of our study was to bring additional information about this possibility. Porcine monocyte derived macrophages (MDMF) and human embryonic kidney (HEK) cells were stimulated with unpurified LF in complex with LPS and with purified LF without bound LPS.

Results: Both cell types were stimulated with unpurified as well as purified LF. On the other hand, neither HEK0 cells not expressing any TLR nor HEK4a cells transfected with TLR4 produced any pro-inflammatory cytokine transcripts after stimulation with purified LF. This suggests that purified LF without LPS stimulates cells via another receptor than TLR4. An alternative, TLR4-independent, pathway was further confirmed by analyses of the NF-kappa-B-inducing kinase (NIK) activation. Western blot analyses showed NIK which activates different NFκB subunits compared to LF-LPS signaling via TLR4. Though, this confirmed an alternative pathway which is used by the purified LF free of LPS. This stimulation of MDMF led to low, but significant amounts of pro-inflammatory cytokines, which can be considered as a positive stimulation of the immune system.

Conclusion: Our results suggest that LF's ability is not only to bind LPS, but LF itself may be a stimulant of pro-inflammatory pathways.

Keywords: Inflammatory cytokines, LPS, NFκB, NIK, TLR4

Background

Lactoferrin (LF) is an 80 kDa iron-binding glycoprotein that plays an important role in the innate immune system [1–7]. It is also a multifunctional protein with a supportive therapeutic potential against bacterial, fungal or viral infections [8–10]. At present, LF is the focus of a variety of research areas due to its anti-inflammatory and anti-cancer properties, and the resulting therapeutical potential [11–14]. It was found that LF interacts with LPS and then activates NFκB via TLR4 pathway [15]. This suggests that immunomodulatory effects of LF could be due, in part, to LPS binding [16, 17]. To date, it is known that bovine LF affects the innate immunity by activating specific events in macrophages through interaction with TLR4-dependent and TLR4-independent pathways. It was found that bovine LF function is not dependent on TLR4 with respect to IL-6 production [18]. Since LF has a broad spectrum of actions, it also uses a wide range of signaling pathways for this purpose. In addition to NFκB, LF affects also the MAP kinase pathway. There, activation of ERK1/2 as well as of p38 stress-activated kinase has been reported [19, 20].

NF-kappa-B-inducing kinase also known as Mitogen-activated protein kinase 14 (NIK) is encoded by the MAP3K14 gene which is a serine/threonine protein kinase [21]. NIK belongs to the group of relevant activators of the NFκB alternative pathway [22]. Although it is known that NIK is necessary for activation of the non-

* Correspondence: faldyna@vri.cz
[1]Veterinary Research Institute, Hudcova 296/70, Brno 621 00, Czech Republic
Full list of author information is available at the end of the article

canonical pathway, it was first identified as a kinase which activates the canonical pathway downstream of TNF and IL-1 receptors. Physiological relevance of NIK in the canonical NFκB activation was examined by Zarnegar and coworkers. It was demonstrated that the role of NIK is in the amplification of signals from the canonical pathway. In concert with findings from NIK over-expression studies, NIK activates the canonical pathway when present at high levels due to stabilization [23].

LF was tested for several applications, incl. as newborn babies supplementation [24] or as a feed supplement for newborn piglets [25, 26]. Although there are several producing systems in which LF can be produced as recombinant [27, 28], bovine colostrum still remains as an important source of bioactive constituents incl. LF [29]. Therefore, the aim of our study was to bring additional information about the pathway used by bovine LF free of LPS to stimulate porcine macrophages to the production of proinflammatory cytokines.

Methods

Preparation of MDMF from CD14$^+$ cells

CD14$^+$ porcine monocytes were isolated from whole blood of six 5–6 week–old pigs as described previously [30]. Peripheral blood mononuclear cells (PBMC) were isolated from the collected whole blood by density gradient centrifugation using Histopaque–1077 (Sigma). CD14$^+$ monocytes were purified from PBMC by staining with mouse–anti–swine CD14 (clone MIL2, AbD Serotec, Oxford, UK, 10 μl per 10^8 cells) and goat-anti-mouse IgG MicroBeads followed by an immunomagnetic separation method (QuadroMACS™ cell separator, Miltenyi Biotec, Gladbach, Germany). The cell subset purity was assessed using flow cytometer LSRFortessa^{TM} (BD Biosciences, San Jose, CA) and was more than 95 % in all cases. Purified CD14$^+$ monocytes were cultured in DMEM medium supplemented with 10 % fetal bovine serum and 1 % antibiotics (Antibiotic Antimycotic Solution 100×: 10,000 units penicillin, 10 mg streptomycin, and 25 μg amphotericin B per mL; Sigma-Aldrich) for 4 days to differentiate into macrophages (MDMF).

HEK cells preparation

HEK 293/hTLR4A-MD2-CD14 (HEK4a) cells expressing human TLR4A, MD2 and CD14 molecules and control HEK 293/null (HEK0) cells were obtained from Invivogen (San Diego, USA). Cells were grown in DMEM (Sigma) according to the manufacturer's recommendation. The cells were seeded onto 24-well plates at the density of 2×10^5 cells per well. A treatment was applied as soon as the cells reached 80 % confluency.

Purification of lactoferrin

In the study, bovine colostral lactoferrin (Sigma-Aldrich, USA). For part of the study, lactoferrin was purified from LPS. This method was previously described [16]. The interaction between LPS and LF was abrogated by NaCl concentrations higher than 0.4 M [15]. Ten mg of LF (Sigma) was dissolved in 0.5 M NaCl of endotoxin-free water (Sigma), and centrifuged with a 100-kDa Centricon® Plus 70 (Millipore, Centricon, Merck KGaA) to remove endotoxin from the LF. The filtrate, an endotoxin free LF in 0.5 NaCl solution, was desalted by centrifugation with the 10-kDa Centricon® Plus 70. The retentate was recovered, and the concentration was quantified by Pierce BCA Protein Assay Kit 23225 (Thermo Scientific). The success of the purification was estimated by *Limulus* Amebocyte Lysate QCL-1000™ (Lonza). Unpurified and purified LF contained 4.10 EU/ml and 0.05 EU/ml, respectively.

Iron saturation determination

The level of iron saturation was estimated according to already described method [31]. Briefly, lactoferrin before and after LPS removal was spectroscopicaly measured at 280 and 466 nm with endotoxin-free water (Sigma-Aldrich, USA) as a blank. UV/VIS spectra were obtained using a microplate reader Synergy H1 (Biotek, USA). The acquired A_{280}/A_{466} ratios were subsequently used to determine the iron saturation applying the correlation curve established by Majka and coworkers [31]. From accurate measurements by ELISA and ICP-mass spectrometry. The estimation of iron saturation was calculated from an equation:

$$A_{280}/A_{466} = a(\text{iron saturation})^b, \text{ where a} = 933.0 \pm 47,\ b = -0.817 \pm 0.056$$

Iron saturation was calculated as 64 and 32 % for unpurified and purified LF, respectively.

Stimulation of HEK and MDMF

In the first experiments, HEK0 and HEK4a cells were treated with 100 μg/ml and 250 μg/ml of lactoferrin to prove the dose-dependent manner. In further experiments, only the dose of 100 μg/ml was used. This dose is sufficient for demonstrable stimulation of cells and is also appropriate because there are significant losses of starting material during purification of lactoferrin. Both types of cells were stimulated with 100 μg/ml unpurified or purified lactoferrin for 6 hours. After stimulation, the cells were harvested. Then, the experiment was stopped by aspiration of the medium and covering cells with RLT buffer with mercaptoethanol (for RT-PCR analysis) or Laemmli buffer (for Western blot analysis). All samples were frozen and stored at −80 °C.

Quantitative RT-PCR

Total RNA from each well was extracted using RNeasy Mini Kit (Qiagen) according to the manufacturer's instructions. mRNA was specifically reverse-transcribed using the M-MLV reverse transcriptase system (Invitrogen, Paisley, UK) in the presence of oligo-dT primers. cDNA was diluted 5× and 0.5 μl was used in qPCR. In qPCR analysis, RNA expression was quantified in triplicate reactions in a final volume of 3 μl in 384-well plates using QuantiTect SYBR Green PCR master mix (Qiagen, Hilden, Germany) following the manufacturer's recommendations, on a LightCycler 480 (Roche Applied Science, http://www.roche.com/). qPCR reactions were prepared with the assistance of Nanodrop II liquid dispenser (Innovadyne Technologies, Rohnert Park, CA). qPCR was performed under the following conditions: denaturation (95 °C for 15 min) and 45 amplification cycles (95 °C for 15 s, 58 °C for 30 s and 72 °C for 30 s). The resulting melting curves were analyzed to test the product specificity. Ten pmol of each primer pair was used per reaction. Primers specific to 4 porcine and 2 human target genes and 2 porcine (HPRT, TBP-1) and 2 human (HPRT, GAPDH) reference genes were used for simultaneous measurements of gene activity (Tables 1 and 2). Among the candidate reference genes, HPRT and GAPDH in the case of MDMF and HEK, respectively, were evaluated as the most constitutively expressed genes in our samples and were selected to adjust mRNA measurements. From the obtained data, relative expression of each target gene was calculated according to the formula [1/(2target gene Ct)]/[1/(2reference gene Ct)].

Western blot analysis

For Western blot analysis, MDMF protein lysates were prepared by adding lysis buffer (Cell Signaling Technology) with Laemmli buffer (0.5 M Tris–HCl pH 6.8, glycerol, 10 % SDS, bromophenol blue, beta-mercaptoethanol, deionized water), sonicated and heated to 95 °C for 5 min.

Table 1 Porcine primers used in the study

Gene	Sequence 5′–3′	Reference
CXCL10/IP10	F: CCC ACA TGT TGA GAT CAT TGC	[55]
	R: CAT CCT TAT CAG TAG TGC CG	
HPRT	F: GAG CTA CTG TAA TGA CCA GTC AAC G	[56]
	R: CCA GTC TCA ATT ATA TCT TCA ACA ATC AA	
IL-8	F: TTC TGC AGC TCT TCG TGA GGC	[57]
	R: GGT GGA AAG GTG TGG AAT GC	
NFκB	F: ACG AGC AGA TGG TGA AGG AG	[57]
	R: TCA TGG ATG ATG GCC AAG T	
TNFα	F: CCC CCA GAA GGA AGA GTT TC	[58]
	R: CGG GCT TAT CTG AGG TTT GA	

Table 2 Human primers used in the study

Gene	Sequence 5′–3′	Reference
GAPDH	F: TCC TAG ATT ATT CTC TGA TTT GGT CGT ATT G	this study
	R: GAA TTT GCC ATG GGT GGA ATC ATA TTG	
IL-8	F: TCC AAA CCT TTC CAC CCC AAA TTT ATC	this study
	R: AGC TCT CTT CCA TCA GAA AGC TTT ACA ATA A	
TNFα	F: CAA TGG CGT GGA GCT GAG AGA TA	this study
	R: CCT TGA AGA GGA CCT GGG ATG AGA T	

Protein lysates were resolved by 10 % SDS-PAGE. After gel electrophoresis, proteins were transferred onto PVDF (polyvinylidene difluoride) membranes. The membranes were blocked in 5 % low-fat dry milk for 1 h at room temperature and followed by incubation with primary antibodies at 4 °C overnight. Primary antibodies used for membrane probing-p-IκB (Ser 32) (Santa Cruz Biotechnology), NIK antibody (Cell Signaling Technology), anti-beta actin antibody (Abcam). Anti-beta actin served as loading control. Next day, the membranes were washed and probed with secondary antibodies for 1 h at room temperature, followed by a washing step. The proteins were visualized on a photographic film, using ECL Plus Western blotting detection reagents (GE Healthcare Life Sciences, UK).

Statistical analysis

Data were analysed using Statistica 12 (StatSoft). The non-parametric test for paired samples (Wilcoxon's signed-rank test) was used to detect significant differences among groups. A p-value of <0.05 was considered significant.

Results

mRNA production of CXCL-10, NFκB inhibitor, IL-8 and TNFα by MDMF after LF treatment

Results in Fig. 1a-d suggest that both purified and unpurified LF stimulated MDMF cells. In the case of unpurified LF (i.e., in a complex with LPS), the TLR4-dependent signaling pathway can be expected. In the case of purified LF, the TLR4-dependent pathway can be excluded. This statement is based on results of the experiment in which a medium containing 1 picogram of LPS per milliliter was used (pure LF medium). The concentration was selected because the same residual amount of LPS was detected by the LAL test in the case of the purified form of LF. The results showed that this amount had no impact on the MDMF activation.

mRNA production of IL-8 and TNFα by HEK0 and HEK4a cells after unpurified LF treatment

Activation of the TLR4-dependent pathway was further excluded in experiments using permanent cell lines

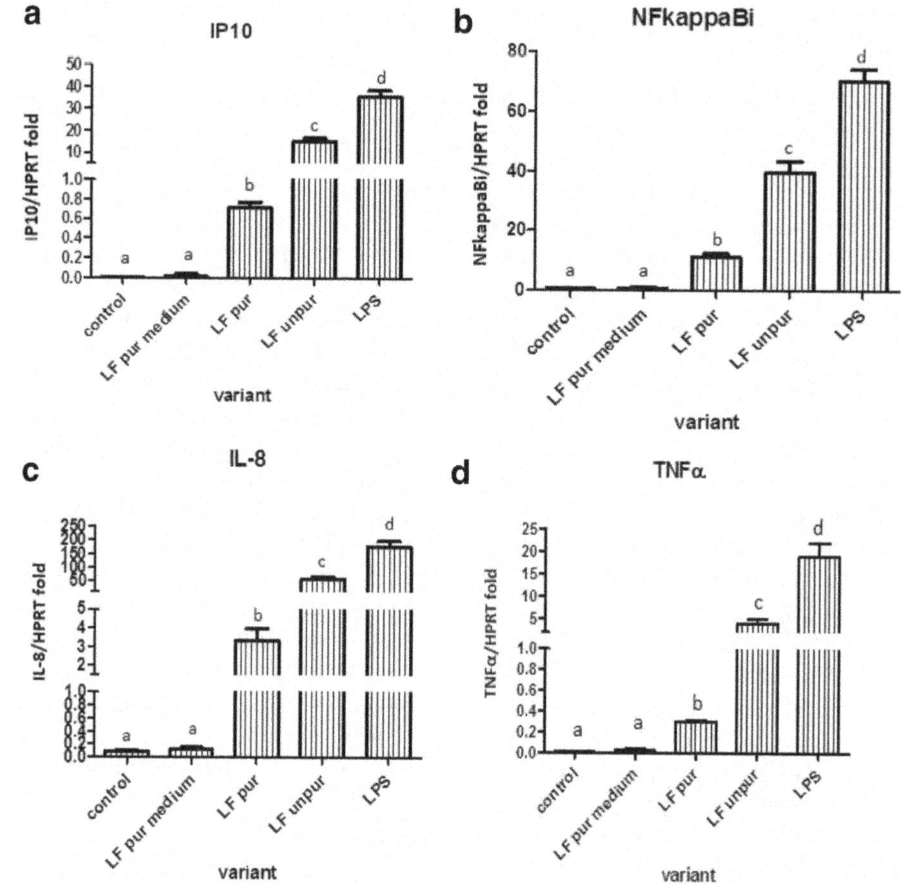

Fig. 1 Expression of mRNA for IP10 (**a**), NFκBi (**b**), IL-8 (**c**) and TNFα (**c**) by porcine monocyte-derived macrophages after stimulation with: 1 μg/ml of lipopolysaccharide (LPS); unpurified lactoferrin (LF unpur) and purified lactoferrin (LF pur) containing 4.10 EU/ml and 0.05 EU/ml, respectively. Cells treated with medium containing 1 pg/ml of LPS (i.e. amount comparable with residual LPS-contamination of LF pur) or cultivated in LPS-free medium were used. Data are shown as mean ± SD, $n = 6$. Different letters show statistical significance when compared with control ($p < 0.05$)

HEK0 and HEK4a that express either no TLR molecules or TLR4a-MD2-CD14 complex. As expected, stimulation of HEK0 with LPS or with two concentrations of unpurified LF (LF-LPS complex) did not increase expression of mRNA for TNFα and IL-8 (Fig. 2a, c). In the case of HEK4a, this response was induced (Fig. 2b, d).

In contrast to results shown in Fig. 2b and d, expression of mRNA for pro-inflammatory cytokines was not increased when purified LF was used (Fig. 3a, b). The production of IL-8 and TNFα was increased only in the case of unpurified LF and LPS, and it was partially decreased by Polymyxin B (PMB).

Western blot

Western blot analysis confirmed that expression of mRNA for pro-inflammatory cytokines in response to LPS alone or unpurified LF (LF-LPS) was conducted via TLR4 and NFκB pathway activation (Fig. 4). Moreover, Western blot analysis also confirmed that MDMF, when stimulated with purified LF, were activated via TLR4-independent pathway. It is clear from the fact that NIK associated with activation of alternative NFκB pathway was present (Fig. 5).

Discussion

The aim of this study was to extend our knowledge about immunomodulatory properties of LF and particularly about the pathway used by LF free of LPS to stimulate macrophages to the production of proinflammatory cytokines.

As expected, MDMF responded to the stimulation with unpurified LF containing LPS by the production of proinflammatory cytokines. To a lesser extent, these cells were stimulated also by purified LPS-free LF. It suggested that this stimulation is LPS independent so does not use the TLR4 associated pathway. This was confirmed when HEK4a cells were treated with purified LPS-free LF. These cells expressing only TLR4 did not increase the production of IL-8 and TNFa transcripts when compared to control. TLR4-dependent activation was observed only in the case when these cells were stimulated with LF-LPS complex in a dose-dependent

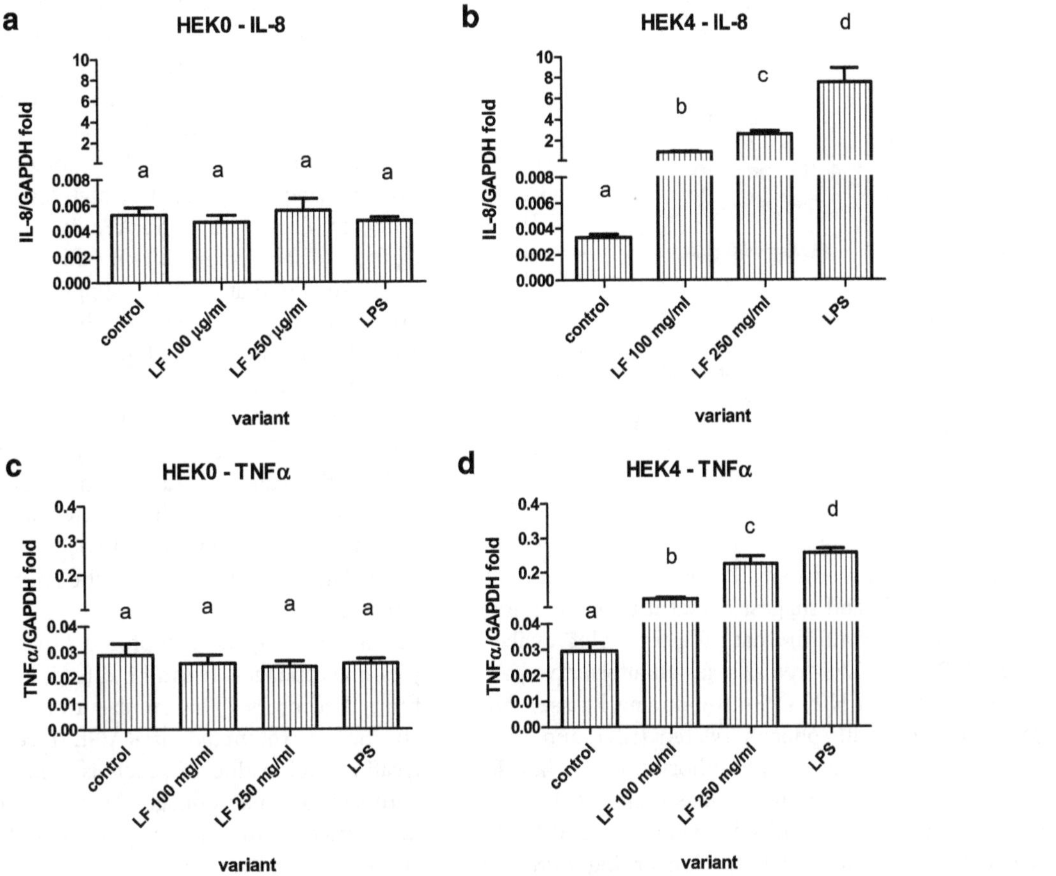

Fig. 2 Expression of mRNA for IL8 (**a** and **b**) and TNFα (**c** and **d**) by HEK cell line expressing no TLR (**a** and **c**) or the same cell line expressing TLR4a-MD2-CD14 (**b** and **d**) after stimulation with two concentrations of lactoferrin (LF) or 1 µg/ml of lipopolysaccharide (LPS). Data are shown as mean ± SD of 5 repetitions. Different letters show statistical significance when compared with control ($p < 0.05$)

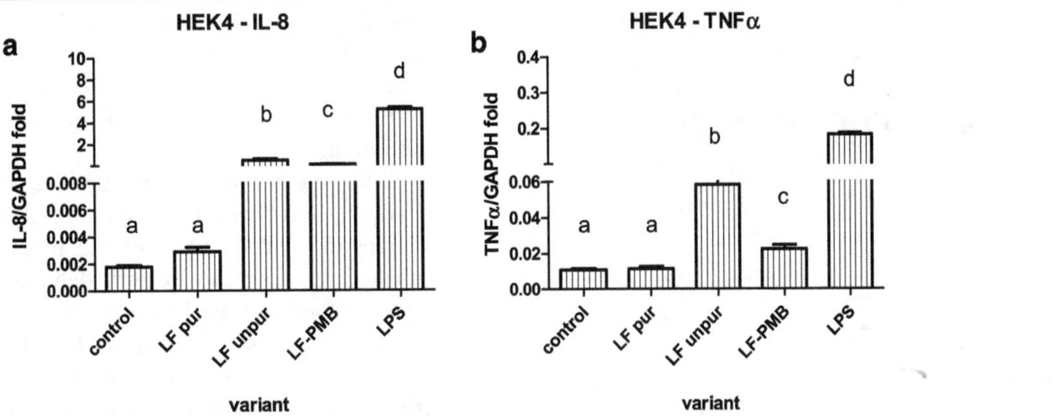

Fig. 3 Expression of mRNA for IL8 (**a**) and TNFα (**b**) by HEK cell line expressing TLR4a-MD2-CD14 after stimulation with 1 µg/ml of lipopolysaccharide (LPS); unpurified lactoferrin (LF unpur) and purified lactoferrin (LF pur) containing 4.10 EU/ml and 0.05 EU/ml, respectively. Cells stimulated with unpurified LF in combination with polymyxine B are marked as LF-PMB. Cells cultivated in LPS-free medium were used. Data are shown as mean ± SD of 5 repetitions. Different letters show statistical significance when compared with control ($p < 0.05$)

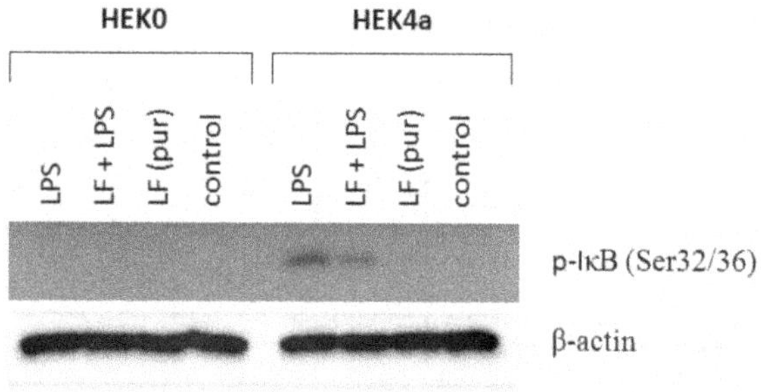

Fig. 4 Presence of phosphorylated form of IκB in HEK cell line expressing no TLR (*left*, HEKO) or the same cell line expressing TLR4a-MD2-CD14 (right, HEK4a) after stimulation with (from *left* o *right*) 1 μg/ml of lipopolysaccharide (LPS); 100 μg/ml of unpurified (LF + LPS) or purified lactoferrin (LF pur) containing 4.10 EU/ml or 0.05 EU/ml, respectively. Cells cultivated in LPS-free medium were used as a control. β-actin served as loading control

manner. This confirmed that LPS-free LF does not activate cells via TLR4 and suggested that it uses an alternative pathway. To confirm that, Western blot analyses were used. Purified LPS-free LF uses an alternative pathway which activates NIK (NF-κB-inducing kinase). NIK forms a complex with phosphorylated IKK1 and IKK2, subsequently leading to the phosphorylation of IκB and translocation of NF-κB to the nucleus [32]. NIK is essential for alternative NFκB pathway, being required for the processing of p100 to p52 [22]. Concerning LPS, it is known that NIK induces phosphorylation of IKK1, that in turn phosphorylates histone H3 [33]. It may be relevant also for LF-LPS (unpurified LF) treatment. Zarnegar and coworkers showed the relevance of NIK in canonical NFκB pathway, when NIK is present at high levels. They demonstrated elevation of canonical NFκB pathway activity on TRAF3−/− mice. In this case NIK was stabilized and present at high levels due to the deficiency of TRAF3 [23]. TRAF3 is an essential negative regulator of alternative NFκB pathway, but in the study it was shown that TRAF3 suppresses canonical NFκB activation and gene expression both in vitro and in vivo. Deregulation of canonical pathway in TRAF3-deficient cells causes NIK accumulation, i.e. TRAF3 inhibition results in coordinated activation of both canonical and alternative pathways [23]. NIK probably lies at a cross road of the canonical and non-canonical pathways of NFκB and it maintains the positive and negative regulatory balance between these two pathways. Maintaining this balance is crucial for the proper physiological responses [32, 34].

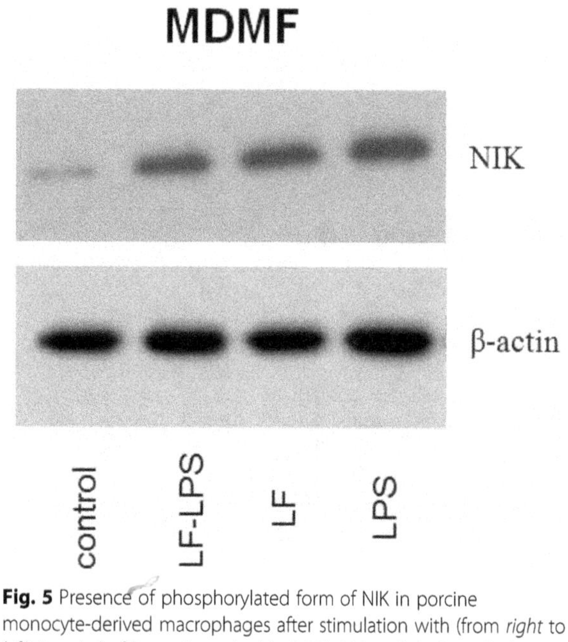

Fig. 5 Presence of phosphorylated form of NIK in porcine monocyte-derived macrophages after stimulation with (from *right* to *left*) 1 μg/ml of lipopolysaccharide (LPS); 100 μg/ml of purified (LF) of unpurified lactoferrin (LF LPS) containing 0.05 EU/ml or 4.10 EU/ml, respectively. Cells cultivated in LPS-free medium were used as a control. β-actin served as loading control

LF has no specific receptor on the surface of monocytes/macrophages. It can bind to different low- or high-affinity receptors. The similar route leading to NFκB activation is conducted via RAGE (receptor for advanced glycation endproducts) which is a multiligand receptor expressed on macrophages, neurons, endothelial cells and a variety of tumor cells [35, 36] and TREM-1 (trigerring receptor expressed on myeloid cells 1) a member of the immunoglobulin superfamily which was found on monocytes and neutrophils [37]. Signals which are conducted via RAGE activate two major pathways. First one comprises CDC42/Rac and the second one encompasses MAPKs that finally lead to NFκB activation [38]. TREM-1 is a positive regulator of inflammation [37]. One of the main mechanisms mediating the expression of TREM-1 is activation of NFκB [39]. TREM-1 modulates the activity and availability of key proteins of the TLR4 signaling cascade [40]. TREM-1 influences TLR4 signaling cascade positively by driving TLR4 into the lipid rafts [41]. In other words, signal conducted via TLR4, RAGE and TREM-1 use connected patways and lead to NFκB activation.

LF has the potential to modify the course of an infection [42]. Being immunomodulator, LF is used as adjunct therapeutic which can alter the immune status of the infected host. Bovine LF is widely used as therapeutic agent or food additive in many animal or human models. In the mouse model, it improves host survival modestly concurrent with the decreased serum cytokines [43] or it protects gut integrity in mouse models of LPS endotoxaemia [44]. Its immunomodulatory properties were also confirmed in the influence of low endotoxine bovine milk LF (<0.2 EU/mg) on expression of presentation molecules on bone marrow derived macrophages

(BMMs) study. LF modulated the activity of macrophages and their ability to present an antigen and to stimulate T-cells through increased surface expression of antigen presentation and co-stimulatory molecules such as CD80 and CD86. BMMs cultured with bovine LF increased the number of MHC II^+ cells [45].

In humans, LF is known to be able to prevent the onset of sepsis in infants [46–48]. LF has a high homology among species [49] thus, heterologous molecules of LF are used as a therapeutic agents. Although, the better protective or immunomodulatory effect of homologous recombinant LF could be expected, it was not proven, concretely in the mouse model of methicillin-resistant *Staphylococcus aureus* challenge. Recombinant mouse LF (rmLF) and recombinant human LF (rhLF) had a high degree of overlap to modulate inflammatory responses. Both recombinant molecules caused a decrease in inflammatory cytokine production in monocytes and an increase of cytokine production in granulocytes during the infection. But the two heterologous recombinant molecules showed some differences in their activities concerning cellular population that were used in the studies-mouse granulocytes demonstrated a different cytokine pattern upon the infection than human granulocytes. LFs were produced under endotoxin free serum conditions, and further affinity purified to remove traces of endotoxin [42]. The data show that endotoxin-free LF has not only immunomodulatory effects but also the therapeutic effects which have different essence than just LPS binding.

A similar study in which low endotoxin human milk LF (<10EU/mg) was used supports a therapeutic significance of LF in vivo. LF treatment reduced TNF-α and IL-6 in serum of LPS-challenged mice during LPS induced endotoxaemia, when it was delivered to naive mice 1 h prior to administration of LPS. The same molecule showed a prophylactic effect of treatment with LF 18 h prior to LPS-induced endotoxaemia when the production of TNF-α but not IL-6 was affected [50].

Conclusion

Our results suggest that LF's ability is not only to bind LPS, but LF itself may be a stimulant of pro-inflammatory pathways. LF's ability to bind LPS from an environment is known very well [16, 51]. This binding is its natural property which is used for example against bacterial infections. Due to this fact, LF often occurs in an unpurified form in a complex with LPS [18]. This kind of "contamination" resulting from one of the natural properties of LF could lead to some overvalued results. It should be considered when studying the production of cytokines or chemokines after LF treatment. Similar situations were found concerning other therapeutic proteins. When testing ovalbumin, it was found that a purified form of this protein did not activate endothelial cells in vitro. If an unpurified commercial ovalbumin containing LPS was used, endothelial cells were activated in early stages of inflammation [52]. False pro-inflammatory effects of commercial alpha-lactalbumin were previously described in RAW 264.7 macrophages due to endotoxin contamination [53]. And, finally, other authors mentioned contamination of commercial LF [18, 54]. This fact should be borne in mind when we perform experiments in which we study pro-inflammatory parameters.

Abbreviations

HEK: Human embryonic kidney; IL: Interleukin; IκB: Inhibitor of NF-kappa-B; LF: Lactoferrin; LPS: Lipopolysaccharide; MDMF: Monocyte-derived macrophages; mRNA: Messenger ribonucleic acid; NFκB: NF-kappa-B; NIK: NF-kappa-B-inducing kinase; RT-PCR: Real-time polymerase chain reaction; TLR: Toll-like receptor; TNF: Tumor necrosis factor

Acknowledgements

Authors would like to thank Ing. Ludmila Faldikova (Veterinary Research Institute, Brno), experienced bioscience literature language advisor, for language correction.

Funding

This study was supported by the Ministry of Agriculture of the Czech Republic (QJ1310258 and RO0516) and Ministry of Education, Youth and Sports of the Czech Republic (LO1218).

Authors' contributions

NZ performed preparation of MDMF, purification of lactoferrin, qRT-PCR. KC carried out western blot analysis. JM was responsible for HEK cell preparation. JG was responsible for work with lactoferrin, incl. iron saturation determination. NZ, JP and MF conceived the study. All authors helped to draft the manuscript, read and approved the final version of the manuscript.

Competing interests

The authors declare that they have no competing interests.

Consent for publication

Not applicable.

Author details

[1]Veterinary Research Institute, Hudcova 296/70, Brno 621 00, Czech Republic. [2]Faculty of Science, Masaryk University, Brno, Czech Republic.

References

1. Buckett NM, Luckas MJ, Gazvani MR, Aird IA, Lewis-Jones DI. Seminal plasma lactoferrin concentrations in normal and abnormal semen samples. J Androl. 1997;18:302–4.
2. Fullard RJ, Snyder C. Protein levels in nonstimulated and stimulated tears of normal human subjects. Invest Opthalmol Vis Sci. 1990;31:1119–26.

3. Hayakawa T, Harada H, Noda A, Kondo T. Lactoferrin in pure pancreatic juice in chronic pancreatitis. Am J Gastroenterol. 1983;78:222–4.
4. Hetherington SV, Spitznagel JK, Quie PG. An enzyme-linked immunoassay (ELISA) for measurement of lactoferrin. J Immunol Methods. 1983;65:183–90.
5. Lonnerdal B, Iyer S. Lactoferrin: Molecular structure and biological function. Annu Rev Nutr. 1995;15:93–110.
6. Ronayne De Ferre PA, Baroni A, Sambucetti ME, López NE, Ceriani Cernadas JM. Lactoferrin levels in term and preterm milk. J Am Coll Nutr. 2000;19:370–3.
7. Shugars DS, Watkins CA, Cowen HJ. Salivary concentration of secretory leukocyte protease inhibitor, an antimicrobial protein, is decreased with advanced age. Gerontology. 2001;47:246–53.
8. Kondori N, Baltzer L, Dolphin GT, Mattsby-Baltzer I. Fungicidal activity of human lactoferrin-derived peptides based on the antimicrobial αβ region. Int J Antimicrob Agents. 2011;37:51–7.
9. Wakabayashi H, Oda H, Yamauchi K, Abe F. Lactoferrin for prevention of common viral infections. J Infect Chemother. 2014;20:666–71.
10. Yin C, Wong JH, Ng TB. Recent studies on the antimicrobial peptides lactoferricin and lactoferrampin. Curr Mol Med. 2014;14:1139–54.
11. Adlerova L, Bartoskova A, Faldyna M. Lactoferrin: a review. Vet Med. 2008;53:457–68.
12. Hung YP, Yang YP, Wang HC, Liao JW, Hsu WL, Chang CC, Chang SC. Bovine lactoferrin and piroxicam as an adjunct treatment for lymphocytic-plasmatic gingivitis stomatitis in cats. Vet J. 2014;202:76–82.
13. Lim LY, Koh PY, Somani S, Al Robaian M, Karim R, Yean YL, Mitchell J, Tate RJ, Edrada-Ebel RA, Blatchford DR, Mullin M, Dufes C. Tumor regression following intravenous administration of lactoferrin- and lactoferricin-bearing dendriplexes. Nanomed Nanotech Biol Med. 2015;11:1445–54.
14. Zemann N, Klein P, Wetzel E, Huettinger F, Huettinger M. Lactoferrin induces growth arrest and nuclear accumulation of Smad-2 in HeLa cells. Biochimie. 2010;92:880–4.
15. Van Berkel PH, Geerts ME, Van Veen HA, Mericskay M, De Boer HA, Nuijens JH. N-terminal stretch Arg2, Arg3, Arg4and Ar g5 of human lactoferrin is essential for binding to heparin, bacterial lipopolysaccharide, human lysozyme and DNA. Biochem J. 1997;328:145–51.
16. Na YJ, Han SB, Kang JS, Yoon YD, Park SK, Kim HM, Yang KH, Joe CO. Lactoferrin works as a new LPS binding protein in inflammatory activation of macrophages. Int Immunopharmacol. 2004;4:1187–99.
17. Puddu P, Carollo MG, Belardelli F, Valenti P, Gessani S. Role of endogenous interferon and LPS in the immunomodulatory effects of bovine lactoferrin in murine peritoneal macrophages. J Leukoc Biol. 2007;82:347–53.
18. Curran CS, Demick KP, Mansfield JM. Lactoferrin activates macrophages via TLR4-dependent and -independent signaling pathways. Cell Immunol. 2006;242:23–30.
19. Oh SM, Hahm DH, Kim IH, Choi SY. Human neutrophil lactoferrin trans-activates the matrixmetalloproteinase 1 gene through stress-activated MAPK signaling modules. J Biol Chem. 2001;276:42575–9.
20. Zhou Y, Zeng Z, Zhang W, Xiong W, Wu M, Tan Y, Yi W, Xiao L, Li X, Huang C, Cao L, Tang K, Li X, Shen S, Li G. Lactotransferrin: a candidate tumor suppressor-deficient expression in human nasopharyngeal carcinoma and inhibition of NPC cell proliferation by modulating the mitogen-activated protein kinase pathway. Int J Cancer. 2008;123:2065–72.
21. Malinin NL, Boldin MP, Kovalenko AV, Wallach D. MAP3K-related kinase involved in NF-kB induction by TNF, CD95 and IL-1. Nature. 1997;385:540–4.
22. Xiao G, Harhaj EV, Sun SC. NF-kappa-B-inducing kinase regulates the processing of NF-kappaB2 p100. Mol Cell. 2001;7:401–9.
23. Zarnegar B, Yamazaki S, He JQ, Cheng G. Control of canonical NF-kappaB activation through the NIK-IKK complex pathway. Proc Natl Acad Sci U S A. 2008;105:3503–8.
24. Pammi M, Abrams SA. Oral lactoferrin for the prevention of sepsis and necrotizing enterocolitis in preterm infants. Cochrane Database Syst Rev. 2015;20, CD007137.
25. Nguyen DN, Li Y, Sangild PT, Bering SB, Chatterton DE. Effects of bovine lactoferrin on the immature porcine intestine. Br J Nutr. 2014;111:321–31.
26. Nguyen DN, Jiang P, Stensballe A, Bendixen E, Sangild PT, Chatterton DE. Bovine lactoferrin regulates cell survival, apoptosis and inflammation in intestinal epithelial cells and preterm pig intestine. J Proteomics. 2016;139:95–102.
27. Choi BK, Actor JK, Rios S, d'Anjou M, Stadheim TA, Warburton S, Giaccone E, Cukan M, Li H, Kull A, Sharkey N, Gollnick P, Kocięba M, Artym J, Zimecki M, Kruzel ML, Wildt S. Recombinant human lactoferrin expressed in glycoengineered *Pichia pastoris*: effect of terminal *N*-acetylneuraminic acid on *in vitro* secondary humoral immune response. Glycoconj J. 2008;25:581–93.
28. Hwang SA, Kruzel ML, Actor JK. CHO expressed recombinant human lactoferrin as an adjuvant for BCG. Int J Immunopathol Pharmacol. 2015;28:452–68.
29. Bagwe S, Tharappel LJ, Kaur G, Buttar HS. Bovine colostrum: an emerging nutraceutical. J Complement Integr Med. 2015;12:175–85.
30. Vicenova M, Nechvatalova K, Chlebova K, Kucerova Z, Leva L, Stepanova H, Faldyna M. Anti-inflammatory activity of biologically active phospholipids endowed with anti-neoplastic potential in *in vivo* and *in vitro* porcine immune systems. BMC Complement Altern Med. 2014;1:339.
31. Majka G, Śpiewak K, Kurpiewska K, Heczko P, Stochel G, Strus M, Brindell M. A high-thoughput method for the quantification of iron saturation in lactoferrin preparations. Anal Bioanal Chem. 2013;405:5191–200.
32. Thu YM, Richmond A. NF-κB inducing kinase: A key regulator in the immune system and in cancer. Cytokine Growth Factor Rev. 2010;21:213–26.
33. Park GY, Wang X, Hu N, Pedchenko TV, Blackwell TS, Christman JV. NIK is involved in nucleosomal regulation by enhancing histone H3 phosphorylation by IKKalpha. J Biol Chem. 2006;281:18684–90.
34. Oeckinghouse A, Hayden MS, Gosh S. Crosstalk in NF-kB signaling pathways. Nature Immunol. 2011;12:695–708.
35. Bierhaus A, Humpert PM, Morcos M, Wendt T, Chavakis T, Arnold B, Stern DM, Nawroth PP. Understanding RAGE, the receptor for advanced glycation end products. J Mol Med. 2005;83:876–86.
36. Veloso CA, Fernandes JS, Volpe CM, Fagundes-Netto FS, Reis JS, Chaves MM, Nogueira-Machado JA. TLR4 and RAGE: similar routes leading to inflammation in type 2 diabetic patients. Diabetes Metab. 2011;37:336–42.
37. Sharif O, Knapp S. From expression to signaling: roles of TREM-1 and TREM-2 in innate immunity and bacterial infection. Immunobiology. 2008;213:701–13.
38. Van Beijnum JR, Buurman WA, Griffioen AW. Convergence and amplification of toll-like receptor (TLR) and receptor for advanced glycation end products (RAGE) signaling pathways via high mobility group B1 (HMGB1). Angiogenesis. 2008;11:91–9.
39. Hosoda H, Kida S, Nagaoka I. Transcriptional regulation of mouse TREM-1 gene in RAW264.7 macrophage-like cells. Life Sci. 2011;89:115–22.
40. Arts RJW, Joosten LAB, van den Meer JWM, Netea MG. TREM-1: intracellular signaling pathways and interaction with pattern recognition receptors. J Leukoc Biol. 2013;93:209–15.
41. Fortin CF, Lesur O, Fulop Jr T. Effects of TREM-1 activation in human neutrophils: activation of signalling pathways, recruitment into lipid rafts and association with TLR4. Int Immunol. 2007;19:41–50.
42. Hwang SA, Kruzel ML, Actor JK. Immunomodulatory effects of recombinant lactoferrin during MRSA infection. Int Immunopharmacol. 2014;20:157–63.
43. Hwang SA, Actor JK. Lactoferrin modulation of BCG-infected dendritic cell functions. Int Immunol. 2009;21:1185–97.
44. Kruzel ML, Harari Y, Chen CY, Castro GA. Lactoferrin protects gut mucosal integrity during endotoxemia induced by lipopolysaccharide in mice. Inflammation. 2000;24:33–44.
45. Hwang SA, Kruzel ML, Actor JK. Influence of bovine lactoferrin on expression of presentation molecules on BCG-infected bone marrow derived macrophages. Biochimie. 2009;91:76–85.
46. Manzoni P, De Luca D, Stronati M, Jacqz-Aigrain E, Ruffinazzi G, Luparia M, Tavella E, Boano E, Castagnola E, Mostert M, Farina D. Prevention of nosocomial infections in neonatal intensive care units. Am J Perinatol. 2013;30:81–8.
47. Ochoa TJ, Pezo A, Cruz K, Chea-Woo E, Cleary TG. Clinical studies of lactoferrin in children. Biochem Cell Biol. 2012;90:457–67.
48. Stronati M, Bollani L, Maragliano R, Ruffinazzi G, Manzoni P, Borghesi A. Neonatal sepsis: new preventive strategies. Minerva Pediatr Febbraio. 2013;65:103–10.
49. Shanbacher FL, Goodman RE, Talhouk RS. Bovine mammary lactoferrin: implications from messenger ribonucleic acid (mRNA) sequence and regulation contrary to other milk proteins. J Dairy Sci. 1992;76:3812–31.
50. Kruzel ML, Harari Y, Mailman D, Actor JK, Zimecki M. Differential effects of prophylactic, concurrent and therapeutic lactoferrin treatment on LPS-induced inflammatory responses in mice. Clin Exp Immunol. 2002;130:25–31.
51. Drago-Serrano ME, de la Garza-Amaya M, Serrano Luna J, Campos-Rodríguez R. Lactoferrin-lipopolysaccharide (LPS) binding as key to antibacterial and antiendotoxic effects. Int Immunopharmacol. 2012;12:1–9.
52. Watanabe J, Miyazaki Y, Zimmerman GA, Albertine KH, McIntyre TM. Endotoxin contamination of ovalbumin suppresses murine immunologic responses and development of airway hyper-reactivity. J Biol Chem. 2003;278:42361–8.

53. Lin IC, Kuo CD. Pro-inflammatory effects of commercial alpha-lactalbumin on RAW 264.7 macrophages is due to endotoxin contamination. Food Chem Toxicol. 2010;48:2642–9.
54. Cohen MS, Rasmussen GT, Serody JS, Britigan BE. Interaction of lactoferrin and lipopolysaccharide (LPS): effects of antioxidant property of lactoferrin and the ability of LPS to prime human neutrophils for enhanced superoxide formation. J Infect Dis. 1992;166:1375–8.
55. Stepanova H, Pavlova B, Stromerova N, Ondrackova P, Stejskal K, Slana I, Zdrahal Z, Pavlik I, Faldyna M. Different immune response of pigs to *Mycobacterium avium* subsp. *avium* and *Mycobacterium avium* subsp. *hominissuis* infection. Vet Microbiol. 2012;159:343–50.
56. Zelnickova P, Matiasovic J, Pavlova B, Kudlackova H, Kovaru F, Faldyna M. Quantitative nitric oxide production by rat, bovine and porcine macrophages. Nitric Oxide. 2008;19:36–41.
57. Kyrova K, Stepanova H, Rychlik I, Faldyna M, Volf J. SPI-1 encoded genes of *Salmonella* Typhimurium influence differential polarization of porcine alveolar macrophages *in vitro*. BMC Vet Res. 2012;8:15.
58. Pavlova B, Volf J, Ondrackova P, Matiasovic J, Stepanova H, Crhanova M, Karasova D, Faldyna M, Rychlik I. SPI-1 type III secretion system of *Salmonella enterica* is required for the suppression of porcine alveolar macrophage cytokine expression. Vet Res. 2011;42:1.

4

Mycoplasma detection by triplex real-time PCR in bronchoalveolar lavage fluid from bovine respiratory disease complex cases

Jan B. W. J. Cornelissen*, Freddy M. de Bree, Fimme J. van der Wal, Engbert A. Kooi, Miriam G. J. Koene, Alex Bossers, Bregtje Smid, Adriaan F. Antonis and Henk J. Wisselink

Abstract

Background: In this study we evaluated the RespoCheck *Mycoplasma* triplex real-time PCR for the detection in bronchoalveolar lavage fluid (BALF) of *Mycoplasma (M.) dispar, M. bovis* and *M. bovirhinis*, all three associated with bovine respiratory disease (BRD). Primers and probes of the RespoCheck *Mycoplasma* triplex real-time PCR are based on the V3/V4 region of the 16S rRNA gene of the three *Mycoplasma* species.

Results: The analytical sensitivity of the RespoCheck triplex real-time PCR was, as determined by spiking experiments of the *Mycoplasma* strains in Phosphate Buffered Saline, 300 colony forming units (cfu)/mL for *M. dispar*, and 30 cfu/mL for *M. bovis* or *M. bovirhinis*. The analytical sensitivity of the RespoCheck *Mycoplasma* triplex real-time PCRwas, as determined on purified DNA, 10 fg DNA per assay for *M. dispar* and 100 fg fo r*M. bovis* and *M. bovirhinis*. The analytical specificity of the RespoCheck *Mycoplasma* triplex real-time PCR was, as determined by testing *Mycoplasmas* strains (n = 17) and other bacterial strains (n = 107), 100, 98.2 and 99.1% for *M. bovis, M. dispar* and *M. bovirhinis* respectively. The RespoCheck Mycoplasma triplex real-time PCR was compared with the PCR/DGGE analysis for *M. bovis, M. dispar* and *M. bovirhinis* respectively by testing 44 BALF samples from calves.

Conclusion: In conclusion, the RespoCheck PCR assay can be a valuable tool for timely and accurate detection of three *Mycoplasma* species associated with in bovine respiratory disease.

Keywords: Bovine *Mycoplasma, M. dispar, M. bovis, M. bovirhinis*, Triplex PCR, RespoCheck, Bovine respiratory disease

Background

Bovine respiratory disease complex (BRDC) is a global problem causing severe economic losses to the cattle farming industry through mortality, loss of production, and treatment costs [1, 2]. It has a complex etiology that involves various pathogens, host factors, and environmental factors. Viruses such as bovine herpes 1 virus (BoHV-1, parainfluenza virus 3 (PBIV-3), bovine respiratory syncytial Virus (BRSV), respiratory bovine coronavirus (BoCoV) and bovine viral diarrhoea virus (BVDV) in conjunction with stress factors have been implicated as causes of respiratory tract infections of cattle by immunosuppression and damage to the respiratory epithelium [3]. A primary viral infection can be followed by an opportunistic secondary infection with bacteria like *Mannheimia haemolytica, Pasteurella multocida, Histophilus somni*, or *Trueperella pyogenes* [2, 4, 5], but these bacteria could also act as primary pathogen. In addition it has become increasingly clear that Mycoplasmas are important contributors to BRD, either as primary pathogens or in co-infection [2, 6–9]. *M. bovis* is the best known *Mycoplasma* species causing respiratory disease [4, 7], but also *M. dispar* and *M. bovirhinis* have been associated with BRD [2, 9–11]. *M. bovis* has not only been identified as a primary or opportunistic pathogen in BRD in beef cattle worldwide, but it has also been implicated in other clinical manifestations in cattle, such as mastitis, otitis, arthritis, and reproductive disorders [7]. *M. bovirhinis* and *M. dispar* are regularly isolated from the nasal cavity of cattle with respiratory disease and are

* Correspondence: jan.cornelissen@wur.nl
Department of Infection Biology, Central Veterinary Institute of Wageningen UR, P.O. Box 658200, AB, Lelystad, The Netherlands

usually regarded as an opportunistic pathogen in respiratory diseases [7, 12].

Bacteriological, serological and histopathological examinations are important tools to detect particular animal-carriers of *Mycoplasma* [13], however, these assays are time-consuming, insensitive and can give false positive results. Bronchoalveolar lavage fluid (BALF) from calves with BRD may contain various potential pathogens, but additional antibiotic use in the affected herds can inhibit cultivation and thereby can cause false-negative test results. In BRD, differential diagnosis of these pathogens with rapid turnaround time procedure is essential to implement appropriate treatment and intervention measures in a timely manner. Rapid detection of these pathogens at the early stage of outbreak can contribute substantially to minimize the spread of infection and increase treatment efficiency. Today quick, highly sensitive and species-specific PCRs are used in the diagnosis of *Mycoplasma*-associated diseases for *M. dispar* [14, 15], *M. bovis* [4, 16] and *M. bovirhinis* [17] in BALF or nasal swabs. Combining a 16S Ribosomal DNA PCR with denaturing gradient gel electrophoresis fingerprinting (PCR/DGGE) enabled the simultaneous detection of mixed *Mycoplasma* populations, however information about the detection limit in clinical samples is limited [18]. Additionally, a DNA microarray assay was developed for the parallel detection of 37 *Mycoplasma* species [19], in which species-specific probes derived from the 23S rRNA and *tuf* genes were used for species differentiation.

Multiplex real-time PCR could be a promising and practical approach to speed up the differential diagnosis from 1 to 2 weeks for traditional culture to 24 h, with limited expenses. This will make diagnostic testing more accessible for veterinary practitioners and thereby improve BRD diagnosis. This report describes the RespoCheck triplex PCR developed by Central Veterinary Institute (CVI, Lelystad, The Netherlands) for detection of three *Mycoplasma* species.

Methods

Strains and growth conditions

M. bovis (ATCC 25025) and *M. bovirhinis* (ATCC 5189985) were purchased from the ATCC (United Kingdom (U.K.), Guernsey, Ireland, Jersey and Liechtenstein) and cultured in Heart Infusion Broth Medium (Difco, Detroit, Mich.). All isolates were grown at 37 °C and 5% CO_2 for seven days in a modified standard mycoplasma broth medium [20] containing 19 g of Heart Infusion Broth, 50 mL of liquid yeast extract (10% [vol/vol]; Oxoid, London, United Kingdom), 2 × 10^6 U of penicillin G (Hoechst, Frankfurt, Germany), and 200 mL of heat-inactivated (56 °C, 30 min) horse serum per liter. Stocks of each isolate were prepared by freezing 1 mL portions of a 10 mL logarithmic-phase broth culture with 15% glycerol at −80 °C. Cultures were titrated on Heart Infusion Agar and were shown to contain 7 × 10^6 cfu/mL for *M. bovis* and 4 × 10^5 cfu/mL for *M. bovirhinis. M. dispar* NCTC 10125) was provided by Helena Windsor *(Mycoplasma* Experience LTD, Bletchingley, UK) with a titre of 1.6 × 10^7 cfu/mL. In addition, DNA from 14 *Mycoplasma* strains (Table 1) were provided by Prof. Konrad Sachse (Friedrich-Loeffler-Institut, Federal Research Institute for Animal Health, Bundesforschungsinstitut für Tiergesundheit, Jena, Germany).

Hundred and seven bacterial isolates, representing 39 different species, were used to evaluate of specificity of the RespoCheck *Mycoplasma* triplex real-time PCR assay (Table 2). These included isolates associated with BRD and isolates associated with other bovine diseases. Prior to testing by PCR, the identity of the isolates was confirmed using MALDI-TOF mass spectrometry (MS Bruker MALDI Biotyper Microflex, version 3.1 with the reference database version 3.1.66 Bruker Daltonics GmbH, Germany).

Field samples and isolation of DNA

Calves (n = 44) with or without BRD (increased respiratory rate and/or dyspnoea) were sampled for diagnostic purposes. Sampling of the calves was granted an exemption from requiring ethics approval by the institutional Animal Experiment Commission "Dier Experimenten Commissie (DEC) Lelystad (2013111.b)" because sampling was performed for diagnostic purposes. BAL samples were obtained as described [21]. Approximately 35–75 ml BAL was obtained from each calf after instillation of 100 ml PBS with 10% Fetal Calf serum (FCS). Foam, large purulent exudates and blood clots were removed from the BALF samples under aseptic conditions. BALF (25 mL) was centrifuged (4600×g, 10 min, 4 °C). Sediment was resuspended in 0.5 mL Dulbecco's minimal essential medium (DMEM) with 5% FCS, carefully added to 1 mL freeze medium (DMEM, 50% FCS and 20% DMSO) and frozen at −80 °C. The BALF supernatants were also stored at −80 °C.

For testing the influence of centrifugation of BALF samples (4600×g, 10 min, 4 °C) on the PCR results we tested three variants of BALF samples: without centrifugation, supernatant and pellet obtained after centrifugation (50 times concentrated). DNA was extracted from 200 µL aliquots of BALF samples. We used the MagNA Pure LC Total Nucleic Acid Isolation Kit (Roche Applied Science), with the Total NA External_lysis" protocol (Version 2.11). With the MagNA Pure LC Total Nucleic Acid Isolation Kit) 32 samples can processed per run. In all runs a positive control (a mix of 1.4 × 10^6 cfu/mL *M. bovis,* 0.5 × 10^7 cfu/mL *M dispar*

Table 1 *Mycoplasma* strains ($n = 17$), which were used as reference material

Species (Type strain)	ID	Ct-values		
		M. bovis	*M. dispar*	*M. bovirhinis*
M. agalactiae (PG2)	R 41[b]	20.9	-	-
M. alkalescens PG 31/D 12	R 18[b]	-	35.7[a]	-
M. bovis PG45	R 9[b]	19.4	-	-
M. bovirhinis PG43	R 12[b]	-	-	25.8
M. bovigenitalium PG11	R 8[b]	-	-	-
M. californicum ST-6	R 26[b]	-	-	-
M. canadense 275C	R 22[b]	-	-	-
M. canis, PG14	R 74[b]	-	-	20.0[a]
M. dispar 462/2.	R 11[b]	-	18.7	-
M. leachii PG50 (former M. bovine group VII)	R 23[b]	-	-	-
M. mycoides subsp. Mycoides PG1 (former Small Colony Type)	R 84[b]	-	-	-
Acholeplasma axanthum S743	R 17[b]	-	33.1[a]	-
A. laidlawii PG8	R 10[b]	-	-	-
A. oculi 19-L	R 62[b]	-	-	-
M. bovis	ATCC 25025	23.9	-	-
M. dispar (NCTC 10125)	ATCC 27140	-	20.3	-
M. bovirhinis	ATCC 5189985	-	-	20.9

[a]Cross-reactions in the RespoCheck triplex Mycoplasma PCR
[b]ID Friedrich-Loeffler-Institut

and 1.3×10^5 cfu/mL *M. bovirhinis*) and a negative water control (NTC) was included.

RespoCheck primers and probes

To enable testing of testing for BRD associated pathogens in a routine setting, real-time PCRs for detection of viral, bacterial and mycoplasma pathogens in bronchoalveolar lavage fluid (BALF) of calves have been set up by the Central Veterinary Institute (Lelystad, The Netherlands) under the name RespoCheck. Primers and probes specific for the bacterial 16S, V3 and V4 regions were based on the Full length, bacterial 16S sequences (50,000 in July 2012) were used from the *nuccore* database at the National Center for Biotechnology Information (NCBI, USA, http://www.ncbi.nlm.nih.gov/nuccore). For *M. bovirhinis* and *M. dispar* the nearly full length 16S sequences were used. These sequences and their taxonomic information were used to build an Insignia-based database [22] from which pathogen-specific sequence regions were extracted with special interest for the V3 and V4 region because these sequences are often targeted for metagenomic next-generation sequencing (NGS) [23]. Using the identified regions, primers and probes were designed with AlleleID 7.8. (Premier Biosoft, palo Alto, USA). The resulting triplex PCR was designated RespoCheck *Mycoplasma* triplex real-time PCR The specificity of the *Mycoplasma* primers and probes was also verified against V3-V4 partial sequences of *M. flocculare, M. ovipneumonia* and *M. hyopneumonia.*

RespoCheck triplex and single real-time PCR

The QuantiFast triplex Kit Real Time-PCR kit (Qiagen) was used for the RespoCheck *Mycoplasma* triplex real-time PCR. The assays were conducted in a 20 µl reaction mix containing 5 µl of the nucleic acid sample, 250 nM of each primer, 100 nM of each MGB probe, 1× QuantiFast triplex Real Time-PCR Master Mix and sterile deionised water. All reactions were conducted with an ABI-7500 with the following cycling parameters: 95 °C for 15 min, followed by 40 cycles of 94 °C for 15 s and 60 °C for 60 s. The machine was set to acquire fluorescence on the FAM, VIC, and NED channels for respectively *M. bovis, M. dispar and M. bovirhinis* All primers and probes were obtained from Life Technologies Europe BV (Bleiswijk, the Netherlands). The final results were analysed using ABI-7500 software (Version 1.4). Samples with a Ct of 40 cycles or less were considered to be positive.

Evaluation of the analytical sensitivity and the analytical specificity

The analytical sensitivity of the RespoCheck triplex PCR was defined as the ability to detect the lowest concentration of *M. bovis, M. dispar* and *M. bovirhinis* expressed as a concentration (cfu/mL) [24]. The analytical

Table 2 Bacterial strains (n = 107), that were used as reference material

Identification (number of isolates tested)	CCUG identification[c]	Source
Acidovorax spp. (3)	NA[c]	CVI collection[a]
Actinomyces	NA	CVI collection[a]
Aerococcus viridans	NA	CVI collection[b]
Bibersteinia trehalosi	*Pasteurella trehalosi*	CCUG 37711
Biberstenia trehalosi 20 AA III 3 E3	NA	CVI collection[a]
Biberstenia trehalosi 21 AA III 3 E4	NA	CVI collection[a]
Brucella abortus	NA	CVI collection[b]
Comamonas kerstersii	NA	CVI collection[a]
Corynebacterium bovis (2)	NA	CVI collection[b]
Corynebacterium pseudotuberculosis	NA	CVI collection[b]
Escherichia coli	NA	CVI collection[b]
Gallibacterium anatis (5)	NA	CVI collection[a]
Hafnia alvei	NA	CVI collection[a]
Histophilus somni	NA	ATCC 22132[e]
Histophilus somni (4)	NA	CVI collection[a]
Klebsiella oxytoca	NA	CVI collection[b]
Klebsiella pneumoniae	NA	CVI collection[b]
Lactobacillus mucosae	NA	CVI collection[a]
Lactococcus garvieae	NA	CVI collection[b]
Lactococcus lactis	NA	CVI collection[b]
Listeria monocytogenes	NA	CVI collection[b]
Mannheimia heamolytica	NA	ATCC 14003
Mannheimia heamolytica	NA	CVI collection[a]
Mannheimia haemolytica	*Mannheimia glucosida*	CCUG 38457-T
Mannheimia granulomatis	*Mannheimia granulomatis*	CCUG 45422-T
Mannheimia granulomatis 25 AA III 3 E8	NA	CVI collection[a]
Mannheimia haemolytica	*Mannheimia ruminalis*	CCUG 38470-T
Mannheimia haemolytica (5)	NA	CVI collection[a]
Mannheimia haemolytica 3 AA III 2 H2	NA	CVI collection[a]
Mannheimia varigena	*Mannheimia varigena*	CCUG 38462-T
Mannheimia varigena 19 AA III 3 E2	NA	CVI collection[a]
Mannheimia varigena 24 AA III 3 E7	NA	CVI collection[a]
Micrococcus luteus	NA	CVI collection[b]
Moraxella bovis	NA	CVI collection[b]
Moraxelle lacunata (2)	NA	CVI collection[a]
Mycobacterium avium subsp. paratuberculosis	NA	CVI collection[b]
Mycobacterium bovis	NA	CVI collection[b]
Mycobacterium tuberculosis	NA	CVI collection[b]
Neisseria zoodegmatis	NA	CVI collection[a]
Pantoea agglomerans Erwina herbicola (n = 13)	NA	CVI collection[a]
Pasteurella multocida	NA	ATCC 15743[e]
Pasteurella multocida	NA	CVI collection[a]
Pasteurella multocida	Bisgaard Taxon 13	CCUG 16497[d]

Table 2 Bacterial strains (n = 107), that were used as reference material *(Continued)*

Pasteurella multocida	Bisgaard Taxon 13	CCUG 16498[d]
Pasteurella multocida	*Pasteurella multocida* ss *gallicida*	CCUG 17978-T[d]
Pasteurella multocida	*Pasteurella multocida* ss *septica*	CCUG 17977-T[d]
Not typable	*Pasteurella aerogenes*	CCUG 27905[d]
Proteus mirabillis	NA	CVI collection[a]
Pseudomonas aeruginosa	NA	CVI collection[b]
Psychrobacter spp.	NA	CVI collection[a]
Salmonella enteritica ssp. *enteritica* serovar Dublin	NA	CVI collection[b]
Salmonella enteritica ssp. *enteritica* serovar Typhinurium	NA	CVI collection[b]
Serratia marcescans	NA	CVI collection[b]
Staphylococcus aureus	NA	CVI collection[b]
Staphylococcus epidermidis	NA	CVI collection[b]
Streptococcus agalactiae	NA	CVI collection[b]
Streptococcus bovis (5)	NA	CVI collection[a]
Streptococcus dysgalactiae	NA	CVI collection[b]
Streptococcus faecalis	NA	CVI collection[b]
Streptococcus hyointestinalis	NA	CVI collection[a]
Streptococcus pluranimalium (5)	NA	CVI collection[a]
Streptococcus pneumoniae	NA	CVI collection[b]
Streptococcus spp. (3)	NA	CVI collection[a]
Streptococcus uberis	NA	CVI collection[b]
Trueperella pyogenes	NA	ATCC 9731[e]
Trueperella pyogenes (5)	NA	CVI collection[a]
Yersinia enterolytica	NA	CVI collection[b]
Yersinia pseudotuberculosis	*Pasteurella lymphangitidis*	CCUG 27188-T[d]

[a]Isolated from lungs of calves
[b]Isolated from other tissues of cattle as lungs
[c]Not applicable
[d]CCUG: Culture Collection University of Götenborg, Sweden
[e]ATCC: American Type Culture Collection, USA
All bacterial strains, except the CCUG strains, were from an in-house strain collection

sensitivity of the single and triplex PCRs for *M. bovis, M. dispar* and *M. bovirhinis* was determined with DNA isolated from 200 μL culture (*M. bovis, M. dispar* and *M. bovirhinis* strain) in a volume of 200 μL elution buffer at a final DNA concentration of 10 ng/ μL. This DNA preparation was tested in seven 10-fold serial dilutions (5 μL per assay) in PBS, resulting in a range with 10 ng down to 1 fg *Mycoplasma* DNA per assay. The Ct was determined for each sample by single and RespoCheck triplex real-time PCR with a threshold of 50% of the Delta Rn value (log). The threshold was manually set at 0.04 in the linear phase of the amplification plot, whereby the Slope and Correlation Coefficient values were 3.22 and 99.99% respectively.

The analytical sensitivity of the *M. bovis, M. dispar* and *M. bovirhinis* single and RespoCheck triplex real-time PCR, was also determined by testing a mixture of *M. bovis* (3×10^6 cfu/mL), *M. dispar* (3×10^6 cfu/mL) and *M. bovirhinis* (3×10^5 cfu/mL) in seven 10-fold serial dilutions in BALF of specific pathogen free (SPF) calves of 3–4 weeks old. Dilution resulted in a series of *M. bovis, M. dispar* and *M. bovirhinis* spiked BALF samples, ranging from 3×10^6 cfu/mL down to 0.3 cfu/mL. Total DNA was isolated from each 200 μl sample with the MAGNA pure isolation kit and the Ct was determined for each sample (5 μl) by both the single and RespoCheck triplex PCR assays. The slope of the curve, the efficiency and the detection limit (for DNA ng/μl; for cells cfu/mL) for each PCR was determined. To determine the analytical specificity of the designed RespoCheck triplex PCR, 17 *Mycoplasma* isolates and 107 bacterial strains (Table 2) were tested.

Diagnostic sensitivity and specificity in BALF samples from calves.

For determining the diagnostic specificity, BALF samples were analysed with the PCR/DGGE method by the

Animal and Plant Health Agency (APHA, Mycoplasma Team, Addlestone Surrey, UK) as earlier described [18, 25]. To determine the analytical sensitivity of the PCR/DGGE analysis, four 10-fold serial dilutions of *M. bovis* (7×10^4 cfu/mL), *M. dispar* (16×10^4 cfu/mL), and *M. bovirhinis* (0.5×10^4 cfu/mL), were prepared in PBS. Samples were sent to the APHA and analysed using the PCR/DGGE method.

Sequencing amplicons

16S rDNA PCR-sequencing was used for confirmation of the results of RespoCheck *Mycoplasma* triplex real-time PCR. 16S rDNA of the DGGE positive /PCR positive (n = 5) and DGGE negative /PCR positive (n = 5) was amplified using the specific *Mycoplasma* primers of the RespoCheck *Mycoplasma* triplex real-time PCR. DNA was sequenced by BaseClear (Leiden, the Netherlands) by an automated DNA sequencer. The nucleotide sequences were compared with GenBank sequences using the Basic Local-Alignment Search Tool(BLAST) of the NCBI-NIH for homology [26]. Pairwise sequence alignments were performed using the Clustal algorithm implemented in the program DNA star (DNASTAR Inc., Madison, WI).

Analyses of sensitivity and specificity

The analytical sensitivity of the RespoCheck triplex PCR was determined by its ability to detect a low concentration of *M. bovis, M. dispar* and *M. bovirhinis* and therefore expressed as a concentration (ng/assay and cfu/mL) [24]. The analytical specificity of the assay was calculated for each target microorganism using the following definition for specificity as the percentage of true negative samples/ the number of true negative samples and the number of false positive samples [27].

Calculation of diagnostic sensitivity, specificity and Cohen's Kappa Coefficient was performed as described [28]. We therefore used the results of the PCR/DGGE analysis as reference standard.

Statistical analyses

Differences in PCR results were analysed for statistical significance by the non-parametric Mann–Whitney U test in the GraphPad Prism version 5.0 software, with $P < 0.05$ considered significant.

Results

Analytical sensitivity and linear detection range of the RespoCheck triplex

The linearity of quantification of the RespoCheck triplex *Mycoplasma* real-time PCR was established through a linear regression plot by plotting the Ct-values against the values of log10 DNA concentration tested per reaction. The *M. dispar* single and RespoCheck triplex real-time PCR showed a linear detection range from 10 ng to 10 fg DNA per assay with a linear correlation (R^2) value of 0.999 (Table 3; Fig. 1.). The *M. bovis* and *M. bovirhinis* single and RespoCheck real-time PCR showed a linear detection range from from 1 ng to 100 fg DNA per assay, with a R^2 value of 0.999 (Table 3; Fig. 1). In BALF

Table 3 Performance of the RespoCheck *Mycoplasma* real-time PCR in which the *M. dispar, M. bovis* and *M. bovirhinis* DNAs were diluted in PBS (A) or cells were spiked in BALF (B)

A

PCR	Agent	Real time PCR				
		R^2	Slope	Efficiency (%)	Linearity (ng)	Detection limit (ng/assay)
Singleplex PCR	*M. dispar*	0.9995	−3.3836	97.49	10 ng-10 fg	10 fg
	M. bovis	0.9966	−3.1137	109.49	10 ng-10 fg	10 fg
	M. bovirhinis	0.9955	−3.5033	92.95	10 ng-10 fg	10 fg
Triplex PCR	*M. dispar*	0.9989	−3.1175	109.3	10 ng-10 fg	10 fg
	M. bovis	0.9955	−3.6240	88.8	10 ng-100 fg	100 fg
	M. bovirhinis	0.9939	−3.3735	97.9	10 ng-100 fg	100 fg

B

PCR	Agent	Real time PCR				
		R^2	Slope	Efficiency (%)	Linearity (CFU/ml; log 10)	Detection limit (CFU/assay)
Singleplex PCR	*M. dispar*	0.995	3.248 ± 0.1276	103.2	6.5–2.5	1–2
	M. bovis	0.995	3.453 ± 0.1178	94.8	6.5–1.5	0.5
	M. bovirhinis	0.981	3.395 ± 0.2698	97.0	5.5–1.5	0.5
Triplex PCR	*M. dispar*	1.000	3.534 ± 0.04608	91.9	6.5–2.5	1–2
	M. bovis	0.993	3.462 ± 0.1440	94.5	6.5–1.5	0.5
	M. bovirhinis	0.965	2.750 ± 0.3014	131.0	5.5–1.5	0.5

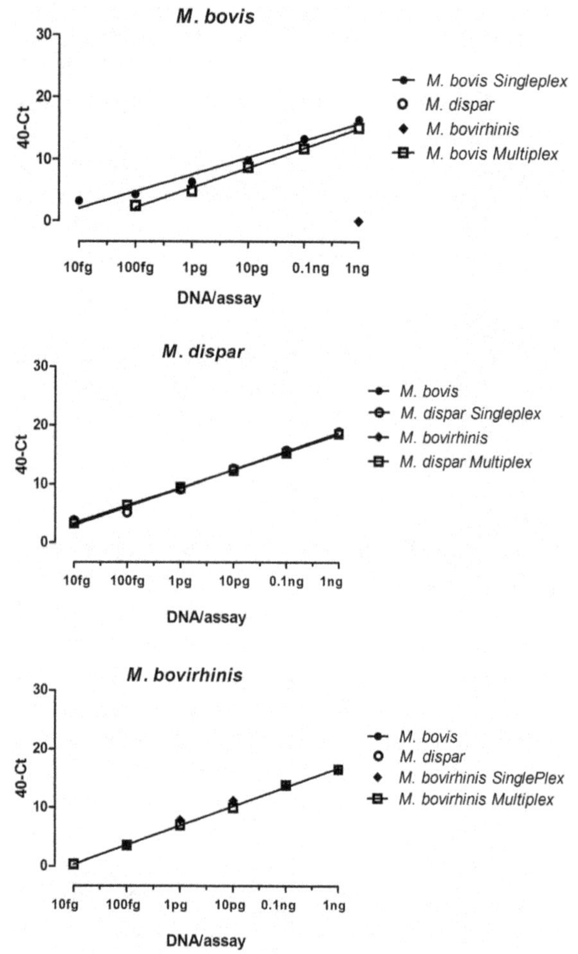

Fig. 1 Analytical sensitivity of *M. bovis, M. dispar* and *M. bovirhinis* in single and RespoCheck triplex PCR assays. Ten-fold serial dilutions of *M. bovis, M. dispar* and *M. bovirhinis* DNA were made in PBS in a range from 1 ng down to 10 fg/assay. The resulting samples were tested in the *M. bovis, M. dispar* and *M. bovirhinis* single and RespoCheck triplex real-time PCR

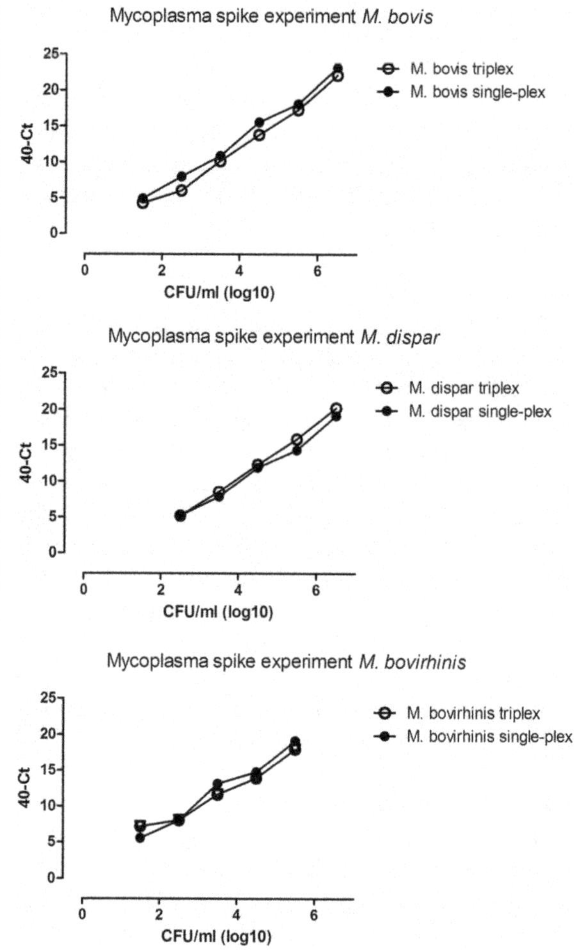

Fig. 2 Analytical sensitivity of *M. bovis, M. dispar* and *M. bovirhinis* single and RespoCheck triplex real-time PCR in spiked BALF samples. Ten-fold serial dilutions of *M. bovis, M. dispar* and *M. bovirhinis* were made in BALF samples in a range from 3×10^6 down to 0.3 cfu/mL. The resulting samples were subjected to DNA isolation and testing in the *M. bovis, M. dispar* and *M. bovirhinis* single and RespoCheck triplex real-time PCR

spiked samples, the detection limit of the RespoCheck triplex real-time PCR was 300 cfu/mL for *M. dispar*, and 30 cfu/mL for *M. bovis* or *M. bovirhinis* (Table 3; Fig. 2). In the RespoCheck *Mycoplasma* real-time PCR, 5 μL was tested and the analytical sensitivity was therefore 1–15 cfu/assay. A good linear correlation ($R^2 > 0.96$) was found between the values of BALF spiked samples and the Ct-values in the RespoCheck *Mycoplasma* triplex or singleplex real-time PCR for the three Mycoplasmas (Table 3).

Analytical specificity of the RespoCheck triplex PCR

RespoCheck Mycoplasma triplex PCR in silico BLAST searchs (http://blast.ncbi.nlm.nih.gov/Blast.cgi) for the specificity of the *M. dispar* amplicon revealed a 100% identity (E-value $1E^{-46}$) for 2 hits for *M. dispar* complete genome sequence. The in silico BLAST search for amplicon of *M. bovis* we found a 99–100% identity (E-values $4E^{-43}$ - $6E^{-45}$) to 31 complete genome or 16S ribosomal partial sequence, and for the amplicon of *M. bovirhinis* we found a 99–100% identity (E-values $5E^{-50}$ - $1E^{-51}$.) For the *M. bovirhinis* amplicon a 97% identity (*3E-40*) was found for 11 hits for *Mycoplasma canis* (Taxid:29,555).

Seventeen *Mycoplasma* strains (Table 1) and 107 bacterial strains (Table 2) were used to calculate the analytical specificity of the RespoCheck *Mycoplasma* triplex PCR. The RespoCheck *Mycoplasma* triplex real-time assay for detecting *M. bovis, M. dispar* and *M. bovirhinis* possessed an analytical specificity of 100% (0 FP), 98.2% (2 FP) and 99.1% (1 FP), respectively. No cross-reactivity in the RespoCheck *Mycoplasma* triplex real-time assay was observed with any of the 107 bacterial strains. The *M. dispar* PCR did however cross-react with *Acholeplasma axanthum*

S743 and *M. alkalescens* PG 31/D 12. In silico sequence data analyses from the 16S V3 genomic DNA region showed no similarity of *M. dispar* specific sequences with the *A. axanthum* S743 and *M. alkalescens* PG 31/D 12 isolates. The *M. bovirhinis* RespoCheck triplex PCR cross-reacted with *M. canis* with a Ct-value of 20.4. In the *M. bovis* RespoCheck triplex PCR we found a cross-reaction with *M. agalactiae*. Based on the almost 100% similarity of *M. canis* PG14 16S rRNA gene and the *M. bovirhinis*16S rRNA, it is not possible to prevent for this cross-reaction.

Diagnostic sensitivity and specificity of the RespoCheck triplex compared with DGGE

To study the influence of centrifugation of the BALF samples on the PCR results we compared the PCR results from the BALF samples before and after centrifugation (10 min at 4600×g). A significant lower Ct-value ($P < 0.05$; non-parametric Wilcoxon statistics) in the RespoCheck *Mycoplasma* triplex real-time PCR was found for *M. bovis* and *M. dispar* in the pellet of the centrifuged BALF samples. Several *M. bovis, M. bovirhinis* and *M. dispar* mix-infections could be detected in one BALF sample with a difference of 10 Ct-values between the three species and were in accordance with the PCR/DGGE analysis (Fig 3). Therefore we used the pellet of the centrifuged BALF samples (50× concentrated) to determine the presence of the three *Mycoplasma* species in 44 BALF samples by real-time PCR.

The calculated diagnostic sensitivity and specificity the RespoCheck triplex PCR is reported in Table 4. As the diagnostic specificity is very low (0.1944, 0.739, 0.3889 for *M. dispar, M. bovis* and *M. bovirhinis* respectively) we analysed the sequence of the produced amplicon of five DGGE negative /PCR positive and five DGGE positive PCR positive samples. The sequence of both products was confirmed as *M. bovis, M. dispar* or *M. bovirhinis,* as all sequences had a high E-value (3e-44) and 100% Query cover (100%) against the homologue sequence using the BLAST of the NCBI-NIH. Comparison of the Ct-values of PCR positive/ DGGE negative and the PCR positive/ DGGE positive samples with a non-parametric Mann Whitney test, showed that the Ct values of *M. dispar* and *M. bovis* were significantly lower, $p = 0.0026$ and 0.0282, respectively. In the *M. bovis* and *M. dispar* PCR, the difference in Ct value between PCR positive/ DGGE positive and PCR positive/ DGGE negative samples is at least 3.2, which indicates a factor of 10 difference in concentration of *M. bovis* and *M. dispar* DNA between these two groups (Fig. 4). As a consequence the diagnostic

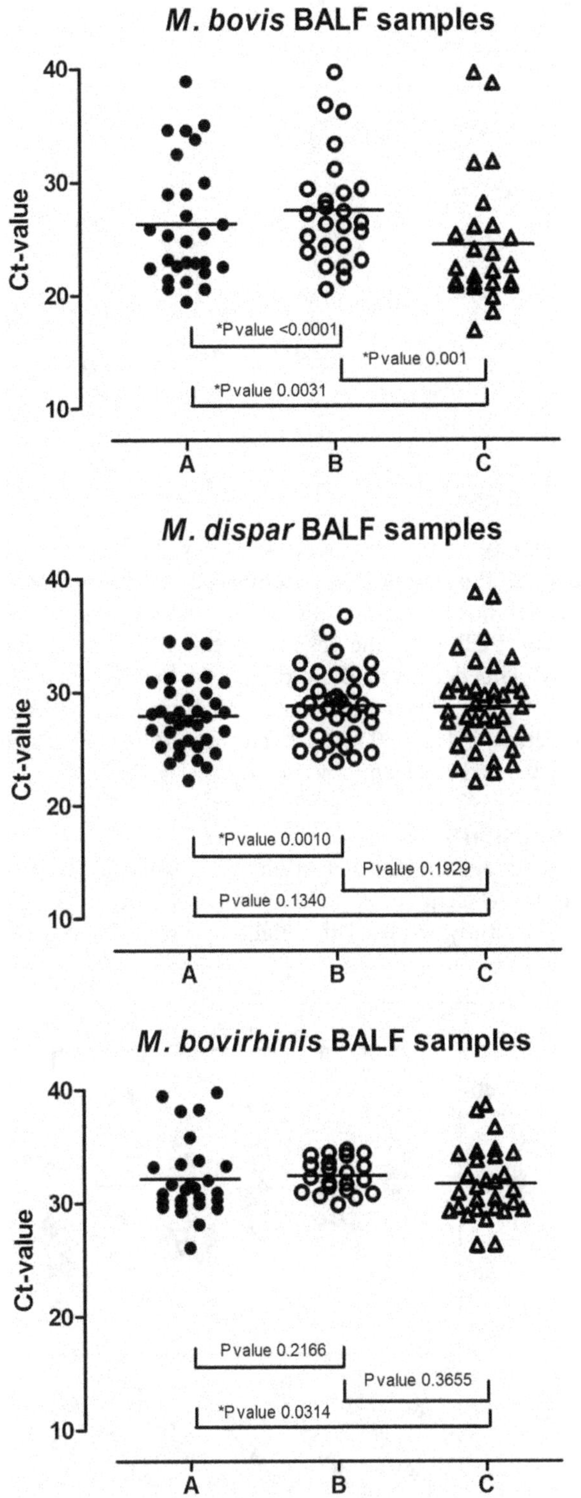

Fig. 3 Ct-values obtained in the RespoCheck *Mycoplasama* triplex real-time PCR on DNA samples derived from BALF samples from *M. bovis, M. dispar and M. bovirhinis* infected calves. PCRs were performed on three variants of BALF samples: without centrifugation (A), supernatant after centrifugation (B) and sediment of after centrifugation (50 times concentrated) (C)

Table 4 Diagnostic sensitivity and specificity of the RespoCheck *Mycoplasma* triplex real-time PCR compared with the PCR/DGGE method

M. dispar	PCR +	PCR -	Total	Diagnostic specificity and sensitivity
DGGE +	8	0	8	Sensitivity =1
DGGE -	29	7	36	Specificity =0.1944 (95% CI: 0.0819–0.3602)
Total	37	7	44	
M. bovis	PCR +	PCR -	Total	Diagnostic specificity and sensitivity
DGGE +	20	1	21	Sensitivity =0.9524 (95% CI: 0.7618–0.9988)
DGGE -	6	17	23	Specificity =0.7391 (95% CI: 0.5159–0.8977)
Total	26	18	44	
M. bovirhinis	PCR +	PCR -	Total	Diagnostic specificity and sensitivity
DGGE +	7	1	8	Sensitivity =0.8750 (95% CI: 0.4735–0.9968)
DGGE -	22	14	36	Specificity =0.3889 (95% CI: 0.2314–0.5654)
Total	29	15	44	

specificity of the RespoCheck triplex PCR is undervalued by this method. We compared the results of the *M. bovis, M. dispar* and *M. bovirhinis* RespoCheck triplex PCR with the results of the PCR/DGGE analysis. The detection limit of the *M. bovis, M. dispar* and *M. bovirhinis* PCR/DGGE analysis was, as determined by APHA, 0.7 × 10^3 cfu/mL, 16 × 10^3 cfu/mL and 0.5 × 10^3 cfu/mL, respectively (Fig. 5).

Discussion

PCR assays for the detection of *Mycoplasmas* generally target sequences on the 16S rRNA gene [29, 30]. In this study we used the highly conserved 16S rRNA sequence to set up the RespoCheck *Mycoplasma* triplex real-time PCR assay for the specific detection of *M. bovis, M. dispar* and *M. bovirhinis* in BALF samples of calves.

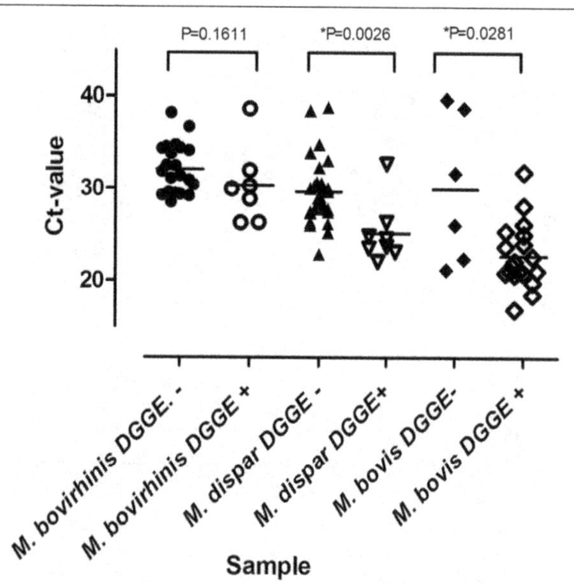

Fig. 4 The Ct-level of DNA derived from BALF samples from *M. bovis, M. dispar* and *M. bovirhinis* infected calves of PCR/DGGE analyses of DGGE APHA negative and positive samples. Significant *P* values are indicated by *

The lowest concentration of *M. dispar* which could be detected with the RespoCheck triplex PCR assay is around 300 cfu/mL. With a copy number of 16S rRNA of one or two (https://rrndb.umms.med.umich.edu/) and with a test volume of 5 µl the lowest concentration which could be detected is around 1–2 cfu/assay. The lowest concentration of *M. bovis* and M. *bovirhinis* which could be detected with the RespoCheck triplex for *M. bovis,* and *M. bovirhinis* is around 0.5 cfu/assay. From the calculated analytical sensitivity of the *M. bovis, M dispar* and *M. bovirhinis* RespoCheck triplex PCR (0.5–2 cfu/assay) we conclude that the RespoCheck triplex PCR has a good analytical sensitivity. It was shown that the use of a pellet from 25 mL BALF after centrifugation instead of not-centrifuged BALF samples increased the analytical sensitivity of the RespoCheck triplex PCR assay. In order to determine the analytical specificity of the RespoCheck triplex PCR we analysed the DNAs from panels of *Mycoplasma* and bacterial strains. In the *M. bovis* RespoCheck *Mycoplasma* triplex real-time PCR we found a cross-reaction with *M. agalactiae.* Phylogenetic analyses on 16S rRNA sequences and comparing the 16S rRNA sequences of *M. bovis* and *M. agalactiae* [25] at NCBI (www.ncbi.nlm.nih.gov), we found a close relationship between *M. agalactiae* and *M. bovis,* with a 99% nucleotide identity between their 16S rRNA sequences. However, *M. bovis* causes calf pneumonia, mastitis, and arthritis in cattle [16, 31], *M. agalactiae* is the causal agent of contagious agalactia in goats and sheep [32]. Although unusual, *M. agalactiae* has been detected from cattle samples [33, 34]. Therefore the cross reactivity for *M. agalactiae* might be a problem for the intended BALF samples in the *M. bovis*

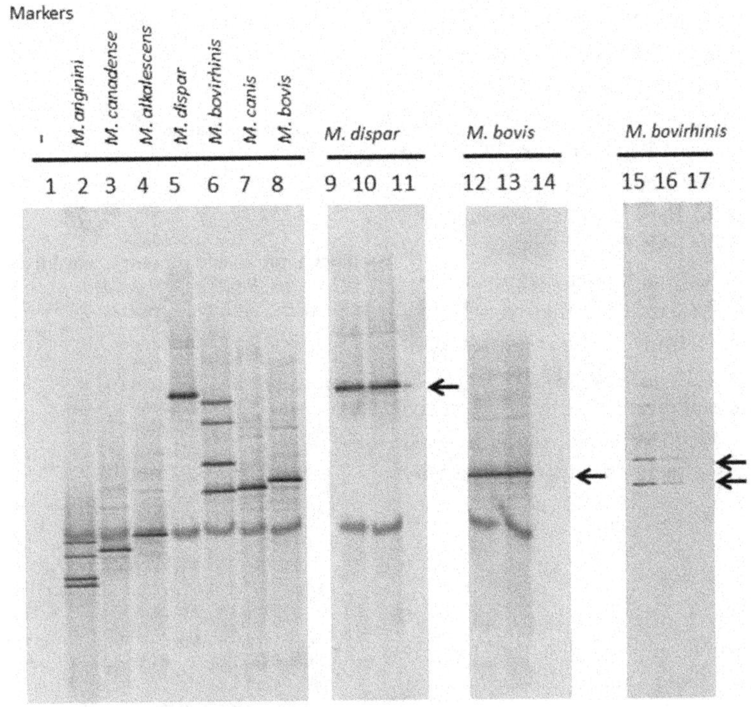

Fig. 5 DGGE fingerprinting profiles of 16S ribosomal DNA fragments obtained after amplification by PCR. Lane 1 contains the negative water control, lanes 2 to 8 contain several Mycoplasma strains as reference (lane and strain designations indicated), lanes 9, 10 and 11 contain three 10-fold serial dilutions of *M. dispar* (starting with 16×10^4 cfu/mL), lanes 12, 13 and 14 contain three 10-fold serial dilutions of *M. bovis* (starting with 7×10^3 cfu/mL) and lanes 15, 16 and 17 contain three 10-fold serial dilution of *M. bovirhinis* (starting with 0.5×10^4 cfu/mL)

PCR. In the *M. bovirhinis* RespoCheck triplex real-time PCR one false positive reaction was obtained on DNA from *M. canis. M. canis* can be isolated from the reproductive tract of dogs, but has not been proved to cause disease in dogs. However, it has been shown to cause clinical signs of pneumonia in experimentally challenged calves [35] and *M. canis* has been isolated from ruminants in Britain [36, 37]. Depending on the incidence of *M. canis* in ruminants, this may give false-positive results in the *M. bovirhinis* RespoCheck triplex real-time PCR. DNA samples from *M. alkalescens* and *A. axanthun* showed high Ct-values (>35) for *M. dispar* in the RespoCheck triplex real-time PCR, and were therefore classified as false-positive (Ct of 40 cycles or less were considered to be positive). *M. alkalescens* and *M. bovigenitalium* are important *Mycoplasmas* that can infect cattle and cause mastitis, arthritis and respiratory disease [17]. However, in the sequence analyses of the PCR-positive and DGGE-negative *M. dispar* BALF samples, we did not find any indication for the presence of *M. alkascens*, underlining the high specificity for *M. dispar* in the RespoCheck triplex real-time PCR.

Monitoring for Mycoplasma species in BALF samples through collection and testing of BALF samples by culture is hampered by the fastidious nutritional requirements, lengthy culture of mycoplasmas, and their susceptibility to growth inhibitors. As a consequence, Mycoplasma culture is time-consuming, costly, and requires specific expertise. Moreover, Mycoplasma species may easily be overgrown by bacterial contaminants or by more rapidly growing *Mollicutes,* notably *Acholeplasmas.* The PCR/DGGE method of the APHA can differentiate 13 bovine *Mycoplasma* species [18] including the target Mycoplasmas of the RespoCheck *Mycoplasma* triplex real-time PCR and in contrary to the RespoCheck can differentiate between *M. bovis* and *M. canis.* Additional the PCR/DGGE is capable of detecting mixed cultures, which would have been difficult to detect by culture methods [18]. Therefore we used this method as a reference for determining the diagnostic sensitivity and specificity of the RespoCheck *Mycoplasma* triplex real-time PCR.

Possibly due to the lower sensitivity of the DGGE analysis compared to the RespoCheck triplex PCR (almost factor 10) and its use as reference method to validate the RespoCheck triplex PCR, the latter test method scores 29, 6 and 22 *M. dispar, M. bovis* and *M. bovirhinis* respectively out of 44 more samples as false-positive and therefore the diagnostic specificity of the RespoCheck triplex PCR is underestimated. The transport and storage conditions or differences in DNA preparation of particularly the more diluted BALF samples for the PCR/DGGE method could have induced a lower sensitivity of the PCR/DGGE analysis. The Ct values of the *M. bovis,* and *M. dispar* PCR positive and DGGE positive samples are significant ($P < 0.05$ Mann Whitney test) lower than the *M. bovis,* and *M. dispar* PCR positive DGGE negative samples, which confirms the difference in the analytical sensitivity between the RespoCheck triplex PCR and DGGE analyse. In the *M. bovis* and *M. dispar* PCR, we found a 10 fold difference in the Ct values between the DGGE positive/ PCR positive and DGGE negative/ PCR positive samples, which indicates a higher diagnostic sensitivity of *M. bovis* and *M. dispar* PCR than the DGGE analyses. Results by DGGE from BALF samples with mixed infections could be reproduced by the triplex PCR, suggesting that there is no significant PCR bias when the triplex PCR is used for Mycoplasma detection in field samples. The PCR has thus a higher analytical sensitivity than the DGGE.

Conclusion

In conclusion, the RespoCheck *Mycoplasma* triplex PCR-test appears to be a sensitive and specific test for the detection of *M. bovis, M. dispar* and *M. bovirhinis* in BALF samples of calves.

Abbreviations

BALF: Bronchoalveolar lavage fluid; BRDC: Bovine respiratory disease complex; PCR/DGGE: PCR with denaturing gradient gel electrophoresis fingerprinting

Acknowledgments

The authors thank Prof. Konrad Sachse (Friedrich-Loeffler-Institut, Federal Research Institute for Animal Health, Jena, Germany) for providing us with DNA from 14 *Mycoplasma* isolates. The authors thank also The Mycoplasma Team (APHA Woodham Lane Addlestone, Surrey, UK) for performing the PCR/DGGE analysis.

This research was part of the project "Development and application of diagnostics" within the public-private partnership "One Health for Food" in The Netherlands.

Funding

The research was funded by the Dutch Ministry of Economic Affairs, Productschap Vee en Vlees / Stichting Brancheorganisatie Kalversector (SBK) and supported by the Van Drie Group, MSD Animal Health and Denkavit.

Authors' contributions

JC authored the manuscript, designed the study, performed the qPCR and analysed the data. AB and FB build an Insignia-based database from which pathogen-specific sequence regions were extracted, designed the oligonucleotide primers and probes and assisted in drafting and editing the manuscript. HW and FW assisted in study design, interpretation of data and editing the manuscript. MK was involved in the microbiological analyses of the samples and participated in the drafting of the manuscript. AA designed and coordinated the field study for the collection of BALF samples and also helped to draft this manuscript. BS conducted and coordinated the field sample collection, performed microbiological analysis of samples and managing the database with results. BK participated in the design of the study and helped in the interpretation of the triplex PCR data. All authors read and critically revised and approved the final manuscript.

Competing interests

The authors declare that they have no competing interests.

Consent for publication

Not applicable.

Ethics approval and consent to participate

Sampling of the calves was granted an exemption from requiring ethics approval by the institutional Animal Experiment Commission "Dier Experimenten Commissie (DEC) Lelystad (2013111.b)" because sampling was performed for diagnostic purposes. Ethics approval is not applicable. Animal handling, including BALF sample collection, was performed or supervised by approved veterinarians Consent was obtained from the farmers for the samples collected at their farm.

References

1. Poulsen KP, McGuirk SM. Respiratory disease of the bovine neonate. Vet Clin North Am Food Anim Pract. 2009;25:121–37. doi:10.1016/j.cvfa.2008.10.007.
2. Griffin D, Chengappa MM, Kuszak J, McVey DS. Bacterial pathogens of the bovine respiratory disease complex. Vet Clin North Am Food Anim Pract. 2010;26:381–94. doi:10.1016/j.cvfa.2010.04.004.
3. Srikumaran S, Kelling CL, Ambagala A. Immune evasion by pathogens of bovine respiratory disease complex. Anim Health Res Rev. 2007;8:215–29. doi:10.1017/s1466252307001326.
4. Bell CJ, Blackburn P, Elliott M, Patterson T, Ellison S, Lahuerta-Marin A, Ball HJ. Investigation of polymerase chain reaction assays to improve detection of bacterial involvement in bovine respiratory disease. J Vet Diagn Investig. 2014;26:631–4. doi:10.1177/1040638714540166.
5. Woolums AR, Ames TR, Baker JC. The bronchopneumonias (respiratory disease complex of cattle, sheep and goats). In: Smith BP, editor. Large Animal Internal Medicine. 4th ed. St Louis: Mosby, Elsevier; 2009. p. 602–43.
6. Nicholas RA, Ayling RD. Mycoplasma bovis: disease, diagnosis, and control. Res Vet Sci. 2003;74:105–12.
7. Nicholas RA. Bovine mycoplasmosis: silent and deadly. Vet Rec. 2011;168: 459–62. doi:10.1136/vr.d2468.
8. Houghton SB, Gourlay RN. Synergism between Mycoplasma bovis and Pasteurella haemolytica in calf pneumonia. Vet Rec. 1983;113:41–2. doi:10.1136/vr.d2468.
9. Caswell JL, Archambault M. Mycoplasma bovis pneumonia in cattle. Anim Health Res Rev. 2007;8:161–86. doi:10.1017/s1466252307001351.
10. Panciera RJ, Confer AW. Pathogenesis and pathology of bovine pneumonia. Vet Clin North Am Food Anim Pract. 2010;26:191–214. doi:10.1016/j.cvfa.2010.04.001.
11. Angen O, Thomsen J, Larsen LE, Larsen J, Kokotovic B, Heegaard PM, Enemark JM. Respiratory disease in calves: microbiological investigations on trans-tracheally aspirated bronchoalveolar fluid and acute phase protein response. Vet Microbiol. 2009;137:165–71. doi:10.1016/j.vetmic.2008.12.024.
12. Hirose K, Kobayashi H, Ito N, Kawasaki Y, Zako M, Kotani K, Ogawa H, Sato H. Isolation of Mycoplasmas from nasal swabs of calves affected with respiratory diseases and antimicrobial susceptibility of their isolates. J Vet Med B Infect Dis Vet Public Health. 2003;50:347–51.
13. Gabinaitiene A, Siugzdaite J, Zilinskas H. Laboratory diagnosis of Mycoplasma infection in young cattle. Pol J Vet Sci. 2011;14:87–93.
14. Marques LM, Buzinhani M, Yamaguti M, Oliveira RC, Ferreira JB, Mettifogo E, Timenetsky J. Use of a polymerase chain reaction for detection of Mycoplasma dispar in the nasal mucus of calves. J Vet Diagn Investig. 2007; 19:103–6.
15. Miles K, McAuliffe L, Ayling RD, Nicholas RA. Rapid detection of Mycoplasma dispar and M. Bovirhinis using allele specific polymerase chain reaction protocols. FEMS Microbiol Lett. 2004;241:103–7. doi:10.1016/j.femsle.2004.10.010.
16. Sachse K, Salam HS, Diller R, Schubert E, Hoffmann B, Hotzel H. Use of a novel real-time PCR technique to monitor and quantitate Mycoplasma bovis infection in cattle herds with mastitis and respiratory disease. Vet J. 2010;186:299–303. doi:10.1016/j.tvjl.2009.10.008.

17. Kobayashi H, Hirose K, Worarach A, Paugtes P, Ito N, Morozumi T, Yamamoto K. In vitro amplification of the 16S rRNA genes from Mycoplasma bovirhinis, Mycoplasma alkalescens and Mycoplasma bovigenitalium by PCR. J Vet Med Sci. 1998;60:1299–303.
18. McAuliffe L, Ellis RJ, Lawes JR, Ayling RD. Nicholas RA 16S rDNA PCR and denaturing gradient gel electrophoresis; a single generic test for detecting and differentiating Mycoplasma species. J Med Microbiol. 2005;54:731–9. doi:10.1099/jmm.0.46058-0.
19. Schnee C, Schulsse S, Hotzel H, Ayling RD, Nicholas RA, Schubert E, Heller M, Ehricht R, Sachse K. A novel rapid DNA microarray assay enables identification of 37 Mycoplasma species and highlights multiple Mycoplasma infections. PLoS One. 2012;7:e33237. doi:10.1371/journal.pone.0033237.
20. Aluotto BB, Wittier RG, Williams CO, Faber JE. Standardized bacteriologic techniques for the characterization of mycoplasma species. Int J Syst Bacteriol. 1970;20:35–58.
21. Fogarty U, Quinn PJ, Hannan J. Bronchopulmonary lavage in the calf: a new technique. Ir Vet J. 1983;37(2):35–8.
22. Phillippy AM, Mason JA, Ayanbule K, Sommer DD, Taviani E, Huq A, Colwell RR, Knight IT, Salzberg SL. Comprehensive DNA signature discovery and validation. PLoS Comput Biol. 2007;3:e98. doi:10.1371/journal.pcbi.0030098.
23. Peer Y van de, Chapelle S, de Wachter S. A quantitative map of nucleotide substitution rates in bacterial RNA. NAR 1996; 24:3381-3391.
24. Saah AJ, Hoover DR. "Sensitivity" and "specificity" reconsidered: the meaning of these terms in analytical and diagnostic settings. Ann Intern Med. 1997;126:91–4.
25. Pettersson B, Uhlén M, Johansson KE. Phylogeny of some mycoplasmas from ruminants based on 16S rRNA sequences and definition of a new cluster within the hominis group. Int J Syst Bacteriol. 1996;46:1093–8. doi:10.1099/00207713-46-4-1093.
26. Altschul SF, Gish W, Miller W, Myers EW, Lipman DJ. Basic local alignment search tool. Mol Biol. 1990;215:403–10. doi:10.1016/s0022-2836(05)80360-2.
27. Martin SW. Estimating disease prevalence and the interpretation of screening test results. Prev Vet Med. 1984;2:463–72.
28. Mackinnon A. A spreadsheet for the calculation of comprehensive statistics for the assessment of diagnostic tests and inter-rater agreement. Comput Biol Med. 2000;30:127–34.
29. Chávez González YR, Ros Bascuñana C, Bölske G, Mattsson JG, Fernández Molina C, Johansson KE. In vitro amplification of the 16S rRNA genes from Mycoplasma bovis and Mycoplasma agalactiae by PCR. Vet Microbiol. 1995; 47:183–90.
30. van Kuppeveld FJ, van der Logt JT, Angulo AF, van Zoest MJ, Quint WG, Niesters HG, Galama JM, Melchers WJ. Genus and species-specific identification of mycoplasmas by 16S rRNA amplification. Appl Environ Microbiol. 1993;59:655.
31. Pfützner H, Sachse K. Mycoplasma bovis as an agent of mastitis, pneumonia, arthritis and genital disorders in cattle. Rev Sci Tech. 1996;15:1477–94.
32. Lambert M. Contagious agalactia of sheep and goats. Rev Sci Technol Off Int Epizoot. 1987;6:6677–711.
33. Catania S, Gobbo F, Schiavon E, Nicholas RAJ. Severe otitis and pneumonia in adult cattle with mixed infection of Mycoplasma bovis and Mycoplasma agalactiae. Vet Rec Case Rep. 2016;4:e000366. doi:10.1136/vetreccr-2016-000366.
34. Bashiruddin JB, Frey J, Königsson MH, Johansson KE, Hotzel H, Diller R, de Santis P, Botelho A, Ayling RD, Nicholas RA, Thiaucourt F, Sachse K. Evaluation of PCR systems for the identification and differentiation of Mycoplasma agalactiae and Mycoplasma bovis: a collaborative trial. Vet J. 2005;169:268–75.
35. Nicholas R, Baker S, Ayling R, Stipkovits L. Mycoplasma infections in growing cattle. Cattle Practice. 2000;8:115–8.
36. Megid R, Nicholas RA, Miles RJ. Biochemical characterization of Mycoplasma bovirhinis, Mycoplasma dispar and recent bovine isolates of Mycoplasma canis. Vet Res Commun. 2001;25:1–12.
37. Ayling RD, Bashiruddin SE, Nicholas RA. Mycoplasma species and related organisms isolated from ruminants in Britain between 1990 and 2000. Vet Rec. 2004;155:413–6.

5

A randomised clinical trial of a metaphylactic treatment with tildipirosin for bovine respiratory disease in veal calves

J. Berman, D. Francoz*, J. Dubuc and S. Buczinski

Abstract

Background: Bovine respiratory disease (BRD) is a major problem in veal calf rearing units. The objective of this randomised clinical trial was to assess the effectiveness of tildipirosin as a metaphylactic treatment in veal calves on the number of BRD treatments, lung consolidation on thoracic ultrasonography (TUS) and average daily gain (ADG). A total of 209 veal calves from a pre-weaning fattening unit were randomly allocated to receive one of two treatments (tildipirosin 4 mg/kg, subcutaneously, n = 109; placebo 0.9% saline, subcutaneously, n = 100) at day 12 after entry in the pre-weaned unit. The calves were followed for a 70-day period. Occurrence of mortality and BRD treatments were recorded during the pre-weaning period. At days 1, 12 and 30, TUS and clinical scores were performed and ADG was measured during the first and second months of feeding.

Results: The use of a metaphylactic treatment of tildipirosin 12 days after arrival of the veal calves was not associated with the number of BRD treatments performed by the producer, ultrasonographic lung consolidation or weight gain (P < 0,05). In this cohort of calves, the proportion of calves treated for BRD by the producer was low at 14% (29/209). However, 13% (26/209) of calves included in the study already had ultrasonographic lung consolidation lesions 12 days after their arrival, which was before treatment time, and 27% (56/209) had lung consolidation at day 30.

Conclusion: In this study population with a low BRD prevalence, we were not able to detect any benefit of tildipirosin as a metaphylactic treatment of BRD at day 12 after arrival based on BRD treatments, TUS, and ADG.

Keywords: Tildipirosin, Lung consolidation, Pneumonia, Ultrasonography, Weight gain, Bronchopneumonia

Background

Bovine respiratory disease (BRD) is a multifactorial disease involving an interaction between stressors, the environment, viruses and bacteria. BRD remains a major disease in all types of bovine production systems especially in veal calf production due to stress factors such as transportation, commingling from different sources, high prevalence of failure of passive transfer and veal calf processing [1]. In this production the mortality rate of BRD is low varying from 1.5% to 2% [1, 2], but the morbidity rate could be important according to the cohorts varying from 7% to 61% [1–3]. At slaughter, prevalence of calves with lung lesions is much higher and varies from 41% to 70% [4, 5]. Discrepancy between the clinical prevalence of BRD (i.e. the proportion of calves treated individually by producers or clinical cases) and the prevalence of lesions at slaughter indicates that many cases remain subclinical and undetected, contributing to long term losses particularly with individual treatment strategies [5].

Consequently, metaphylactic antimicrobial therapy is commonly used in the veal calf industry [6]. Metaphylactic antimicrobial therapy has been defined by Nickell and White as a mass treatment of animal populations currently experiencing any level of disease before the onset of blatant illness [6, 7]. The administration of metaphylactic treatment is usually instituted before the main peak of BRD clinical incidence [6] that occurs about three weeks after arrival at the fattening unit [1]. In Europe, metaphylactic antimicrobial treatments in veal calves are mostly administered by the oral route. By this

* Correspondence: david.francoz@umontreal.ca
Département des sciences cliniques, Faculté de médecine vétérinaire, Université de Montréal, Saint-Hyacinthe, QC, J2S 7C6, Canada

route, the majority of cases, the antimicrobial drug used is underdosed compared to the labelled dose essentially because of underestimation of weight [8]. This raises concerns about the effectiveness of oral metaphylactic antimicrobial treatments and the development of antimicrobial resistance [8, 9]. In order to reduce the use of antimicrobial drugs and resistance development, parenteral administration of targeted antimicrobial drugs is reported as an alternative [9]. In veal calves, tulathromycin is the only antimicrobial drug for which a study on the effectiveness of its parenteral administration has been published [9]. No differences in growth performance or clinical signs compared to oral treatment with tetracycline were found, but a difference between the treated control group and the placebo was observed [9]. Tildipirosin (Zuprevo; Merck Animal Health, Kirkland, Québec, Canada), is another macrolide antimicrobial drug rapidly and extensively distributed to the respiratory tract followed by slow elimination [10]. In vitro, Tildipirosin is reported to be efficacious against *Pasteurella multocida, Mannheimia haemolytica* and *Histophilus somni* [11]. Its effectiveness against *Mycoplasma bovis* in vitro is reported but the CMI are higher than those for the others agents [12]. Tildipirosin is already approved for treatment and control of BRD in cattle in different countries. However, to our knowledge, the effectiveness of tildipirosin metaphylactic treatment, including its effect on growth performance in high-BRD risk calves, has never been studied in veal calves.

The main objective of this study was to assess the effectiveness of tildipirosin as a metaphylactic antimicrobial treatment on the average daily gain (ADG) in veal calves. Secondary objectives were to evaluate the effect of this metaphylactic treatment on the number of calves treated for BRD by the producer during the pre-weaning period and on the number of calves with lung consolidation on thoracic ultrasonography (TUS). The first hypothesis was that treated calves would have a greater ADG than non-treated calves. The second hypothesis was that calves treated with tildipirosin two weeks after arrival would have fewer BRD treatments and fewer ultrasonographic lung lesions than non-treated calves.

Methods

A double-blinded randomised clinical trial was conducted between the months of July and October of 2014. The experimental protocol was approved by the Comité d'éthique et d'utilisation des animaux de l'Université de Montréal (protocol: 14-Rech_1727).

Population and study design

The study was performed in one commercial pre-weaning veal calf fattening unit in St-Hélène de Bagot, Québec (Délimax, Veaux Lourds Ltée, Saint-Hyacinthe, Québec, Canada). Calves coming from local auction markets were enrolled at the approximate age of seven days of life (exact age was not reported for each calf) on arrival at the farm. A sample size estimation of 200 calves (100 calves per treatment group) was based on detecting a difference in treated versus control calves of 4 kg of weight at the end of the pre-weaning period (105 kg in control vs 109 kg in treated calves; assuming a population variance (s^2) of 100) as well as a difference of 15% in the proportion of calves treated for BRD (25% in control vs 10% or less in treated calves) with a statistical power of 80% and an alpha error of 5%. The difference of 4 kg was defined as the minimal difference financially beneficial for the producer considering the cost of tildipirosin. The proportions used in this study were estimated based on previous data of BRD reported in veal calves by the owner and the literature [1], and the differences were based on those observed in feedlots [11].

On arrival at the farm (day 0, D0), calves were immediately intra-nasally vaccinated against infectious bovine rhinotracheitis and para-influenza 3 virus (Nasalgen® IP, Merck Animal Health, Kirkland, Québec, Canada) and were housed in duckboard individual pens and fed with milk replacer (21% crude protein, 18.5% fat) until a quantity of 25 to 28 kg at the time of weaning period 70 days after arrival. Each calf was identified using an ear tag, and calf pen allocation was also recorded. Calves were randomly distributed between two groups using the RAND function in Excel (Microsoft, Redmond, WA, USA) software. One group of calves received tildipirosin (TILD) at the labelled dose subcutaneously (SQ) (4 mg/kg). The second group received an equivalent volume of 0.9% saline SQ (PLAC). The metaphylactic antimicrobial and placebo treatments were given at day 12 (D12), which was one week before the main peak of clinical incidence according to historical health observations in this fattening unit (Annie Dubuc, Dr. Frédéric Beaulac, Délimax veaux lourds Ltée, personal communication) that is similar of the peak reported in veal calves in the litterature [1].

The commercial solution of tildipirosin (18%) and sterile saline (0.9%) were specifically prepared for the trial, using opaque vials, and they were kept locked away by a pharmacist. The bottles were labelled with either an A or a B, and the corresponding keys were kept by the pharmacist during the entire study period. The bottle identifications were only revealed to the researcher at the end of the data collection period.

Data collection

The different data collected and their times of collection are presented in Fig. 1.

The first visit was performed by a veterinarian and an animal health technician one day after arrival (D1).

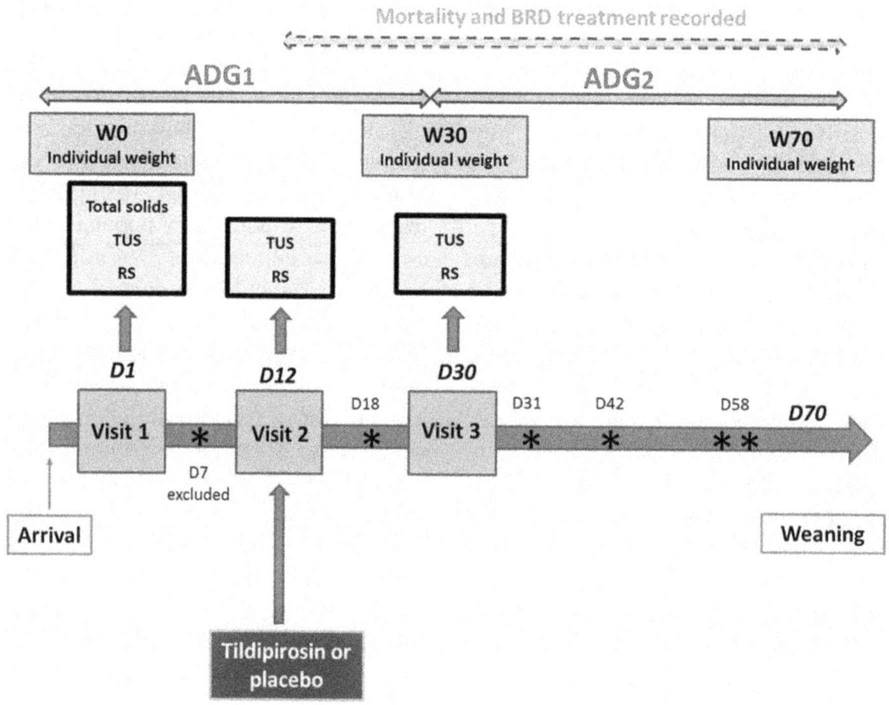

Fig. 1 Timeframe of the study period with the different data collected and their times of collection as well as calf deaths during the study. Legends. D: Day; W: Weight; TUS: Systematic thoracic ultrasonography; RS: Respiratory score; ADG1: Average daily gain for the first month of feeding; ADG2: Average daily gain for the second month of feeding; (*) represents dead calves; *dotted arrow* represents the mortality and BRD treatment recorded and integrated in the statistical analysis

Calves were bled from the jugular vein for assessment of serum total solids. Blood samples were centrifuged at 3500 rpm for 10 min at room temperature. The serum total solids were then determined using a manual handheld refractometer (Atago SUR-NE, Tokyo, Japan). Failure of passive transfer (FPT) was defined when total solids were less than or equal to 52 g/l [13].

Throughout the study period, the producer was allowed to treat the calves according to on-farm protocols but did not participate in the research project visits. Individual medical treatments or events, with a focus on BRD treatments, and mortalities were recorded during the entire pre-weaning period but no necropsies were performed. The number of BRD treatments was used to estimate the apparent prevalence by the producer.

On D1, D12 (the day of metaphylactic treatment) and day 30 (D30), a respiratory score (RS) was also performed on each calf using the Calf Respiratory Scoring Chart [14] as previously described [15]. An RS of five or more has been suggested as a BRD clinical case [14]. All clinical examinations were performed by two different operators blinded to TUS results and treatment allocation.

On D1, D12 and D30, TUS was performed with a 7.5 MHz linear probe (Imago, ECM, Angoulème, France) on the right and left thorax by two different operators blinded to respiratory score results and treatment allocation. The area screened was the mid to ventral portion of the right and left lung as well as the parenchyma cranial to the heart [16]. Thoracic ultrasonography was used to detect calves with lung consolidation. The maximal depth of consolidation (DEPTH) was recorded. The DEPTH (cm) was calculated by a manual count of the 1 cm squares using the 1 cm^2 grid of the ultrasound unit. As previously described, lung consolidation was considered significant when at least one point of DEPTH greater than or equal to 3 cm was observed [16].

During the study period, the calves were weighed by a blind research assistant three times: at arrival (D0, W0), at D30 (W2) and just before leaving the fattening unit at day 70 (W3). Individual calf weights were determined by a digital scale. Average daily gain was defined for the first month of feeding (ADG1; difference between W2 and W0 divided by 30 days), for the second month of feeding (ADG2; difference between W3 and W2 divided by 40 days) and for the total of the pre-weaning period (ADG tot; difference between W3 and W0 divided by 70 days).

Statistical analyses

The calf was the unit of interest in this study. As recommended by Sargeant [17], intention-to-treat (ITT) analyses were used. This analysis compares the group exactly as randomised. This approach is considered the gold standard if there are losses to follow-up in a randomised controlled trial because randomisation is maintained and adverse effects, as well as lack of compliance, is likely to occur in "real world" use of the intervention [17]. For the only calf that died before the third visit at D30, the medians of ultrasonographic lung consolidation, of clinical score and of ADG, were assigned [18]. For the calves that died after D30, a last observation carried forward approach [18] was performed for ADG, i.e. the ADG during the second period of the dead calves (ADG2) was considered similar to the one observed during the first period (ADG1; ADG1 = ADG2). Two calves had no weight at D30 to calculate ADG1 and ADG2, and ADG tot was attributed to calves.

Statistical analyses were performed using SAS (version 9.3, SAS Institute Inc. Cary, NC, USA). Descriptive statistics were first calculated for the two treatment groups using the MEANS and FREQ procedures in SAS. A T-test was used to compare both treatment groups concerning the continuous variables (total solids at arrival, weight, ADG), with statistical significance set at $P < 0.05$. Fisher's exact test was used to compare both treatment groups concerning the dichotomized variables (breed, sex, FPT, TUS, RS, mortality, BRD treatments by the producer after administration of TILD), with statistical significance set at $P < 0.05$.

Secondly, logistic regression analyses were performed to assess the impact of treatment group on the presence of significant lung consolidation on D30 (GLIMMIX procedure with LOGIT link in SAS). Univariable model building was performed using Chi-squared or Fisher's exact tests (FREQ procedure). The covariates with $P < 0.25$ were retained for building a final multivariable linear regression model using a backward elimination strategy. The final model was obtained when all covariates had $P < 0.05$.

Finally, ADG1, ADG2 and ADG tot were considered as dependent variables in univariable linear regression models (MIXED procedure in SAS). All the potential covariates were first assessed using univariable analyses with ADG as the dependent variable. The covariates with $P < 0.25$ were retained for building a final multivariable linear regression model using a backward elimination strategy. The final model was obtained when all covariates had $P < 0.05$. The treatment group was forced in every multivariable model since it was the main study interest.

Results

Two hundred and ten dairy calves, with an average weight of 51.7 kg (SD = 6.4 kg) were enrolled in the study. One calf died at day seven, before metaphylactic treatment, and was therefore excluded from the study. At D12, a total of 209 calves received either tildipirosin (TILD, n = 109) or the placebo (PLAC, n = 100). No adverse drug effect was observed.

Descriptive statistics at arrival (D1) and at the moment of treatment administration (D12) are presented in Table 1. Five calves died during the whole pre-weaning period.

Table 1 Descriptive statistics at days 1 (D1) and 12 (D12) of the 209 calves randomly assigned to receive a metaphylactic antimicrobial treatment of tildipirosin (TILD) or placebo (PLAC) at D12

	PLAC (n = 100)		TILD (n = 109)	
Nominal variables*	Number	Proportion (%)	Number	Proportion (%)
Male	91	91	103	95
Holstein breed	89	89	99	93
RS ≥ 5[a]				
D1	1	1	0	0
D12	3	3	5	5
DEPTH ≥ 3 cm[b]				
D1	1	1	2	2
D12**	8	8	18	17
Failure of passive transfer[c]	55	55	59	54
Continuous variables*	Mean (range)	SD	Mean (range)	SD
Serum total solids (g/L)	54.3 (38–83)	7.9	53.3 (41–80)	7.9
Weight at arrival (kg)	47.6 (34.5–63.5)	5.9	46.7 (33.1–65.3)	6.8

[a]Calf respiratory score assessing rectal temperature, cough, nasal and ocular discharge as well as ear drop. Calf is considered clinically sick when score ≥ 5
[b]Ultrasonographic assessment of lung consolidation maximal depth using thoracic ultrasonography. Calf is considered sick by ultrasonography when depth ≥ 3
[c]Serum total solids less than 52 g/l
*Used of Fisher's test or chi-squared test for the nominal variables and t-test for the continuous variables
**$P = 0.06$; no significant difference observed for all data

The times of death during the study period are described in Fig. 1. All calves who died during the study were in the PLAC group (Fisher's exact test; $P = 0.02$). Among the dead calves, four had lung consolidation with DEPTH > 3 cm at the last TUS before death, and two among this four had been treated for BRD. No differences were observed for the number of calves treated for BRD during the entire pre-weaning period (TILD group: 17/109 (16%) vs PLAC group 18/100 (18%) (Fisher's exact test; $P = 0.64$). Between D12 and D30, three antimicrobial treatments for another indication of BRD were used. They were not included in the statistical analysis.

At D30, 29 (27%) calves had ultrasonographic lung consolidation in the PLAC group versus 27 (27%) calves in the TILD group (Chi-squared test; $P = 0.91$). Among these 56 calves, 14 already had ultrasonographic lung consolidation at D12 (9 in PLAC group and 5 in TILD group). Data concerning the proportion of subclinical and clinical calves are presented in Additional file 1: Table S1. In the univariable analyses, the treatment group was not associated with the presence of ultrasonographic lung consolidation at D30 ($P = 0.89$). The final multivariable model showed that, when accounting for treatment group (TILD vs PLAC), ultrasonographic lung consolidation at D30 was not different between groups ($P = 0.78$). Presence of lung consolidation at D12 and FPT were associated with the presence of consolidation at D30 in this model ($P < 0.05$; Table 2).

Results of the univariable analyses reported no association between ADG1 and potential covariates ($P \geq 0.25$). Lung consolidations at D30 and BRD treatments by producer were both associated with ADG2 and with ADG tot ($P < 0.25$; Table 3). The final multivariable model showed that, after adjusting for treatment groups (TILD vs PLAC), BRD treatments by producer ($P < 0.01$) was negatively associated with ADG2 and ADG tot. The effect of the metaphylactic treatment was not significant ($P = 0.07$; Table 4).

Table 2 Results from the multivariable logistic regression analysis showing variables that were associated with the presence of lung consolidation at D30

Variables		Coefficient	SE	OR	95% CI	*P*-value
Treatment	PLAC	Referent	-	-	-	-
	TILD	−0.09	0.33	0.91	0.47–1.76	0.78
DEPTH d12	<3 cm	Referent	-	-	-	-
	≥3 cm	1.47	0.45	4.35	1.79–10.6	<0.01
Total solids	>52 g/L	Referent	-	-	-	-
	≤52 g/L	0.85	0.34	2.35	1.19–4.65	0.02

PLAC placebo treated calves, *TILD* tildipirosin treated calves, *DEPTH D12* maximal depth of lung consolidation found during thoracic ultrasonography at day 12 after arrival (i.e. time of injection of TILD or PLAC).

Discussion

The use of a metaphylactic treatment of TILD 12 days after arrival of the veal calves was not associated with the number of BRD treatments performed by the producer, ultrasonographic lung consolidation or weight gain. In this cohort of calves, the proportion of calves treated for BRD by the producer was low 14% (29/209). However, 13% (26/209) of calves included in the study already had ultrasonographic lung consolidation lesions 12 days after their arrival, which was before treatment time, and 27% (56/209) had lung consolidation at D30.

One of the major limitations of this study was the absence of identification of the etiological agent involved. In vivo, Tildipirosin was proven efficacious against *Mannheimia hemolytica, Pasteurella multocida* and *Histophilus somni* [11] but evidences of the effectiveness against others agents especially *Mycoplasma bovis* is limited [19]. Moreover, no necropsy was performed. Thus, despites the fact that four dead calves on five had lung consolidation with DEPTH > 3 cm at the last TUS before death and two among this four had been treated for BRD, we cannot concluded on the real cause of death and proved a superiority of TILD (no dead calves) vs control (five dead calves) on mortality. However, since only 2 out of 4 of the dead calves were diagnosed by the owner as having BRD, these results underline the already reported poor sensitivity of owners to diagnose BRD in cattle [20, 21].

The second major limitation of this study was the low apparent BRD clinical and lung consolidation prevalence in this cohort of calves with a proportion of calves treated for BRD by the producer of 14% (29/209) and lung consolidation of 27% (56/209). This may have limited our ability to find a difference between both treatment groups. Indeed, we calculated our estimated sample size based on finding a difference of 15% in calves treated for BRD. With a proportion of 15%, that means that we should have had a proportion of 0% in the TILD to detect a significant difference, which is almost impossible in a fattening unit. Therefore, even if there was a difference, it could have been missed by the low prevalence in both groups.

Concerning the conclusion from TUS, the absence of difference between both groups have to be interpreted with caution because despite the randomisation process on arrival there was an important, but not significant, difference in the proportion of lung consolidation at D12 between the treatment and the control groups (18%; TILD vs 8% PLAC; $P = 0.06$). When we consider the model without the animal already consolidated at D12, the final multivariable model showed that TILD ($P = 0.04$) was positively associated with ADG2 (Additional file 1: Table S2, S3, S4 and S5). So, probably this difference could have impacted the effect of the metaphylactic treatment but

Table 3 Results of the univariable analyses assessing the association between the average daily gain during the 1st month of feeding (ADG1), the 2nd month of feeding (ADG2) and during whole the pre-weaning period (ADG tot), and potential covariates measured during the study

		ADG 1			*ADG 2*			*ADG tot*		
Variables		LSM (lbs/d)	SEM	*P*-value	LSM (lbs/d)	SEM	*P*-value	LSM (lbs/d)	SEM	*P*-value
Treatment	PLAC	1.32	0.05	0.37	2.22	0.05	0.37	1.90	0.04	0.97
	TILD	1.28	0.05		2.27	0.04		1.90	0.04	
Sex	Female	1.42	0.13	0.33	2.19	0.12	0.68	1.91	0.10	0.95
	Male	1.29	0.04		2.24	0.03		1.90	0.03	
DEPTH D1	≥ 3 cm	1.39	0.28	0.71	2.55	0.26	0.25	2.13	0.23	0.35
	< 3 cm	1.31	0.03		2.26	0.03		1.92	0.03	
DEPTH D12	≥ 3 cm	1.17	0.10	0.25	2.30	0.09	0.58	1.89	0.08	0.78
	< 3 cm	1.29	0.04		2.26	0.03		1.91	0.03	
DEPTH D30	≥ 3 cm	1.28	0.07	0.93	2.34	0.06	0.13	1.97	0.05	0.17
	< 3 cm	1.29	0.04		2.24	0.04		1.89	0.08	
Total solids	≤ 52 g/l	1.29	0.05	0.95	2.25	0.04	0.98	1.91	0.04	0.83
	>52 g/l	1.30	0.05		2.25	0.05		1.90	0.04	
BRD treatments	0	NA			2.29	0.03	< 0.01	1.95	0.03	< 0.01
	≥ 1	NA			1.85	0.15		1.42	0.22	

LSM Least Squares Means, *SEM* Standard Error of the Mean, *DEPTH D1* maximal depth of lung consolidation found during thoracic ultrasonography at arrival, *DEPTH D30* maximal depth of lung consolidation found during thoracic ultrasonography at day 30 after arrival. BRD treatments represent the proportion of calves treated for bovine respiratory disease by the producer during the feeding period. NA: not available. The breed (not presented here) was not significantly associated with ADG1 ($P = 0.30$) or ADG2 ($P = 0.26$).

most likely it could have limited the ability to find a difference in favour of the treatment group.

One hypothesis that can also explain the absence of effectiveness of TILD in this study is the inadequate time of administration. In the feedlot industry, reduction of morbidity and the negative impact of BRD on ADG is more important when the metaphylactic treatment is given before the morbidity peak [6], which occurs earlier in feedlot (about one week after arrival). In this study, the moment of TILD administration was determined according to the main expected peak of BRD clinical incidence at about three weeks after arrival (diagnosed by the producer or veterinarian) that is similar with what is reported in veal calves in the literature [1]. In this study, using TUS allowed to diagnose lung lesions associated with clinical and subclinical BRD and provided an accurate antemortem assessment of lung health [15, 16]. This tool show a sensitivity and a specificity higher than clinical signs and can detect the disease earlier [15, 16]. In this study, during the first 12 days after arrival, the proportion of calves with ultrasonographic lung consolidation increased from 1% (3/209) at D1 to 12% (26/209) at D12 (see more details in the Additional file 1: Table S1). This finding suggests that BRD develops earlier after arrival and remains undetected for a greater or lesser period before detection of clinical signs. This finding emphasizes that the true peak of BRD in veal calf units differs from the apparent, clinical peak of BRD. Moreover, when we consider the model without the animal already sick at D12 (animal with lung consolidation and RS ≥ 5), the final multivariable model showed that TILD ($P = 0.03$) was positively associated with ADG2 (Additional file 1: Table S6 and S7). Thus, the treatment of tildipirosin should probably be efficient as a prophylactic treatment, when animals are not sick yet. With this knowledge, metaphylactic antimicrobial treatment should be given earlier during the pre-weaning period (before D12) or at arrival [6, 9].

Intention-to-treat analyses were used in this clinical trial. In randomised controlled trials (RCT), if there are losses in follow-up, there are three possible approaches to

Table 4 Results of the multivariable linear regression analysis showing variables that have an impact on average daily gain during the second month of feeding (ADG 2)

Variables		Estimate	SE	*P*-value	LSM (lbs/d)	SEM
	Intercept	1.7305	0.19	<0.0001		
Treatment	PLAC	Referent			2.06	0.08
	TILD	0.12	0.07	0.07	1.94	0.08
DEPTH D30	<3 cm	Referent			1.97	0.07
	≥3 cm	−0.07	0.09	0.47	2.05	0.09
BRD treatments	0	Referent	0.18	<0.01	2.29	0.05
	≥1	−0.41	0.25		1.87	0.16

LSM Least Squares Means, *SEM* Standard Error of the Mean, BRD treatments represent the proportion of calves treated for bovine respiratory disease by the producer during the feeding period.

the analysis: comparison of the groups exactly as randomised (ITT), analysis only of individuals who complied completely with the intervention (compliers-only analysis) or analysis of individuals based on the intervention that was received, regardless of whether the individual was initially randomly allocated to that intervention group [17, 18]. The model ITT is considered as the gold standard for statistical analyses in RCT because it has the advantage of taking into account effects that may be related to the intervention and better reflects real life [17]. In this present study, six out of the seven calves were missing data in the PLAC, and the missing data were mainly due to dead calves indicating a beneficial effect of TILD. This beneficial effect would be hidden with the approach of compliers-only analysis where dead calves are excluded from statistical analysis. Unfortunately, there is no clear consensus on how to deal with missing data. As previously described, we used the methods of "imputation of the median (or mean) of the calf's own group" or "last observation carried forward" [18]. Consequently, as recommended by Wright and Sim [18], and in order to be sure not to under- or overestimate the treatment effect, we also analysed the data with the compliers-only analysis, i.e. by excluding calves with missing data. As for the model ITT, treatment was not significantly associated with BRD treatments, TUS or weight gain ($P = 0.89$).

In this cohort of veal calves, metaphylactic TILD treatment had no effect on ADG during the first month of feeding, second month of feeding or whole pre-weaning period ($P > 0.05$). To the author's knowledge, this is the first study to assess the effectiveness of tildipirosin on growth performance in veal calves. Indeed, the only previous study that assessed the effectiveness of tildipirosin over the long term was the label study for the homologation of the product as a metaphylactic treatment in feedlots, and only the clinic effectiveness was taken into account [11]. In our study, apparent prevalence of BRD was estimated by the number of calves treated for BRD by the producer, and no differences between the TILD and PLAC groups were observed ($P > 0.05$). The absence of difference in this study compare with the homologation study, could be explained by the relatively low proportion of calves treated for BRD in this cohort of calves (PLAC group: 15%) in comparison to what was expected (25%). In this study, the proportion of ultrasonographic lung consolidation at D30 was the same in the TILD and PLAC groups. In a previous study, crossbred dairy heifers were first treated with TILD, challenged with *Mannheimia haemolytica* 10 days later and necropsied three days post challenge [20]. Results of this study showed that lung lesions, mainly lung consolidation, observed at necropsy were less extensive in the group treated with TILD than the control group [20]. Explanations for this difference could be that lung lesions were not evaluated in the same way (presence or absence of lung consolidation vs extension of lung consolidation) and that different or multiple BRD pathogens could have been implicated in the present study.

Finally, absence of effectiveness of tildipirosin on weight gain, the number of BRD treatment and lung consolidation in this cohort underlines the fact that metaphylactic treatments must be based on cohort risk assessment. In feedlot cattle and in veal calves, it was shown that low-risk cattle did not benefit from a metaphylactic treatment [6]. Consequently, it is very likely that this fattening unit with such a low BRD prevalence did not benefit from a metaphylactic antimicrobial treatment.

Conclusion

In this particular study, the metaphylactic treatment of veal calves at D12 after arrival at a fattening unit did not demonstrate any beneficial impact on growth performance. Additionally, it was not associated with a decrease in the proportion of calves treated for BRD or with lung consolidation on TUS. These results should be confirmed in a veal calf population with a greater BRD prevalence (high risk population). They also underline the need to improve the capacity to identify cohorts of cattle with low or high risk of BRD in order to determine which ones could benefit or not of a metaphylactic antimicrobial treatment. It would also be important to evaluate the effectiveness of metaphylactic treatment administered upon arrival of veal calves in the fattening unit rather than D12 as in the current study.

Additional file

Additional file 1: Table S1. Descriptive data of disease status of the 209 calves randomly assigned to receive a metaphylactic antimicrobial treatment of tildipirosin (TILD) or placebo (PLAC) at D12. **Table S2.** Results from the multivariable logistic regression analysis showing variables that were associated with the presence of lung consolidation at D30, excluding calves already ill at D12 (RS ≥ 5). **Table S3.** Results from the multivariable logistic regression analysis showing variables that were associated with the presence of lung consolidation at D30, excluding calves already ill at D12 ((RS ≥ 5) and consolidated. **Table S4.** Results of the univariable analyses assessing the association between the average daily gain during the 1st month of feeding (ADG1), the 2nd month of feeding (ADG2) and during whole the pre-weaning period (ADG tot), and potential covariates measured during the study. **Table S5.** Results of the multivariable linear regression analysis showing variables that have an impact on average daily gain during the second month of feeding (ADG 2). **Table S6.** Results of the univariable analyses assessing the association between the average daily gain during the 1st month of feeding (ADG1), the 2nd month of feeding (ADG2) and during whole the pre-weaning period (ADG tot), and potential covariates measured during the study. **Table S7.** Results of the multivariable linear regression analysis showing variables that have an impact on average daily gain during the second month of feeding (ADG 2)

Abbreviations

BRD: Bovine respiratory disease; DEPTH: Maximal depth of consolidation; FPT: Failure of passive transfer; PLAC: Placebo; RCT: Randomised controlled trials; RS: Respiratory scoring; SQ: Subcutanously; TILD: Tildipirosin; TUS: Thoracic ultrasonography

Acknowledgements
The authors thank J. Ménard, N. Tison, A. AbdAllah, C. Faure, S. Jolivet, G. Boisclair from Merck Health Animal, A. Dubuc from Délimax, and SecRepro for their technical help and financial support during the study.

Funding
This study was conducted at the Université de Montréal and was funded by Merck Animal Health Canada. The funding body wanted to assess the effectiveness of tildipirosin (Zuprevo; Merck Animal Health, Kirkland, Québec, Canada) as a metaphylactic treatment for bovine respiratory disease in veal calves. Otherwise, the funding body had no input on study design nor in data collection, analysis or interpretation. No member of the funding body reviewed this manuscript prior to its submission.

Author's contributions
JB collected of the clinical and ultrasonographic data, contributed in their analyses and wrote the manuscript. DF contributed to the study design, participated in the analyses of the data and revised the manuscript critically. JD contributed to the study design, performed the statistical analyses and revised the manuscript critically. SB had the idea of performing the study, designed and coordinated the study. He participated in the collection of the clinical and ultrasonographic data and participated in their analyses and revised critically the manuscript. All authors read and approved the manuscript for publication.

Competing interests
The authors declare that they have no competing interests. This study was funded by Merck Animal Health Canada. It is a standard requirement, with no implications impropriety on our part.

Consent for publication
Not applicable

References

1. Pardon B, Hostens M, Duchateau L, Dewulf J, De Bleecker K, Deprez P. Impact of respiratory disease, diarrhea, otitis and arthritis on mortality and carcass traits in white veal calves. BMC Vet Res. 2013;9:1.
2. Brscic M, Leruste H, Heutinck LF, Bokkers EA, Wolthuis-Fillerup M, Stockhofe N, et al. Prevalence of respiratory disorders in veal calves and potential risk factors. J Dairy Sci. 2012;95:2753–64.
3. Pardon B, Alliët J, Boone R, Roelandt S, Valgaeren B, Deprez P. Prediction of respiratory disease and diarrhea in veal calves based on immunoglobulin levels and the serostatus for respiratory pathogens measured at arrival. Prev Vet Med. 2015;120:169–76.
4. Van der Mei J, Van den Ingh TS. Lung and pleural lesions of veal calves at slaughter and their relationship with carcass weight. Vet Quart. 1987;9:203–7.
5. Leruste H, Brscic M, Heutinck LF, Visser EK, Wolthuis-Fillerup M, Bokkers EA, et al. The relationship between clinical signs of respiratory system disorders and lung lesions at slaughter in veal calves. Prev Vet Med. 2012;105:93–100.
6. Nickell JS, White BJ. Metaphylactic antimicrobial therapy for bovine respiratory disease in stocker and feedlot cattle. Vet Clin N Am-Food A. 2010;26:285–301.
7. Young C. Antimicrobial metaphylaxis for undifferentiated bovine respiratory disease. Comp Cont Educ Pract Vet. 1995;17:133–42.
8. Pardon B, Catry B, Dewulf J, Persoons D, Hostens M, De Bleecker K, et al. Prospective study on quantitative and qualitative antimicrobial and anti-inflammatory drug use in white veal calves. J Antimicrob Chem. 2012;67: 1027–38.
9. Rérat M, Albini S, Jaquier V, Hüssy D. Bovine respiratory disease: efficacy of different prophylactic treatments in veal calves and antimicrobial resistance of isolated Pasteurellaceae. Prev Vet Med. 2012;103:265–73.
10. Menge M, Rose M, Bohland C, Zschiesche E, Kilp S, Metz W, et al. Pharmacokinetics of tildipirosin in bovine plasma, lung tissue, and bronchial fluid (from live, nonanesthetized cattle). J Vet Pharma Thera. 2012;35:550–9.
11. U.S. Food and Drug Administration. Freedom of information summary original new drug application of U.S. Food and Drug Administration. 2012. http://www.fda.gov/downloads/AnimalVeterinary/Products/ApprovedAnimalDrugProducts/FOIADrugSummaries/UCM314826.pdf. Accessed May 14 2015.
12. Gautier-Bouchardon AV, Ferré S, Le Grand D, Paoli A, Gay E, Poumarat F. Overall decrease in the susceptibility of Mycoplasma bovis to antimicrobials over the past 30 years in France. PLoS One. 2014;9:e87672.
13. Godden S. Colostrum management for dairy calves. Vet Clin N Am-Food A. 2008;24:19–39.
14. McGuirk SM. Disease management of dairy calves and heifers. Vet Clin N Am-Food A. 2008;24:139–53.
15. Buczinski S, Ollivett TL, Dendukuri N. Bayesian estimation of the accuracy of the calf respiratory scoring chart and ultrasonography for the diagnosis of bovine respiratory disease in pre-weaned dairy calves. Prev Vet Med. 2015; 119:227–31.
16. Ollivett TL, Caswell JL, Nydam DV, Duffield T, Leslie KE, Hewson J, et al. Thoracic ultrasonography and Bronchoalveolar lavage fluid analysis in Holstein calves with subclinical lung lesions. J Vet Intern Med. 2015;29:1728–34.
17. Sargeant JM, Kelton DF, O'Connor AM. Randomized controlled trials and challenge trials: design and criterion for validity. Zoonoses Pub Health. 2014; 61(S1):18–27.
18. Wright CC, Sim J. Intention-to-treat approach to data from randomized controlled trials: a sensitivity analysis. J Clin Epidemio. 2003;56:833–42.
19. Bartram DJ, Moyaert H, Vanimisetti BH, Ramage CP, Reddick D, Stegemann MR. Comparative efficacy of tulathromycin and tildipirosin for the treatment of experimental Mycoplasma bovis infection in calves. Vet Med Sci. 2016;2:170–8.
20. Amrine DE, White BJ, Larson RL, Mosier DA. Pulmonary lesions and clinical disease response to challenge 10 days following administration of tildipirosin or tulathromycin. J Anim Sci. 2014;92:311–9.

6

Toxin-neutralizing antibodies protect against *Clostridium perfringens*-induced necrosis in an intestinal loop model for bovine necrohemorrhagic enteritis

Evy Goossens[1], Stefanie Verherstraeten[1], Bonnie R. Valgaeren[2], Bart Pardon[2], Leen Timbermont[1], Stijn Schauvliege[3], Diego Rodrigo-Mocholí[3], Freddy Haesebrouck[1], Richard Ducatelle[1], Piet R. Deprez[2] and Filip Van Immerseel[1*]

Abstract

Background: Bovine necrohemorrhagic enteritis is caused by *Clostridium perfringens* type A. Due to the rapid progress and fatal outcome of the disease, vaccination would be of high value. In this study, *C. perfringens* toxins, either as native toxins or after formaldehyde inactivation, were evaluated as possible vaccine antigens. We determined whether antisera raised in calves against these toxins were able to protect against *C. perfringens* challenge in an intestinal loop model for bovine necrohemorrhagic enteritis.

Results: Alpha toxin and perfringolysin O were identified as the most immunogenic proteins in the vaccine preparations. All vaccines evoked a high antibody response against the causative toxins, alpha toxin and perfringolysin O, as detected by ELISA. All antibodies were able to inhibit the activity of alpha toxin and perfringolysin O in vitro. However, the antibodies raised against the native toxins were more inhibitory to the *C. perfringens*-induced cytotoxicity (as tested on bovine endothelial cells) and only these antibodies protected against *C. perfringens* challenge in the intestinal loop model.

Conclusion: Although immunization of calves with both native and formaldehyde inactivated toxins resulted in high antibody titers against alpha toxin and perfringolysin O, only antibodies raised against native toxins protect against *C. perfringens* challenge in an intestinal loop model for bovine necrohemorrhagic enteritis.

Keywords: Bovine necrohemorrhagic enteritis, *Clostridium perfringens*, Neutralizing antibodies, Alpha toxin, Perfringolysin O

Background

The ubiquitous, spore forming, Gram-positive bacterium *Clostridium perfringens* is considered to be the most widespread pathogenic bacterium in the world [1–4]. It can cause a wide range of diseases including, amongst others, gas gangrene in man and necrohemorrhagic enteritis in suckling and veal calves [5–8]. Most of these diseases follow a very rapid, often fatal course. Therefore, curative treatment is difficult and control must rely on preventive measures, including vaccination. Virulence properties of different *C. perfringens* strains are largely determined by their ability to secrete a variety of proteinaceous toxins and enzymes, which can cause different forms of tissue damage [2–4, 9]. Alpha toxin and perfringolysin O have been identified as the principal toxins involved in the pathogenesis of both *C. perfringens*-induced gas gangrene and bovine necrohemorrhagic enteritis [10, 11]. These toxins exert different effects in both diseases. Bovine necrohemorrhagic enteritis is characterized by congestion of the capillaries, hemorrhages and inflammation. This is in contrast to gas gangrene, where these toxins lead to tissue necrosis, thrombosis and lack of leukocyte infiltration at the site of infection [10–12]. It is well known that humoral antibodies against secreted proteinaceous virulence factors

* Correspondence: Filip.VanImmerseel@UGent.be
[1]Department of Pathology, Bacteriology and Avian Diseases, Faculty of Veterinary Medicine, Ghent University, Salisburylaan 133, B-9820 Merelbeke, Belgium
Full list of author information is available at the end of the article

of *C. perfringens* can be protective, as shown in different animal models. As the enzymes and toxins of *C. perfringens* are highly destructive to tissues, vaccines against a variety of clostridial diseases have been developed using the denatured proteins [13–15]. Despite the usefulness of formaldehyde toxoids for other *C. perfringens*-associated diseases, there is controversy about the efficacy of such vaccines for gas gangrene, as opposed to crude toxin preparations [8, 16–18]. In addition, multivalent clostridial vaccines based on formaldehyde inactivated exotoxins derived from culture supernatant are commercially available for domestic livestock, including bovines, but no studies on their efficacy for necrohemorrhagic enteritis in calves are available.

The objective of the present study was to evaluate whether antibodies against *C. perfringens* toxins could protect against the development of necrotic lesions in the intestine. Therefore, calves were immunized with native *C. perfringens* toxins. To evaluate whether we could eliminate the undesired toxin activity, but conserve the immune-protective potential, a previously described, modified formaldehyde treatment was also tested [19]. Also a commercial formaldehyde inactivated multivalent clostridial vaccine was used. As necrohemorrhagic enteritis in veal calves is an unpredictable event and experimental reproduction of the disease is difficult, the neutralizing activity of the antibodies was evaluated in a previously developed intestinal loop model [20]. To further unravel the mechanism of protection, the inhibitory effect of the evoked antibodies on *C. perfringens*-induced cytotoxicity to bovine endothelial cells was evaluated and the toxin-neutralizing capacity against alpha toxin and perfringolysin O was analyzed.

Results

Western blot analysis

The proteins in the *C. perfringens* toxin preparation were visualized by SDS-PAGE (Fig. 1a). In the vaccinated calves, the production of circulating antibodies against *C. perfringens* supernatant and the *C. perfringens* toxin preparation was analyzed by western blot in three separate experiments (Fig. 1). No immune reaction was detected in the sera before immunization (data not shown). Sera obtained from calves six weeks after initial vaccination with either native toxins or the L-lysine protected, formaldehyde inactivated toxins, revealed immunoreactivity towards two proteins. Immune sera from calves vaccinated with the commercial formaldehyde inactivated clostridial vaccine showed immunoreactivity towards more proteins. The two proteins that were immunoreactive with antisera raised against all vaccine preparations were further identified as alpha toxin and perfringolysin O by MALDI analysis.

ELISA

In the vaccinated calves, the production of circulating antibodies directed against alpha toxin and perfringolysin O was also monitored by ELISA. No antibodies against alpha toxin or perfringolysin O were detected in the sera before immunization. In all calves a strong antibody response against both alpha toxin and perfringolysin O was detected 6 weeks after initial immunization. Calves vaccinated with the native *C. perfringens* toxins showed the highest antibody titers, whereas vaccination with formaldehyde inactivated toxins (either L-lysine protected or commercial inactivation) resulted in a more variable immune response (Table 1).

Protective effect of antisera against *C. perfringens*-induced necrosis in an intestinal loop model

The potential of the antisera, derived after vaccination of calves with the respective vaccines, to inhibit *C. perfringens*-induced necrosis, was evaluated in an intestinal loop assay. All positive control loops inoculated with *C. perfringens*

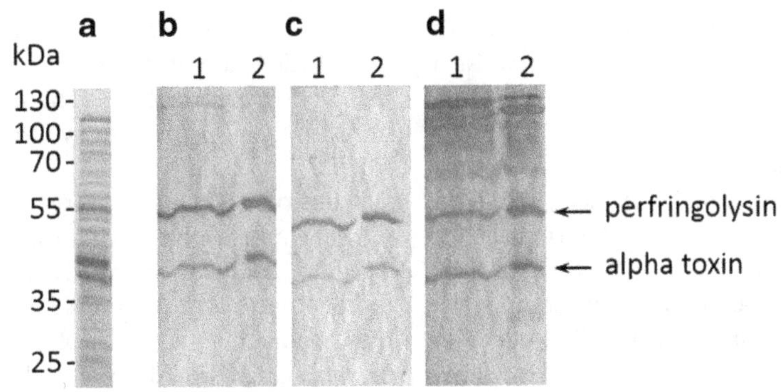

Fig. 1 Western blot analysis of the immune sera. **a** SDS-PAGE of the *C. perfringens* toxin preparation after Coomassie staining. **b-d** Representative Western blots showing the immunoreactivity towards crude *C. perfringens* supernatants (lane 1) and the *C. perfringens* toxin preparation (lane 2). The immune sera from calves vaccinated with native toxins (**b**) and formaldehyde inactivated, L-lysine protected *C. perfringens* toxins (**c**) detect only two proteins, whereas the commercial formaldehyde inactivated multivalent clostridial vaccine (**d**) reacts with multiple proteins. The blots shown are representative pictures of one out of three experiments

Table 1 Calves were immunized with either a *C. perfringens* toxin preparation (native toxins), L-lysine protected, formaldehyde inactivated *C. perfringens* toxins (L-lysine/formaldehyde toxoid) or a commercial multivalent formaldehyde inactivated clostridial vaccine

Vaccine	Anti-alpha toxin titer	Anti-perfringolysin O titer
Native toxins	64.44 ± 0.22	25600 ± 0
L-lysine/formaldehyde toxoid	24.26 ± 2.96	16000 ± 9600
Commercial formaldehyde vaccine	45.14 ± 20.42	4800 ± 1600

The anti-alpha toxin and perfringolysin O response was measured by ELISA. The data represent antibody titers (mean ± standard error of the means), six weeks after initial immunization

developed necrosis. Injection of loops with *C. perfringens* together with sera from naive calves (pre-immune sera) also resulted in a high percentage of necrotic loops. Injection of *C. perfringens* together with antisera raised against native toxins resulted in significantly fewer necrotic loops as compared to the positive control loops ($p < 0.001$) and the loops injected with the pre-immune sera ($p < 0.01$). Antisera raised against formaldehyde inactivated toxoid (either L-lysine protected or commercial) were unable to significantly neutralize the necrosis-inducing activity of *C. perfringens* (Figs. 2 and 3).

Neutralization of alpha toxin and perfringolysin O activity in vitro

The inhibitory capacity of the sera towards alpha toxin and perfringolysin O activities was further examined using recombinant toxins. All antisera decreased the activity of alpha toxin in vitro (Table 2). Up to a final dilution of 409.8 the antisera against the native toxins neutralized 50 % of the alpha toxin activity. To the contrary, in order to obtain the same inhibition of alpha toxin activity, antisera against L-lysine protected, formaldehyde inactivated toxoid or against the commercial formaldehyde inactivated clostridial vaccine could only be diluted up to final dilutions of 80.47 or 22.39, respectively.

The hemolytic activity of perfringolysin O towards equine erythrocytes in vitro was decreased by all antisera (Table 2). Up to a final dilution of 18 either antiserum inhibited the perfringolysin O activity completely. This neutralizing ability of the sera was observed up to a final dilution of 48 for the anti-native toxins antisera, a final dilution of 72 for the anti-L-lysine protected, formaldehyde inactivated toxoid or a final dilution of 18 when the antiserum obtained after vaccination with the commercial vaccine was used. The pre-immune sera had no effect on the alpha toxin or perfringolysin O activity in vitro.

Neutralization of *C. perfringens* cytotoxicity to bovine endothelial cells

To determine whether the cytotoxic activity of *C. perfringens* could be inhibited by the antisera to the vaccines, *C. perfringens* supernatants were incubated with serial dilutions of the antisera. Exposure of the endothelial cells to untreated *C. perfringens* supernatant resulted in 100 % cell death. More than 80 % cell viability could be measured by pre-incubation of the *C. perfringens* supernatant with a 32-fold dilution of the native toxins antiserum. At this concentration, neither the antisera raised against L-lysine protected, formaldehyde inactivated toxins nor the antisera raised against the commercial formaldehyde inactivated clostridial vaccine had an effect on the cytotoxicity (Table 3). The pre-immune sera had no effect on the *C. perfringens* cytotoxicity.

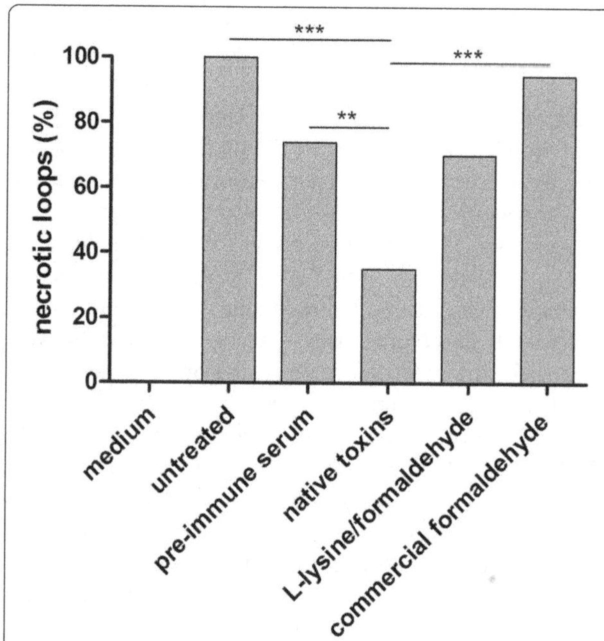

Fig. 2 Neutralization of the lesion-inducing potential of *C. perfringens*. The graph represents the percentage of ligated intestinal loops in which necrotic lesions were present after 5 h of incubation with sterile culture medium ($n = 20$), *C. perfringens* alone (untreated, $n = 20$) or *C. perfringens* in combination with naive sera (pre-immune serum, $n = 60$), antiserum to *C. perfringens* toxins (native toxins, $n = 20$), formaldehyde inactivated, L-lysine protected *C. perfringens* toxins (L-lysine/formaldehyde, $n = 20$) and commercial formaldehyde inactivated multivalent clostridial vaccine (commercial formaldehyde, $n = 20$). ** $0.001 \leq p < 0.01$ or *** $p < 0.001$

Discussion

Necrohemorrhagic enteritis caused by *C. perfringens* in suckling and veal calves is characterized by sudden death. Due to the very rapid course of the disease, curative treatment is not possible and therefore, protection by vaccination would be of high value. The virulence of *C. perfringens* is due to the many extracellular toxins it produces. In this study we showed that toxin neutralizing antibodies protect against *C. perfringens*-induced necrotic lesions in an intestinal loop assay and are able to prevent endothelial damage. Western blot analysis

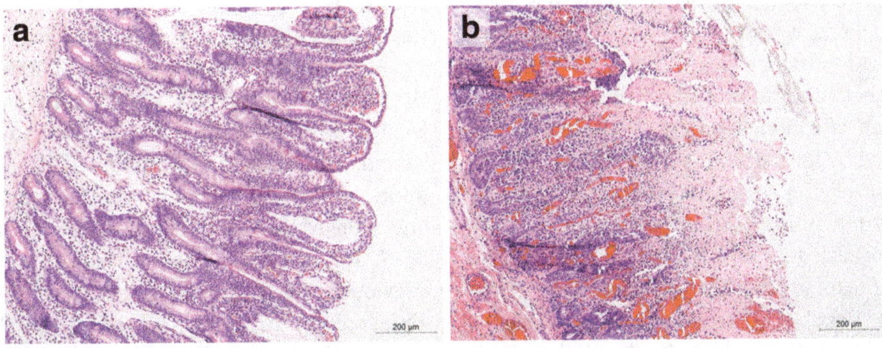

Fig. 3 *C. perfringens*-induced necrosis in experimentally infected intestinal loops in calves. **a** Representative histological section from an intestinal loop without necrotic lesions. This loop was injected with *C. perfringens* in combination with antiserum to native *C. perfringens* toxins. **b** Representative section from an intestinal loop from the same calf, showing hemorrhage and extensive necrosis of the villi. This loops was injected with *C. perfringens* in combination with naïve immune serum

revealed antibodies towards alpha toxin and perfringolysin O as the most abundant antibodies in the immune sera from calves vaccinated against *C. perfringens* toxins.

We previously reported congestion and leakage of the capillaries as an early event in the pathogenesis of necrohemorrhagic enteritis as shown in an intestinal loop assay [20]. Furthermore we showed that alpha toxin and perfringolysin O may exert their effect by directly targeting the endothelial cells [10]. This points towards endothelial damage as a key event in the pathogenesis of bovine necrohemorrhagic enteritis. Indeed, in the present study antisera which protected against *C. perfringens*-induced cytotoxicity to bovine endothelial cells also offered protection against *C. perfringens*-associated necrosis in an intestinal loop assay. Moreover, the protective antisera were shown to inhibit the activity of alpha toxin and perfringolysin O, which further underscores the roles of these toxins in the pathogenesis of bovine necrohemorrhagic enteritis. It can, however, not be ruled out that antibodies induced against other substances present in the vaccines also played a role in the protection observed in the intestinal loop model.

Formaldehyde inactivation of *C. perfringens* toxins diminished their capacity to induce protective antibodies. Antisera raised against L-lysine protected, formaldehyde inactivated *C. perfringens* toxins were also not protective in the intestinal loop model. This result is in disagreement with previous studies showing high antigenicity, low toxicity, and protection in mice that were immunized with L-lysine protected, formaldehyde inactivated toxoid and subsequently challenged with lethal doses of *C. perfringens* [19, 21]. In the present study we demonstrated that vaccinations with *C. perfringens* toxins, either in their native forms or as formaldehyde inactivated toxoids, all resulted in high antibody responses as detected by ELISA. However, only serum derived from animals immunized with the native toxins offered protection against necrosis in an intestinal loop assay. There is thus a discrepancy between the antibody titers against formaldehyde inactivated *C. perfringens* toxins measured by ELISA and the protective capacity of these antibodies in the intestinal loop model. Nevertheless, the value of vaccines based on

Table 2 In vitro neutralization of biological activities of alpha toxin and perfringolysin O. Calves were immunized with either a *C. perfringens* toxin preparation (native toxins), L-lysine protected, formaldehyde inactivated *C. perfringens* toxins (L-lysine/formaldehyde toxoid) or a commercial multivalent formaldehyde inactivated clostridial vaccine

	Inhibitory capacity (Mean ± SEM)	
Antiserum	Alpha toxin activity[a]	PFO activity[b]
Native toxins	409.8 ± 5.75	48.0 ± 0.0
L-lysine/formaldehyde toxoid	80.47 ± 46.93	72.0 ± 24.0
Commercial formaldehyde vaccine	22.39 ± 2.17	18.0 ± 6.0

[a]Neutralization of 10 µg/ml alpha toxin. The inhibitory capacity of the antiserum is expressed as the dilution that gives 50 % inhibition of the alpha toxin activity
[b]Neutralization of 2 µg/ml perfringolysin O. The inhibitory capacity of the antiserum is expressed as the highest dilution that inhibited perfringolysin O-induced hemolysis
Alpha toxin activity was determined by measuring its lecithinase activity on egg yolk lipoproteins. Perfringolysin O (PFO) activity was determined by measuring the hemolysis of horse erythrocytes

Table 3 In vitro neutralization of *C. perfringens* cytotoxicity. Calves were immunized with either a *C. perfringens* toxin preparation (native toxins), L-lysine protected, formaldehyde inactivated *C. perfringens* toxins (L-lysine/formaldehyde toxoid) or a commercial multivalent formaldehyde inactivated clostridial vaccine

Antiserum	Inhibitory capacity (Mean ± SEM)
Native toxins	32.00 ± 0.0
L-lysine/formaldehyde toxoid	9.00 ± 7.0
Commercial formaldehyde vaccine	4.00 ± 0.0

The cytotoxicity of *C. perfringens* supernatant to primary bovine endothelial cells was measured using a neutral red uptake (NRU) assay. The inhibitory capacity of the antiserum is expressed as the highest dilution that yields 80 % cell viability

formaldehyde inactivated *C. perfringens* toxins has been demonstrated for diseases associated with toxins other than alpha toxin and perfringolysin O [22–25]. This suggests that the protective immunogenicity of other *C. perfringens* toxins, such as, amongst others, NetB and epsilon toxin, is not affected by formaldehyde inactivation.

Although the use of *C. perfringens* native toxins represents an efficient strategy for vaccine development, active toxins cannot be regarded as safe. Therefore methods for the development of toxoids other than formaldehyde inactivation are needed. Possible strategies include the use of genetically modified toxoids based on site-directed mutants with reduced toxic activity or the use of immunologically active fragments of the essential toxins. Immunization with the carboxy-terminal domain of alpha toxin has previously been shown to provide protection in a mouse model against *C. perfringens* gas gangrene and may be a good candidate for development of a vaccine against bovine necrohemorrhagic enteritis [21, 26]. The identification of the structural elements responsible for membrane interaction of perfringolysin O provides opportunities for the development of non-toxic site-directed mutants as alternatives for native perfringolysin O [27].

In order to obtain the ultimate evidence that vaccination against *C. perfringens* toxins protects against bovine necrohemorrhagic enteritis, field trials need to be performed. However, since necrohemorrhagic enteritis is a low incidence disease, this would be a huge cost and more evidence concerning the immune-protective potential of the antisera is needed before considering this type of trial. Unfortunately, no in vivo model to validate the protective immune-potential of the candidate vaccines against bovine necrohemorrhagic enteritis is available. Niilo and colleagues were able to induce a mild diarrhea in cattle inoculated intraduodenally or per os with *C. perfringens* type A cultures, but no necrohemorrhagic enteritis was established [28]. Also we were unable to develop a reliable model of bovine necrohemorrhagic enteritis after per os or intraduodenal administration of *C. perfringens* type A cultures (unpublished results).

Conclusion

This study showed that toxin-neutralizing antibodies protect against *C. perfringens* challenge in an intestinal loop model for bovine necrohemorrhagic enteritis. Immunization of calves with either native or formaldehyde inactivated toxins resulted in a strong immune response against alpha toxin and perfringolysin O, but only antibodies raised against native toxins were protective in the intestinal loop model. Therefore it seems that, at least for alpha toxin mediated diseases, antibody titers detected by ELISA are not a guarantee for protection, even if protection against the disease is antibody mediated.

Methods

Vaccine preparation and immunization

C. perfringens toxin preparation (P4039, Sigma-Aldrich, Bornem, Belgium) was either used as native toxin or treated with formaldehyde to generate a formaldehyde toxoid. Inactivation was obtained by adding a combination of 0.4 % formaldehyde solution (Sigma-Aldrich) and 0.05 M L-lysine (Sigma-Aldrich) and incubation at 37 °C for two days. The addition of 0.05 M L-lysine has previously been shown to preserve the antigenicity of alpha toxin during toxoid formation [19]. Inactivation of alpha toxin was confirmed by spotting 5 μl drops on 2 % egg yolk Columbia agar plates (Oxoid, Wesel, Germany), followed by incubation for 16 h at 37 °C [29]. Native and formaldehyde inactivated toxins were formulated with the adjuvant Quil A (Brenntag Biosector, Frederikssund, Denmark) at a final concentration of 350 μg antigen and 750 μg Quil A in 1.5 ml phosphate buffered saline (PBS) per animal and filter-sterilized using a 0.2 μm filter. A standard formalin inactivated multivalent commercial vaccine was used according to the manufacturer's instructions (Covexin 10®, Zoetis, Louvain-la-Neuve, Belgium).

For immunization six 2-months old male Holstein Friesian calves were used. The calves were purchased from a local tradesman which collects dairy calves from herds in Eastern Flanders. They were housed on straw and received water and hay *at libitum*, and concentrates adjusted to the body weight.

For each antigen, two calves were immunized subcutaneously in the neck. The calves received a primer vaccination at the age of two months, with booster immunizations 14 and 28 days later. No strong adverse reactions were observed. Although no fever (>39.5 °C) was induced, all calves experienced a mild hyperthermia for two days following the vaccination. As described in the drug information leaflet of the commercial vaccine, localized swelling occurred at the site of injection. This effect was more pronounced in the calves vaccinated with the commercial formaldehyde inactivated clostridial vaccine (7–10 cm diameter) as compared to the calves vaccinated with either native toxins or the L-lysine protected, formaldehyde inactivated toxins (0–6 cm diameter). Blood samples were taken before primer vaccination and two weeks after the final booster vaccination.

SDS-PAGE and Western Blot

The proteins present in the toxin preparation were visualized on a 12 % SDS-PAGE followed by Coomassie Briliant Blue staining (Sigma-Aldrich). For the Western Blot analysis, 16 μl of cell-free supernatants of the *C. perfringens* strain JIR325 (10x concentrated using Vivaspin, Sartorium Stedim Biotech GmbH, Goettingen, Germany) or 6 μg of the *C. perfringens* toxin preparation were loaded on a 12 % SDS-PAGE. The proteins from the gel were transferred to nitrocellulose membranes of 0.45 µm pore size. Non-specific binding to the blots was blocked with 5 % skimmed milk powder in PBS, followed by overnight incubation at 4 °C with a 1/500 dilution of the immune sera collected two weeks after the final booster immunization. For this incubation step, the sera of the 2 animals that were vaccinated with a given vaccine preparation were pooled. Blots were washed with 0.1 % tween 20 in PBS and incubated for 1 h at room temperature with horseradish peroxidase-labelled rabbit-anti-bovine IgG (Sigma-Aldrich). Blots were developed with CN/DAB substrate kit (Thermo Fisher Scientific, Rockford, USA). The test was performed in triplicate. The specific immunoreactive protein bands were identified in the parallel-run Coomassie stained gel followed by MALDI analysis.

Enzyme-linked immunosorbent assay

The immune response following vaccination was also measured by ELISA using serum samples two weeks after the final booster immunization.

Alpha toxin-specific antibody levels were determined by the end-point dilution method using a blocking ELISA (*Clostridium perfringens* alpha toxin serological ELISA kit, Bio-X Diagnostics, Jemelle, Belgium). For each ELISA, sera were used at a dilution 1:50 and assays were performed in duplicate. The specific antibody level of the immune serum was expressed as the percent inhibition (% inhib) by means of the following formula: % inhib = [(OD neg – OD sample)/OD neg] * 100.

Perfringolysin O-specific antibody levels were measured using an indirect ELISA. Briefly, 96-well microtitre plates (Nunc MaxiSorp, Thermo Fisher Scientific) were coated with 20 µg recombinant perfringolysin O [30]. Non-specific binding was blocked with 1 % (w/v) bovine serum albumin (Sigma-Aldrich) in PBS. Two-fold dilutions of the sera ranging from a dilution of 1:50 to 1:51200 were added to the plates (100 µl of each dilution/well; in duplicate) and incubated for 2 h at 37 °C. Plates were washed with 0.1 % (v/v) Tween 20 in PBS and incubated for 1 h 30 min at 37 °C with horseradish peroxidase-labelled rabbit-anti-bovine IgG (Sigma-Aldrich). Bound conjugate was detected using the substrate 3,3′,5,5′-Tetramethylbenzidine (TMB) (Sigma-Aldrich). The reaction was blocked with H_2SO_4 and the absorbance was measured at 450 nm using a microplate reader (Multiscan MS, Thermo Labsystems, Helsinki, Finland). The end-point titer is expressed as the reciprocal of the last dilution that gave a reading of 0.1U above background (precolostral neonatal bovine serum).

Intestinal loop model

To study the protection against *C. perfringens*-induced necrosis provided by the antisera from calves vaccinated with the vaccine preparations, four intestinal loop experiments were performed. Intestinal loop experiments were performed according to a previously described protocol using 4 healthy male Holstein Friesian calves, purchased from a local tradesman which collects dairy calves from herds in Eastern Flanders [20]. Briefly, the calves were anesthetized and the small intestine was exteriorized. Per calf 80 intestinal loops of approximately 10 cm were ligated in the jejunum and a 5 cm space was left between the loops. Only half of the loops were injected, thus each time leaving one intervening loop to avoid leakage between sampled loops. For each vaccine preparation individual pre- and post-vaccination sera of 2 calves were used in two intestinal loop experiments. Intestinal loops were inoculated with 20 ml of a wild-type strain (JIR325) in combination with 10 ml of 25 % commercial milk replacer suspended in sterile NaCl solution, resulting in a total volume of 30 ml which was the same across all treatments and control loops. Prior to inoculation pre- or post-immune serum derived from calves immunized with the different vaccine preparations was added to the NaCl solution containing milk replacer, to obtain a final concentration of 6 % serum (v/v). In each calf five intestinal loops per test serum were injected. Also an equal number of control loops without addition of serum were injected either with *C. perfringens* (positive control) or with sterile bacterial growth medium (negative control). After injection of the loops, the abdomen was closed and the calves were maintained under anesthesia. At 5-h post-inoculation intestinal biopsy samples were taken, after which the animals were euthanized. Samples were fixed in 4 % phosphate buffered formaldehyde. They were embedded in paraffin wax, sectioned and stained with hematoxylin-eosin. The sections were evaluated in a blinded manner by a board certified pathologist for presence of tissue necrosis (0 – absence of necrosis, 1 = necrotic lesions present).

In vitro neutralization and cytotoxicity tests

Neutralization of alpha toxin activity on egg yolk lipoproteins in vitro

The alpha toxin activity was determined by its effect on egg yolk lipoproteins as previously described [31]. Therefore, fresh egg yolk was centrifuged (10,000 × g for 20 min at 4 °C) and diluted 1:10 in PBS. The ability of the sera to neutralize the alpha toxin activity was assessed by pre-incubating a two-fold dilution series of the sera (two wells per dilution) with a constant amount of alpha toxin (10 µg/ml recombinant alpha toxin in PBS solution) for 30 min at 37 °C prior to the addition of 10 % egg yolk emulsion. Recombinant alpha toxin was expressed in *E. coli* using the pBAD TOPO® TA Expression Kit (Invitrogen, Paisley, UK) followed by purification onto a Ni-sepharose column (His Gravitrap, GE Healthcare, Buckinghamshire, UK). After incubation of the plates at 37 °C for 1 h, the A_{620} was determined. Alpha toxin activity was indicated by the development of turbidity which results in an increase in absorbance. The inhibitory capacity of the antiserum was determined by applying a Hill function to the concentration-response data (GraphPad Prism 5, GraphPad Software, San Diego, CA, USA) and expressed as the dilution that inhibited 50 % of the alpha toxin activity. The test was performed in duplicate.

Neutralization of perfringolysin O activity in vitro

Perfringolysin O (PFO) activity was determined by measuring the hemolysis of horse erythrocytes using a doubling dilution assay as previously described [32]. The PFO titer is the reciprocal of the last dilution which showed complete hemolysis. Similar to the inhibition of the alpha toxin activities, the ability of sera to neutralize the PFO activity was assessed by pre-incubating a two-fold dilution series of the sera (two wells per dilution) with a constant amount of perfringolysin O (2 µg/ml recombinant perfringolysin O). Recombinant perfringolysin O was produced as previously described [33]. The inhibitory capacity of the antiserum was expressed as the highest dilution that inhibited perfringolysin O induced hemolysis. The test was performed in duplicate.

Endothelial cell cytotoxicity assay

Primary bovine umbilical vein endothelial cells (BUVEC) were isolated from umbilical cord veins by an adaptation of the method of Jaffe et al. as performed previously [10, 34]. The toxicity of *C. perfringens* supernatant towards cultured bovine endothelial cells has been reported previously [10]. The ability of the antisera to neutralize the *C. perfringens* cytotoxicity towards BUVECs was determined using a Neutral Red Uptake assay (NRU) [35]. Therefore, a two-fold dilution series of the sera (100 % - 0.4 %) prepared in serum free cell culture medium was pre-incubated for 30 min at 37 °C with an equal amount of *C. perfringens* supernatant. Cells were treated with 100 µl of these supernatant-serum mixtures. The inhibitory capacity of the antiserum was expressed as the highest dilution that yielded 80 % cell viability. As a positive control, cells were treated with *C. perfringens* supernatant which was pre-incubated for 30 min with serum free cell culture medium. Untreated cells, incubated with serum free cell culture medium served as a negative control. The test was performed in duplicate.

Statistical analysis

The 20 loops tested for each condition provided enough statistical power to detect a 40 % reduction in the development of necrotic lesions in the intestinal loop assay (95 % confidence, 80 % power) (Winepiscope 2.0).

The protective effect of the antisera in the intestinal loop assay as compared to the pre-immune sera and the untreated control loops were determined by a Fisher's exact test (GraphPad Prism 5 software). Differences between groups were considered significant at $p < 0.05$.

Abbreviations

BUVEC, bovine umbilical vein endothelial cells; ELISA, enzyme-linked immunosorbent assay; NRU assay, neutral red uptake assay; PBS, phosphate buffered saline; PFO, perfringolysin O; TMB, 3,3',5,5'-Tetramethylbenzidine

Acknowledgements

We are grateful to Christian Puttevils and Delphine Ameye for their skillful technical assistance. The authors thank Prof. Bart Devreese for the MALDI analysis. Rodney Tweten kindly provided the pTrcHisA plasmid encoding native perfringolysin O.

Funding

This work was supported by the Agency for Innovation by Science and Technology under contract number 090910.

Authors' contributions

Study design: EG, SV, BV, LT, BP, FH, RD, PD, FVI; Animal experiments: EG, SV, BV, DRM, SS; in vitro experiments: EG, SV; Preparation of the manuscript: EG, BP, FH, RD, PD, FVI. All authors have read and approved the final version of the manuscript.

Competing interests

The authors declare that they have no competing interests.

Consent for publication

Not applicable.

Author details

[1]Department of Pathology, Bacteriology and Avian Diseases, Faculty of Veterinary Medicine, Ghent University, Salisburylaan 133, B-9820 Merelbeke, Belgium. [2]Department of Internal Medicine and Clinical Biology of Large Animals, Faculty of Veterinary Medicine, Ghent University, Salisburylaan 133, B-9820 Merelbeke, Belgium. [3]Department of Surgery and Anesthesia of Domestic Animals, Faculty of Veterinary Medicine, Ghent University, Salisburylaan 133, B-9820 Merelbeke, Belgium.

References

1. Matches JR, Liston J, Curran D. *Clostridium perfringens* in the environment. Appl Microbiol. 1974;28(4):655–60.
2. Rood JI. Virulence genes of *Clostridium perfringens*. Annu Rev Microbiol. 1998;52:333–60.
3. Petit L, Gibert M, Popoff MR. *Clostridium perfringens*: toxinotype and genotype. Trends Microbiol. 1999;7(3):104–10.
4. Songer JG. Clostridial enteric diseases of domestic animals. Clin Microbiol Rev. 1996;9(2):216–34.
5. Manteca C, Daube G, Pirson V, Limbourg B, Kaeckenbeeck A, Mainil JG. Bacterial intestinal flora associated with enterotoxaemia in Belgian Blue calves. Vet Microbiol. 2001;81(1):21–32.
6. Lebrun M, Mainil JG, Linden A. Cattle enterotoxaemia and *Clostridium perfringens*: description, diagnosis and prophylaxis. Vet Rec. 2010;167(1):13–22.
7. Muylaert A, Lebrun M, Duprez JN, Labrozzo S, Theys H, Taminiau B, Mainil J. Enterotoxaemia-like syndrome and *Clostridium perfringens* in veal calves. Vet Rec. 2010;167(2):64–5.
8. Titball RW. Gas gangrene: an open and closed case. Microbiology. 2005; 151(Pt 9):2821–8.

9. Songer JG, Uzal FA. Clostridial enteric infections in pigs. J Vet Diagn Invest. 2005;17(6):528–36.
10. Verherstraeten S, Goossens E, Valgaeren B, Pardon B, Timbermont L, Vermeulen K, Schauvliege S, Haesebrouck F, Ducatelle R, Deprez P, et al. The synergistic necrohemorrhagic action of *Clostridium perfringens* perfringolysin and alpha toxin in the bovine intestine and against bovine endothelial cells. Vet Res. 2013;44(1):45.
11. Awad MM, Ellemor DM, Boyd RL, Emmins JJ, Rood JI. Synergistic effects of alpha-toxin and perfringolysin O in *Clostridium perfringens*-mediated gas gangrene. Infect Immun. 2001;69(12):7904–10.
12. Stevens DL, Tweten RK, Awad MM, Rood JI, Bryant AE. Clostridial gas gangrene: evidence that alpha and theta toxins differentially modulate the immune response and induce acute tissue necrosis. J Infect Dis. 1997;176(1):189–95.
13. Chandran D, Naidu SS, Sugumar P, Rani GS, Vijayan SP, Mathur D, Garg LC, Srinivasan VA. Development of a recombinant epsilon toxoid vaccine against enterotoxemia and its use as a combination vaccine with live attenuated sheep pox virus against enterotoxemia and sheep pox. Clin Vaccine Immunol. 2010;17(6):1013–6.
14. Lobato FC, Lima CG, Assis RA, Pires PS, Silva RO, Salvarani FM, Carmo AO, Contigli C, Kalapothakis E. Potency against enterotoxemia of a recombinant *Clostridium perfringens* type D epsilon toxoid in ruminants. Vaccine. 2010; 28(38):6125–7.
15. Zeng J, Deng G, Wang J, Zhou J, Liu X, Xie Q, Wang Y. Potential protective immunogenicity of recombinant *Clostridium perfringens* alpha-beta2-beta1 fusion toxin in mice, sows and cows. Vaccine. 2011;29(33):5459–66.
16. Altemeier WA, Furste WL, Culbertson WR, Wadsworth CL, Tytell AA, Logan MA, Tytell AG. Toxoid Immunization in Experimental Gas Gangrene: A Preliminary Report. Ann Surg. 1947;126(4):509–21.
17. Boyd NA, Walker PD, Thomson RO. The prevention of experimental *Clostridium novyi* gas gangrene in high-velocity missile wounds by passive immunisation. J Med Microbiol. 1972;5(4):459–65.
18. Maclennan JD. The histotoxic clostridial infections of man. Bacteriol Rev. 1962;26:177–276.
19. Ito A. Alpha toxoid of *Clostridium perfringens*. I. Purification and toxoiding of alpha toxin of *C. perfringens*. Jpn J Med Sci Biol. 1968;21(6):379–91.
20. Valgaeren B, Pardon B, Goossens E, Verherstraeten S, Schauvliege S, Timbermont L, Ducatelle R, Deprez P, Van Immerseel F. Lesion Development in a New Intestinal Loop Model Indicates the Involvement of a Shared *Clostridium perfringens* Virulence Factor in Haemorrhagic Enteritis in Calves. J Comp Pathol. 2013;149(1):103–12.
21. Stevens DL, Titball RW, Jepson M, Bayer CR, Hayes-Schroer SM, Bryant AE. Immunization with the C-Domain of alpha -Toxin prevents lethal infection, localizes tissue injury, and promotes host response to challenge with *Clostridium perfringens*. J Infect Dis. 2004;190(4):767–73.
22. Green DS, Green ML, Hillyer MH, Morgan KL. Injection site reactions and antibody responses in sheep and goats after the use of multivalent clostridial vaccines. Vet Rec. 1987;120(18):435–9.
23. Uzal FA, Kelly WR. Protection of goats against experimental enterotoxaemia by vaccination with *Clostridium perfringens* type D epsilon toxoid. Vet Rec. 1998;142(26):722–5.
24. Fernandes da Costa SP, Mot D, Bokori-Brown M, Savva CG, Basak AK, Van Immerseel F, Titball RW. Protection against avian necrotic enteritis after immunisation with NetB genetic or formaldehyde toxoids. Vaccine. 2013; 31(37):4003–8.
25. Thachil AJ, McComb B, Early MM, Heeder C, Nagaraja KV. *Clostridium perfringens* and *Clostridium septicum* toxoid to control cellulitis in turkeys. J Appl Poultry Res. 2012;21(2):358–66.
26. Williamson ED, Titball RW. A genetically engineered vaccine against the alpha-toxin of *Clostridium perfringens* protects mice against experimental gas gangrene. Vaccine. 1993;11(12):1253–8.
27. Soltani CE, Hotze EM, Johnson AE, Tweten RK. Structural elements of the cholesterol-dependent cytolysins that are responsible for their cholesterol-sensitive membrane interactions. Proc Natl Acad Sci U S A. 2007;104(51):20226–31.
28. Niilo L, Moffatt RE, Avery RJ. Bovine "Enterotoxemia". II. Experimental Reproduction of the Disease. Can Vet J. 1963;4(11):288–98.
29. Rigby GJ. An egg-yolk agar diffusion assay for monitoring phospholipase C in cultures of *Clostridium welchii*. J Appl Bacteriol. 1981;50(1):11–9.
30. Tweten RK. Cloning and expression in *Escherichia coli* of the perfringolysin O (theta-toxin) gene from *Clostridium perfringens* and characterization of the gene product. Infect Immun. 1988;56(12):3228–34.
31. Logan AJ, Williamson ED, Titball RW, Percival DA, Shuttleworth AD, Conlan JW, Kelly DC. Epitope mapping of the alpha-toxin of *Clostridium perfringens*. Infect Immun. 1991;59(12):4338–42.
32. Awad MM, Bryant AE, Stevens DL, Rood JI. Virulence studies on chromosomal alpha-toxin and theta-toxin mutants constructed by allelic exchange provide genetic evidence for the essential role of alpha-toxin in *Clostridium perfringens*-mediated gas gangrene. Mol Microbiol. 1995;15(2):191–202.
33. Valgaeren BR, Pardon B, Goossens E, Verherstraeten S, Roelandt S, Timbermont L, Vekens NV, Stuyvaert S, Gille L, Van Driessche L et al. Veal Calves Produce Less Antibodies against C. Perfringens Alpha Toxin Compared to Beef Calves. Toxins. 2015;7(7):2586–97.
34. Jaffe EA, Nachman RL, Becker CG, Minick CR. Culture of human endothelial cells derived from umbilical veins. Identification by morphologic and immunologic criteria. J Clin Invest. 1973;52(11):2745–56.
35. Repetto G, del Peso A, Zurita JL. Neutral red uptake assay for the estimation of cell viability/cytotoxicity. Nat Protoc. 2008;3(7):1125–31.

7

PCR-diagnosis of *Anaplasma marginale* in cattle populations of Ecuador and its molecular identification through sequencing of ribosomal 16S fragments

Leandro Tana-Hernández[1], Katherine Navarrete-Arroyo[1], Jorge Ron-Román[2], Armando Reyna-Bello[1,3] and María Augusta Chávez-Larrea[1,4*]

Abstract

Background: Bovine anaplasmosis is an endemic disease in tropical and subtropical areas. It is caused by a bacterium named *Anaplasma marginale*, and represents an economic problem for cattle farmers due to the losses it generates, such as: mortalities, reduced production, quarantine measures, treatments and control of vectors. The method most often used to diagnose this haemotrophic bacterium is direct examination on blood smear, which sensitivity and specificity are limited compared to other methods such as PCR. The present study aimed at investigating the presence of *A. marginale* in dairy cattle of Luz de América commune, province of Santo Domingo de los Tsachilas. Two PCRs were used to amplify specific regions of the *Rickettsia* for its molecular identification.

Results: At first, 151 blood samples were tested: *msp5* specific gene of *A. marginale* was identified in 130 samples, meaning 86.1% of them were infected by the rickettsia. Two positive samples were further randomly selected to confirm the presence of *A. marginale* through amplification, cloning and sequencing of the conserved region of gene 16S rRNA. The analysis of sequences obtained through cloning revealed a 100% identity between both samples and those registered in GenBank for *A. marginale*.

Conclusion: This is the first report and molecular identification of *A. marginale* in the bovine population of Ecuador and its prevalence was high at the level of farms and animals. These results demonstrate the importance of proceeding to evaluate and characterize bovine Anaplasmosis in Ecuador in order to establish control measures and reduce their impact.

Keywords: Bovine Anaplasmosis, *Anaplasma marginale*, *msp5*, 16S rRNA, Bovine disease, Ecuador

Background

Bovine anaplasmosis is a vector-borne disease caused by the rickettsia *Anaplasma marginale* (*A. marginale*). The disease is mainly characterized by fever, anaemia, weight loss, pale mucous membranes and sometimes death of affected animals. Its distribution includes the whole American continent, Asia, Africa, Europa and Australia, and generates reduced production, increased costs and hinders genetic improvement through the difficulty of introducing susceptible animals in endemic herds [1–3].

The distribution of the bacteria depends on the presence/absence of vectors, which are arthropods belonging to the Family Ixodidae; the most important are genera *Dermacentor* and *Rhipicephalus*. In Latin America, the tick of major distribution is *Rhipicephalus microplus* (*R. microplus*), which is incriminated as vector of anaplasmosis [4]. However, the epidemiological importance of ticks in the eco-epidemiology of the disease in Latin America is controversial, as the transmission of the

* Correspondence: machavez@espe.edu.ec
[1]Grupo de Investigación en Sanidad Animal y Humana (GISAH), Carrera Ingeniería en Biotecnología, Departamento de Ciencias de la Vida y la Agricultura, Universidad de las Fuerzas Armadas ESPE, Sangolquí, Ecuador
[4]Carrera de Ingeniería en Biotecnología, Departamento de Ciencias de la Vida y la Agricultura, Universidad de las Fuerzas Armadas – ESPE, Av. General Rumiñahui s/n, P.O.BOX: 171-5-231, Sangolquí, Ecuador

rickettsia by blood sucking insects, such as horseflies, would be more important [4, 5].

Most animals positive for anaplasmosis are permanently infected, with a rickettsemia ranging from 10^4 to 10^7 [6]. They are responsible for an epidemiological status known as 'enzootic stability' in herds [4]. As the majority of animals in this condition have a low rickettsemia and do not show clinical signs, diagnosis is difficult [1, 6, 7].

Serological and molecular diagnoses are the only methods allowing the detection of *A. marginale*, as their sensitivity and specificity are high [8]. Polymerase Chain Reaction (PCR), based on amplification of DNA fragment, has been recommended to detect infection in animals to be commercialized and/or moved internationally [9]. A positive result to the Enzyme Linked Immunosorbent Assay (ELISA) only confirms the presence of the pathogen at some time, as it detects antibodies; it does not necessarily mean the pathogen is present by the time the test is performed [8]. Culture and isolation of the causing agent are frequently used as gold standard methods for diagnosing other bacterial diseases; they are not applicable in the case of *A. marginale* as it cannot be cultured [10]. The unique gold standard remains xenodiagnosis which is not very convenient [8, 9]. For such reason, PCR recommended as confirmatory test for the diagnosis of bovine anaplasmosis [11]. In addition to PCR, the sequencing of 16S of ribosomal RNA (16S rRNA) gene allows identifying genus and species of these microorganisms; it has thus been used concomitantly with groESL operon for further classification [10].

Despite the tropical distribution of bovine anaplasmosis, little has been done to confirm the presence of the disease in Ecuadorian cattle, even though horseflies and *R. microplus* ticks are present in Ecuador [12, 13]. However, neither spatiotemporal distribution, risk factors associated with the disease nor characterization of the causing agent has been studied deeply in the country. The scarce scientific information available is mainly found in theses performed in universities and in non-indexed journals. For example a 68%-prevalence was recently reported by Muñoz and collaborators after using blood smear [14]. On the other hand, a prevalence of 85.5% was estimated by nested PCR in cattle sampled in the province of Los Rios, Quevedo canton [15], while Soto reported a prevalence of 91.7% by using a commercial ELISA [16].

The rickettsia was even identified through sequencing of 16S rRNA gene in *R. microplus* ticks collected on two cows [13]. In view of these preliminary results, one could think the disease is endemic in cattle of Ecuador. In order to collect a better information on the presence of bovine anaplasmosis, a PCR based on the detection of the DNA fragment 605 bp of Mayor Surface Protein 5 (*msp5*) [17] was standardized and tested in animals located in an area where vectors are present. Subsequently, a sequencing of previously cloned 16S rRNA obtained from two samples collected at random was performed for confirmation.

Methods

Type of study and geographic localization

In December 2014, a transversal study including two levels of sampling (herd and animal) was carried out. Blood samples were collected from cattle (n = 151) belonging to 15 dairy herds gathered in province of Santo Domingo de los Tsáchilas, humid tropical region at an altitude of 655 m above sea level [18] (Figs. 1 and 2).

Collection of blood samples

Blood samples were collected at random, in 4 cm^3-tubes with EDTA, through coccygeal vein puncture with 0.75 mm × 25 mm Vacutainer® needles. Samples were then stored at −20 °C until further use and analysis.

DNA extraction

Genomic DNA was extracted from 1 mL of blood using a protocol previously described by J Sambrook and DW Russell [19]. Extracted DNA was eluted in 100 µl Tris-borate EDTA (TBE) and stored at −80 °C until further analysis. DNA concentration was determined using a NanoDrop ND-1000 (NanoDrop Technologies Inc., Wilmington, USA) b at 260 nm.

PCR of msp5 and 16S rRNA genes

In order to perform the PCR allowing the detection of *msp5* gene, 100 ng of each sampled DNA were used. The PCR primer and technique used has been described previously [17, 20, 21] with some modifications. A total volume of 25 µL was composed of 2.5 PCR buffer, 0.5 U Platinum® Taq DNA Polymerase (Invitrogen™), 1 µM of each primer (Ana 19A/Ana 19B or RYaF16S/RYaR16S, Invitrogen™, as shown in Table 1 Invitrogen™), 0.2 mM of each nucleotide (dATP, dTTP, dCTP and dGTP; Invitrogen™) and 2.5 mM of $MgCl_2$. The program used for *msp*5 gene in the thermal cycler (TC-512 TECHNE) consisted in: 5 min-incubation at 94 °C, followed by 35 cycles of 45 s at 94 °C, 30 s at 64 °C and 1 min at 72 °C. A final 10 min-extension at 72 °C ended the program. For 16S rRNA PCR, 68 °C was used for hybridization and the extension time was changed 1.5 min. PCR products were analyzed on 1.5% agarose gel in TBE (45 mM Tris-borate, 1 mM EDTA) and visualized with SYBER® Safe DNA in a UV-transilluminator.

Cloning

Two µL of each PCR sample (An-SD-1 and An-SD-18) were ligated to TOPO TA vector (TOPO TA Cloning®

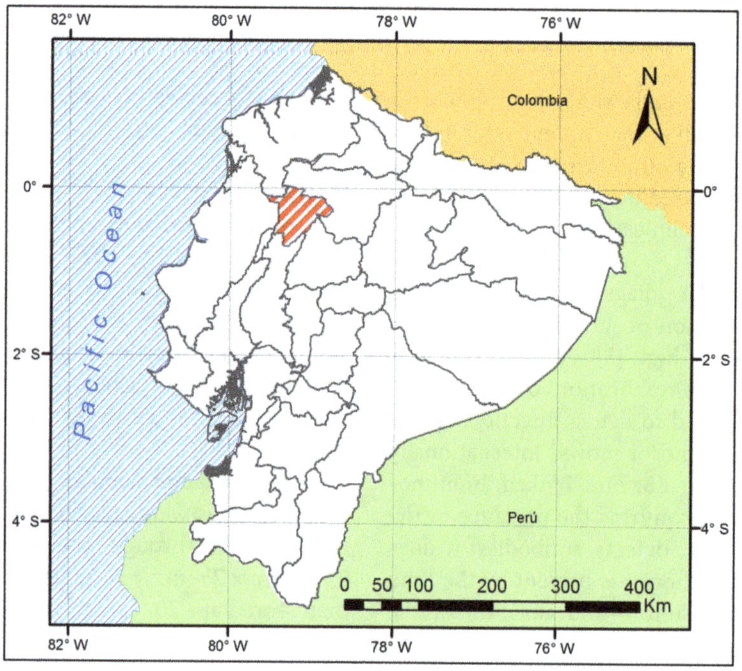

Fig. 1 Location study area, province of Santo Domingo de los Tsáchilas in red lines

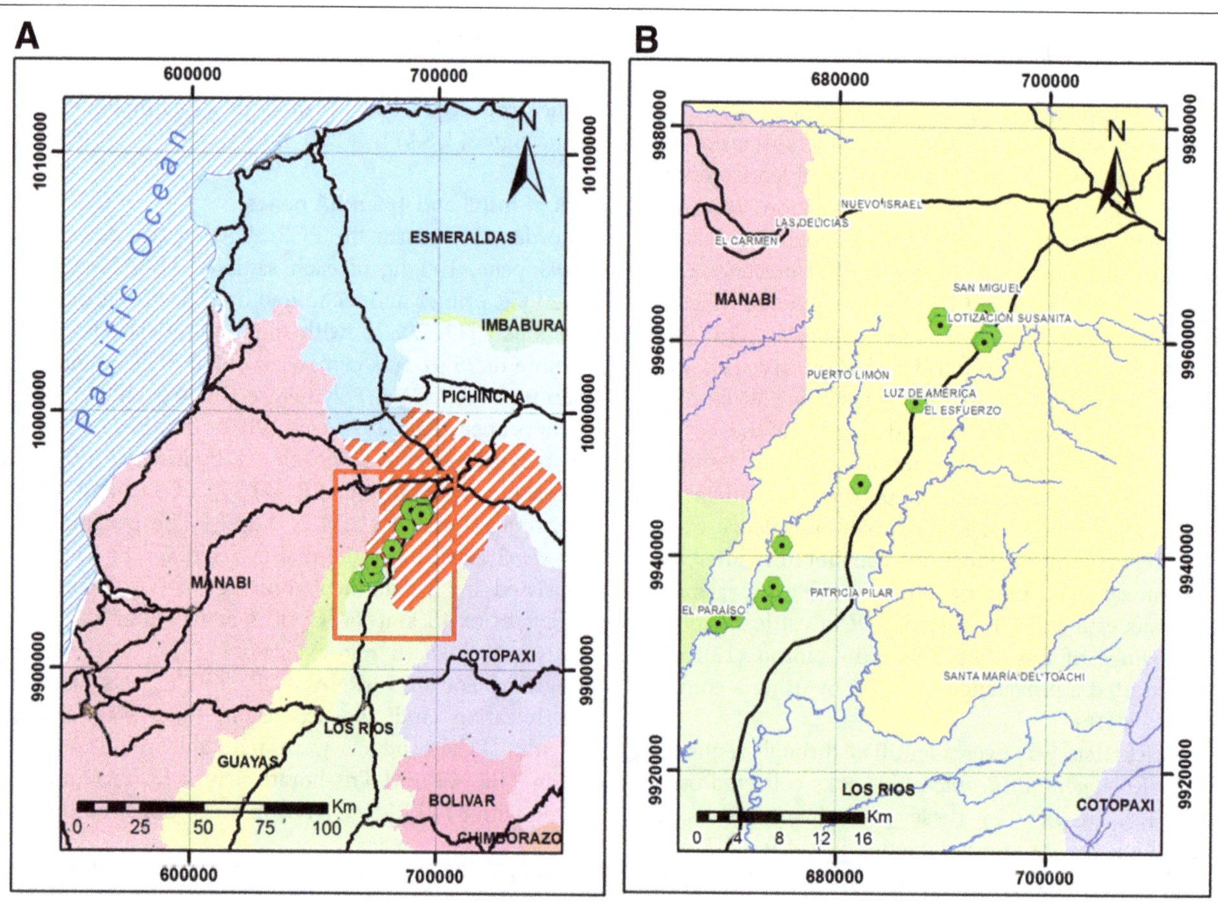

Fig. 2 Sampling area in panel **a** and farm location in Panel **b**

Table 1 Sequence of primers used for amplification of *msp5* gene and 16S rRNA

Primer	5'-3' Sequence	Temp. hybrid	Target	Reference and target
Ana 19A	GTGTTCCTGGGGTACTCCTA	64 °C	MSP5	Reyna-Bello et al., 1998
Ana 19B	TGATCTGGTCAGCCCCAGCT			
RYaF16S	TAACACATGCAAGTCGAACGG	68 °C		Chacón 2012
RYaR16S	ACCCAACCTTAAATGGTCGC			
M13 F M13 R	ACATTCCAGCAGCAGTTC GAG CACGTGAATCCTCAATTTTGT	63 °C	Fragment inserted into the plasmid	Invitrogen™

kit, Invitrogen™), following manufacturer's instructions. Subsequently, electroporation was performed with the Gene Pulser Xcell™ (BIORAD™), according to the protocol described by the manufacturer, for the transformation of *E.coli* with ligated plasmid. Once the cells were electroporated, 1 mL SOC medium (composed of 2% Tryptone peptone, 0.5% yeast extract, 100 mM NaCl, 2.5 mM KCL, 100 mM $MgCl_2$ and 100 mM $MgSO_4$ at pH 7) was added and placed one hour for stirring at 225 rpm in a 15 mL-centrifuge tube placed horizontally. Later, 50 μL of cloning were spread on LBP Petri dishes (prepared according to Sambrook and Russell, [19]) containing 50 μg/mL Kanamycin [SIGMA™], 40 μL of 40 mg/mL X-Gal (Invitrogen™) and 40 μL 100 mM-IPTG (Invitrogen™); it was further stored 24 h at 37 °C.

Purification of plasmid DNA of transformed dH5α *E. coli*

In all transformations of *E.coli*, 3 colonies were selected by colony PCR using the protocol described above using primers Ana 19A and B. These colonies were seeded in 10 mL LB culture medium [19] in the presence of 20 μg/mL Kanamycin (SIGMA™) for 24 h. Afterwards, cells were recovered through centrifugation of 3 mL from the culture at 10,000 xg for 10 min; plasmid DNA was further extracted from the pellet by a PureLink kit (K210010, Invitrogen™,USA).

Sequencing

Extracted plasmid DNAs were sent to Macrogen™ sequencing company (Seoul, South Korea), in order to sequence the cloned fragments, using forward and reverse M13 universal primers.

Sequence analysis

Cloning sequences were analyzed, cut and aligned in order to establish the consensus sequences with the MEGA6 Software; later, these consensus sequences were submitted to the BLAST database (NCBI Blastn) for identification and similarity.

Statistical analyses

At first, the disease prevalence was estimated, both at farm and individual levels. A farm was categorized as positive if at least one PCR-*msp5* tested positive among all samples analyzed. On the other hand, the influence of gender as possible risk factor was assessed by estimating Odds Ratios (ORs). Statistical analyses were performed with Epi Info 7™ and R software.

Results

Our transversal study was performed in December 2014; 151 bovine blood samples were collected in 15 dairy farms located in the commune Luz de América, province of Santo Domingo de los Tsáchilas, in Ecuador's coastal region. The sample size reached 49.5% of all animals present in the farms (N = 305). No sampling bias was to be considered, as the maximal proportion reached 25.2% (N = 38) in SD-4 farm. Males represented 11.9% (N = 18) of animals sampled.

Prevalence reached 100% at herd level and 86.1% at individual level (130 positive animals). Herd prevalence ranged from 40.0% (2/5) in SD-7 farm, to 100% in farms SD-4 (38/38), SD-8 (14/14), SD-9 (3/3) and SD-13 (5/5). No significant difference was observed between males and females (P > 0.05; OR = 1.94 for females; CI = 0.57–6.62) when considering *msp5* PCR results (males = 77.0% and females = 87.2%). Table 2 summarises the distribution of samples and results of *msp5* PCR analysis.

Out of all samples positive by *msp5* PCR (N = 130), two samples (An-SD-1 y An-SD-18) were selected in order to identify the rickettsia with 16S rRNAr PCR and further cloning and sequencing. Once consensus sequences were obtained for each clone, they were compared: sequences had 1383 bp and showed a 100% identity. After consulting Blast Database (NCBI Blastn), a 100% similarity was identified with GenBank (access number CP000030.1), registered for St Maries strain of *A. marginale* by Brayton and collaborators in 2005 [22].

Discussion

Bovine anaplasmosis is a disease commonly reported on the five continents. The rickettsia is transmitted through biological and mechanical vectors, but also iatrogenically. The disease is considered as endemic in tropical and subtropical areas and responsible for economic losses in cattle herds [2, 7]. Nevertheless, no report of *A.*

Table 2 Summary of results obtained per farm by PCR for the detection of *A. marginale* msp5 gene

Parameter	Animals present in the farm (N)	Animals sampled			Positive PCRmsp5	
		N	% per farm	% sampled	N	%
Sex						
Male		18		11.9	14	77.8
Female		133		88.1	116	87.2
Farm						
SD-1	47	12	25.5	8.0	8	66.7
SD-2	15	6	40.0	4.0	5	83.3
SD-3	10	10	100.0	6.6	8	80.0
SD-4	38	38	100.0	25.2	38	100.0
SD-5	20	9	45.0	6.0	8	88.9
SD-6	7	7	100.0	4.6	5	71.4
SD-7	5	5	100.0	3.3	2	40.0
SD-8	20	14	70.0	9.3	14	100.0
SD-9	3	3	100.0	2.0	3	100.0
SD-10	3	3	100.0	2.0	2	66.7
SD-11	35	12	34.3	8.0	10	83.3
SD-12	14	4	28.6	2.7	3	75.0
SD-13	40	5	12.5	3.3	5	100.0
SD-14	20	12	60.0	8.0	10	83.3
SD-15	28	11	39.3	7.3	9	81.8
TOTAL	305	151	49.5	100.0	130	86.1

N = number

marginale in Ecuadorian cattle has ever been published in indexed journals to date. Only few studies were issued in local journals and dissertations.

Our transversal study aimed at assessing the prevalence of bovine anaplasmosis, both at herd and individual levels in a district of Ecuador coastal region. The presence of *A. marginale* was assessed by using a standardized PCR for *msp5* gene amplification, according to a procedure previously described [20].

All herds were identified as being positive, along with a high proportion of infected animals per herd from 40 to 100%. Our result confirms the fact that anaplasmosis is endemic in the area of study with 86,1%, as it is the case in other tropical and subtropical regions of Latin America [23]. A second PCR was developed to amplify the rickettsia 16S rRNAr gene in two samples positive by *msp5* PCR. After cloning and sequencing, the 1400 bp fragment showed a 100% homology with St. Maries strain of *A. marginale* [22]. Such result allow us to conclude that *A. marginale* is present in Ecuadorian cattle; the sequence we operated is commonly used to perform taxonomy among bacteria of the Order Rickettsiales, Family Anaplasmataceae [10, 24].

It is important to mention that no farm provided more than 50% of samples. In this study, there was a total of 151 animals, 18 (11.9%) were males while 133 (88.1%) were females due to the dairy characteristic of farms. However, sex was not identified as a risk factor, which confirms previous results [25].

Significant differences were highlighted between farms for what PCR/*msp5* results are concerned: 40% of results were positive in SD-7 farm vs. 100% in SD-4, SD-10 and SD-12 farms. It is not surprising, the longer animals remain in a farm, the higher the risk of transmission. The localization of farms has probably influenced the distribution of the disease. In the area of study, 86.1% of animals were positive for *A. marginale* without clinical signs (anaemia, fever or more than 1% *A. marginale* in smear) suggesting the circulation of *A. marginale* in apparently healthy animals or persistent infection [6]. And due to the high prevalence found, the concept of enzootic stability can also be used [4].

The PCR usually detects DNA traces from previous infections, however, in the case of bovine anaplasmosis this is not true because once *A. marginale* infects an animal, it becomes infected for years [6] allowing us to infer that a positive PCR animal is still infected with *A. marginale.*

On the other hand infection by blood parasites such as *Trypanosoma vivax*, *Babesia bovis* or *Babesia bigemina*, frequently identified in Latin America, causes a clinical picture quite similar, worthless from a diagnostic point of view (Reyna-Bello, 2014). PCR test is seen as an alternative for these cases, or for persistently infected animals [9, 20, 26]. The gene encoding *msp5* has been used in numerous PCR trials [20, 27], due to the fact that, as a single copy, it is highly conserved in all *A. marginale* isolates [9, 28].

In endemic areas, clinical cases are rarely observed, most often in naïve animals recently introduced (and coming from free areas). Animals persistently infected can be responsible for outbreaks in a naïve herd, when moved to a disease-free area where vectors are present [2]. The high prevalence of anaplasmosis estimated in the area of study and for previous reports of 85.5% in cattle sampled in Quevedo canton, province of Los Ríos [15] and 91.7% in cattle slaughtered in Quito Metropolitan Slaughterhouse [16], places the lowlands of Ecuador in enzootic stability for bovine anaplasmosis.

The moderate prevalence (68%) reported by Muñoz and collaborators in cattle of Zamora, Canton [14] is probably related to the lower sensitivity of the test used, i.e. blood smear, compared to PCR or ELISA [9]. Another study performed in the Galápagos Islands reported a 64.1% prevalence estimated by a commercial competitive ELISA (VMRD Inc.™) in a sample of 184 animals [29].

In neighbouring countries, anaplasmosis was reported to be present at prevalence rates similar to our study. In Colombia, for example, a 90.3% prevalence obtained by agglutination test was reported [30]; the performance of an ELISA using *msp5* recombinant protein as antigen revealed a prevalence ranging between 47% [17] and 94% [31, 32] in Venezuela. In Costa Rica, a prevalence ranging from 20.0 to 72.0% was estimated by using an ELISA/MSP5r kit (VMRD Inc.™) [25] while a prevalence of 15% was estimated in Texas, with the same kit [33]. Indeed, such differences in prevalence have been mentioned by other authors [1].

The distribution of the rickettsia depends on the presence of vectors, which are arthropods belonging to the Ixodidae family; the most important belong to the genera *Dermacentor* and *Rhipicephalus*. In Latin America, the tick most widely distributed is *Rhipicephalus microplus*, wich is known as vector of *A. marginale* [4]. Nevertheless, in Latin America, the epidemiological importance of ticks as vectors is controversial: the transmission by blood sucking dipterans, such as horseflies, would play a more important role [4, 5]. Distribution and prevalence of anaplasmosis are directly related with the epidemiological role played by the different vectors. *Dermacentor albipictus* was mentioned as the main vector of the disease in Texas [33]. In Latin America, the predominating tick is *Rhipicephalus (Boophilus) microplus* (*R. microplus*), which does not seem to be an efficient vector for *A. marginale* [4, 7]. Transmission of anaplasmosis was mentioned to be mainly operated by blood sucking dipterous, iatrogenic or vertical transmission in Latin America [7, 34]. Vertical transmission would depend on the type of strain involved but infecting until 25% of calves and sometimes causing their death [35, 36]. On the other hand, Scoles and collaborators (2008) reported that *Dermacentor andersoni* was more efficient in transmitting *A. marginale* than horseflies. Nevertheless, the relationship between pathogen and vector has not been studied widely in Latin America yet (Baldacchino et al., 2014). While it is true, *R. microplus* happens to be an efficient vector, but to be species-specific (one sole host) limits its vector capacity in Latin America and Africa, where it is the predominating species [23].

In this regard, differences between *A. marginale* strains have been described in the world; some strains of Florida and Mississippi are not transmitted by ticks [1].

A study carried out in Costa Rica pointed out the presence of horseflies as the major risk factor for bovine anaplasmosis, not the ticks presence [25]. It is possible that, like what happened when *T. vivax* 'travelled' from Africa to Latin America more than 100 years ago, it adapted from a transmission by Tse Tse fly in Africa to a horsefly transmission in America, after gene deletion in its kinetoplast [37]. *Anaplasma marginale* might also have adapted itself to vectors existing in Central and South America, which would explain such a high prevalence of the disease in the region.

Conclusion

The high prevalence observed at farm and animal levels, as well as the molecular characterization of *A. marginale* in Ecuador, would allow to clarify the epidemiological situation of this hemotrophic, but also to better focus its diagnosis, treatment and control in Ecuador.

Abbreviations

16S rRNA: 16S of ribosomal RNA; *A. marginale*: *Anaplasma marginale*; ELISA: Enzyme Linked Immunosorbent Assay; MSP5: Mayor Surface Protein 5; ORs: Odds Ratios; PCR: Polymerase Chain Reaction; *R. microplus*: *Rhipicephalus microplus*

Acknowledgements

We would like to thank all farmers who participated in this study.

Funding

This work was supported by the 'Secretaría de Educación Superior, Ciencia, Tecnología e Innovación de la República del Ecuador' (SENESCYT), in the framework of Dr. Armando Reyna-Bello's 'Proyecto Prometeo", and the "Universidad de las Fuerzas Armadas – ESPE', through funding by the 'Unidad de Gestión de la Investigación – UGI'.

Authors' contributions

LT, KN, JR, MCH and ARB collected cattle samples, LT, KN carried out the PCR and LT drafted the manuscript. JR, carried out the statistical analysis and participated in preparation of the manuscript. ARB and MACH develop the idea of this research and help with the preparation and revision of the manuscript. All authors have read and approved the final version of this manuscript.

Ethics approval and consent to participate

The study was conducted according to the ethical rules of the Universidad de las Fuerzas Armadas ESPE and was approved by its ethics committee (reference number 2014-PIC-045), and the cattle samples were obtained under veterinary supervision. The cattle were owned by farms located in the Luz de America Community at Sto. Domingo de los Tsáchillas Province. Before sampling, the study objectives, schedule and work methodology were explained to the farmers. 15 farmers were involved in the study and all 15 gave their verbal consent for their cattle to be used in this study. This consent included the serological sampling of the cattle and we also provided sanitary-epidemiologic information about the farms and also about the animals. After our study, each farmer received the results and our recommendations from his farm.

Consent for publication

Not applicable

Competing interests

Authors declare that they have no competing interests.

Author details

[1]Grupo de Investigación en Sanidad Animal y Humana (GISAH), Carrera Ingeniería en Biotecnología, Departamento de Ciencias de la Vida y la Agricultura, Universidad de las Fuerzas Armadas ESPE, Sangolquí, Ecuador. [2]Grupo de Investigación en Sanidad Animal y Humana (GISAH), Carrera

Ingeniería Agropecuaria, Departamento de Ciencias de la Vida y la Agricultura, Universidad de las Fuerzas Armadas ESPE, Sangolquí, Ecuador. [3]Investigador-Prometeo, Departamento de Ciencias de la Vida, Carrera Ingeniería en Biotecnología, Universidad de las Fuerzas Armadas ESPE, Sangolquí, Ecuador. [4]Carrera de Ingeniería en Biotecnología, Departamento de Ciencias de la Vida y la Agricultura, Universidad de las Fuerzas Armadas – ESPE, Av. General Rumiñahui s/n, P.O.BOX: 171-5-231, Sangolquí, Ecuador.

References

1. Suarez CE, Noh S. Emerging perspectives in the research of bovine babesiosis and anaplasmosis. Vet Parasitol. 2011;180(1–2):109–25.
2. Reyna-Bello A: Anaplasmosis bovina: logros y retos inmediatos. In: Logros & Desafíos de la Ganadería Doble Propósito. Volume Cap LXXIV, edn. Edited by C González-Stagnaro NM, E Soto Belloso: Fundación GIRARZ. Ediciones Astro Data S.A. Maracaibo, Venezuela; 2014: 703–710.
3. Zivkovic Z, Nijhof AM, de la Fuente J, Kocan KM, Jongejan F. Experimental transmission of *Anaplasma marginale* by male Dermacentor reticulatus. BMC Vet Res. 2007;3:32.
4. Guglielmone AA. Epidemiology of babesiosis and anaplasmosis in south and central America. Vet Parasitol. 1995;57(1–3):109–19.
5. Coronado A: Is *Boophilus microplus* the main vector of *Anaplasma marginale*? *Revista Científica, FCV-LUZ* 2001, XI(5):408–411.
6. Brown WC, Barbet AF. Persistent infections and immunity in ruminants to arthropod-borne bacteria in the family Anaplasmataceae. Annual review of animal biosciences. 2016;4:177–97.
7. Kocan KM, de la Fuente J, Blouin EF, Coetzee JF, Ewing SA. The natural history of *Anaplasma marginale*. Vet Parasitol. 2010;167(2–4):95–107.
8. Aubry P, Geale DW. A review of bovine anaplasmosis. Transbound Emerg Dis. 2011;58(1):1–30.
9. Corona B, Dasiel Obregón D, YA YAI, Ifonso P, Vega E, Díaz A, Martinez S. Tendencies in diagnostic of bovine anaplasmosis. Rev Salud Anim. 2014; 36(2):73–9.
10. Dumler JS, Barbet AF, Bekker CP, Dasch GA, Palmer GH, Ray SC, Rikihisa Y, Rurangirwa FR. Reorganization of genera in the families Rickettsiaceae and Anaplasmataceae in the order Rickettsiales: unification of some species of Ehrlichia with Anaplasma, Cowdria with Ehrlichia and Ehrlichia with Neorickettsia, descriptions of six new species combinations and designation of Ehrlichia equi and 'HGE agent' as subjective synonyms of Ehrlichia phagocytophila. Int J Syst Evol Microbiol. 2001;51(Pt 6):2145–65.
11. OIE.: Bovine anaplasmosis. In: *OIE Terrestrial Manual.* edn. Edited by OIE; 2012: 589–600.
12. Cárdenas RE, Jaime Buestán J, Dangles O. Diversity and distribution models of horse fl ies (Diptera: Tabanidae) from Ecuador. Ann soc entomol Fr. 2009; 45(4):511–28.
13. Pesquera C, Portillo A, Palomar AM, Oteo JA. Investigation of tick-borne bacteria (Rickettsia spp., Anaplasma spp., Ehrlichia spp. and Borrelia spp.) in ticks collected from Andean tapirs, cattle and vegetation from a protected area in Ecuador. Parasit Vectors. 2015;8:46.
14. Muñoz TR, Ayora P, Jiménez V. Prevalencia de Anplasma marginale mediante extendidos sanguíneos en el cantón Zamora, provincia de Zamora Chinchipe. Centro de Biotecnología. 2014;3(1):44–51.
15. Escobar A, Cevallos O, Patricio Villarreal P, Carranza M, Carranza H, Pinargote E. Prevalence and detection by nested PCR of *Anaplasma marginale* in cattle and tick in the center of the coast of Ecuador. Ciencia y Tecnología. 2015;8(1):11–7.
16. Soto K: Determinación de la prevalencia de Anaplasmosis en el ganado bovino faenado en la Empresa Metropolitana de Rastro de Quito (EMRQ) mediate la aplicación de las técnicas de diagnóstico: Microscopía de Frotis Sanguíneos, Reacción en cadena de la polimerasa (PCR) y Rnsayo Inmunoenzimático Competitivo (cELISA). Escuela Politécnica del Ejército; 2010.
17. Reyna-Bello A, Cloeckaert A, Vizcaino N, Gonzatti MI, Aso PM, Dubray G, Zygmunt MS. Evaluation of an enzyme-linked immunosorbent assay using recombinant major surface protein 5 for serological diagnosis of bovine anaplasmosis in Venezuela. Clin Diagn Lab Immunol. 1998;5(2):259–62.
18. GAD., Domingo., S. 2016, Ecuador, R.d., ed. (http://www.santodomingo.gob.ec/).
19. Sambrook J, Russell DW. Molecular cloning: a laboratory manual, 3. EDN. 2001;
20. Tavares-Marques L, Reyna-Bello A. Estandarización de la técnica de PCR para el diagnóstico de la anaplasmosis bovina y ovina. Agronomía Trop. 2006; 56(4):501–12.
21. Chacón: Identificación Molecular de organismos Rickettsiales en caninos de diferentes regiones de Venezuela. In: Facultada de Ciencias, Escuela de Biología. Licenciatura: Universidad central de Venezuela; 2013: 122.
22. Brayton KA, Kappmeyer LS, Herndon DR, Dark MJ, Tibbals DL, Palmer GH, McGuire TC, Knowles DP, Jr.: Complete genome sequencing of Anaplasma marginale reveals that the surface is skewed to two superfamilies of outer membrane proteins. Proc Natl Acad Sci U S A 2005, 102(3):844–849.
23. Kocan KM, de la Fuente J, Guglielmone AA, Melendez RD. Antigens and alternatives for control of *Anaplasma marginale* infection in cattle. Clin Microbiol Rev. 2003;16(4):698–712.
24. Chakravorty S, Helb D, Burday M, Connell N, Alland D. A detailed analysis of 16S ribosomal RNA gene segments for the diagnosis of pathogenic bacteria. J Microbiol Methods. 2007;69(2):330–9.
25. Oliveira JB, Montoya J, Romero JJ, Urbina A, Soto-Barrientos N, Melo ES, Ramos CA, Araujo FR. Epidemiology of bovine anaplasmosis in dairy herds from Costa Rica. Vet Parasitol. 2011;177(3–4):359–65.
26. Figueroa JV, Chieves LP, Johnson GS, Buening GM. Multiplex polymerase chain reaction based assay for the detection of Babesia bigemina, Babesia bovis and *Anaplasma marginale* DNA in bovine blood. Vet Parasitol. 1993; 50(1–2):69–81.
27. Torioni de Echaide S, Knowles DP, TC MG, Palmer GH, Suarez CE, McElwain TF. Detection of cattle naturally infected with *Anaplasma marginale* in a region of endemicity by nested PCR and a competitive enzyme-linked immunosorbent assay using recombinant major surface protein 5. Journal of clinical microbiology. 1998;36(3):777–82.
28. Visser ES, McGuire TC, Palmer GH, Davis WC, Shkap V, Pipano E, Knowles DP Jr. The *Anaplasma marginale* msp5 gene encodes a 19-kilodalton protein conserved in all recognized Anaplasma species. Infect Immun. 1992;60(12): 5139–44.
29. Monroy MF: Determinación de la seroprevalencia de *Anaplasma marginale*, a través del Ensayo por Inmunoabsorción Ligado a Enzimas (ELISA) en la población Bovina de la provincia de Galápagos-Ecuador. San Francsco de Quito; 2015.
30. Patarrollo JH, Villa O, Diazgranados H. Epidemiology of cattle anaplasmosis in Colombia: I prevalence and distribution of agglutinating antibodies. Trop Anita Hlth Prod. 1978;10:171–4.
31. Eleizalde M, Caballero H, Reyna-Bello A. Evaluación y mejoramiento del ensayo inmunoenzimático (elisa) para el diagnóstico de la anaplasmosis bovina, utilizando la msp5 recombinante como antígeno. Revista Científica, FCV-LUZ. 2007;17(4):349–56.
32. Eleizalde MC, Reyna-Bello A. Mechanisms of antigenic variation in *Anaplasma marginale*. Rev Fac Cs Vets UCV. 2014;55(2):112–23.
33. Hairgrove TB, Craig TM, Budke CM, Rodgers SJ, Gill RJ. Seroprevalence of *Anaplasma marginale* in Texas cattle. Preventive veterinary medicine. 2014; 116(1–2):188–92.
34. Baldacchino F, Desquesnes M, Mihok S, Foil LD, Duvallet G, Jittapalapong S. Tabanids: neglected subjects of research, but important vectors of disease agents! Infection, genetics and evolution: journal of molecular epidemiology and evolutionary genetics in infectious diseases. 2014;28:596–615.
35. Grau HE, Cunha Filho NA, Pappen FG, Farias NA. Transplacental transmission of *Anaplasma marginale* in beef cattle chronically infected in southern Brazil. Revista brasileira de parasitologia veterinaria = Brazilian journal of veterinary parasitology: Orgao Oficial do Colegio Brasileiro de Parasitologia Veterinaria. 2013;22(2):189–93.
36. Costa SC, de Magalhaes VC, de Oliveira UV, Carvalho FS, de Almeida CP, Machado RZ, Munhoz AD. Transplacental transmission of bovine tick-borne pathogens: frequency, co-infections and fatal neonatal anaplasmosis in a region of enzootic stability in the northeast of Brazil. Ticks and tick-borne diseases. 2016;7(2):270–5.
37. Greif G, Rodriguez M, Reyna-Bello A, Robello C, Alvarez-Valin F. Kinetoplast adaptations in American strains from Trypanosoma vivax. Mutat Res. 2015; 773:69–82.

8

Search for OIE-listed ruminant mycoplasma diseases in Afghanistan

W. Bahir[1], O. Omar[1], R.S. Rosales[2], M. Hlusek[2], G. Ziay[1], W. Schauwers[3], A.M. Whatmore[2] and R.A.J. Nicholas[4*]

Abstract

Background: Little is known about the occurrence of important diseases of ruminants in Afghanistan because of the conflict affecting the country over the last 40 years. To address this discrepancy, ruminant herds in Afghanistan were screened for OIE-listed mycoplasma diseases, contagious bovine (CBPP) and caprine pleuropneumonias (CCPP).

Results: Of the 825 samples from 24 provinces tested for serological evidence of CBPP caused by *Mycoplasma mycoides* subsp.*mycoides*, 20 (3.4%) had ELISA values greater than the positive threshold of 50% though all were less than 55%. Repeat testing of these suspect sera gave values below 50. A smaller number of sera (330) from cattle in nine provinces were also tested by the rapid latex agglutination test (LAT) for CBPP, 10 of which were considered suspect. However, no positive bands were seen when immunoblotting was carried out on all sera that gave suspect results. Serological evidence of *Mycoplasma bovis* was detected in half of 28 herds in eight provinces. The cause of CCPP, *M. capricolum* subsp. *capripneumoniae* was not detected in any of the 107 nasal swabs and lung tissue collected from goats in seven provinces though sample handling and storage were not optimal. However, strong serological evidence was detected in goat herds in several villages near Kabul some of which were over 50% seropositive by LAT and ELISAs for CCPP; immunoblotting confirmed positive results on a selection of these sera.

Conclusions: The data presented here provide a first assessment of the occurrence of the two OIE listed mycoplasma diseases in Afghanistan. From the results of the testing bovine sera from the majority of provinces there is no evidence of the presence of CBPP in Afghanistan. However the samples tested represented only 0.03% of the cattle population so a larger survey is required to confirm these findings. Serological, but not bacterial, evidence was produced during this investigation to show that CCPP is highly likely to be present in parts of Afghanistan.

Background

Little was known about the current prevalence of animal diseases in Afghanistan until the establishment of the Central Veterinary and Diagnostic Research Laboratories (CVDRL) in 2009 which occupy newly constructed facilities in Kabul, funded by the EU and FAO. Despite considerable challenges within the country, the CVDRL, linked to a network of six regional and twelve provincial laboratories, has established diagnostic facilities for a range of bacterial, viral and protozoal diseases. The formation of an OIE twinning programme with the Mycoplasma Group, Animal and Plant Health Agency (APHA) in the UK provided an opportunity to investigate the presence of important transboundary diseases in particular the OIE-listed contagious bovine (CBPP) and caprine (CCPP) pleuropneumonias caused by *Mycoplasma species.*

Today, CBPP, caused by *Mycoplasma mycoides* subsp. *mycoides*, is believed to be confined to sub-Saharan Africa where it affects at least 26 countries [1]. Little targeted surveillance of the disease has taken place in Asia over the last few decades but freedom from CBPP was declared in India in 1990, China in 1996 and Pakistan in 1997 [1]. There have been no reports of CBPP in Afghanistan or its other neighbours, Iran, Turkmenistan, Tajikistan and Uzbekistan although it is unlikely that these countries routinely monitor for the disease.

The presence of CCPP, caused by *Mycoplasma capricolum* subsp. *capripneumoniae*, has long been suspected in goats in Afghanistan because of reports of outbreaks of disease with similar clinical signs though no laboratory confirmation had been made. Iran reports large numbers of cases annually and there have been

* Correspondence: robin.a.j.nicholas@gmail.com
[4]Consultant, The Oaks, Nutshell Lane, Farnham, Surrey GU9 0HG, UK
Full list of author information is available at the end of the article

detections of CCPP in China, Pakistan and Tajikistan over the last decade [2] so it is highly likely that the disease is present in Afghanistan though the on-going hostilities makes investigation difficult.

The aim of this work was to screen for these two OIE listed diseases in Afghanistan with the support of the OIE reference laboratory CIRAD, Montpellier. It also provided an opportunity to look for other mycoplasmas present in ruminants in particular *Mycoplasma bovis*, an increasingly important cause of bovine mycoplasmosis occasionally mistaken for CBPP [3], and *M. ovipneumoniae,* a frequent isolate from small ruminants affected with atypical pneumonia [4].

Materials and methods

Sample collection

Sera and clinical samples were taken randomly from ruminants in most provinces of Afghanistan. Sera for CBPP testing originated from a serum bank established during a national *Brucella* survey carried out in 2012. Samples for CCPP testing were obtained during a joint *Pasteurella/ Mannheimia/ Mycoplasma* survey in 2013 which covered seven provinces close to Kabul but did not achieve all its objectives because of problems with sample collection and storage from some provinces. In most cases sera were transported successfully and without delay to the CVDRL in cool boxes at approximately 4 °C and stored at -80 °C until all sera was collected and then tested together. All swab and tissue samples were transported as for the sera but tested immediately on arrival. It should also be noted that due to logistical and other problems clinical reports of disease were not comprehensive or considered reliable so were not included in this study.

Cattle samples

In total, 825 samples from 79 villages comprising 142 herds were tested. The sera had been collected from 24 of the 34 provinces of Afghanistan (Fig 1); sample sizes varied from 1 to 32 animals per village averaging 7. About 90% of cattle sampled were female and all were over 2 years old.

Goat samples

A total of 325 serum samples were obtained from herds in seven provinces in 2013; 88 nasal swabs and 19 lung tissues were taken from adult goats in an additional four provinces where clinical signs suggested the presence of CCPP. Sample size averaged 15 goats per village which sometimes consisted of more than one herd but animals mostly shared the same grazing area.

Fig. 1 Map of Afghanistan showing (shaded) provinces sampled for CBPP

Laboratory tests for CBPP and bovine mycoplasmosis:
The following tests were performed: commercial competitive (c) ELISA (Idexx, France) and latex agglutination test (Bovi-LAT, APHA, UK) for CBPP and indirect ELISA for *M. bovis* [5].

The cELISA was performed as stated by the manufacturer with a cut-off established at 50%; results below this cut-off are considered seronegative for CBPP. Training of staff from CVDRL was conducted by APHA, Weybridge during the OIE twinning programme. The World Reference Centre for CBPP, CIRAD, France conducted quality assurance testing of the CVDRL staff performing the cELISA and also kindly provided the kits; 32 sera of unknown status were sent to CDVRL for QC testing. Results matched exactly those of CIRAD.

Training and reagents for the indirect ELISA for *M. bovis* were provided by the APHA, UK. The test is run routinely at APHA where a cut-off point of 0.3 had been established values greater than this are considered positive.

The latex agglutination test (Bovi-LAT) was obtained from APHA and was performed according to their instructions. Results are scored subjectively after 2 min as: no agglutination: negative; slight agglutination: + suspect positive; agglutination within 2 min: ++ positive; agglutination with 1 min +++: strong positive.

Western blotting was carried out using the type strain PG1 as antigen. Strips were prepared by APHA, Weybridge [6] and dispatched to CDVRL. Sera were diluted 1/50 and applied to the antigen coated strips. For the western blot to be valid, the positive control should develop a distinctive pattern of bands [7]. Samples are compared to the positive control. A positive result is determined by four or more bands of equal size to the positive control. A sample is negative if the result does not follow the distinctive pattern or is not of equal size to the positive control.

Laboratory tests for CCPP
Mycoplasma culture was carried out on nasal swabs and lung tissue from selected goats in seven provinces [8]. Specific *Mycoplasma* medium was obtained from Mycoplasma Experience Ltd. (Reigate, UK) which was capable of supporting the growth a range of ruminant mycoplasmas including the fastidious *M. c. capripneumoniae.* The following PCRs were used to identify mycoplasmas following 24 h growth in mycoplasma medium: first, a generic mycoplasma PCR was carried out with primers described by Tang et al. [9] using *M. c. capripneumoniae, M. ovipneumoniae* and *M. arginini* as positive controls. If the samples matched the bands of either the *M. ovipneumoniae* or *M. arginini* controls then identification was achieved for these mycoplasmas; however as this PCR does not differentiate the *M mycoides* cluster (which includes *M c. capripneumoniae)* then a *M. mycoides* cluster PCR [10] and specific PCR for *M. c. capripneumoniae* [11] were carried out.

The latex agglutination test (Capri-LAT) was obtained from APHA and was performed according to their instructions. Results were interpreted as for CBPP.

The competitive ELISA (Idexx) was carried out according to the manufacturer's instructions. In addition, an indirect test for antibodies to CCPP was carried out with concentrated culture-grown antigen prepared at APHA [12]. and sent to CDVRL where a chequerboard titration was carried out to establish optimum concentrations of antigen and peroxidase-labelled IgG conjugate. Antigen and conjugate was diluted 1/200 and 1/2000 respectively. A cut-off point of the mean optical density plus 4 standard deviations was established as 0.3 after testing 100 goat sera from the UK, where CCPP has never been recorded.

The immunoblotting method was adapted from that previously described using the F38 strain of *M. c. capripneumoniae* [6]. Field sera were diluted 1/50 and applied to the antigen-coated strips which had been prepared at APHA and sent to CDVRL for testing. Results were interpreted as for CBPP.

Results

CBPP

Of the 825 samples tested by cELISA, 20 (3.4%) had values greater than 50% though all were less than 55% (Table 1). Repeat testing of these suspect sera gave values of below 50. Immunoblotting was carried out on all sera that gave suspect results in the cELISA. No bands of the correct size for CBPP were seen; indeed bands on the strips were either absent, faint or indistinct. However because of the shortages of reagents, none of these suspect sera were tested by LAT or *M bovis* ELISA.

LAT was carried out on 330 sera from 29 herds in 9 provinces, 10 sera had scores of 1+, which is considered suspect (Table 1). However these sera were tested by immunoblotting and found to be negative.

The indirect ELISA for *M bovis* was carried out on 272 sera from 28 herds in eight provinces. In all, 14 herds contained seropositive cattle, seven of which were over 50% seropositive (Table 1). No seropositive cattle were seen in the provinces of Bamiyan, Kanduz and Balkh though numbers of cattle tested were relatively few. No clinical reports were received from the herds tested.

CCPP

Mycoplasma c.capripneumoniae was not detected in any of the 107 nasal swabs and lung tissue cultured and tested by PCR for mycoplasmas. *M. ovipneumoniae* and

Table 1 Serological survey for CBPP and bovine mycoplasmosis in cattle herds in 24 provinces of Afghanistan

Province	No. of herds	No. of samples	No of positives cELISA	No of positives Bovi-LAT	No of positives *M bovis* ELISA
Bamyan	5	35	0	(1)[b]	0 (0)[c]
Kunar	4	32	7[a]	-	-
Kabisa	2	16	3[a]	-	-
Parwan	4	36	0	0	13 (4)[c]
Badakhshan	1	7	0	-	-
Balkh	3	30	0	0	0
Laghman	3	28	0	(1)[b]	18 (3)[c]
Ghor	3	35	0	(6)[b]	30 (3)[c]
Khandahar	2	39	0	(1)[b]	20 (2)[c]
Kanduz	2	18	0	(1)[b]	0
Nangharhar	1	31	0	0	-
Zabul	6	78	2[a]	0	20 (2)[c]
Patkya	6	64	2[a]	-	-
Khost	6	113	1[a]	-	-
Helmand	3	84	0	-	-
Wardak	6	29	1[a]	-	-
Logar	4	33	1[a]	-	-
Daykondi	5	26	1[a]	-	-
Farah	4	29	1[a]	-	-
Herat	2	19	1[a]	-	-
Badgis	4	33	0	-	-
Samangan	1	7	0	-	-
Uruzgan	1	3	0	-	-
Fariab	1	0	0	-	-
Total	79	825	20	0	101 (14)

[a]Negative on repeat testing and by western blot
[b]Sera giving suspect results on LAT but negative by western blot
[c]Herds seropositive for *M bovis*
- not tested

M. arginini were detected in 55 and 9 samples respectively (data not shown). A total of 16 of 19 nasal swabs were positive for *M. ovipneumoniae* in one village in Kunar Province but no clinical reports were available on goat health.

Of the 21 villages sampled in seven provinces, seven were seropositive by cELISA, 8 were seropositive by LAT and 12 by iELISA (Table 2). Five of the villages were positive by all tests; two of these, Starbik and Arbabdara, were from Kabul province and were 47 and 40% by cELISA, 47% and 83% seropositive by LAT and 73% and 63% seropositive respectively by iELISA. A selection of 10 sera from different herds, positive by indirect ELISA, were tested by immunoblot. Of these six gave strong bands of the same size as the control serum.

Discussion

The data presented here provide a first assessment of the occurrence of the two OIE listed mycoplasma diseases in Afghanistan. From the results of the testing of cattle sera from the majority of provinces there is little evidence of the presence of CBPP in Afghanistan. The sample tested represented only 0.03% of the cattle population but has a good statistical chance of detecting disease at a prevalence of at least 1%. One province, Kunar, contained a high number of suspect ELISA-positive samples (nearly 20%) spread amongst four herds on first testing but these were found to be negative on retesting and by immunoblotting. Publicity including talks and posters have been generated by the CVDRL staff during the twinning process with APHA describing the typical clinical signs and lesions of CBPP and will help the local veterinary staff to recognise the disease should it be present.

However from a smaller survey of eight provinces, it is clear that *M. bovis* is present with most herds examined containing seropositive animals, some up to 90% positive. Only the five herds tested in Bamiyan Province

Table 2 Serological testing for CCPP of goat herds in 8 provinces in Afghanistan

Province	Villages	No of samples	No of positives Capri-LAT	No of positives iELISA	No of positives cELISA
Bamyan	Akhzarat	10	3	0	2
Kunar	Shahedan	9	0	0	2
Kabul	Mohammady	3	1	1	0
	Dasht said	7	0	0	0
	Starbik	30	14	22	14
	Arbabdara	30	25	19	12
	Khoshab	31	3	4	1
	Qasaban	25	0	0	0
Kabisa	Qasaba	18	0	0	0
	Dara Dasht	6	0	1	0
Parwan	Kampany	4	0	0	0
	Parwan	4	0	0	0
	Sabat	27	0	0	0
	Polemoghulan	30	0	0	0
Badakhshan	Said Khil	16	0	2	0
Balkh	Sanjed Dara	19	0	3	0
	Barek Ab	13	0	4	0
	Qala-e-Dana	11	0	3	0
	Dahandara	12	4	1	7
	Gozar Marpol	8	2	2	1
	Arghawan	12	1	2	0[a]
Total 7	21	325	53 (8)[b]	64 (12)[b]	38 (7)[b]

[a]Only 2 samples tested by cELISA
[b]number of positive villages

were negative. This finding is not surprising given that only one cattle-rearing country to date, New Zealand, remains *M bovis*-free. Unfortunately, the presence of *M bovis* could confound attempts to declare Afghanistan free of CBPP as gross lesions can sometimes be very similar [13]. Furthermore, serological cross-reactions have been reported between the two mycoplasmas during the CBPP eradication campaign in Portugal [3] although there was little evidence of it in the present study as only two false positives were seen during the testing of 14 herds which were seropositive to *M bovis*. Interestingly the false positive rate of the cELISA for CBPP was just over 3% using the strict cut-off of 50% on first testing but 0% if 55% is used. All sera gave values below 50% on repeat testing. The use of the highly specific immunoblotting test provided confirmation of the negative status of all suspect sera.

The presence of CCPP in Afghanistan has long been suspected due to its presence in neighbouring countries and from clinical reports from the field. Indeed the main reason for the twinning programme was to provide reagents, equipment and training to enable detection of this most fastidious of mycoplasmas. To this effect serological, but not bacterial, evidence was produced during this investigation to show that CCPP is highly likely to be present in parts of Afghanistan.

Of the goat herds in seven provinces, strong serological evidence was seen in Kabul and Balkh where many sera were positive by ELISAs and LAT; in a smaller sample both the cELISA and LAT were positive in several goats in Bamyan province. Some differences were seen in the sensitivities of the tests with cELISA detecting fewer positive sera than the other tests but this is to be expected as the cELISA is generally considered more specific [14]. Unfortunately the causative mycoplasma *M. c.capripneumoniae* could not be detected or isolated from nasal or lung samples probably due to poor sample handling, storage and time taken for the samples to reach the CVDRL in Kabul. Predictably, the less fastidious mycoplasmas such as *M. ovipneumoniae*, an occasional cause of respiratory disease in small ruminants, and the opportunist *M. arginini* were detected, mostly in nasal swabs from goats in most provinces especially in a herd in Kunar where there had been no serological evidence of CCPP.

Abbreviations

(c) ELISA: (competitive) enzyme-linked immunosorbent assay; CBPP: contagious bovine pleuropneumonia; CCPP: contagious caprine pleuropneumonia; LAT: latex agglutination test; PCR: polymerase chain reaction

Acknowledgements

Work was undertaken as part of an OIE twinning project between APHA and CVRDL with funding from the UK International Biological Security Programme (projects RDOR1018 and RDOG1301). We would like to thank Dr. Keith Hamilton, formerly of the OIE, for setting up the twinning programme, and the Director and staff of the Pendik Veterinary Control Institute, Istanbul for hosting meetings and training sessions. We are also grateful to Dr. Francois Thiaucourt (CIRAD) for kindly providing serology kits for CBPP and conducting the QA tests.

Authors' contributions

WB and OO planned, carried out the surveys and performed the assays; RR and MH performed the training enabling the Afghan staff to conduct the study; WS supervised the planning, survey and assays and ensuring quality control; GZ and AW directed the project in which the studies were carried and approved the final version. RN was in charge of mycoplasma training and wrote the paper with considerable input from WB, RR, MH and AW. All authors read and approved the final manuscript.

Competing interests

The authors declare that they have no competing interests.

Consent for publication

Not applicable.

Ethics approval and consent to participate

Samples were collected for diagnostic purposes from farms throughout Afghanistan or came from serum banks taken for other survey purposes held at the laboratory in Kabul with permission of the owners. A local ethics committee, the CVDRL Ethics and Technical Committee, comprising Epidemiology Department adviser (Dr.Maken Ali), head of Laboratory (Dr. Ziay Ghulam), laboratory consultant (Willy Schauwers) and AHDP Team leader (Dr.Susan Chadima) waived the need for formal approval in this particular case.

Author details

[1]Central Veterinary and Diagnostic Research Laboratories, Kabul, Afghanistan. [2]Animal and Plant Health Agency, Woodham Lane, Addlestone, Surrey KT15 3NB, UK. [3]Landell Mills, Trowbridge, Wilts BA14 8HE, UK. [4]Consultant, The Oaks, Nutshell Lane, Farnham, Surrey GU9 0HG, UK.

References

1. OIE. World animal health, Animal health situation and disease control measures by country/territory by alphabetical order, vol. 2. Paris: World Animal Health Organisation; 2012.
2. Nicholas RAJ, Churchward C. Contagious caprine pleuropneumonia: new aspects of an old disease. Transbound Emerg Dis. 2012;59:189–96.
3. Gonçalves R, Regalla J, Ayling RD, Nicholas RAJ. Impact of *Mycoplasma bovis* infection on serosurveillance for contagious bovine pleuropneumonia. Vet Rec. 2009;163:632–3.
4. Nicholas RAJ. Improvements in the diagnosis and control of diseases of small ruminants caused by mycoplasmas. Small Rum Res. 2000;45:145–9.
5. Byrne WJ, Ball HJ, Brice N, Baker SE, Ayling RD, Nicholas RAJ. An application of an indirect ELISA to milk samples to identify cows with *Mycoplasma bovis* mastitis. Vet Rec. 2000;146:368–9.
6. Nicholas RAJ, Santini FG, Clark KM, Palmer NMA, De Santis P, Bashiruddin JB. A comparison of serological tests and gross lung pathology for the detection of contagious bovine pleuropneumonia in two groups of Italian cattle. Vet Rec. 1996;139:89–93.
7. OIE. Contagious bovine pleuropneumonia. In: Manual of diagnostic tests and vaccines for terrestrial animals. Paris: World Organisation for Animal Health; 2008. p. 718.
8. Nicholas R, Baker SE. Recovery of mycoplasma from animals. In: Miles R, Nicholas R, editors. Mycoplasma protocols. New Jersey: Humana Press; 1998. p. 37–44.
9. Tang J, Hu M, Lee S, Roblin R. A polymerase chain reaction based method for detecting Mycoplasma/Acholeplasma contaminants in cell culture. J Microbiol Methods. 2000;39:121–6.
10. Bashiruddin JB. In: Miles R, Nicholas R, editors. PCR and RFLP methods for specific detection and identification of Mycoplasma mycoides subsp. mycoides SC, In Mycoplasma Protocols. New Jersey: Humana Press; 1998. p. 167–78.
11. Woubit S, Lorenzon S, Peyraud A, Manso-Silván L, Thiaucourt F. A specific PCR for the identification of *Mycoplasma capricolum* subsp. *capripneumoniae*, the causative agent of contagious caprine pleuropneumonia (CCPP). Vet Microbiol. 2004;104:125–32.
12. Tekliogiorghis T. Natural outbreaks of suspected contagious caprine pleuropneumonia in Eritrea. Royal Veterinary College, London: Thesis for the Degree of Master of Science; 1997.
13. Nicholas R, Ayling R, McAuliffe L. Contagious bovine pleuropneumonia. Mycoplasma diseases of ruminants. Wallingford: CABI Publishing; 2008. p. 69–97.
14. Thiaucourt F, Bolske G, Libeau G, Le Goff C, Lefevre P-C. The use of monoclonal antibodies in the diagnosis and control of contagious caprine pleuropneumonia (CCPP). Vet Microbiol. 1994;41:191–203.

Biopanning of polypeptides binding to bovine ephemeral fever virus G_1 protein from phage display peptide library

Peili Hou, Guimin Zhao, Chengqiang He, Hongmei Wang* and Hongbin He*

Abstract

Background: The bovine ephemeral fever virus (BEFV) glycoprotein neutralization site 1 (also referred as G_1 protein) , is a critical protein responsible for virus infectivity and eliciting immune-protection, however, binding peptides of BEFV G_1 protein are still unclear. Thus, the aim of the present study was to screen specific polypeptides, which bind BEFV G_1 protein with high-affinity and inhibit BEFV replication.

Methods: The purified BEFV G_1 was coated and then reacted with the M13-based Ph.D.-7 phage random display library. The peptides for target binding were automated sequenced after four rounds of enrichment biopanning. The amino acid sequences of polypeptide displayed on positive clones were deduced and the affinity of positive polypeptides with BEFV G_1 was assayed by ELISA. Then the roles of specific G_1-binding peptides in the context of BEFV infection were analyzed.

Results: The results showed that 27 specific peptide ligands displaying 11 different amino acid sequences were obtained, and the T18 and T25 clone had a higher affinity to G_1 protein than the other clones. Then their antiviral roles of two phage clones (T25 and T18) showed that both phage polypeptide T25 and T18 exerted inhibition on BEFV replication compared to control group. Moreover, synthetic peptide based on T18 (HSIRYDF) and T25 (YSLRSDY) alone or combined use on BEFV replication showed that the synthetic peptides could effectively inhibit the formation of cytopathic plaque and significantly inhibit BEFV RNA replication in a dose-dependent manner.

Conclusion: Two antiviral peptide ligands binding to bovine ephemeral fever virus G_1 protein from phage display peptide library were identified, which may provide a potential research tool for diagnostic reagents and novel antiviral agents.

Keywords: Bovine ephemeral fever virus, Phage display random peptide library, G_1 binding peptide, Antiviral peptide

Background

Bovine ephemeral fever (BEF) caused by bovine ephemeral fever virus (BEFV), also known as 3 day fever, is an acute febrile infection of cattle and water buffaloes. The typical clinical symptoms of BEF are characterized by the rapid onset and rapid recovery of clinical signs, such as fever, depression, muscle stiffness, oral and nasal discharges, joint pain and lamenes [1–3].The disease is more frequently distributed in a vast expanse of the world including tropical, subtropical and temperate regions of Africa, the Middle East, Australia and Asia, and leads to serious economic losses to the cattle industry through loss of milk production, significant impacts on trade [4–6]. To date, no specific treatment and drug is available for BEFV.

BEFV belongs to the genus *Ephemerovirus* in the family *Rhabdoviridae*, and consists of a single stranded, negative-sense RNA genome with a lipid envelope and five structural proteins [7, 8]. The G protein plays an important role in inducing neutralizing antibodies, and it is also closely associated with cell tropism, pathogenicity, fusion and interaction with host cellular receptors [5, 9–12]. The

* Correspondence: hongmeiwang@sdnu.edu.cn; hongbinhe@sdnu.edu.cn
Key Laboratory of Animal Resistant Biology of Shandong, Ruminant Disease Research Center, College of Life Sciences, Shandong Normal University, No. 88 East Wenhua Road, Jinan City, Shandong Province, China

protein G consists of five major neutralisation sites (G_1,G_2,G_{3a},G_{3b},G_4) defined by BEFV G protein specific monoclonal antibodies (MAbs) in competition ELISAs, and among these sites, site G_1 composed of two minimal B cell epitopes is a linear neutralisation site involving in eliciting neutralizing antibodies. It is reported that site G_1 is predicted to face the viral membrane in both the pre-fusion and post-fusion forms of the trimer and may be approachable to antibodies [13–15]. Therefore, G_1 protein can be used as a potential target for antiviral drug screening.

Phage display, a well-established powerful and popular technology, has been extensively used in many fields, including antibody engineering [16, 17], ligand screening [18, 19], peptide drug discovery and manufacture [20, 21], disease molecular diagnostic analysis [22], biosensing [23] and vaccine research and development [24]. The high capacity and abundance of random phage display library makes it appropriate for high-throughput screening of peptide ligands that specifically bind with the given targets. In virology, several peptides have been identified as potential antivirals against classical swine fever virus [25], transmissible gastroenteritis virus [26], influenza virus [27], newcastle disease virus [28], human immunodeficiency virus (HIV) [29] and hepatitis B virus [30] using phage display technology. However, there is no report about screening of polypeptides binding to BEFV by phage display library.

In this study, we identified specific peptides that were capable of binding to the G_1 protein of BEFV from the 7-mer phage display library using the immobilized BEFV G_1 protein as a target, and we analyzed their roles in the context of BEFV infection in vitro, which is of great significance for potential diagnosis and treatment of BEF.

Methods

Virus, cells and plasmid

Ph.D.-7 phage-displayed peptide library was purchased from New England Biolabs (Ipswich, MA, USA). The titer of this peptide library was up to 2×10^{13} plaque forming units (pfu/mL), containing about 2.8×10^9 independent random peptide sequences. BEFV(Shandong/China/2011) were isolated and stored by the Ruminant Disease Research Center, Shandong Normal University, Jinan, Shandong Province, China, and the 50% tissue culture infected dose ($TCID_{50}$) of BEFV determined by Reed-Muench method were $5.0 \times 10^{5.0} TCID_{50}$/mL. Anti-BEFV antibody extracted from cattle was preserved in our laboratory. More concretely, clinical serum samples were collected from detected dairy cattle cases of BEFV infections in China. The antibodies against BEFV were identified by the microneutralization assay according to the instructions of the Agricultural Industry Criteria of the People's Republic of China (publication no. NY/T543–2002). The anti-BEFV antibodies in this study were applied to analyze the bioactivity of purified recombinant G_1 protein in Western Blot. BHK-21 cells were incubated in Dulbecco's modified Eagle's medium (DMEM, Invitrogen, Carlsbad, CA, USA), supplemented with 10% fetal bovine serum (FBS, Gibco, USA) at 37 °C, 5%CO_2. Prokaryotic expression vector pET-32a(+) plasmid were maintained in our laboratory. The G genes of Shandong/China/2011 (GenBank: JX234571.1) was previously described [31], and the pMD18-T-G was used as template for PCR amplifying the G_1 gene using sense primer G1-F:5'-CGGGATCCAGAGCTTGGTGTGAATAC-3′ and anti-sense primer G1-R:5'-CCGCTCGAGCCAACCTACAACAGCAGATA-3′. Underlined parts of primers reprensent *Bam* HI and *Xho* I restriction sites, respectively. The amplified G_1 gene region and the points linked to the primers used were shown in Supplement Figure one with the size of 420 bp. The PCR product was cloned into the pET-32a(+) vector and the plasmids with G_1 gene was sequenced by BGI (BGI, China).

Expression and purification of BEFV recombinant G_1 protein in *E. coli*

The recombinant plasmid pET-32a(+)-G_1 was transformed into high efficiency chemical competent cells *E.coli* BL21(DE3) pLysS (TransGen Biotech Company, Beijing, China). BL21 cells were cultured in LB medium containing 100 mg/L Ampicillin. The LB medium was shaken at 37°C until the OD_{600} reached 0.6–0.8, then isopropyl-d-thiogalactoside (IPTG) was added to a final concentration of 1 mmol/L to induce G_1 expression. The BL21 cells were further incubated for 6 h at 37 °C, and harvested by centrifugation at 8000×g for 5 min at 4 °C.The precipitation of BL21 cells were resuspended in lysis buffer including 10 mmol/L phosphate buffered saline (PBS, pH 7.4), imidazole (10 mmol/L), lysozyme (10 g/L), trypsin inhibitor (10 g/L), pepsin inhibitor (1.0 g/L) and PMSF (1.0 g/L) for sonication on ice. The cell lysate was centrifuged at 12,000×g for 20 min at 4 °C. The supernatant was collected and poured in a 10 mL Ni^{2+}-Sepharose 6 Fast Flow column pre-equilibrated with lysis buffer. The column was then washed with imidazole (10, 50 and 100 mmol/L) in PBS (10 mmol/L, pH 7.4) to remove nonspecifically bound proteins, and eluted with 250 mmol/L imidazole. Meanwhile, the designated pET32a(+) vector control protein was also expressed and purified in *E. coli* in the same condition. The yield protein concentration was measured using the Pierce BCA-200 Protein Assay Kit. To verify the reactivity of purified G_1 protein, we conducted western blotting using purified G_1 protein and pET32a(+) vector control protein as antigen by incubating with anti-BEFV positive antibody (diluted at 1:1000) for 2 h, and then with goat anti-cattle IgG conjugated to peroxidase (diluted at 1:2000, Abcam,ab102150,USA). Full immunoblots was visualized by enhanced chemiluminescence immunoassay (ECL kit, Thermo Fisher Scientific Inc., USA).

Biopanning and enrichment analysis of phage library using BEFV G_1 as target

The Ph.D.-7 peptide library (NEB, New England Biolabs, USA, E8100S) was used to screen for G_1 binding peptides according to the manufacturer's instructions with slight modifications. In short, 96-well plates were coated with 100 μL/well of the recombinant G_1 protein at a concentration of 100 μg/mL, biopanning was carried out by incubating the phage display library (10^{11} phages/mL). After washing away the unbound phage with Tris-buffered saline (TBS + 0.1% Tween 20), a bound phage was eluted in TBS and titrated as described in the standard protocol, and then subjected to the next round of panning. The second, third and fourth rounds of panning were done under more stringent conditions using less amount of the target (100 μL/well of a solution at 75,50,25 μg/mL of the recombinant G_1 protein, and shorter incubation time (1.5 h, 1 h and 0.5 h in the 2, 3 and 4th round, respectively), and washed with higher concentrations of Tween 20-TBS (0.25%, 0.5% and 0.5%, respectively) for longer time (10 × 2 min, 10 × 3 min and 10 × 4 min). Inbetween each round of panning, the titer of the panning phages in binding buffer (here referred to as input) and that in the elution buffer (here referred to as output) was determined, and the ratio of output to input was analyzed to assess the enrichment efficiency. After four rounds of panning, individual clones were randomly picked and amplified in *E. coli* ER2738 strain to prepare their DNA.

Sequencing of positive phage clones

Individual positive phage clones were amplified, and phage single-stranded DNA was extracted and purified using M13 Phage DNA Rapid Extraction Kit (Spin-column, Signalway Biotechnology, USA), DNA was sequenced by Sangon sequencing (Sangon Biotech Co.,Ltd., Shanghai, China) using the-96 gIII sequencing primer 5′-CCCTCATAGTTAGCG TAACG-3′ provided by the Ph.D.-7 peptide library kit. Amino acid sequences were deduced from phage display peptide DNA sequences by ExPASy Translate tool (http://web.expasy.org/translate/). The polypeptides used in the study were synthesis and purified by the Beijing Science and Technology Co., Ltd. and an irrelevant control peptide (AEMLELS) acquired from another screening strategy was synthesized by China Peptides Co., Ltd. (Shanghai, China).

Binding analysis of individual phage using ELISA

The phage displayed peptide clones with different amino acid sequences were subjected to ELISA. Briefly, ELISA plates were coated with the G_1 protein at a concentration of 10 μg/well diluted in 0.1 mol/L $NaHCO_3$ (pH 8.6) overnight at 4°C. Meanwhile, the purified pET32a(+) vector control protein coated with the same concentration was used as control group. All ELISA plates were blocked with 200 μL blocking buffer (5% BSA) for 2 h at room temperature. Ten-fold serial dilutions of the selected phages were added to both coated plates starting with 10^{12} phages in the first well. Plates were incubated for 2 h at room temperature and then washed 6 times with TBST (TBS containing 0.5% Tween-20). Bound phage was then detected with 200 μL 1:1000 diluted horseradish peroxidase (HRP)-conjugated anti-M13 antibody (GE Healthcare, 27–9421-01, USA) in blocking buffer. Finally, the peroxidase activity was rapidly detected with substrate solution TMB upon addition of a sulfuric acid stop solution. The absorbance of the reaction was determined at 450 nm with an ELISA microplate reader (Shanghai Utrao Medical Instrument Co., Ltd., Shanghai, China).

Antiviral activity of positive phage clones and synthetic peptides

BHK-21 cells (1 × 10^5/well) were plated in 12-well culture plates to 90% confluence. Amplified positive phage clones at concentrations of 10^{10}(pfu/mL) and synthetic peptides at concentrations of 0, 10, 20, 40, 80, 160, and 320 μg/mL were incubated with 100TCID$_{50}$/0.1 mL BEFV at 37°C for 1 h, respectively, and then a mixture of virus and amplified positive phages or various concentrations of synthetic peptides transferred onto the cell monolayers for adsorption for 2 h. The medium was then replaced with maintenance medium containing 2%FBS in 1 mL DMEM and incubated for a further 36 h. The treated cells and their supernatants were collected followed by three cycles of alternate freezing and thawing and stored at −80 °C before use.

Plaque reduction assay

Vero cells (1 × 10^5 cells/well) were seeded into 12-well plates for 12 h maintained in Dulbecco's modified Eagle's medium (DMEM, Invitrogen, USA), supplemented with 10% inactivated fetal bovine serum (FBS, Gibco, USA) and a mix of the antibiotics Penicillin (100 IU) and Streptomycin (100 μg/mL) at 37°C, 5%CO_2. Synthetic peptides at concentrations of 0, 10, 20, 40, 80, 160, and 320 μg/mL were incubated with 100TCID$_{50}$/0.1 mL BEFV at 37°C for 1 h, and then transferred onto the cell monolayers. After 2 h of incubation, discarding the supernatant, the cells were incubated under normal culture conditions at 37°C for 36 h. The monolayers were spread out with 2% low melting agarose, and formation of cytopathic plaque was visualized by staining the cell monolayers with 1%(*W*/*V*) neutral red solution (Sigma-Aldrich, USA). The effects of peptides on BEFV infection were evaluated by counting the plaques and calculating the titer (pfu/mL) and inhibitory rate.

Real-time reverse transcription polymerase chain reaction (RT-PCR)

Total RNA was extracted from 200 μL treated samples using QIAamp® Viral RNA Mini Kit (QIAGEN, Valencia, CA, USA) and eluted into final volumes of 50 μL. Reverse transcription was performed using PrimeScript RT reagent Kit (TaKaRa, Japan) according to manufacturer's protocols. The SYBR Green-based quantitative real-time RT-PCR was carried out using a Premix Ex Taq kit according to the manufacturer's protocols (TaKaRa, Japan) in a total volume of 25 μL on a Roche LightCycler 480 Real Time PCR System (Roche Applied Science, Germany). The sequences of primer sets used in real-time RT-PCR with specific primers for BEFV N gene and endogenous glyceraldehyde 3-phosphate dehydrogenase (GAPDH) gene were listed in Table 1. Real-time RT-PCR was carried out as follows: 95 °C for 5 min, followed by 40 cycles of 94 °C for 20 s, 60 °C for 30 s and 30 s at 72°C. The threshold cycle (Ct) value related to RNA levels of each target gene was normalized to GAPDH expression and the relative expression of each sample was analysed by the $2^{(-\Delta\Delta Ct)}$ method.

Statistical analysis

Each treated sample was run in triplicate. Statistical analysis was performed by one-way analysis of variance and values were considered significant when $p < 0.05$. Figures were performed using the GraphPad™ Prism 5.0 software (Graphpad Software Inc., San Diego, CA, USA).

Results

Purification of recombinant BEFV G_1 protein in *E. coli*

The G_1 gene region of 420 bp (Additional file 1: Figure S1) was amplified by PCR and cloned into pET32a(+) vector. G_1 protein was expressed in *E.coli* strain BL21 (DE3) and it mainly existed in the supernatant of the bacterial lysate, which indicated that G_1 gene had been efficiently expressed in prokaryotic expression system in a soluble form. Large amounts of highly expressed protein were purified using Ni^{2+}-sepharose 6 Fast Flow affinity media. The results of SDS-PAGE indicated that we obtained the highly-pure recombinant G_1 protein with an expected size of ~ 37 kDa. Meanwhile, the expected thioredoxin protein encoded by empty pET-32a (+) vector alone was also clearly visible with the size of ~ 21 kDa (Fig. 1a). Western blot were performed to analyze the bioactivity of purified recombinant G_1 protein. The result indicated that purification of recombinant

Table 1 The sequences of primers used in real-time RT-PCR

Gene		Sequences (5'-3')	Amplicon size(bp)
BEFV N	N-F	TCCTTCACCATGTACTGCAC	138
	N-R	ACCTTGTGGCACTCTCAAC	
GAPDH	GAPDH-F	TCATGACCACAGTCCATGCC	144
	GAPDH-R	GGATGACCTTGCCCACAGCC	

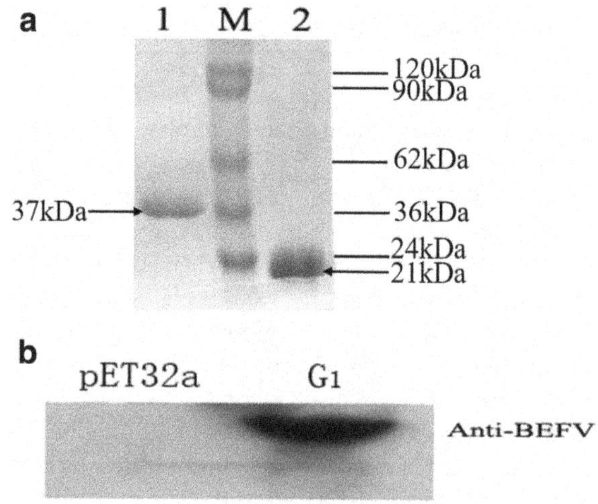

Fig. 1 SDS-PAGE and Western blot analysis of purified recombinant G_1 protein. Large amounts of highly expressed recombinant G_1 protein were purified using Ni^{2+}-sepharose 6 Fast Flow affinity media under native conditions. **a** SDS-PAGE analyses of purified recombinant G_1 protein and pET32a vector protein. M, molecular protein marker; Lane 1, purified recombinant G_1 protein; Lane 2, pET32a vector protein. **b** Western blot analysis of purified recombinant G_1 protein and pET32a vector protein react with BEFV antibody. The left lane: vector protein control; the right lane: purified recombinant G_1 protein

G_1 protein could immunoreact with anti-BEFV antibody (Fig. 1b). Therefore, those data were indicated that the purified G_1 used in this study was biologically active. The concentration of purified G_1 and control pET-32a (+) vector protein determined by the absorbance at 280 nm were 1.65 mg/mL and 2.85 mg/mL.

Identification of phage-displayed peptides with BEFV G_1 protein

In this study, a Ph.D.-7™ Phage Display Peptide Library was used to screen high-affinity ligands that could selectively bind to G_1 protein. Following four rounds of biopanning, the amounts of input phages and output phages were determined by titration, and the properties of phage recovery efficiency (output phages/input phages) in each round were calculated. As shown in Table 2, the recovery efficiency was 195-fold higher (from 7.2×10^{-3} to 3.7×10^{-5}) after four rounds of panning than after the first round, suggesting that the phages specifically bound to G_1 protein were successfully enriched.

DNA sequencing of the selected phage clones

After the 4th panning round, a total of 27 phage clones were randomly selected and amplified. The DNA of selected positive phage clones was extracted and sequenced. The DNA sequencing reports showed that 11 deduced peptides were finally obtained among the 27 phages and presented in Table 3. Alignment of the peptides revealed probably putative G_1 binding conservative

Table 2 Enrichment of phage peptides from library by BEFV G_1 protein biopanning

Item	First round biopanning	Second round biopanning	Third round biopanning	Fourth round biopanning
G1(μg)	10	7.5	5	2.5
Concentration of Tween20 (*v/v*)	0.1%	0.25%	0.5%	0.5%
Input/PFU (pfu/ml)	1×10^{11}	1×10^{11}	1×10^{11}	1×10^{11}
Output/PFU (pfu/ml)	3.7×10^{6}	5.2×10^{7}	8.7×10^{7}	7.2×10^{8}
Recovery (Output/input)	3.7×10^{-5}	5.2×10^{-4}	8.7×10^{-4}	7.2×10^{-3}

consensus motifs. Phages 4, 11, 12, 23 and 25 had a consensus sequence YSLRSDY named T25, which bear YSLR similarity with other 11 phage clones. The overlapping motif sequence derived from Phages 3 and 18 was HSIRYDF named T18.

Binding analysis of individual phage with G_1 using ELISA

In order to test the binding ability of the selected phage clones, 11 independent phage clone groups encoding different amino acid sequences were selected and detected by ELISA using the purified recombinant G_1 protein as target. As shown in Fig. 2, the absorbance obtained from each phage combined with the target G_1 protein was significantly higher than that of the control pET-32a (+) vector protein. Among these G_1 binding peptides, T18 and T25 showed the higher binding ability to G_1 protein when compared to the other clones. Although there is no criterion for judging the positive clones, it is suggested that the T18 and T25 clone might have a higher affinity to G_1 protein than the other clones.

Inhibition of peptides on BEFV replication

To further determine whether the phages that displayed G_1-binding peptides have the potential to inhibit BEFV infection, we performed virus-blocking assays in BHK-21 cells. Briefly, 100 $TCID_{50}$/0.1 mL BEFV was incubated with individual clone with a titer of 10^9pfu/0.1 mL that displayed G_1-binding peptide or irrelevant phage, respectively, and then inoculated into BHK-21 cells. Meanwhile, the non-phage treated BEFV was included as a control. After incubation for 36 h, the SYBR Green-based real-time RT-PCR was performed to determine the RNA copies of BEFV in different phage-treated cells. The results showed that T25 and T18 peptides significantly exerted antiviral activity, and the phage polypeptide T25 could reduce BEFV RNA copy number by 2 times, while T18 decrease by 1.5 times compared to control group. Obviously, T25 was the greatest antiviral peptide among all these selected peptides (Fig. 3).

To further investigate the antiviral activity of selected polypeptide (T18 and T25), firstly, we synthesized the peptides borne by phages T18 and T25 and conducted the dose-dependent effect of synthetic peptides T18 and T25 on inhibition of BEFV replication by virus plaque assay. At indicated concentration points, the reduce plaque assay showed the synthetic peptides could significantly inhibit the formation of cytopathic plaque at a concentration of 80 μg/mL, while the group of 320 μg/mL nearly abolished plaque formation in Vero cells, and the reduce plaque result of serial diluted synthetic peptides T25 was shown in Fig. 4a. Then antiviral activities of synthetic peptides T18 and T25 were determined by calculating the percentage of inhibitory rate compared against the controls. The results suggested that the similar

Table 3 Sequencing results of positive phage colonies with specifically binding to BEFV G_1 protein

Groups	Phage clones	Phage displayed peptide DNA sequence	Phage displayed peptide amino acid sequence	Poly-peptide names
1	4, 11, 12, 23, 25	TATTCTCTTCGTTCTGATTAT	YSLRSDY	T25
2	5, 6, 20	TATTCTCTTCGTCAGGATTGG	YSLRQDW	T20
3	1, 13, 19	TATAGTTTGCGTACTGATTGG	YSLRTDW	T19
4	16, 22, 26	TATTCTCTTCGTCAGGAGAGG	YSLRQER	T26
5	14, 24	TATAGTTTGCGTGCGGATCGT	YSLRADR	T24
6	10, 17	GCTCTGAGTAGTCTGCGTAAT	ALSSLRN	T17
7	3, 18	CATTCTATTCGGTATGATTTT	HSIRYDF	T18
8	2, 15	CATAGTATTCGTGTTGATTGG	HSIRVDW	T15
9	21	GATTGGATTTTTCCTGCGTTT	DWIFPAF	T21
10	9, 27	AAGGTTTGGATTGTTCCTTCT	KVWIVPS	T27
11	7, 8	AAGGTTTGGTTGCTTCATTCT	KVWLLHS	T8

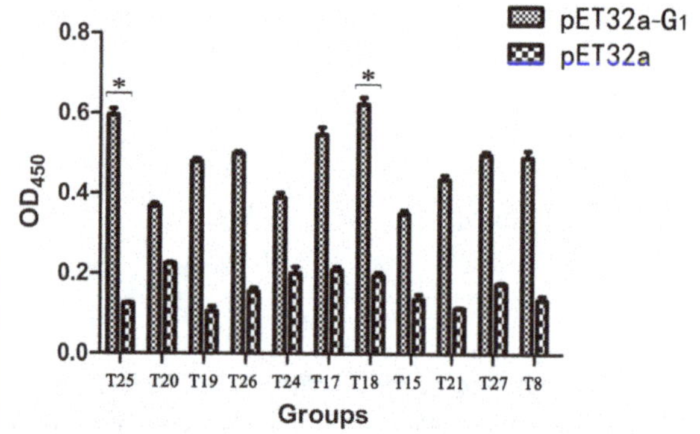

Fig. 2 ELISA analysis of binding activities between the selected phages and BEFV G_1 protein. 11 selected phages labeled T25,T20,T19,T26,T24,T17,T18,T15,T21,T27 and T8 were incubated with recombinant protein G_1 and pET32a vector control protein in ELISA plates to determine individual binding activities as measured by OD_{450} values. The OD_{450} readings were performed in triplicates. Statistical significance ($p < 0.05$) is noted by "*" compared to control groups

trend was at work on the inhibition of T18 and T25 on BEFV infection. It turned out that 80 μg/mL, 160 μg/mL and 320 μg/mL of synthetic peptides could significantly interfere with BEFV formation of cytopathic plaque ($p < 0.05$) (Fig. 4b).

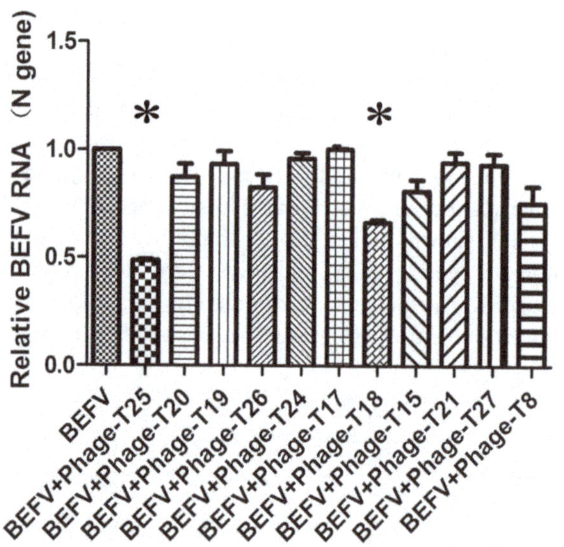

Fig. 3 Inhibitory effects of selected phage on BEFV detected by real-time RT-PCR. 11 selected phages labeled Phage T25,T20,T19,T26,T24,T17,T18,T15,T21,T27 and T8 were incubated with 100TCID$_{50}$/0.1 mL, respectively. The SYBR Green-based quantitative real-time PCR was used to determine BEFV RNA with specific primers for N gene. The relative expression of BEFV was normalized to GAPDH expression and analysed by the $2^{(-\Delta\Delta C_t)}$ method. Each treated sample was run in triplicate. Statistical analysis was p erformed by one-way analysis of variance and values were considered significant when $p < 0.05$. Figures were performed using the GraphPad™ Prism 5.0 software. Statistical significance ($p < 0.05$) was noted by "*" compared to control groups

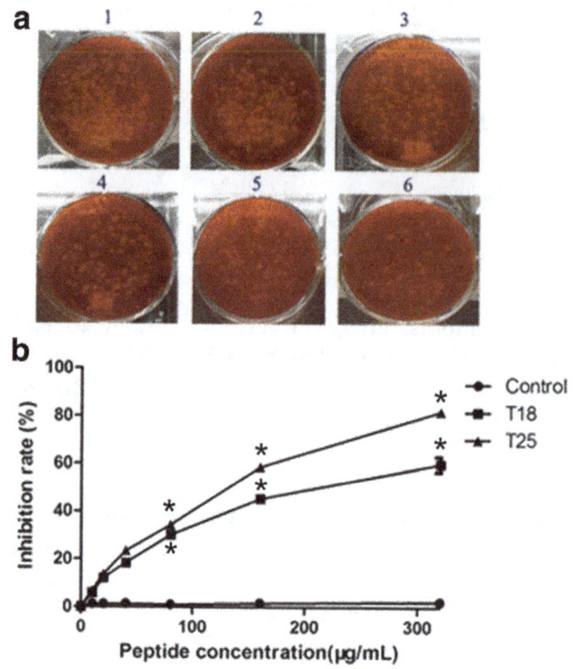

Fig. 4 Inhibition of peptide T18 and T25 to plaque production on Vero cells. **a** Inhibition ability of synthetic peptides T25 was determined by the reduce plaque assay. 100 TCID50/0.1 mL BEFV and 2-fold serial diluted synthetic peptides ranging from 10 μg/mL to 320 μg/mL were pre-inoculated and then inoculated Vero cells.1, BEFV; 2, BEFV + 20 μg/mL;3, BEFV + 40 μg/mL; 4, BEFV + 80 μg/mL; 5, BEFV + 160 μg/mL; 6, BEFV + 320 μg/mL. **b** Synthetic peptides T18 and T25 at 2-fold serial dilution concentrations of 10-320 μg/mL were incubated with 100 TCID$_{50}$/0.1 mL BEFV, respectively. The synthetic irrelevant peptide (AEMLELS) acted as control. Antiviral activities of the peptides were determined by calculating the percentage of inhibitory rate compared against the controls using the following f ormula; IN (%) = T/C × 100 [42], where, T is the mean of the number of plaque treated with different concentration of peptides. C is the mean of the number of plaque treated with corresponding synthetic irrelevant peptides that did not react with G_1 protein. Data presented were the mean OD values (±SD) of triplicate samples. Figures were performed using the GraphPad™ Prism 5.0 software. Statistical significance ($p < 0.05$) was noted by "*" compared to control groups

We further analyzed dose dependence of antiviral activity of the synthetic peptide T18 and T25 alone, and compared the synergistic effects of combined use on BEFV replication. As shown in Fig. 5, the results of real time RT-PCR showed that 320 μg/mL, 160 μg/mL and 80 μg/mL of synthetic peptides alone and combined use could significantly inhibit BEFV RNA replication ($P < 0.05$), and both T18 and T25 exerted inhibition on BEFV replication of BHK-21cells in a dose-dependent manner. However, the effects of peptides T18 and T25 combined were similar to that of peptide T18 or T25 alone ($P > 0.5$) (Fig. 5).

Discussion

The study of the molecular mechanism of virus infection is the basic and vital approach for prevention and control

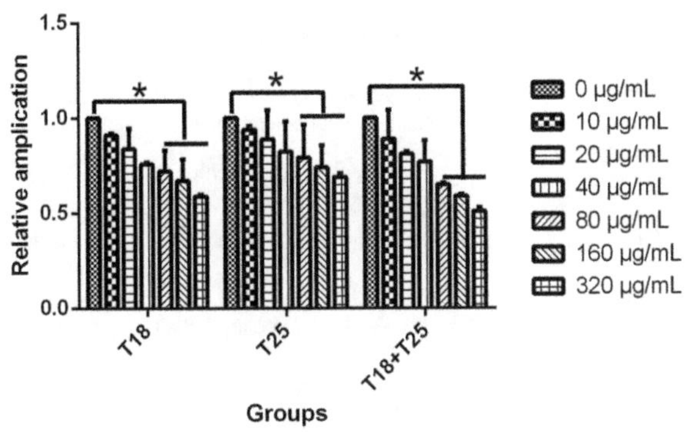

Fig. 5 Antiviral activities of synthetic polypeptides on BEFV infection detected by real-time RT-PCR. Synthetic polypeptides T18, T25 and T18 + T25 were incubated with 100TCID$_{50}$/0.1 mL, respectively. The SYBR Green-based quantitative real-time PCR was used to determine BEFV RNA with specific primers for N gene. The relative expression of BEFV was normalized to GAPDH expression and analysed by the $2^{(-\Delta\Delta Ct)}$ method. Each treated sample was run in triplicate. Statistical analysis was performed by one-way analysis of variance and values were considered significant when $p < 0.05$. Figures were performed using the GraphPad™ Prism 5.0 software

of the disease. As for BEFV infection, it has been proposed that the attachment of virus to the host cell membrane via the interaction of the viral envelope G protein with cell surface receptors is considered to be the initial step of BEFV infection [32]. And G_1 protein is considered to be the main protective antigen of BEFV that only reacts with sera against BEFV, currently some molecular and serological diagnostic methods to detect BEFV were established based on G_1 protein [12, 33]. In this study, we aimed at utilizing G_1 protein as target protein for screening its ligands and preliminary investigating the role of binding peptides during BEFV infection. The highly effective prokaryotic system is exemplified an optimal choice for heterologous protein, and the BEFV G_1 protein was purified in a moderately form of soluble (Fig. 1a), ensuring the target protein's native configuration in vitro. Furthermore it can specifically react with anti-BEFV antibodies as an antigen protein (Fig. 1b). Taken together, the G_1 glycoprotein used in this study is biologically active.

In recent years, using purified protein or whole virus particle as the targets of phage peptide library screening provides an ideal approach to identify multifunctional peptides [22, 34–36]. In this study, we used a phage-displayed heptapeptide library to identify G_1 protein interactors, and the conditions of the selection were strictly limited to enrichment specific G_1-binding phages(Table 2). After four rounds of panning, 11 deduced peptide amino acid sequences from 27 selected candidate phage clones exhibited high affinity to G_1 protein according to ELISA (Fig. 2). And these peptide sequences showed partial consensus motifs (Table 3), indicating that the screening process was successful. Meanwhile, partial consensus motifs which is possibly typical functional motifs of more complex proteins would be promising to discover the ligands of G_1 glycoprotein and analyze their roles in the context of BEFV infection. In addition, there has no report about BEFV receptor, and the consensus motif may mimic the discontinuous binding site as receptor agonists/antagonists or regulate the biological function of receptors [37–39].

Phage-display technology provides an effective way of identifying peptides with affinities to specific proteins for diagnostic analysis and the discovery of mimic antibodies [22, 40]. As the phages bearing the specific affinity peptides to BEFV G_1 protein, they have the potential to develop a phage-mediated diagnostic assay to BEFV. It is reported that the BEFV G_1 protein located at the C-terminal stalk of the G protein is one target of virus-neutralizing antibodies, and G_1 protein is a linear neutralization site (Y487-K503) that comprises two minimal B cell epitopes [5, 11].The potential significant achievement of our work is the discovery of mimic antibodies, synthetic peptide in vitro could be used as a specific antibody mimic to detect BEFV in virus-infected cells.

Peptide ligands targeting a specific protein surface obtained by phage-display technology are widely used as therapeutic agents by interfering with protein-protein interactions [20, 21, 25–29, 41]. Interestingly, the phage clone T18 and T25 exhibited binding affinity to G_1 protein and antiviral activity against BEFV (Fig. 3), whereas other phages with high affinity to G_1 protein did not effectively inhibit the replication of BEFV. This means that the binding affinity of other phages do not necessarily connection with antiviral activity. In addition, a possible explanation of antiviral activity is that the G_1 protein binding phage ligand mediates an inhibitory effect of BEFV infection by competitive binding to receptor binding sites or neutralizing epitopes and thus inhibits the virus attachment/entry. In order to exclude the effect of phage itself on protein binding and antiviral activity against BEFV, we synthesized T18 and T25 peptides in vitro and conducted the reduce plaque assay in vero cells that significantly formed cytopathic plaque during BEFV infection. The results showed that the synthetic peptides could significantly inhibit the formation of cytopathic plaque(Fig. 4) and inhibit BEFV replication in a dose dependent way (Fig. 5). Altogether, T18 and T25 peptides are potentially potent antiviral molecules worthy of further development and test. This study provides an inexpensive way to discover functional motif for BEFV infection using the phage-display technology, which is possible to find the potential protein to mediate the interaction between BEFV and the host.

Conclusion

In summary, we successfully isolated G_1 protein binding peptides from phage display library. Antiviral effect assay showed that G_1-binding peptides were dose-dependently effective against BEFV infection at the cellular level, which might have potential applications in the diagnosis and treatment of BEFV.

Additional file

Additional file 1: Figure S1. Schematic drawing of BEFV G and the points linked to the primers of cloned G_1 used. (A) The number of amino acids region about full length G gene and Region G_1 (from 390 aa–529 aa) were indicated. (B) Presentation of G_1 nucleotides sized and gene sequence amplification locus and positions of the forward (G1-F) and reversed (G1-R) primers used in this study.

Abbreviations

BEF: Bovine ephemeral fever; BEFV: Bovine ephemeral fever virus; BHK-21: Baby Hamster Syrian Kidney; BSA: Bovine serum albumin; DMEM: Dulbecco's modified Eagle's medium; ELISA: Enzyme linked immunosorbent assay; FBS: Fetal bovine serum; GAPDH: Glyceraldehyde 3-phosphate dehydrogenase; IPTG: Isopropyl-d-thiogalactoside; NEB: New England Biolabs; PBS: phosphate buffered saline; PMSF: Phenylmethanesulfonyl fluoride; RT-PCR: Reverse transcription polymerase chain reaction; TBS: Tris-buffered saline; $TCID_{50}$: 50% tissue culture infected dose

Acknowledgments

Not applicable.

Funding

This work was partially supported by grants from National Natural Science Fund of China (31,502,064, 31,672,556), Taishan Scholar and Distinguished Experts (H. H.), Shandong Major Agricultural Application Technology Innovation Project (H.H.,H.W), Primary Research & Developement Plan of Shandong Province(2015GNC113006, 2016GNC113006).

Authors' contributions

The study was conceived, designed and critically revised by HHB and WHM. Data collection, study execution and the drafted manuscript were done by HPL and ZGM. Laboratory data analysis was performed by HCQ. All authorship gave final approval and agreed to be accountable for all aspects of the work in ensuring that questions related to the accuracy or integrity of any part of the work.

Consent for publication

Not applicable.

Competing interests

The authors declared no potential conflicts of interest with respect to the research, authorship, and/or publication of this article.

References

1. St George TD. Bovine ephemeral fever: a review. Trop Anim Health Prod. 1988;20:194–202.
2. Uren MF. Bovine ephemeral fever. Aust Vet J. 1989;66:233–6.
3. Wang FI, Hsu AM, Huang KJ. Bovine ephemeral fever in Taiwan. J Vet Diagn Investig. 2001;13:462–7.
4. Walker PJ. Bovine ephemeral fever in Australia and the world. Curr Top Microbiol Immunol. 2005;292:57–80.
5. Walker PJ, Klement E. Epidemiology and control of bovine ephemeral fever. Vet Res. 2015;46:124.
6. Yeruham I, Van Ham M, Bar D, Yadin H, Tiomkin D. Economic aspects of the 1999 outbreak of bovine ephemeral fever in dairy cattle herds in the Jordan Valley in Israel. Vet Rec. 2003;153:180–2.
7. Nandi S, Negi BS. Bovine ephemeral fever: a review. Comp Immunol Microbiol Infect Dis. 1999;22:81–91.
8. Walker PJ, Byrne KA, Cybinski DH, Doolan DL, Wang YH. Proteins of bovine ephemeral fever virus. J Gen Virol. 1991;72(Pt 1):67–74.
9. Cybinski DH, Walker PJ, Byrne KA, Zakrzewski H. Mapping of antigenic sites on the bovine ephemeral fever virus glycoprotein using monoclonal antibodies. J Gen Virol. 1990;71(Pt 9):2065–72.
10. Hertig C, Pye AD, Hyatt AD, Davis SS, McWilliam SM, Heine HG, Walker PJ, Boyle DB. Vaccinia virus-expressed bovine ephemeral fever virus G but not G(NS) glycoprotein induces neutralizing antibodies and protects against experimental infection. J Gen Virol. 1996;77(Pt 4):631–40.
11. Kongsuwan K, Cybinski DH, Cooper J, Walker PJ. Location of neutralizing epitopes on the G protein of bovine ephemeral fever rhabdovirus. J Gen Virol. 1998;79(Pt 11):2573–81.
12. Zheng FY, Lin GZ, Qiu CQ, Zhou JZ, Cao XA, Gong XW. Serological detection of bovine ephemeral fever virus using an indirect ELISA based on antigenic site G1 expressed in Pichia Pastoris. Vet J. 2010;185:211–5.
13. Hsieh YC, Wang SY, Lee YF, Chen SH, Mak PO, Chu CY. DNA sequence analysis of glycoprotein G gene of bovine ephemeral fever virus and development of a double oil emulsion vaccine against bovine ephemeral fever. J Vet Med Sci. 2006;68:543–8.
14. Zheng FY, Lin GZ, Qiu CQ. Expression, purification and antigenic characterization of the Epitope-G1 gene of bovine ephemeral fever virus in Escherichia Coli. Wei sheng wu xue bao. 2007;47:498–502.
15. Ting LJ, Lee MS, Lee SH, Tsai HJ, Lee F. Relationships of bovine ephemeral fever epizootics to population immunity and virus variation. Vet Microbiol. 2014;173:241–8.
16. Wang X, Zhong P, Luo PP, Wang KC. Antibody engineering using phage display with a coiled-coil heterodimeric fv antibody fragment. PLoS One. 2011;6:e19023.
17. Frei JC, Lai JR. Protein and antibody engineering by phage display. Methods Enzymol. 2016;580:45–87.
18. Odermatt A, Audige A, Frick C, Vogt B, Frey BM, Frey FJ, Mazzucchelli L. Identification of receptor ligands by screening phage-display peptide libraries ex vivo on microdissected kidney tubules. J Am Soc Nephrol. 2001; 12:308–16.
19. Mohrluder J, Stangler T, Hoffmann Y, Wiesehan K, Mataruga A, Willbold D. Identification of calreticulin as a ligand of GABARAP by phage display screening of a peptide library. FEBS J. 2007;274:5543–55.
20. Liu JK, Teng Q, Garrity-Moses M, Federici T, Tanase D, Imperiale MJ, Boulis NM. A novel peptide defined through phage display for therapeutic protein and vector neuronal targeting. Neurobiol Dis. 2005;19:407–18.
21. Molek P, Strukelj B, Bratkovic T. Peptide phage display as a tool for drug discovery: targeting membrane receptors. Molecules. 2011;16:857–87.
22. Wang X, Li G, Ren Y, Ren X. Phages bearing affinity peptides to bovine rotavirus differentiate the virus from other viruses. PLoS One. 2011;6:e28667.
23. Wu J, Park JP, Dooley K, Cropek DM, West AC, Banta S. Rapid development of new protein biosensors utilizing peptides obtained via phage display. PLoS One. 2011;6:e24948.
24. Toledo-Machado CM, Bueno LL, Menezes-Souza D, Machado-de-Avila RA, Nguyen C, Granier C, Bartholomeu DC, Chavez-Olortegui C, Fujiwara RT. Use of phage display technology in development of canine visceral leishmaniasis vaccine using synthetic peptide trapped in sphingomyelin/cholesterol liposomes. Parasit Vectors. 2015;8:133.

25. Yin L, Luo Y, Liang B, Wang F, Du M, Petrenko VA, Qiu HJ, Liu A. Specific ligands for classical swine fever virus screened from landscape phage display library. Antivir Res. 2014;109:68–71.
26. Zou H, Zarlenga DS, Sestak K, Suo S, Ren X. Transmissible gastroenteritis virus: identification of M protein-binding peptide ligands with antiviral and diagnostic potential. Antivir Res. 2013;99:383–90.
27. Rajik M, Jahanshiri F, Omar AR, Ideris A, Hassan SS, Yusoff K. Identification and characterisation of a novel anti-viral peptide against avian influenza virus H9N2. Virol J. 2009;6:74.
28. Chia SL, Tan WS, Shaari K, Abdul Rahman N, Yusoff K, Satyanarayanajois SD. Structural analysis of peptides that interact with Newcastle disease virus. Peptides. 2006;27:1217–25.
29. Welch BD, Francis JN, Redman JS, Paul S, Weinstock MT, Reeves JD, Lie YS, Whitby FG, Eckert DM, Hill CP, et al. Design of a potent D-peptide HIV-1 entry inhibitor with a strong barrier to resistance. J Virol. 2010;84:11235–44.
30. Ho KL, Yusoff K, Seow HF, Tan WS. Selection of high affinity ligands to hepatitis B core antigen from a phage-displayed cyclic peptide library. J Med Virol. 2003;69:27–32.
31. He CQ, Liu YX, Wang HM, Hou PL, He HB, Ding NZ. New genetic mechanism, origin and population dynamic of bovine ephemeral fever virus. Vet Microbiol. 2016;182:50–6.
32. Albertini AA, Baquero E, Ferlin A, Gaudin Y. Molecular and cellular aspects of rhabdovirus entry. Viruses. 2012;4:117–39.
33. Zheng FY, Lin GZ, Qiu CQ, Zhou JZ, Cao XA, Gong XW. Development and application of G1-ELISA for detection of antibodies against bovine ephemeral fever virus. Res Vet Sci. 2009;87:211–2.
34. Lim KS, Lee DY, Valencia GM, Won YW, Bull DA. Cell surface-engineering to embed targeting ligands or tracking agents on the cell membrane. Biochem Biophys Res Commun. 2017;482:1042–7.
35. Ladner RC, Sato AK, Gorzelany J, de Souza M. Phage display-derived peptides as therapeutic alternatives to antibodies. Drug Discov Today. 2004;9(12):525–9.
36. Hamzeh-Mivehroud M, Alizadeh AA, Morris MB, Church WB, Dastmalchi S. Phage display as a technology delivering on the promise of peptide drug discovery. Drug Discov Today. 2013;18:1144–57.
37. Deshayes K, Schaffer ML, Skelton NJ, Nakamura GR, Kadkhodayan S, Sidhu SS. Rapid identification of small binding motifs with high-throughput phage display: discovery of peptidic antagonists of IGF-1 function. Chem Biol. 2002; 9:495–505.
38. Belloc CG, Longhi SA, Pena C, Blank VA, Miranda ME, Retegui LA. Identification, synthesis and properties of a consensus peptide recognized by a monoclonal antibody directed to various type I cytokine receptors. Life Sci. 2007;81:553–9.
39. de Souza CA, Teixeira PC, Faria RX, Krylova O, Pohl P, Alves LA. A consensus segment in the M2 domain of the hP2X(7) receptor shows ion channel activity in planar lipid bilayers and in biological membranes. Biochim Biophys Acta. 1818;2012:64–71.
40. Wang C, Sun X, Suo S, Ren Y, Li X, Herrler G, Thiel V, Ren X. Phages bearing affinity peptides to severe acute respiratory syndromes-associated coronavirus differentiate this virus from other viruses. J Clin Virol. 2013;57: 305–10.
41. Ren X, Liu B, Yin J, Zhang H, Li G. Phage displayed peptides recognizing porcine aminopeptidase N inhibit transmissible gastroenteritis coronavirus infection in vitro. Virology. 2011;410:299–306.
42. Moghaddam E, Teoh B-T, Sam S-S, Lani R, Hassandarvish P, Chik Z, Yueh A, Abubakar S, Zandi K. Baicalin, a metabolite of baicalein with antiviral activity against dengue virus. Sci Rep. 2014;4:5452.

Bovine mastitis: prevalence, risk factors and isolation of *Staphylococcus aureus* in dairy herds at Hawassa milk shed, South Ethiopia

Rahmeto Abebe[1*], Hagere Hatiya[2], Mesele Abera[1], Bekele Megersa[1] and Kassahun Asmare[1]

Abstract

Background: Mastitis is a disease of major economic importance in dairy industry worldwide. It is of particular concern in developing countries like Ethiopia, where milk and milk products are scarce. The objectives of the study were to estimate the prevalence of mastitis, identify the cow-and herd-level potential risk factors and isolate *Staphylococcus aureus*, one of etiological agents for contagious mastitis, from cows positive for mastitis. A total of 529 lactating cows selected randomly from 95 herds were screened by California mastitis test (CMT) for sub-clinical mastitis. Also 172 milk samples collected from CMT positive cows were cultured for isolation of *S. aureus*.

Results: Based on CMT result and clinical examination, the prevalence of mastitis at herd-level was 74.7% (95% CI: 64.5, 82.8). The corresponding cow-level prevalence was 62.6% (95% CI: 58.3, 66.7), of which 59.2 and 3.4% were sub-clinical and clinical mastitis cases, respectively. *S. aureus* was isolated from 51.2% of the milk samples cultured and 73.2% of the herds affected with mastitis. In the multivariable logistic regression model, the herd-level factors significantly associated ($p < 0.05$) with the presence of mastitis were herd size, bedding material, and milking mastitic cows last, while at cow-level, breed, parity, stage of lactation, udder and leg hygiene, and teat end shape were noted to have a significant effect on mastitis occurrence.

Conclusion: The very high prevalence of mastitis, more importantly the sub-clinical one, in the herds examined revealed the huge potential economic loss the sector suffers. Perhaps this was attributed to lack of implementation of the routine mastitis prevention and control practices by all of the herd owners. The findings of this study warrants the need for strategic approach including dairy extension that focus on enhancing dairy farmers' awareness and practice of hygienic milking, regular screening for sub-clinical mastitis, dry cow therapy and culling of chronically infected cows.

Keywords: CMT, Dairy herds, Mastitis, Risk factors, *S. aureus*

Background

Mastitis is the most widespread and costly disease in dairy cattle occurring throughout the world. It is of particular concern for farmers in developing countries like Ethiopia. Costs due to mastitis include reduced milk production, condemnation of milk due to antibiotic residues, veterinary costs, culling of chronically infected cows and occasional deaths [1]. Moreover, mastitis has a serious zoonotic potential associated with shedding of bacteria and their toxins in the milk [2].

* Correspondence: rahmetoabe@gmail.com
[1]School of Veterinary Medicine, Hawassa University, P.O.Box 05, Hawassa, Ethiopia
Full list of author information is available at the end of the article

Mastitis is caused by a wide spectrum of pathogens and, epidemiologically categorized in to contagious and environmental mastitis [3]. Contagious pathogens are those for which udders of infected cows serve as the major reservoir. They spread from cow to cow, primarily during milking, and tend to result in chronic sub-clinical infections with flare-ups of clinical episodes. Contagious pathogens include: *Staphylococcus aureus, Streptococcus agalactiae, Mycoplasma* spp. and *Corynebacterium bovis* [4]. On the other hand, environmental mastitis can be defined broadly as those intra-mammary infections caused by pathogens whose primary reservoir is the environment in which the cow lives [5]. Environmental pathogens include *E. coli,* Klebsiella spp., *Strept.*

dysgalactiae and *Strept. uberis* and the majority of infections caused by these pathogens are clinical and of short duration [6].

Mastitis can also be classified as either clinical or sub-clinical. Clinical mastitis is characterized by sudden onset, alterations of milk composition and appearance, decreased milk production, and the presence of the cardinal signs of inflammation in infected mammary quarters. It is readily apparent and easily detected. In contrast, no visible signs are seen either on the udder or in the milk in case of sub-clinical mastitis, but the milk production decreases and the somatic cell count increases. It is more common and has serious impact in older lactating animals than in first lactation heifers [7]. Because of the lack of any overt manifestation, the diagnosis of sub-clinical mastitis is a challenge in dairy animal management and in veterinary practice [8].

Ethiopia has the largest cattle population in Africa, which is estimated at 53.99 million animals of which 7.2 million are primarily held for milk production [9]. Despite these large numbers, milk production often does not satisfy the country's requirements due to a multitude of factors amongst which the occurrence of mastitis has great prominence. Over the years a number of researchers have studied the occurrence of mastitis in dairy herds in Ethiopia. According to the most recent published studies, the cow-level mastitis prevalence estimate falls within the range of 23.2 and 81.1% for the country [10–16]. Consequently, the disease has been documented as one of the major constraint of the dairy sector that needs attention.

Most of the previous studies in Ethiopia were concentrated on the investigation of the prevalence and few risk factors for mastitis at cow level and no or little effort has been made to assess the prevalence, management and hygienic practices at herd/farm-level. A focused study on contagious mastitis with emphasis on subclinical type is lacking. Besides, teat morphology, which is inherited anatomical feature of the cow that may affect mastitis occurrence [17] and could serve as a marker trait for selection to reduce mastitis in dairy cattle [18], is less investigated. It is therefore important to assess the effect of the variable on clinical and sub-clinical mastitis under Ethiopian condition and recommend preventive measures to reduce loss attributed to the disease. Moreover, given the huge economic relevance due to lack of clinical visibility and subsequent effects, investigation of sub-clinical mastitis at herd-level is of paramount importance for designing feasible prevention and control strategy.

Thus, this study was aimed to estimate the prevalence of clinical and sub-clinical mastitis, identify the major risk factors and isolate *S. aureus,* one of the etiological agents for contagious mastitis, from dairy cows in intensive and semi intensive management in southern Ethiopia.

Methods

Study area

This study was conducted between December 2014 and May 2015 in dairy herds found at Hawassa milk shed. The milk shed includes Hawassa city and its adjacent towns which supply milk for the big market at Hawassa. For this study, Hawassa city and two of the towns (Wondo Genet and Arsi Negelle) that are supplying milk for Hawassa were purposively selected because of their relatively larger potential for dairy cattle population. Hawassa is the capital city of the Southern Nations, Nationalities and Peoples' Regional State, and located at 275 km south of Addis Ababa, the capital. The city is situated at an elevation of 1708 m above sea level and located at 7^0 3' north latitude and 38^0 29' east longitudes. The annual rain fall and temperature varies from 800 to 1000 mm and 20.1–25 °C, respectively. Wondo Genet is located at 30 km west of Hawassa. It is situated about 1723 m above sea level at latitude of $7^0$5' N and longitude of 38^0 37' E. The town receives an average annual rainfall of 1372 mm. The mean annual temperature is 19 °C. Arsi Negelle town is found in the West Arsi zone of the Oromia regional state at a distance of 225 km from Addis Ababa. The town is situated about 2043 m above sea level at latitude of 7^0 21' N and longitude of 38^0 42' E. The average annual temperature of the area varies from 10 to 25 °C while rainfall varies between 500 and 1000 mm.

Study population and farm type

A total of 95 dairy herds selected randomly from the dairy farms present in the study areas were included in the study. The herd size of the selected farms varied from four to 112 cattle of which two to 57 were lactating cows. A total of 529 lactating cows were selected. Most of them (84.3%) were cross-breeds (Holstein-Friesian x Zebu) whilst few were local (Borana and Arsi breeds) (15.7%). With regard to management, 35 (36.8%) of the herds were managed intensively while 60 (63.2%) herds were semi-intensive. The intensively managed cattle were kept in doors and received concentrate feeds in addition to hay and crop residues (such as corn stalks, wheat/barley straw and other leftovers from grain threshing). On the other hand, the semi-intensively managed cattle grazed freely on pasture but received supplementary feeds in the morning and evening when they were milked. All cows were hand milked twice daily, in the morning and evening. The milk yield of the cows ranged from four to 14 l per day for cross breeds while two to four liters for local breeds.

Study design and sample size

A cross-sectional study design was followed to address the objective of the study. The sample size was estimated following the method described by Thrusfield [19] for simple random sampling with 95% confidence level and 5% absolute precision, considering an optimum expected herd level prevalence of 50%. Accordingly, the sample size was computed to be 384 herds. This sample size would be acceptable if the size of the study population were large in relation to the sample. However, since the number of dairy farms in the region registered and known by the respective district veterinary departments was smaller (only 126) than the calculated sample size, an adjustment was made based on the formula recommended for a small population size [19].

$$n_{\text{adj}} = \frac{N \times n}{N + n}$$

Where n is the sample size based on an infinite population and N is the size of the study population. Thus, the required sample size, n_{adj}, was calculated to be 95. Consequently, proportional allocation was made for each district in the milk shed and 40 herds each was considered for Hawassa and Arsi Negele and the remaining 15 allotted for Wondo Genet. In each district, respective herd sampling frame was constructed in collaboration with district veterinary department and herds were picked randomly using computer generated random numbers. Before data collection, a support letter written by the respective district veterinary department was sent to the identified farms requesting them to collaborate in the study. Fortunately, all of the farm owners who were randomly selected showed willingness to participate in the study. In each herd a minimum of 10% of the lactating cows were picked at random following the same procedure used for herd selection. Eventually a total of 529 lactating cows were included in the study.

Data collection

A structured questionnaire with close ended questions was used to collect data about various herd- and animal-level factors thought to influence the occurrence of mastitis in the dairy herds. The herd-level factors assessed were study area, herd size, herd management system, use of bedding material, milking practices (machine/hand), frequency of milking, cleaning and drying methods of udder, post-milking teat dipping, hygiene of cow barns, milking mastitic cow last, dry cow therapy, and culling of chronically infected animals. Those farms with good drainage system and used to clean cow manure daily were considered as "good hygiene" while those with no drainage system and absence of daily manure cleaning was regarded as "poor hygienic status." For the purpose of this study and ease of data analysis, the herd size was categorized as ≤10 cows and >10 cows. This is due to the fact that under Ethiopian context, farms with ≤10 cows are small scale semi-intensive farms that are run as side business. Moreover, these farms usually do not have feed reserves sufficient for a year and operated mainly by family labor. In contrast, farms with more than ten cows are medium or large scale intensive farms that have feed reserves sufficient for a year, operated by hired labor and owned by farmers whose primary occupation is dairying. Thus, the herds were categorized in that way just to evaluate the effect of herd management difference on mastitis occurrence. Pertinent to cow-level factors, biodata including lactation (age, parity & stage), breed, milk yield, udder and leg hygiene, teat end morphology, udder position, and previous mastitis history were recorded.

The udder and leg hygiene of each cow was assessed and scored based on a four-point scale (1–4) as described by Schreiner and Ruegg [20]. Accordingly, an udder hygiene score (UHS) or leg hygiene score (LHS) of '1' was referred to no contamination of the skin of the rear of the udder or the hind limb between the hock and coronary band; '2' slightly dirty (2–10% of the area covered in dirt; '3' moderately dirty (10–30% of the area covered in dirt); and '4' caked-on dirt (>30% of these areas completely covered in dirt).

Each cow was clinically observed for the manifestation of general clinical signs related to udder and teats and presence of any gross abnormalities like fibrosis, inflammatory swellings, pain, visible injury or lesion, atrophy of the tissue and teat blindness. The milk sample was observed for any change regarding color, odor and consistency. The presence of clots, flakes, blood and other consistency changes were part of the indicators for clinical mastitis along with udder and teat morphological changes. Udder position was recorded as normal or pendulous while teat end shape was categorized as pointed, round or flat. Though all the researchers were involved in questionnaire and data collection format designing, to avoid any bias, all the field data collection process (questionnaire, hygiene scores, CMT, milk sample collection, etc.) was done by one of the researchers. During the study period each farm was visited only once.

California mastitis test

The California mastitis test (CMT) was used as a screening test for sub-clinical mastitis. It was carried out according to the procedure described by Quinn et al. [21]. A squirt of milk, about 2 ml from each quarter was placed in each of four shallow cups in the CMT paddle. An equal amount of the commercial CMT reagent was added to each cup. A gentle circular motion was applied to the mixtures in a horizontal plane for 15 s. The CMT results were scored as 0 (negative), trace, 1(weak positive), 2 (distinct positive) and 3 (strong positive) based on gel formation. All CMT scores of 0 and trace were considered as negative while CMT scores of 1, 2, and 3 were considered indicators of sub clinical mastitis. Positive cows were defined as having at least one quarter with CMT score of 1 + .

Isolation and identification of *Staphylococcus aureus*

For identification of the causative microorganisms, milk samples were collected from about 50% of CMT positive cows, randomly selected from all mastitis cases, assuming that the causative organisms within a herd are similar and also to reduce the time, labor and cost burdens. In case where only one mastitis case found in a farm, that positive cow was directly sampled. CMT negative cows (0 and trace scores) were not included in milk sampling. During sampling, milk was collected from all quarters of CMT positive cows at the same time as the CMT was done.

Milk sampling was carried out following aseptic procedures as described by National Mastitis Council (NMC) [22]. The time chosen for milk sample collection was before milking. Udder and especially teats were cleaned and dried before sample collection. Each teat end was scrubbed vigorously with a pledged of cotton moistened with 70% ethyl alcohol. Recontamination of teats during scrubbing was avoided by scrubbing carefully, the teats on the far side of the udder first, then those on the near side. A separate pledged of cotton was used for each teat. Teats towards sample collection were sampled first and then the far ones. The first few streams of milk were discarded and approximately 10 ml of milk was collected into horizontally held sterile universal sample collection bottle. After collection, the samples were labeled and placed in icebox and transported to the microbiology laboratory of the School of Veterinary Medicine, Hawassa University. The samples were immediately cultured or stored at 4 °C for a maximum of 24 h until cultured on standard bacteriological media.

To comply with specific objective of the study, bacteriological examination was focused only on identification and isolation *S. aureus*, which is the most common cause of contagious mastitis in dairy herds worldwide. The isolation of *S. aureus* was carried out according to the standard microbiological procedures described by NMC [22]. One standard loop (0.01 ml) of milk sample was streaked on 7% sheep blood agar (Oxoid, Hampshire, England) using the quadrant streaking method for each cow. In case of refrigerated milk samples, as bacteria might be concentrated in the cream layer and held with in clumps of fat globules, dispersion of fat and bacteria was accomplished by warming the samples at 25 °C for 15 min before plating on blood agar. The inoculated plates were then incubated aerobically at 37 °C for 24 to 48 h. When growth was not observed after incubation for 24 to 48 h, the milk sample was re-inoculated on an enriched tryptone soya broth (Oxoid, Hampshire, England) to amplify the bacterial growth. The plates were examined for growth, morphologic features such as colony size, shape, color and hemolytic characteristics. Presumptive colonies were selected and sub cultured on nutrient agar (Oxoid, Hampshire, England) and incubated aerobically at 37 °C for 24–48 h to get a pure culture. After incubation, colonies were identified according to their Gram reaction (Gram-positive or Gram-negative), cellular morphology (coccus or rod) and arrangements of the bacteria and the catalase test. The Gram and catalase-positive cocci were characterized for mannitol fermentation on mannitol salt agar (Oxoid, Hampshire, England), which was followed by tube coagulase test. Samples were considered positive for *S. aureus* when at least one colony was identified as *S. aureus.*

Data analysis

The data were recorded in Microsoft Excel spread sheet and coded before statistical analysis. All the statistical analyses were performed using Stata 11 statistical software (StataCorp, 4905 Lakeway Drive, College Station, Texas). The prevalence of mastitis was calculated as the proportion of mastitis-positive cows (clinical and subclinical) against the total number of animals investigated. A herd was denoted as positive for mastitis if at least a single animal with clinical mastitis or CMT positive result was detected. The association between the dependent variable, cow mastitis status (0 = negative and 1 = positive) and categorical independent variables was assessed using univariable logistic regression analyses. The independent variables evaluated were management type, herd size, use of bedding material, milking mastitic cow last, breed, age, parity, stage of lactation, milk yield, udder and leg hygiene, udder position, teat end morphology, udder washing, udder drying after washing, and previous mastitis history. All variables with *p-value* < 0.25 in the initial univariable analysis were checked for multicollinearity using Kruskal gamma statistics and those variables whose gamma value ranged between −0.6 and +0.6 were considered in a multivariable logistic regression analysis. The final model was built in backward stepwise elimination procedure in reference to log-likelihood ratio and wald statistics. Finally the model was assessed for goodness-of-fit using the Hosmer-Lemeshow method and the predictive ability using the receiver operating characteristic (ROC) curve [23]. In this analysis statistical significance was set at $p < 0.05$.

Results

Management and hygienic practices in the dairy farms

During the course of the study, data on various management and hygienic practices implemented by the dairy farms were collected from a total of 95 farms. A summary of data on the practices assessed was presented in Table 1.

Prevalence of mastitis

Table 2 illustrates the prevalence of clinical and sub-clinical mastitis at herd, cow and quarter level. Based on CMT results and clinical examination, 74.7% of the herds and 62.6% of the cows were positive for mastitis. All quarters of cows (2119) were checked for the presence of gross abnormalities and it was found that 90 (4.3%) teats were blind while 15 (0.7%) of them had various types of lesions. Subsequently, CMT was conducted on milk samples from 2026 quarters out of which 729 (36%) were positive for mastitis. Sub-clinical mastitis was the predominant type of mastitis observed at herd, cow and quarter levels. Of the total cows with subclinical mastitis, 35.2, 49.5 and 15.3% were weak, distinct and strong CMT positives, respectively.

Mastitis risk factors

The prevalence of mastitis at each udder quarter has been shown in Table 3. Overall, the hind quarters were more affected than the fore quarters. The proportion of right hind quarters with positive CMT result was significantly ($p < 0.001$) higher than that of left or right fore quarters but no significant difference was noted between left and right hind quarters ($p > 0.05$).

Several herd and cow-level factors described in the data analysis section were considered in the univariable logistic regression analysis for the presence of mastitis. Among those factors, management type, herd size, use of bedding material, milking mastitic cows last, breed, age, parity, stage of lactation, milk yield, udder and leg hygiene, udder position, and teat end morphology were found to be significantly ($p < 0.05$) associated with the presence of mastitis. On the other hand, udder washing, udder drying after washing and previous mastitis history did not have significant effect ($p > 0.05$) on the occurrence of mastitis (Table 4).

Among the factors considered in the initial univariable analysis, age of the cows was dropped from further analysis due to multi-collinearity with parity (gamma = 0.99), udder position (gamma = 0.97) and previous history of mastitis (gamma = 0.91). With regard to mastitis,

Table 1 Management and hygienic practices implemented by the dairy farms selected for mastitis study at Hawassa milk shed, South Ethiopia

Variable	Category	No. of herds	Distribution (%)
Herd size	≤10 cows	60	63.2
	>10 cows	35	36.8
Management type	Semi-intensive	60	63.2
	Intensive	35	36.8
Housing	Stall barn	23	24.2
	Group barn	72	75.8
Bedding material	Yes	39	41.1
	No	56	58.9
Pre/post teat dipping	Yes	-	-
	No	95	100
Culling chronically infected cows	Yes	-	-
	No	95	100
Dry cow therapy	Yes	-	-
	No	95	100
Udder washing before milking	Whole udder	56	58.9
	Teats only	39	41.1
Drying of udder after washing	Yes	41	43.2
	No	54	56.8
Milking mastitic cow last	Yes	31	32.6
	No	64	67.4
Hand prewashing	With soap	44	46.3
	Without soap	51	53.7
Hygiene of the farm	Good	19	20
	Poor	76	80

Table 2 Prevalence of mastitis in dairy cows at Hawassa milk shed, South Ethiopia

Observation	Overall mastitis				Clinical mastitis	Subclinical mastitis
	No examined	No pos	Prevalence (%)	95% CI	No (%)	No (%)
Herd-level	95	71	74.7	64.5 – 82.8	10 (10.5)	71 (74.7)
Cow-level	529	331	62.6	58.3 – 66.7	18 (3.4)	313 (59.2)
Quarter-level	2026	729	36	33.9 – 38.1	26 (1.3)	703 (34.7)

parity was considered the more relevant variable than age based on biological plausibility and thus retained. Also milk yield was excluded due to multi-collinearity with herd size (gamma = 0.63) and breed (gamma = 1.00). Likewise, udder position and previous history of mastitis were eliminated from further analysis due to collinearity with parity (gamma = 0.99 and 0.92, respectively). All the cows with four or more parities had pendulous udder and previous history of mastitis. Thus, the variables subjected to multivariable logistic regression analysis were management type, herd size, breed, use of bedding material, milking mastitic cows last, parity, stage of lactation, udder & leg hygiene, and teat end shape. Accordingly, the final logistic regression model revealed that with the exception of management type ($p > 0.05$) all the variables entered remained significant predictors of mastitis in the cows ($p < 0.05$). The Hosmer-Lemeshow goodness-of- fit test suggested that the model fit the data ($\chi^2 = 1.7$; $p = 0.99$) (Table 5).

Bacterial isolation

During the course of the study, a total of 172 milk samples collected from 18 clinical and 154 sub-clinical mastitic cows in 71 herds were cultured on blood agar media. Accordingly, growth of different groups of bacteria was observed in 170 (98.8%) samples. However, since the objective was to isolate *S. aureus*, further bacteriological tests were conducted by taking only those colonies presumptive of this species and leaving others. Finally, *S. aureus* was detected in 51.2% (88/172) of the milk samples cultured. The proportion of isolation from milk of clinically and sub-clinically affected cows was 22.2 and 54.5%, respectively. At herd level, *S. aureus* was isolated from 73.2% of the herds affected with mastitis (Table 6).

Table 3 Analysis of the prevalence of mastitis in dairy cows at quarter-level

Quarter	No. examined	No. positive	Prevalence (%)	χ^2	p
Right hind	504	249	49.4	Ref	
Left hind	501	223	44.2	2.7	0.101
Right fore	507	132	26.0	59.0	<0.001
Left fore	511	125	24.5	67.6	<0.001

Discussion

This is one of the few bovine mastitis studies in Ethiopia which involved a large sample size of lactating cows (n = 529) selected from 95 dairy herds. The study revealed that 74.7% of the herds observed had atleast a cow suffering from mastitis. Due to lack of similarly designed studies with in Ethiopia, it was difficult to compare the herd-level results with other mastitis studies. However, in line with similar studies abroad, the prevalence reported is considerably higher than that of Tanzania (21.7%) [24], Zimbabwe (49.3%) [25] and South Wales (29%) [26].

The overall cow-level prevalence was 62.6% and it was largely accounted from sub-clinical mastitis (59.2%) and the smallest proportion (3.4%) from clinical mastitis. The present finding is within the range of cow-level mastitis prevalence (23.2–81.1%) recorded by most recent published studies in the country [10–16]. Some of the earlier findings and our observation is higher relative to the available reports from other African countries whose dairy management is more or less similar to ours, i.e., 51.6% in Tanzania [24], 42.9% in Egypt [27], 38.89% in Sudan [28], 21.1% in Zimbabwe [25], and 51.8% in Rewanda [29]. In contrast, substantially higher prevalence was reported from Uganda (87.9%) [30] and Nigeria (85.3%) [31]. This entails how serious the problem is, in the dairy industry sector of the continent that warrant due attention.

Indeed both the herd- and cow-level mastitis prevalence observed in the current study are very high. According to Levesque [32], within-herd mastitis prevalence of 40% or over must sound an alarm to the producer. Thus, it is obvious that the high mastitis prevalence seen in the current study area is a serious problem that not only reduces milk production but adversely affects the quality of milk and leads to economic losses and public health hazard. Perhaps the prevailing high prevalence of mastitis could either be due to lack of implementation of regular mastitis prevention and/or control strategies other than treating clinical cases. The fact that none of the dairy herds implementing routine mastitis prevention practices such as post-milking teat disinfection, wearing of gloves during milking, treatment of mastitis during non-lactating period, culling of chronically infected animals and others substantiate the assumption. Moreover, none of the farmers were doing

Table 4 Univariable logistic regression analysis of the association of cow-level mastitis with different risk factors

Variable	Number	No. pos	Prevalence (%)	p
Management type				<0.001
Semi-intensive	200	92	46.0	
Intensive	329	239	72.6	
Herd size				<0.001
≤ 10	200	92	46.0	
> 10	329	239	72.6	
Bedding material				<0.001
Yes	145	61	42.1	
No	384	270	70.3	
Udder washing				0.707
Whole udder	291	180	61.9	
Teats only	238	151	63.5	
Udder drying after washing				0.774
Yes	295	183	62.0	
No	234	148	63.3	
Milking mastitic cow last				0.001
Yes	202	109	54.0	
No	327	222	67.9	
Breed				<0.001
Local	83	33	39.8	
Cross	446	298	66.8	
Age				<0.001
≤ 5 years	227	85	37.4	
> 5 years	302	246	81.5	
Parity				<0.001
1	66	17	25.8	
2	145	60	41.4	
3	115	75	65.2	
≥ 4	203	179	88.2	
Stage of lactation				<0.001
1 – 2 months	198	158	79.8	
3 – 6 months	186	80	43.0	
≥ 7 months	145	93	64.1	
Udder & leg hygiene score				<0.001
Slightly dirty	241	120	49.8	
Moderately dirty	203	136	67.0	
Very dirty	85	75	88.2	
Udder position				<0.001
Normal	315	162	51.4	
Pendulous	214	169	79.0	
Teat end shape				<0.001
Pointed	175	74	42.3	
Round	197	126	64.0	
Flat	157	131	83.4	
History of mastitis				0.174
No	189	111	58.7	
Yes	340	220	64.7	

Table 4 Univariable logistic regression analysis of the association of cow-level mastitis with different risk factors *(Continued)*

CMT or other tests routinely to screen their cows for sub-clinical mastitis. On the other hand, the risk factors contributing to mastitis were more predominant. These were large herd size, cross breeding local with exotic breeds, not milking mastitic cows last, absence of bedding material, a higher proportion of multiparous older cows with 4 or more parities, early stage of lactation, dirty udder and leg, and teat end morphology,

The finding of sub-clinical mastitis as the predominant form both at cow (59.2%) and herd level (74.7%) is a very clear evidence of huge economic loss the sector is suffering from. In fact, our finding corroborate with earlier reports of 54.4 to 73.3% [11, 14, 15]. Likewise, the preponderance of sub-clinical mastitis and its serious economic relevance compared to clinical mastitis was underscored elsewhere [8, 25, 27, 33, 34]. According to Seegers et al. [1], the sub-clinical form is 15 to 40 times more prevalent than the clinical form, and usually precedes the clinical form and is of long duration and high economic relevance. Thus, it is important to emphasize that the sub-clinically affected animals remain a continuing source of infection for herd mates. The dominance of sub-clinical mastitis in the herds is most likely attributed to the little attention paid by the farmers to this form of mastitis, as the infected animal shows no obvious clinical symptoms and secrets apparently "normal" milk. Lack of implementation of regular mastitis monitoring program such as CMT or other testing by all of the farms also contributed to the observed high prevalence of sub-clinical mastitis in our case. Perhaps this fact may justify the lack of awareness on the invisible losses from sub-clinical mastitis.

The observation of higher mastitis prevalence in the hind quarters is similar to earlier studies [10, 16, 34]. Although it is difficult to give a plausible explanation for this observation, this might be due to the lower position of the hind quarters in relation to fore ones, which make them more prone to contamination or injuries leading to mastitis. Greater milk yield produced by the hind quarters may also be a reason as risk of mastitis tends to increase with increase in milk production [35]. In contrast to the present finding, a significantly higher prevalence was reported in fore than hind quarters in Nigeria [31].

Table 5 Multivariable logistic regression analysis of the association of cow-level mastitis with different risk factors

Variable	Category	OR	SE	z	*p*	95% CI for OR
Herd size						
	≤10 cows	Ref				
	>10 cows	3.1	1.0	3.74	<0.001	1.7 – 5.7
Breed						
	Local	Ref				
	Cross	16.4	8.3	5.55	<0.001	6.1 – 44.1
Milking mastitic cow last						
	Yes	Ref				
	No	2.0	0.6	2.53	0.011	1.2 – 3.5
Bedding material						
	Yes	Ref				
	No	2.8	0.9	3.24	0.001	1.5 – 5.3
Parity						
	1	Ref				
	2	1.2	0.5	0.48	0.629	0.5 – 2.9
	3	3.2	1.5	2.46	0.014	1.3 – 8.1
	≥4	24.8	12.2	6.50	<0.001	9.4 – 65.2
Stage of lactation						
	Early (1 – 2 months)	Ref				
	Mid (3 – 6 months)	0.3	0.1	−4.17	<0.001	0.1 – 0.5
	Late (≥7 months)	0.7	0.2	−0.99	0.323	0.4 – 1.4
Udder & leg hygiene						
	Slightly dirty	Ref				
	Moderately dirty	2.6	0.8	3.27	0.001	1.5 – 4.7
	Very dirty	8.7	4.0	4.73	<0.001	3.5 – 21.3
Teat end shape						
	Pointed	Ref				
	Round	3.2	1.0	3.72	<0.001	1.7 – 5.9
	Flat	7.6	2.8	5.57	<0.001	3.7 – 15.7
Constant				−7.44	<0.001	

In the present study, the presence of mastitis was significantly influenced by herd size. The odds of finding a cow with positive CMT result was 3.1 times higher in herds with more than ten cows than in herds with ten or fewer cows. It was observed that all the herds with more than ten cows were intensively managed and thus, the higher odds of mastitis occurrence might be associated with increased exposure of the cows for mastitis pathogens in their environment due to high stocking density, dirty ground, infected utensils, poor ventilation and high humidity. The study further revealed that lack of bedding material in the cow barns was significantly associated with the presence of mastitis. Cows in herds that did not use bedding material were 2.8 times more likely to have mastitis than those that did use. It appeared that the floor was a potential source for mastitis organisms, particularly the environmental pathogens, to enter the udder through the teat orifice. In connection to this, Radostitis et al. [4] stated that the use of adequate amounts of good bedding material will reduce incidence

Table 6 Isolation of *S. aureus* in mastitic milk samples from dairy herds at Hawassa milk shed

Clinical mastitis		Sub-clinical mastitis		Total cultured	Total positive (%)
No. samples cultured	No. pos (%)	No. samples cultured	No. pos (%)		
18	4(22.2%)	154	84(54.5%)	172	88 (51.2%)

of mastitis. In the current study however, no attempt was made to isolate environmental pathogens and thus, difficult to explain what proportion of the herds had these pathogens. Hence, this is considered as one of the limitations of the study that should be addressed in future studies.

Cows in herds that did not milk mastitic cows last were significantly more likely to have mastitis than those that did that. It is obvious that, failure to milk mastitic cows last would favor spread of mastitis pathogens between cows by milker's hands resulting in contagious mastitis [8]. It is only 31 of the 95 herds that practiced milking mastitic cows last. The current finding complies with reports of Belayneh et al. [12] in Ethiopia.

Increasing parity number was also one of the predictors noted to associate with the presence of mastitis. Such scenario has also been documented in a number of studies [12, 14, 25, 30, 36]. The likelihood of mastitis was 24.8 times higher in multiparous cows having four or more calvings compared with primiparous cows. This partly, might be associated with the position of udder in older cows. In the current study it was noticed that all of the older cows particularly those with four or more parities had pendulous udder and previous history of mastitis. It has also been stated that cows with the most pendulous quarters appear to be the most susceptible to mammary infections, the pendulous udder exposes the teat and udder to injury and pathogens easily adhere to the teat and gain access to the gland tissue [34].

Cows in the early lactation were significantly less likely to have mastitis than cows in the mid lactation stage albeit no significant difference was appreciated between early and late lactation stages. Evidence on the trend of prevalence increment at the early stage was also produced in previous studies [14, 16, 27, 29]. Perhaps this could be linked to the fact that diapedesis of neutrophils into the mammary gland take longer time in recently calved cows [4], and increased oxidative stress and reduced antioxidant defense mechanisms during early lactation [37]. Moreover, absence of dry cow therapy regime could possibly be among the major factors contributing to higher prevalence at early lactation. In contrast to the present finding, Belayneh et al. [12] reported higher prevalence of mastitis in late stage of lactation while Mureithi and Njuguna [36] in Kenya found a significantly higher prevalence in the mid stage. The variations in the effect of stages of lactation among different studies could be related to disparities in age, parity and breed of the sampled animals. However, the actual source of variation needs to be sorted out in the future research.

The status of cows' udder and leg hygiene was also noted as part of the risk factors that enhance the occurrence of mastitis. Based on the udder and leg hygiene score used, all the examined cows had slightly to very dirty udders and legs, and there was a marked increase in the detection of mastitis as the level of dirtiness increases. Thus, the odds of mastitis presence was 2.6 or 8.7 times higher in cows with moderate or very dirty udder and legs compared to cows with slightly dirty udder and legs. This dirtiness of cows' udder and legs was the result of poor hygiene of the dairy farms. Of the total farms investigated, 80% were categorized as poor hygienic due to lack of waste drainage system and accumulation of manure and urine. In agreement to the present study, a significant association between poor udder hygiene and increased risk of mastitis has also been reported by other studies [29, 30, 36].

In this study, a strong association was observed between teat end shape and CMT positive result. Cows with flat teat or round teat ends were 7.6 or 3.2 times more likely to have mastitis than cows with pointed teat ends. This result may be explained by the presence of wider streak canals in teats with rounded or flat ends, which have been shown to allow more pathogen penetration [38]. The current finding is also in agreement with Bharti et al. [39] who reported a significantly higher mean somatic cell counts in cows with flat teat-end shape.

This study also revealed a significant association between breed and the presence of mastitis. The likelihood of mastitis was 16.4 times higher in Holstein-Friesian x zebu crosses than pure local zebu cattle. This shows that pure local breeds are more resistant to contracting mastitis than the European. Breed difference in susceptibility to mastitis has also been reported by other studies [29, 31, 40].

According to the microbiological finding of the study, *S. aureus* was isolated from 51.2% mastitic milk samples and 73.2% herds affected with mastitis. The cow-level finding is comparable to a previous report of 51.7% from northern part of Ethiopia [16] and also other mastitis studies in the country which recorded *S. aureus* as the predominant agent [11–15]. Apart from Ethiopia, *S aureus* has also been reported as the chief aetiological agent of mastitis in cattle by many studies from African and Asian countries [8]. The high prevalence of *S. aureus* in this study could be associated with absence of post milking teat dipping, lack of culling of chronically infected cows, absence of dry cow therapy and the invariable hand milking practice among the dairy herds. *S. aureus* and other contagious microorganisms are usually found on the udder or teat surface of infected cows and are the primary source of infection between uninfected and infected udder quarters, usually during milking. Even though in all the observed herds milkers' used to wash their hands before milking, they do it only before milking the first cow. Hence it is evident that the causative organisms could be transmitted easily from infected to uninfected udder quarters or cows through the milkers' hands. Cure rate of *S. aureus* infections with antibiotic therapy during lactation is very low and many infected animals become chronic cases and have to be culled [41]. Thus the fight against contagious mastitis should encompass dry cow therapy as major component of mastitis control program. Unfortunately, none of the dairy farmers have the practice of culling of chronically infected animals, dry cow therapy and post-milking teat disinfection and this has created a good opportunity for the establishment of the organism in the dairy herds.

Conclusions

The present study has shown that mastitis, particularly sub-clinical type, is a widely prevalent disease in the dairy farms of Hawassa milk shed at both herd- and cow-level. Lack of implementation of the routine mastitis prevention and control practices by all of the farms observed and preponderance of the risk factors noted are the main reasons for the observed high prevalence of mastitis in the milk shed. Accordingly, multiparous cross-breed cows at early stage of lactation which had dirty udder and legs and round or flat teat ends were at higher risk of contracting mastitis. Besides, larger herds, failure to milk mastitic cows last and keeping cows in the absence of bedding material were also noted to increase the exposure of cows for mastitis pathogens. The present study has also revealed that *S. aureus* is an important cause of mastitis which was isolated from more than half of mastitic cows tested and 73.2% of herds affected with mastitis. Moreover, the isolation of *S. aureus* in observed proportion indicates the higher public health risk due to consumption of raw milk and its products.

Therefore the current study warrants the need for applying feasible mastitis intervention strategy with special emphasis on contagious and sub-clinical mastitis. These include strong dairy extension service that focused on awareness creation and hygienic milking practice. Besides, the animal health service delivery need to focus on regular screening of dairy cows for subclinical mastitis and treating of the cases both in lactation and dry period, and provision of advice to cull chronically infected animals. Since the current study on etiological agents was limited to contagious ones, future studies in the milk shed should include isolation of environmental pathogens. Last but not least the rational use of antibiotic and regular antibiogram surveillance should be made part and parcel of the approach.

Additional file

Additional file 1: Raw data from bovine mastitis study at Hawassa milk shed, South Ethiopia. The additional file contains the entire herd and animal-level raw data collected from 529 lactating cows at Hawassa milk shed, South Ethiopia and on which the conclusion of the manuscript rely. The data will be freely available at the journal's webpage to any scientist wishing to use them for noncommercial purposes.

Acknowledgments

The authors express special thanks to owners or managers of the dairy farms included in the study for their collaboration during the study period. Members of the veterinary departments in the surveyed districts are also very much appreciated for their cooperation in the field work. Authors are thankful to the editor and an anonymous reviewer for their constructive comments that helped to improve the quality of this paper.

Funding

This study was funded by Hawassa University Research and Technology Transfer Vice President Office by the operational fund obtained from the Federal Democratic Republic of Ethiopia.

Authors' contributions

HT collected all the required data. HT and MA performed the laboratory works. KA and RA designed the study, and analyzed and interpreted the data. RA drafted the manuscript. BM and KA critically and substantially revised the manuscript. All authors read and approved the final manuscript.

Competing interests

The authors declare that they have no competing interests.

Consent for publication

Not applicable.

Author details

[1]School of Veterinary Medicine, Hawassa University, P.O.Box 05, Hawassa, Ethiopia. [2]Animal and Fisheries Resources Development Bureau of Southern Nations, Nationalities and Peoples' Regional State, P.O.Box 309, Hawassa, Ethiopia.

References

1. Seegers H, Fouricho C, Beaudeau F. Production effects related to mastitis and mastitis economics in dairy cattle herds. Vet Rec. 2003;34:475–91.
2. González RN, Wilson DJ. Mycoplasmal mastitis in dairy herds. Vet Clin North Am Food Anim Pract. 2003;19(1):199–221.
3. Cervinkova D, Vlkova H, Borodacova I, Makovcova J, Babak V, Lorencova A, Vrtkova I, Marosevic D, Jaglic Z. Prevalence of mastitis pathogens in milk from clinically healthy cows. Vet Med. 2013;58(11):567–75.
4. Radostits OM, Gay CC, Hinchcliff KW, Constable PD. Veterinary Medicine: A text book of the disease of cattle, horses, sheep, pigs and goats. 10th ed. London: Elsevier Ltd; 2007.
5. Smith KL, Todhunter DA, Schoenberger PS. Environmental mastitis: cause, prevalence, prevention. J Dairy Sci. 1985;68:1531–53.
6. Harmon RJ. Symposium - Mastitis and Genetic Evaluation for Somatic Cell Count - Physiology of Mastitis and Factors Affecting Somatic Cell Counts. J Dairy Sci. 1994;77:2103–12.
7. Khan MZ, Khan A. Basic facts of mastitis in dairy animals: A review. Pakistan Vet J. 2006;26:204–8.
8. Food and Agriculture Organization (FAO). Impact of mastitis in small scale dairy production systems. Animal Production and Health Working Paper. No. 13. Rome; 2014.
9. Central Statistical Agency of Ethiopia (CSA). Report on livestock and livestock characteristics (private peasant holdings). Agricultural sample

survey 2011/12. 2012. http://www.csa.gov.et/newcsaweb/images/general/news/livestock%20report%202005%20ec_2012_13.pdf.
10. Girma S, Mammo A, Bogele K, Sori T, Tadesse F, Jibat T. Study on prevalence of bovine mastitis and its major causative agents in West Harerghe zone, Doba district, Ethiopia. J Vet Med Anim Health. 2012;4:116–23.
11. Abera M, Demie B, Aragaw K, Regassa F, Regassa A. Isolation and identification of *Staphylococcus aureus* from bovine mastitic milk and their drug resistance patterns in Adama town, Ethiopia. Afr J Dairy Farm Milk Prod. 2013;1:19–23.
12. Belayneh R, Belihu K, Wubete A. Dairy cows mastitis survey in Adama town, Ethiopia. J Vet Med Anim Health. 2013;5:281–7.
13. Gebremichael L, Deressa B, Begna F, Mekuria A. Study on prevalence of bovine mastitis in lactating cows and associated risk factors in and around Areka town, Southern of Ethiopia. Afr J Microbiol Res. 2013;7:5051–6.
14. Zeryehun T, Aya T, Bayecha R. Study on prevalence, bacterial pathogens and associated risk factors of bovine mastitis in small-holder dairy farms in and around Addis Ababa, Ethiopia. J Anim Plant Sci. 2013;23:50–5.
15. Duguma A, Tolosa T, Yohannes A. Prevalence of clinical and sub-clinical mastitis on cross breed dairy cows at Holleta Agricultural Research Center, Central Ethiopia. J Vet Med Anim Health. 2014;6:13–7.
16. Zenebe N, Habtamu T, Endale B. Study on bovine mastitis and associated risk factors in Adigrat, Northern Ethiopia. Afr J Microbiol Res. 2014;8:327–31.
17. Seykora AJ, McDaniel BT. Heritability of teat traits and their relationships with milk yield, somatic cell count, and percent two-minute milk. J Dairy Sci. 1985;68:2670–83.
18. Nakov D, Hristov S, Andonov S, Trajchev N. Udder related risk factors for clinical mastitis in dairy cows. Vet Arhiv. 2014;84:111–27.
19. Thrusfield M. Veterinary Epidemiology. 3rd ed. Oxford: Blackwell science Ltd; 2005. p. 183–9.
20. Schreiner DA, Ruegg PL. Relationship between udder and leg hygiene scores and subclinical mastitis. J Dairy Sci. 2003;86:3460–5.
21. Quinn PJ, Carter ME, Markey B, Carter GR. Clinical Veterinary Microbiology. 1st ed. London: Elsevier Ltd; 1999.
22. National Mastitis Council (NMC). Microbiological procedures for the diagnosis of udder infection. 3rd ed. Arlington: National Mastitis Council Inc; 2004.
23. Dohoo I, Martin W, Stryhn H. Veterinary Epidemiologic Research. 2nd ed. Charlottetown: AVC Inc; 2009. p. 27–407.
24. Mdegela RH, Ryoba R, Karimuribo ED, Phiri EJ, Løken T, Reksen O, Mtengeti E, Urio NA. Prevalence of clinical and subclinical mastitis and quality of milk in smallholder dairy farms in Tanzania. J S Afr Vet Assoc. 2009;80:163–8.
25. Katsande S, Matope G, Ndengu M, Pfukenyi DM. Prevalence of mastitis in dairy cows from smallholder farms in Zimbabwe. Onderstepoort J Vet Res. 2013;80:523.
26. Plozza K, Lievaart JJ, Pottsb G, Barkema HW. Sub clinical mastitis and associated risk factors on dairy farms in New South Wales. Aust Vet J. 2011;89:41–6.
27. Elbably MA, Emeash HH, Asmaa NM. Risk factors associated with mastitis occurrence in dairy herds in Beni-Suef Governorate. Worlds Vet J. 2013;3:05–10.
28. Hussein SA. Prevalence and bacterial etiology of sub clinical mastitis in dairy cows in Al-Sulaimaniyah District. Kufa J Vet Med Sci. 2012;3:190–203.
29. Iraguha B, Hamudikuwanda H, Mushonga B. Bovine mastitis prevalence and associated risk factors in dairy cows in Nyagatare District, Rwanda. J S Afr Vet Assoc. 2015;86:1228.
30. Abrahmsén M, Persson Y, Kanyima BM, Båge R. Prevalence of subclinical mastitis in dairy farms in urban and peri-urban areas of Kampala, Uganda. Trop Anim Health Prod. 2014;46:99–105.
31. Shittu M, Abdullahi J, Jibril A, Mohammed AA, Fasina FO. Sub-clinical mastitis and associated risk factors on lactating cows in the Savannah Region of Nigeria. BMC Vet Res. 2012;8:134.
32. Levesque P. Moins de mammite meilleure lait. Québec: Institut de Technologie Agroalimentaire, Campus de La Pocatière; 2004.
33. Kaliwal BB, Kurjogi MM. Prevalence and antimicrobial susceptibility of bacteria isolated from bovine mastitis. Adv Appl Sci Res. 2011;2:229–35.
34. Awale MM, Dudhatra GB, Avinash K, Chauhan BN, Kamani DR. Bovine Mastitis: A Threat to Economy. Open Access Sci Rep. 2012;1:295.
35. Kocak O. Influence of mastitis on milk yields in Holstein Cows. Acta Vet Brno. 2006;75:507–13.
36. Mureithi DK, Njuguna MN. Prevalence of subclinical mastitis and associated risk factors in dairy farms in urban and peri-urban areas of Thika Sub County, Kenya. Livest Res Rural Dev. 2016;28:13.
37. Sharma N, Singh NK, Singh OP, Pandey V, Verma PK. Oxidative stress and antioxidant status during transition period in dairy cows. Asian-Aust J Anim Sci. 2011;24:479–84.
38. Appleman RD. Quantifying the genetic effects on the anatomy of streak canal. In: Proceedings of the VI International Conference on Cattle Disease. 1970. p. 104–8.
39. Bharti P, Bhakat C, Pankaj PK, Bhat SA, Prakash MA, Thul MR, Japheth KP. Relationship of udder and teat conformation with intra-mammary infection in crossbred cows under hot-humid climate. Vet World. 2015;8(7):898–901.
40. Sanotharan N, Pagthinathan M, Nafees MSM. Prevalence of bovine subclinical mastitis and its association with bacteria and risk factors in milking cows of Batticaloa district in Sri Lanka. Int J Sci Res Innov Technol. 2016;3:2313–3759.
41. Kulkarni AG, Kaliwal BB. Bovine mastitis: A review. Int J Recent Sci Res. 2013;4:542–8.

11

Assessment of yeast *Saccharomyces cerevisiae* component binding to *Mycobacterium avium* subspecies *paratuberculosis* using bovine epithelial cells

Ziwei Li[1], Qiumei You[1], Faisury Ossa[2], Philip Mead[1], Margaret Quinton[1] and Niel A. Karrow[1*]

Abstract

Background: Since yeast *Saccharomyces cerevisiae* and its components are being used for the prevention and treatment of enteric diseases in different species, they may also be useful for preventing Johne's disease, a chronic inflammatory bowel disease of ruminants caused by *Mycobacterium avium* spp. *paratuberculosis* (MAP). This study aimed to identify potential yeast derivatives that may be used to help prevent MAP infection. The adherence of mCherry-labeled MAP to bovine mammary epithelial cell line (MAC-T cells) and bovine primary epithelial cells (BECs) co-cultured with yeast cell wall components (CWCs) from four different yeast strains (A, B, C and D) and two forms of dead yeast from strain A was investigated.

Results: The CWCs from all four yeast strains and the other two forms of dead yeast from strain A reduced MAP adhesion to MAC-T cells and BECs in a concentration-dependent manner after 6-h of exposure, with the dead yeast having the greatest effect.

Conclusions: The following in vitro binding studies suggest that dead yeast and its' CWCs may be useful for reducing risk of MAP infection.

Keywords: *Saccharomyces cerevisiae, Mycobacterium avium* spp. *paratuberculosis* (MAP), Epithelial cell adhesion

Background

Dairy producers have been using commercially available yeast probiotics and their components as feed supplements for nearly two decades based on claims that these products will improve animal production, promote health, and reduce the need for antibiotic use. Studies demonstrate that supplementing the ruminant diet with specific strains of *S. cerevisiae* improves feed intake [1, 2], weight gain [3], and fiber digestion [4, 5]. It has also been reported that live yeast stabilizes rumen pH [6, 7], and the number of anaerobic cellulolytic bacteria [8, 9]. In addition to having nutritional value, there is evidence that yeast probiotics and their components, such as mannan-oligosaccharides (MOS), can adhere to enteric pathogens including *Campylobacter jejuni* [10], *Escherichia coli* serotypes (O2 and O88) [11], and *Salmonella* [12, 13], thereby reducing their ability to attach to and invade host cells [14]. Given these properties, it is possible that dietary supplementation with yeast probiotics and/or their components may help protect calves that are vulnerable to Johne's disease (JD).

Neonatal calves and calves less than six months of age are most susceptible to MAP infection, the causative agent of JD [15, 16], in part, due to their underdeveloped immune system [17]. MAP invades and subsequently damages the small intestine, causing diarrhea, weight loss and severe dehydration, and also reduces milk production [18]. Clinical signs of JD

* Correspondence: nkarrow@uoguelph.ca
[1]Department of Animal Biosciences, University of Guelph, Guelph N1G 2W1, ON, Canada
Full list of author information is available at the end of the article

may take 2 to 5 years to develop [19], and its long subclinical stage facilitates continuous exposure of non-infected animals within an infected herd. MAP has a thick lipid-rich cell wall containing mycolic acids, which in part contributes to its resilience to chemical and physical degradation [20–22]. Currently, there is no satisfactory treatment for JD, since antibiotics only help to contain the disease, and vaccines only help to reduce disease incidence but they do not eliminate the disease completely [23]. Presently, the best way to control JD is through good management practices and early detection using different commercial available diagnostic tests [24], however, these diagnostic tests all have limitations and are costly.

Given these limitations, alternative strategies need to be explored to help reduce calf exposure to MAP and stimulate the host immune system to help combat MAP infection. We hypothesized that yeast and its CWCs may help reduce MAP adhesion to bovine epithelial cells, thereby reducing the risk of JD.

Methods

MAP isolation

The mCherry-labeled MAP used in the present study was developed by Mead [25] using the clinical isolate Gc86 strain previously isolated in the laboratory of Dr. Lucy Mutharia by Melinda Raymond (Department of Molecular and Cellular Biology, University of Guelph).

Preparation of MAP infection stock

Liquid nitrogen frozen mCherry MAP Gc86 was thawed at 37 °C and was used to inoculate 5 ml of 7H9 broth (Difco laboratories, Franklin Lakes, NJ, USA) supplemented with 10 % oleic acid-albumin-dextrose-catalase (OADC; Sigma-Aldrich, St. Louis, MO, USA), 0.25 % v/v Tyloxopol (Sigma-Aldrich), 50 μg/ml kanamycin (Allied Laboratories, Wichita, KS, USA) and 2 μg/ml of mycobactin J (Allied Laboratories). The cultures were incubated at 37 °C, and once they reached a fluorescent intensity (FI) of 45000, equivalent to $OD_{600} = 0.8$ [25], 5 ml aliquots were sub-cultured into 100 ml of media in a 250 ml sterile culture flask and incubated at 37 °C. When the cultures reached the logarithmic stage of growth (FI = 40000–50000 equivalent to $OD_{600} = 0.6–0.9$), cells were centrifuged at 2000 × g for 30 min. The cells were re-suspended to reach FI = 60000 equivalent to $OD_{600} = 1.0$ using Fig. 1, then to establish Colony Forming Units (CFU/ml) using Fig. 2 [25]. Quantification of fluorescence was based on the specific emission (587 nm) and excitation wavelengths (610 nm) for mCherry using the Wallac-1420 VICTOR3 Multilabel Counter (Perkin Elmer, Woodbridge, ON, Canada).

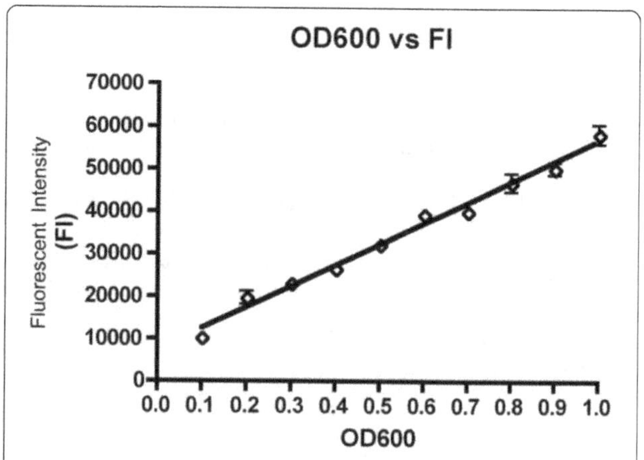

Fig. 1 OD_{600} versus fluorescent intensity of mCherry MAP Gc86 [25]

Bovine mammary epithelial cell line (MAC-T cells) and culture conditions

The MAC-T cells were cultured according to the reference [26]. MAC-T cells were cultured in T75 tissue culture flasks (Corning, Tewksbury, MA, USA) at 37 °C with 5 % CO_2, containing Dulbecco's Modified Eagle Medium (DMEM; Invitrogen, Burlington, ON, Canada) supplemented with 4.0 mM L-glutamine, 10 % heat inactivated fetal bovine serum (FBS; Invitrogen), 25 mM HEPES buffer (Invitrogen), 0.25 μg/ml amphotericin B (Invitrogen), 1 % Penicillin/Streptomycin (100 unit/ml of Penicillin and 100 μg/ml Streptomycin; Invitrogen), 1 mM Sodium Pyruvate (Invitrogen), and 5 μg/ml insulin-transferrin-selenium (Invitrogen).

Bovine intestinal epithelial cell (BEC) and culture conditions

Bovine ileum was aseptically harvested from a neonatal Holstein calf that was euthanized under the approval of the University of Guelph Animal Care Committee. The ileum (15 cm) was flushed with tap water and placed in ice cold HBSS (Invitrogen) supplemented with 5x100 U/ml

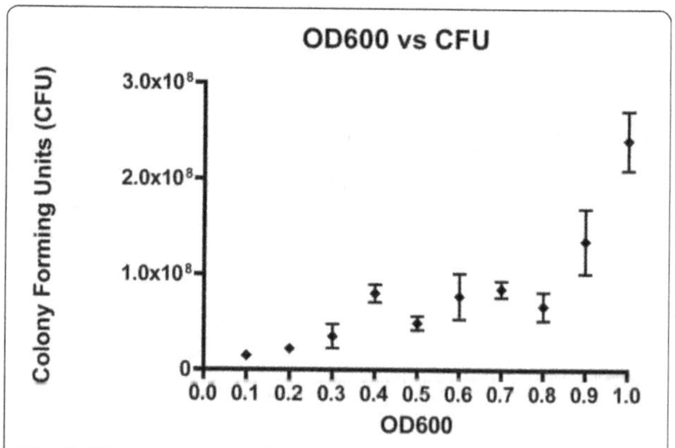

Fig. 2 OD_{600} versus CFU of mCherry MAP Gc86 [25]

Penicillin/Streptomycin (Invitrogen), 2.5 μg/ml Fungizone (Invitrogen), and 25 μg/ml Gentamicin. After the tissue was washed with HBSS several times, the mucosa was scrapped off the underlying epithelium, which was then cut into small pieces, and washed three times in HBSS by centrifugation at 140 × g for 5 min at 4 °C to remove mucus. The tissue pellet (6–10 g) was then digested by shaking for 30 min in a 37 °C water bath in 30 ml of DMEM containing 150 U/ml collagenase I, 100 ~ 150 U/ml Penicillin/Streptomycin, 2.5 μg/ml Gentamicin and 2.5 μg/ml Fungizone. Following this, the epithelium was mechanically dispersed using a 25 ml serological pipette by drawing the digested tissue through it several times. DMEM containing 10 % FBS (approximately 30 ml) was added to stop enzymatic digestion, and the crypts were concentrated by centrifuged at 80 × g for 5 min. To isolate crypts, the cell pellet was diluted with DMEM and enriched 3–5 times by centrifugation over a 2 % D-sorbitol gradient for 5 min at 50 × g. The isolated crypts were washed with DMEM, and then cultured in the cell culture flask using DMEM supplemented with high glucose, 10 % heat inactivated FBS, 2.5 % HEPES buffer, 1 % Fungizone, 1 % Penicillin/Streptomycin, 1 % sodium pyruvate, 1 % insulin, 1 % non-essential amino acid solution, Gentamicin (25 μg/ml) and epidermal growth factor (30 ng/ml; EGF; Sigma-Aldrich); each crypt consists of 200–300 epithelial cells [27]. BECs were isolated as described by Kaushik et al. [28], with slight modification. Primary BECs were confirmed by fluorescent microscopy using the monoclonal anti-cytokeratin antibody pan (mixture) (Sigma-Aldrich), which reacts specifically with a wide variety of epithelial tissues, and the secondary antibody anti-mouse IgG FITC-conjugated antibody (Sigma-Aldrich). The primary BECs were subsequently frozen in cryopreservation media containing 10 % DMSO, 40 % DMEM and 50 % FBS at -80 °C for storage.

In preparation for the adhesion assay, frozen BECs were thawed at 37 °C and washed with DMEM centrifuged at 300 × g for 6 min. Following this, the pellet was re-suspended with DMEM supplemented with high glucose, 1 % heat inactivated FBS, 2.5 % HEPES buffer, 1 % Fungizone, 1 % Penicillin/Streptomycin, 1 % sodium pyruvate, 1 % insulin, 1 % non-essential amino acid solution, Gentamicin (25 μg/ml) and EGF (30 ng/ml) according to Bridger et al. [29], and cultured for at least a week to reach confluence.

Assessment of epithelial cell viability

After reaching 80-100 % confluence in flask, epithelial cells were washed with warm phosphate buffered saline (PBS; Sigma-Aldrich), dislodged with TrypLE Express (Invitrogen) for 5 min, and counted with 0.4 % trypan blue (Invitrogen) using a hemocytometer chamber slide.

Cells were seeded into black 96-well flat bottom plates (50,000 cells per well) within cell culture medium that did not contain Fungizone and other antibiotics, and incubated at 37 °C with 5 % CO_2 overnight (17 h for MAC-T cells and 21 h for BECs). Cells were then exposed to a range of concentrations (0.25, 0.5, 1,2, 4, 6, 8 and 16 mg/ml) of yeast CWCs from strains A, B, C and D of *S. cerevisiae* and two forms of dead yeast from strain A provided by Lallemand Inc., QC, Canada. The cell culture plate was centrifuged briefly at 200 × g for 3 min to ensure interaction between epithelial cells and yeast derivatives before incubation at 37 °C with 5 % CO_2. In a preliminary trial, exposures of 2, 4, 6 and 24 h were assessed, and 6 h was deemed to be suitable when considering binding efficacy, epithelial cell proliferation and viability, and MAP proliferation. After a 6-h period, the epithelial cells were washed with warm PBS, then incubated with calcein AM diluted in culture media (Invitrogen) at room temperature for 30 min to stain live cells. The number of live cells was estimated by measuring the fluorescence of calcein AM (excitation 494/emission 517 nm) using a 1420 Victor2 Multilabel Counter (Beckman Coulter, Inc. California). The cell viability was calculated by using the formula: Percent of cell viability = (X/Y) × 100, where X is the fluorescence value in each well containing yeast derivative-treated cells and Y is the mean value of the fluorescence reading of all control wells.

Assessment of yeast and CWC adhesive properties using MAP

The MAC-T cells/BECs were seeded into 96-black well flat bottom plates at 50,000 cells per well with cell culture medium that did not contain Fungizone and other antibiotics, and MAC-T cells and BECs were incubated for 17 and 21 h, respectively, at 37 °C with 5 % CO_2. The cells were then exposed to a range of concentrations of yeast derivatives with MAP at a 10:1 (CFU: cell) multiplicity of infection. The cell culture plates were centrifuged at 200 × g for 3 min to ensure interaction between MAP and epithelial cells before incubation at 37 °C with 5 % CO_2 for 6 h. Following this, the plate was washed with warm PBS, and the adhesion of MAP was estimated by measuring the fluorescence intensity of mCherry-MAP (excitation 587/emission 610 nm) using a 1420 Victor2 Multilabel Counter. The adhesion of MAP to epithelial cells was calculated using the formula: Adhesion of mCherry-MAP = the fluorescence value in each well of treatments with MAP infection (A) - the mean value of the fluorescence of all wells of the corresponding control group with same exposure concentration of yeast or yeast CWCs (B).

Statistical analysis

The BEC and MAC-T cell viability were all analyzed separately as randomized complete block designs, in which the three independent trials represented the random blocks in each analysis. There were 6 replicates for each treatment within each block. The model for each of experiment included blocks as a random effect, and yeast derivatives and concentration level plus their interactions as fixed effects. All data were log-transformed prior to analysis in order to stabilize variances. Separate residual variances for each yeast derivative were incorporated in the model. Linear and quadratic orthogonal polynomial contrasts across concentration were used to assess changes in viability using the mixed procedure SAS 9.4 (SAS Institute Inc., 2012).

For the adhesion data, MAP adhesion to BECs or MAC-T cells was analyzed separately as a randomized complete block design with three independent trials representing each block. All data were log-transformed prior to analysis in order to stabilize variances. Concentration-dependent responses were compared between control and treatments by one-way ANOVA using Dunnett's test for statistical significance with the mixed procedure from SAS 9.4 (SAS Institute Inc.).

Graphs were generated using the Graphpad Prism version 4.00 (GraphPad Software, 2003, San Diego California, USA), and all data were presented as least squares means of the log transformed data from the percentage of cell viability and adhesion of MAP. A p-value ≤ 0.05 was considered statistically significant.

Results

Effect of yeast derivatives on the viability of MAC-T cells and BEC

MAC-T cell viability was significantly reduced by CWCs from all strains as indicated by linear contrasts ($p < 0.01$, Fig. 3a-d). In contrast, MAC-T cell viability was significantly increased by inactive yeast from strain A ($p < 0.01$) as indicated by quadratic contrasts, and no cytotoxicity was observed (Fig. 3e). The autolyzed yeast also reduced MAC-T viability at the highest concentration (16 mg/ml)

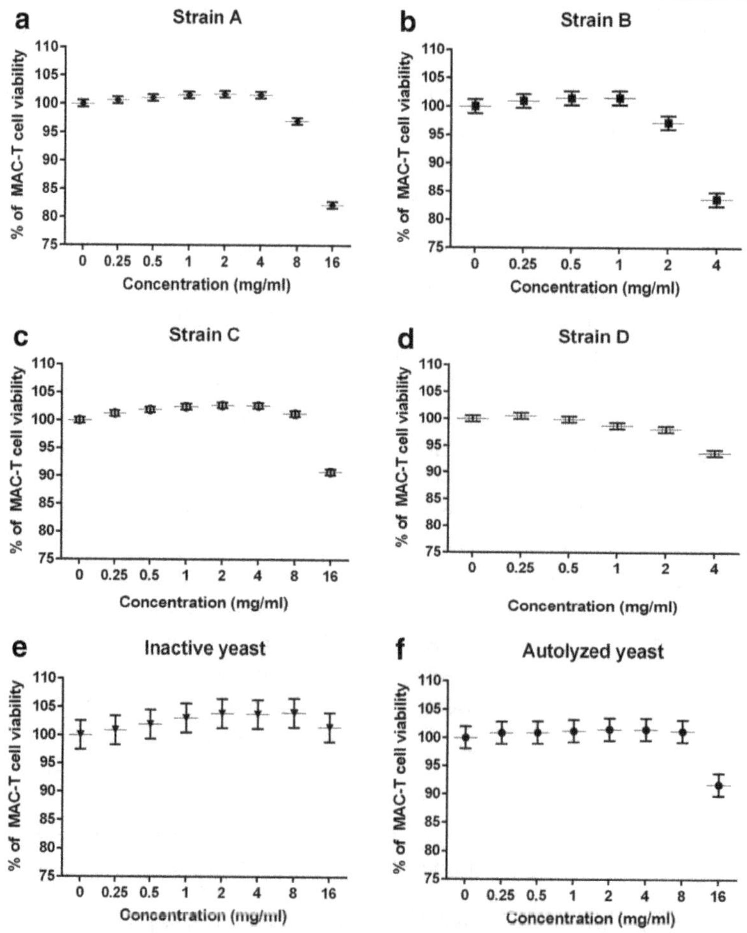

Fig. 3 Viability of MAC-T cells following 6-h exposure to CWCs from yeast strain A (**a**), B (**b**), C (**c**) and D (**d**), and two forms of dead yeast from strain A (inactive yeast (**e**); autolyzed yeast (**f**)). Data are presented as least square mean +/- standard error

and this resulted in a significant linear contrast across concentration ($p < 0.01$, Fig. 3f).

BEC viability was significantly reduced by CWCs from all strains as indicated by linear contrasts ($p < 0.01$, Fig. 4a-d). BEC viability was also significantly increased by inactive yeast from strain A as indicated by quadratic contrasts, and cytotoxicity was not observed for either form of dead yeast from strain A ($p < 0.01$, Fig. 4e-f).

Effect of yeast derivatives on MAP binding to MAC-T cell and BEC

CWCs from the four yeast strains concentration-dependently reduced MAP binding to MAC-T cells under cultured conditions for 6 h. There was significant reduction in the number of MAP adherent to MAC-T cells after 6-h exposure to CWCs from strain A and C at 0.5–8 mg/ml ($p < 0.05$, Fig. 5a and 5c). A significant reduction of MAP adhesion was also seen when MAC-T cells were cultured with all concentrations of CWCs from strain B and D ($p < 0.05$, Fig. 5b and 5d). Similarly, the number of MAP adhering to MAC-T cells appeared to decrease as the concentration of both forms of dead yeast from strain A increased; the significant reduction can be seen from concentrations 2 mg/ml and higher for both inactive and autolyzed yeast ($p < 0.05$, Fig. 5e-f).

CWCs from strain A and D also concentration-dependently reduced MAP binding to BECs during the 6-h exposure period. A significant reduction in MAP binding was observed from strain A at concentrations 1-8 mg/ml, and from strain D at concentrations 0.5-1 mg/ml ($p < 0.05$, Fig. 6a and 6d). However, there was no significant difference in MAP binding in response to CWCs from strain B and C (Fig. 6b and 6c). Additionally, a concentration-dependent reduction of MAP binding to BECs was observed with both inactive and autolyzed yeast from strain A, with significant differences appearing between 2-16 mg/ml ($p < 0.05$, Fig. 6e-f).

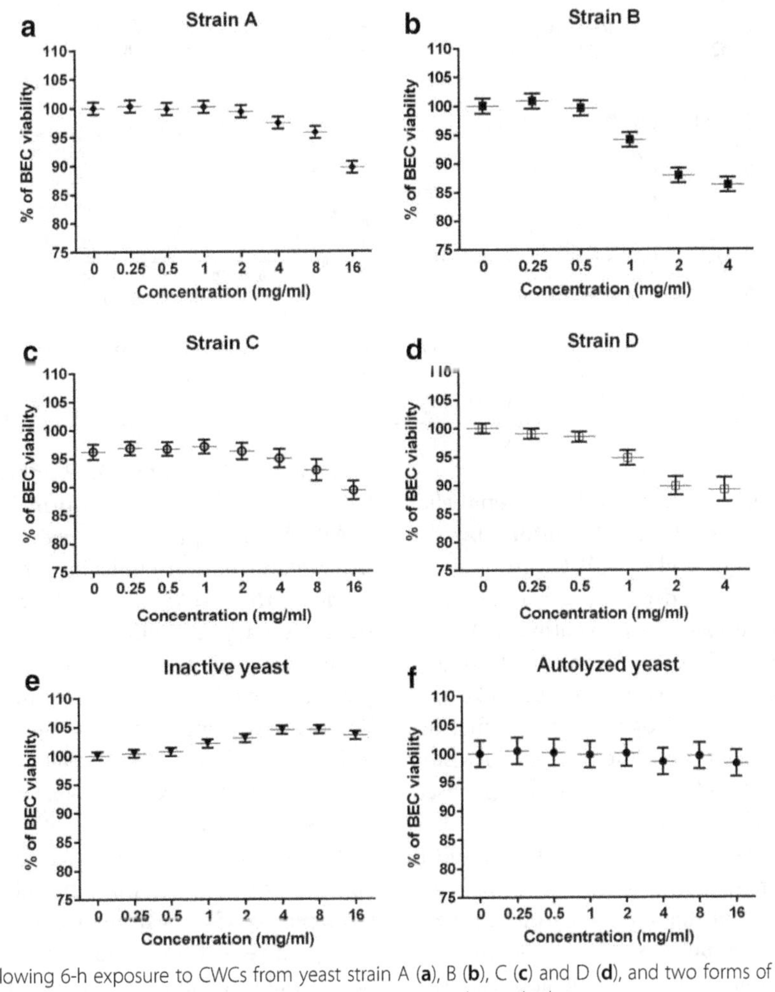

Fig. 4 Viability of BECs following 6-h exposure to CWCs from yeast strain A (**a**), B (**b**), C (**c**) and D (**d**), and two forms of dead yeast from strain A (inactive yeast (**e**); autolyzed yeast (**f**)). Data are presented as least square mean +/- standard error

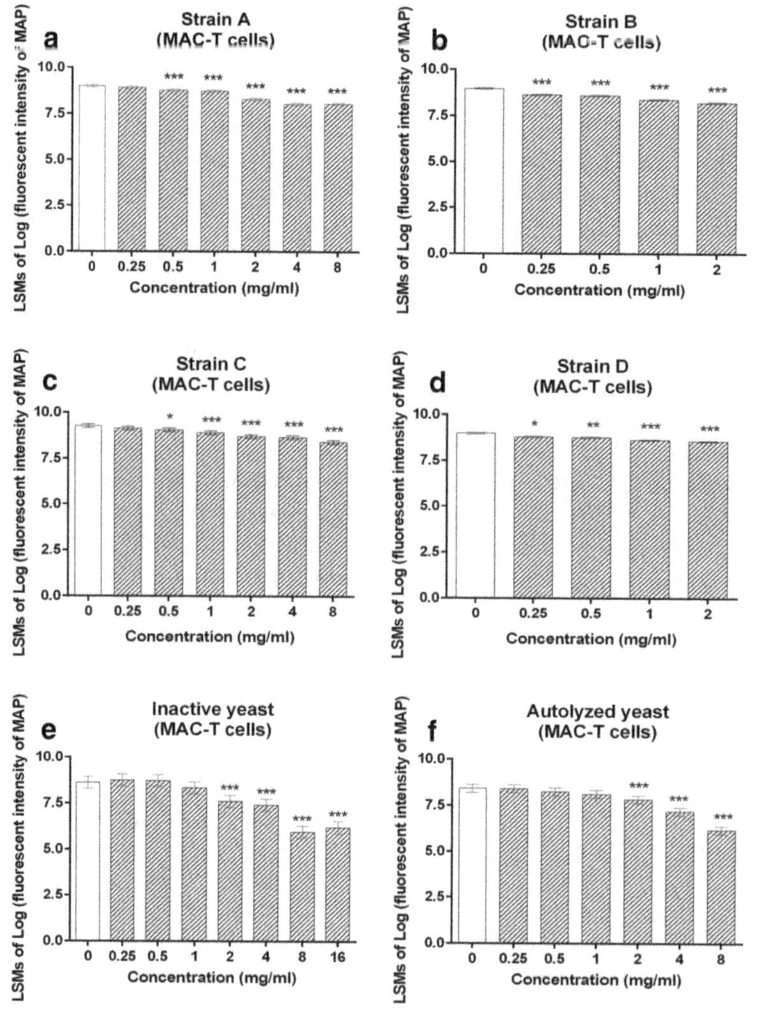

Fig. 5 Binding of mCherry-MAP to MAC-T cells in the presence of yeast CWCs from yeast strain A (**a**), B (**b**), C (**c**) and D (**d**), and two forms of dead yeast from strain A (inactive yeast (**e**); autolyzed yeast (**f**)) after 6-h exposure. Data are presented as least square mean +/- standard error. Significant differences relative to the control are indicated at $p < 0.05$ (*), 0.01 (**), and 0.001 (***)

Discussion

In this study, the adherence of MAP to bovine epithelial cells was significantly reduced in cell culture by the addition of yeast *S. cerevisiae* and its' CWCs. In terms of effectiveness, the two forms of dead yeast from strain A had the most significant impact on preventing MAP adhesion to MAC-T cells and BECs, with maximal binding for the autolyzed and inactive yeast being around 26 and 30 %, respectively. To our knowledge, this is the first investigation of the adhesive properties of *S. cerevisiae* components against MAP. The MAC-T cells were used in this study because of ease of culture and lack of a suitable bovine small intestinal epithelial cell line, and it is challenging to obtain sufficient numbers of primary intestinal epithelial cells for high through-put *in vitro* screening of bioactive compounds. One concern with using MAC-T cells is that they may not appropriately model MAP binding to intestinal epithelial cells; however, in this study MAP adhesion appeared similar between MAC-T cells and primary BECs. Additionally, MAP has also previously been detected within mammary gland of sub-clinically infected cows [30], which suggests that MAP is able to disseminate and invade the mammary epithelium.

Another potential concern with assessing adhesion of yeast and its CWCs to MAP under cell culture conditions is that it may influence epithelial cell viability. The viability of both MAC-T cells and BECs was therefore tested after a 6-h exposure period with the yeast and its CWCs. The results varied with the epithelial cell type and yeast component. MAC-T cell viability for example, appeared to be lower than that of BECs within the same treatment of CWCs from strain A at 16 mg/ml, and the autolyzed yeast from strain A at 16 mg/ml. The decrease of cell viability might be related to changes in intracellular homeostasis and apoptosis-regulated gene expression.

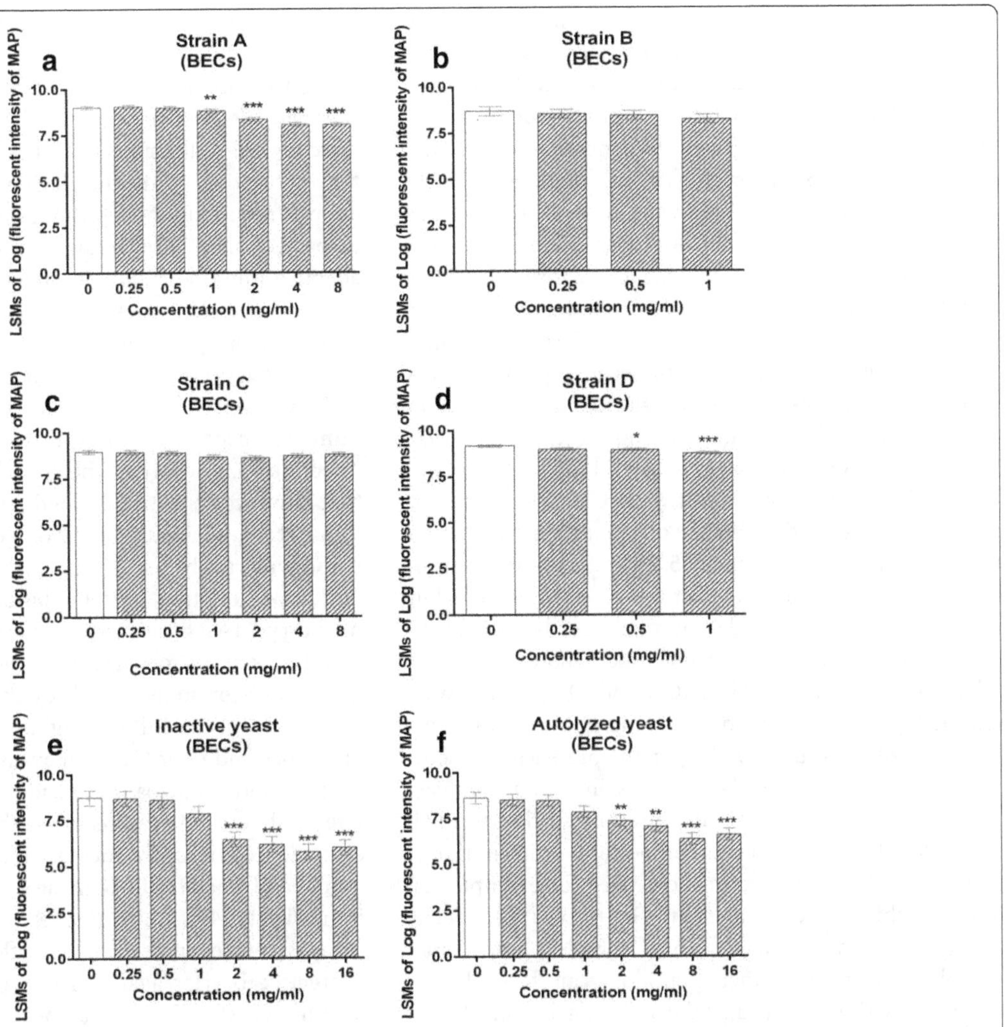

Fig. 6 Binding of mCherry-MAP to BECs in the presence of yeast CWCs from yeast strain A (**a**), B (**b**), C (**c**) and D (**d**), and two forms of dead yeast from strain A (inactive yeast (**e**); autolyzed yeast (**f**)) after 6-h exposure. Data are presented as least square mean +/- standard error. Significant differences relative to the control are indicated at $p < 0.05$ (*), 0.01 (**), and 0.001 (***)

Higher concentrations of yeast derivatives for example, may induce cell apoptosis via altering intracellular calcium levels and the ratio of Bax and bcl-2 genes, as was demonstrated in human breast cancer cells exposed to heat-killed *S. cerevisiae* [31]. Bax and bcl-2 are two discrete members of a gene family involved in the regulation of cellular apoptosis; Bax protein is known as an apoptosis-promoting factor, whereas bcl-2 protein as an apoptosis-suppressing factor [32–34]. Future studies will explore whether or not yeast components from this study induced epithelial cell apoptosis.

It was interesting to note that not all of the yeast components assessed in this study reduced the cell viability. Inactive yeast for example, actually increased the viability of both MAC-T cells and BECs in a quadratic manner. It is possible that inactive yeast stimulated cell proliferation during the 6-h exposure period. Increased cell proliferation was observed with lymphocytes isolated from weaned pigs [35]; in this study, lymphocytes were stimulated with different concentrations of β-glucans from a variety of sources, and *S. cerevisiae* β-glucan significantly induced the cell proliferation compared with the negative control.

Another potential concern about the following observations is that the reduction of MAP binding to epithelial cells was partially attributed to a decrease of epithelial cell viability. CWCs from strain B for instance, reduced MAC-T cell viability and MAP binding at the same concentrations (2 and 4 mg/ml); thus, the reduction of MAP binding could have, in part, been attributed to a decrease in the number of viable epithelial cells. However, a reduction in MAP binding to epithelial cells was also observed at lower non-cytotoxic concentration of CWC from strain B and with the two forms of dead yeast from strain A without a decrease of cell viability.

In addition to the observed decrease in MAP binding to epithelial cells in the presence of *S. cerevisiae* components, reduced MAP infection has also been demonstrated with mice fed the probiotic *Lactobacillus animalis* [36]. Thus, both studies provide evidence that risk of MAP infection may be reduced in ruminants by probiotic/prebiotic use. Similar observations have also been observed with other enteropathgenic bacteria such as *E. coli, Salmonella* Typhimurium and *Salmonella* Typhi when *S. cerevisiae* was used as an adhesive agent [37–39]. Although these later studies were carried out in vivo, making comparisons with the present study difficult, these studies and our study suggest that *S. cerevisiae* may be used, in milk replacer for example, to reduce calf exposure to Gram-negative pathogens and MAP. This hypothesis should be validated in vivo, since increased yeast binding to MAP may not necessarily translate to lower infection load. Ganan et al. [10] for example, demonstrated that while whole yeast, yeast CWC, and the mannose fraction blocked adherence of *Campylobacter jejuni* to Caco-2 cells by approximately 55, 85 and 55 %, respectively, only the mannose fraction effectively reduced microbial invasion. In the present study, the whole cell lysate (inactive yeast and autolyzed yeast) were much more effective at inhibiting MAP binding to MAC-T and BECs than CWCs, and at higher concentrations (>2 mg/ml), the adherence was reduced by around 10–30 % in the presence of inactive yeast (Figs. 5e and 6e). Differences in binding efficacy between the Ganan et al. and present study are likely attributed to the different components within the *C. jejuni* and MAP cell membranes. MAP for example, has a much thicker cell wall than Gram-negative species (*C. jejuni*), and it is made up of a hydrophobic outer mycolate layer and inner peptidoglycan layer bound together by the polysaccharide, arabinogalactan [40]. In contrast, the cell membrane of Gram-negative species is made up of an outer lipopolysaccharide layer and a thin inner peptidoglycan layer [41].

The present in vitro binding assay that involved mCherry-MAP and bovine epithelial cells offers a potentially high through-put platform for screening the adhesive properties of probiotics/prebiotics and their bioactive components that may be useful for controlling JD. Another potential concern about this in vitro study, however, is that the yeast components may influence other mucosal cell types such as goblet cells and dendritic cells (DCs) that were not taken into consideration with this cell culture model. For example, in previous in vivo studies, dietary supplementation with MOS derived from *S. cerevisiae* increased the number of cells secreting acid mucins in posterior gut of sea bass [42]. Increased mucus secretion induced by MOS may help prevent pathogenic bacteria adhesion via the anti-adhesive action of mucins [43], as well as via the bulk physical properties of mucus that help to clear bacteria [44]. However, a reduction in the number of goblet cells that produce mucus has also been reported in the ileal sections of chicken poults fed 0.02 % *S. cerevisiae var boulardii* [45].

Yeast components can also have interactions with DCs. Mannan derived from the *S. cerevisiae* or *Candida albicans* cell wall for example, has been reported to induce the maturation of murine DCs *in vivo* [46]. DC maturation is required for the priming of T helper 1 (Th1) or T helper 2 (Th2) cells, both of which are involved in the adaptive immune response. Sheng et al. [46] also showed that two different types of mannan, oxidized mannan and reduced mannan, differentially stimulated Th1/Th2 cytokine production from murine bone-marrow-derived DCs. Th1 cytokines play a key role during the cell-mediated immune response, which is required to control MAP infection [47]; whereas, the stimulation of Th2 cytokines that support an antibody-mediated immune response is thought to be insufficient for controlling intracellular pathogens such as MAP [48]. Given the potential limitation of the present binding assay with using mCherry-MAP, *in vivo* studies are also needed to validate the benefits of promising yeast and yeast CWCs that have been identified through *in vitro* screening.

In addition to influencing host mucosal cells, yeast probiotics and their CWCs may also influence commensal bacteria and protozoa that reside within ruminant gastrointestinal tract. It was reported that MOS treatment for example, significantly increased the diversity and band number of bacteria in both ileal and colonic fermentum from piglets [49], which implies that MOS could stabilize the gastrointestinal tract microflora that may act as a protective barrier against MAP infection. Additionally, feeding yeast culture was previously reported to increase the number of protozoa in the rumen [50], since the yeast culture is used as a protein and energy source by protozoa [51, 52]. Given that two commonly occurring environmental protozoa, *Acanthamoeba castellanii and A. polyphaga*, have been reported to be the vectors for MAP [53], it is still remains to be determined if the oral administration of yeast or yeast CWCs will help reduce risk of MAP infection.

Conclusions

The results from the present study demonstrate that the adhesion of MAP to epithelial cells was significantly reduced *in vitro* by *S. cerevisiae* and its' CWC in a strain-dependent manner. Since MAP is an intracellular pathogen, this anti-adhesive action could potentially reduce uptake of MAP by epithelial cells. However, since the specific mechanism of reduced MAP adhesion remains to be elucidated, further characterization and investigation is warranted to determine potential mechanisms of action, the stability of the binding between yeast components and MAP, and the efficacy of the tested yeast components *in vivo*.

Abbreviations

BEC: Bovine primary intestinal epithelial cell; CFU: Colony forming units; CWC: Cell wall component; JD: Johne's disease; MAC-T cell: Bovine mammary epithelial cell line; MAP: *Mycobacterium avium* subspecies *paratuberculosis*; MOS: Mannan oligosaccharides.

Competing interests

The authors declare that they have no competing interests.

Author's contributions

ZL research design, conduct research, manuscript preparation; QY sample preparation, manuscript revision; FO industry collaborator, manuscript revision; PM research design; MQ statistical analysis; NK research design, manuscript revision. All authors read and approved the final manuscript.

Acknowledgements

This research was financially supported by collaborative grants with Lallemand Inc. and the Natural Science and Engineering Research Council of Canada (NSERC) Engage and Engage Plus programs, and the MITACS Accelerate program.

Author details

[1]Department of Animal Biosciences, University of Guelph, Guelph N1G 2W1, ON, Canada. [2]Lallemand Inc., Montréal H4P 2R2, QC, Canada.

References

1. Williams PEV, Tait CA, Innes GM, Newbold CJ. Effects of the inclusion of yeast culture (*Saccharomyces cerevisiae* plus growth medium) in the diet of dairy cows on milk yield and forage degradation and fermentation patterns in the rumen of steers. J Anim Sci. 1991;69(7):3016–26.
2. Robinson PH, Garrett JE. Effect of yeast culture (*Saccharomyces cerevisiae*) on adaptation of cows to postpartum diets and on lactational performance. J Anim Sci. 1999;77(4):988–99.
3. Salama AAK, Caja G, Garín D, Albanell E, Such X, Casals R. Effects of adding a mixture of malate and yeast culture (*Saccharomyces cerevisiae*) on milk production of Murciano-Granadina dairy goats. Anim Res. 2002;51:295–303.
4. Wohlt JE, Corcione TT, Zajac PK. Effect of yeast on feed intake and performance of cows fed diets based on corn silage during early lactation. J Dairy Sci. 1998;81:1345–52.
5. Kamel HEM, Sekine J, El-Waziry AM, Yacout MHM. Effect of *Saccharomyces cerevisiae* on the synchronization of organic matter and nitrogen degradation kinetics and microbial nitrogen synthesis in sheep fed Berseem hay (*Trifolium alexandrinum*). Small Rumin Res. 2004;52:211–6.
6. Doreau M, Jouany JP. Effect of a *Saccharomyces cerevisiae* culture on nutrient digestion in lactating dairy cows. J Dairy Sci. 1998;81:3214–21.
7. Jouany JP, Mathieu F, Senaud J, Bohatier J, Bertin G, Mercier M. The effect of *Saccharomyces cerevisiae* and *Aspergillus oryzae* on the digestion of the cell wall fraction of a mixed diet in defaunated and refaunated sheep rumen. Reprod Nutr Dev. 1998;38:401–16.
8. Mosoni P, Chaucheyras-Durand F, Béra-Maillet C, Forano E. Quantification by real-time PCR of cellulolytic bacteria in the rumen of sheep after supplementation of a forage diet with readily fermentable carbohydrates: effect of a yeast additive. J Appl Microbiol. 2007;103:2676–85.
9. Silberberg M, Chaucheyras-Durand F, Mialon MM, Monteils V, Mosoni P, Morgavi DP, Martin C. Repeated acidosis challenges and live yeast supplementation shape rumen microbiota and fermentations and modulate inflammatory status in sheep. Animal. 2013;7:1910–20.
10. Ganan M, Carrascosa AV, de Pascual-Teresa S, Martinez-Rodriguez AJ. Inhibition by yeast-derived mannoproteins of adherence to and invasion of Caco-2 cells by *Campylobacter jejuni*. J Food Prot. 2009;72(1):55–9.
11. Baurhoo B, Letellier A, Zhao X, Ruiz-Feria CA. Cecal populations of lactobacilli and bifidobacteria and Escherichia coli populations after in vivo Escherichia coli challenge in birds fed diets with purified lignin or mannanoligosaccharides. Poult Sci. 2007;86(12):2509–16.
12. Fernandez F, Hinton M, Gils BV. Dietary mannan-oligosaccharides and their effect on chicken caecal microflora in relation to Salmonella Enteritidis colonization. Avian Pathol. 2002;31(1):49–58.
13. Posadas SJ, Caz V, Caballero I, Cendejas E, Quilez I, Largo C, Elvira M, De Miguel E. Effects of mannoprotein E1 in liquid diet on inflammatory response and TLR5 expression in the gut of rats infected by *Salmonella typhimurium*. BMC Gastroenterol. 2010;10(1):58.
14. Ganner A, Stoiber C, Wieder D, Schatzmayr G. Quantitative in vitro assay to evaluate the capability of yeast cell wall fractions from *Trichosporon mycotoxinivorans* to selectively bind Gram-negative pathogens. J Microbiol Methods. 2010;83(2):168–74.
15. Hines 2nd ME, Kreeger JM, Herron AJ. Mycobacterial infections of animals: pathology and pathogenesis. Lab Anim Sci. 1995;45(4):334–51.
16. Veterinary Laboratories Agency (VLA). Surveillance Report. Johne's disease continues to be the most common cause of bovine enteric disease. Vet Rec. 2008;163(6):171–4.
17. Windsor PA, Whittington RJ. Evidence for age susceptibility of cattle to Johne's disease. Vet J. 2010;184(1):37–44.
18. Whitlock RH, Buergelt C. Preclinical and clinical manifestations of paratuberculosis (including pathology). Vet Clin North Am Food Anim Pract. 1996;12(2):345–56.
19. Coussens PM. *Mycobacterium paratuberculosis* and the bovine immune system. Anim Health Res Rev. 2001;2(02):141–62.
20. Duffield BJ, Young DA. Survival of *Mycobacterium bovis* in defined environmental conditions. Vet Microbiol. 1985;10:193–7.
21. Desikan KV. Extended studies on the viability of *Mycobacterium leprae* outside the human body. Lepr Rev. 1995;66:287–95.
22. Jackson R, Morris RS. A study of the environmental survival of Mycobacterium bovis on a farm in New Zealand. New Zealand Vet J. 1995;43(7):346–52.
23. Rosseels V, Huygen K. Vaccination against paratuberculosis. Exp Rev Vaccines. 2008;7(6):817–32.
24. National Center for Animal Health Programs-APHIS. Johne's disease in cattle. http://www.aphis.usda.gov/publications/animal_health/content/printable_version/faq_johnes_disease08.pdf (2008). Accessed Nov 2008.
25. Q. You, P, Mead, S. Oh, S. Mallikarjunappa, L. Mutheria, and N.A. Karrow 2014. Construction of a reporter Mycobacterium avium subsp. paratuberculosis (Map) strain and infection of monocyte-derived macrophages from cows homozygous for SNP -298 A > G in the macrophage migration inhibitory factor (MIF) gene. Proceedings of the 10th World Congress on Genetics Applied to Livestock Production. P 541. Vancouver, B.C. August 17–22.
26. Huynh HT, Robitaille G, Turner JD. Establishment of bovine mammary epithelial cells (MAC-T): an in vitro model for bovine lactation. Exp Cell Res. 1991;197(2):191–9.
27. Föllmann W, Weber S, Birkner S. Primary cell cultures of bovine colon epithelium: isolation and cell culture of colonocytes. Toxicol In Vitro. 2000;14(5):435–45.
28. Kaushik RS, Begg AA, Wilson HL, Aich P, Abrahamsen MS, Potter A, et al. Establishment of fetal bovine intestinal epithelial cell cultures susceptible to bovine rotavirus infection. J Virol Methods. 2008;148(1–2):182–96.
29. Bridger PS, Mohr M, Stamm I, Fröhlich J, Föllmann W, Birkner S, et al. Primary bovine colonic cells: a model to study strain-specific responses to *Escherichia coli*. Vet Immunol Immunopathol. 2010;137(1–2):54–63.
30. Streeter RN, Hoffsis GF, Bech-Nielsen S, Shulaw WP, Rings DM. Isolation of *Mycobacterium paratuberculosis* from colostrum and milk of subclinically infected cows. Am J Vet Res. 1995;56(10):1322–4.
31. Ghoneum M, Matsuura M, Braga M, Gollapudi S. *S. cerevisiae* induces apoptosis in human metastatic breast cancer cells by altering intracellular Ca2+ and the ratio of Bax and Bcl-2. Int J Oncol. 2008;33(3):533–9.
32. Oltval ZN, Milliman CL, Korsmeyer SJ. Bcl-2 heterodimerizes in vivo with a conserved homolog, Bax, that accelerates programed cell death. Cell. 1993;74(4):609–19.
33. Hockenbery D, Nuñez G, Milliman C, Schreiber RD, Korsmeyer SJ. Bcl-2 is an inner mitochondrial membrane protein that blocks programmed cell death. Nature. 1990;348(6299):334–6.
34. Yang E, Korsmeyer SJ. Molecular thanatopsis: a discourse on the BCL2 family and cell death. Blood. 1996;88:386–401.
35. Sonck E, Stuyven E, Goddeeris B, Cox E. The effect of beta-glucans on porcine leukocytes. Vet Immunol Immunopathol. 2010;135(3–4):199–207.
36. Karunasena E, Kurkure PC, Lackey RD, McMahon KW, Kiernan EP, Graham S, Alabady MS, Campos DL, Tatum OL, Brashears MM. Effects of the probiotic *Lactobacillus animalis* in murine *Mycobacterium avium* subspecies *paratuberculosis* infection. BMC Microbiol. 2013;13:8.

37. Tiago FC, Martins FS, Souza EL, Pimenta PF, Araujo HR, Castro IM, Brandão RL, Nicoli JR. Adhesion to the yeast cell surface as a mechanism for trapping pathogenic bacteria by *Saccharomyces* probiotics. J Med Microbiol. 2012;61(9):1194–207.
38. Becker PM, Galletti S. Food and feed components for gut health-promoting adhesion of *E. coli* and *Salmonella enterica*. J Sci Food Agric. 2008;88(11):2026–35.
39. Chaucheyras-Durand F, Ossa F, Habouzit C, Castex M, Henri D. Effect of live yeast and yeast derivatives on adhesion capacity of *E.coli* strains to intestinal epithelial cells. In: Symposium on Gut Health in Production of Food Animals, College Station, Texas, US. 2012. https://ca.linkedin.com/in/faisuryossa. Accessed 3–5 Dec 2012.
40. Ghuysen JM, Hakenbeck R. Bacteria Cell Wall. Amsterdam: Elsevier Science; 1994. p. 1–606.
41. Archibald AR, Hancock IC, Harwood CR. Bacillus subtilis and other gram-positive bacteria: biochemistry, physiology, and molecular genetics. Washington DC: ASM; 1993. p. 381–410.
42. Torrecillas S, Makol A, Caballero MJ, Montero D, Ginés R, Sweetman J, Izquierdo M. Improved feed utilization, intestinal mucus production and immune parameters in sea bass (Dicentrarchus labrax) fed mannan oligosaccharides (MOS). Aquac Nutr. 2011;17(2):223–33.
43. Carlstedt I, Davies JR. Glycoconjugates facing the outside world. Biochem Soc Trans. 1997;25(1):214–9.
44. Sandberg T, Nestor M, Pahlson C, Shi L, Caldwell KD. Mucin as surface protectant against bacterial adhesion. In: Abstract of papers of the American Chemical Society. Amer Chemical Soc. 2000. http://uu.diva-portal.org/smash/record.jsf?pid=diva2%3A66296&dswid=-5673. Accessed 17 Oct 2008
45. Bradley GL, Savage TF, Timm KI. The effects of supplementing diets with *Saccharomyces cerevisiae* var. *boulardii* on male poult performance and ileal morphology. Poult Sci. 1994;73(11):1766–70.
46. Sheng KC, Pouniotis DS, Wright MD, Tang CK, Lazoura E, Pietersz GA, Apostolopoulos V. Mannan derivatives induce phenotypic and functional maturation of mouse dendritic cells. Immunology. 2006;118(3):372–83.
47. Pieters J. Entry and survival of pathogenic mycobacteriain macrophages. Microbes Infect. 2001;3(3):249–55.
48. Alonso-Hearn M, Patel D, Danelishvili L, Meunier-Goddik L, Bermudez LE. The *Mycobacterium avium* subsp. *paratuberculosis* MAP3464 gene encodes an oxidoreductase involved in invasion of bovine epithelial cells through the activation of host cell Cdc42. Infect Immun. 2008;76(1):170–8.
49. Hang S, Zhu W. Gut bacterial and Lactobacilli communities of weaning piglets in response to mannan oligosaccharide and sugar beet pulp in vitro fermentation. J Integr Agr. 2012;11:122–33.
50. Arakaki LC, Stahringer RC, Garrett JE, Dehority BA. The effects of feeding monensin and yeast culture, alone or in combination, on the concentration and genetic composition of rumen protozoa in steers fed on low-quality pasture supplemented with increasing levels of concentrate. Anim Feed Sci Technol. 2000;84(1):121–7.
51. Dehority BA. Protozoa of the digestive tract of herbivorous mammals. Int J Trop Insect Sci. 1986;7(3):279–96.
52. Dehority BA, Orpin CG. Development of, and natural fluctuations in, rumen microbial populations. In: Hobson PN, Stewart CS, editors. The Rumen Microbial Ecosystem. Netherlands: Springer; 1997. p. 196–245.
53. Whan L, Grant IR, Rowe MT. Interaction between Mycobacterium avium subsp. paratuberculosis and environmental protozoa. BMC Microbiol. 2006;6:63.

12

Temporary carriage of bovine coronavirus and bovine respiratory syncytial virus by fomites and human nasal mucosa after exposure to infected calves

Veslemøy Sunniva Oma[1*], Thea Klem[1], Madeleine Tråvén[2], Stefan Alenius[2], Britt Gjerset[3], Mette Myrmel[4] and Maria Stokstad[1]

Abstract

Background: In order to prevent spread of the endemic pathogens bovine coronavirus (BCoV) and bovine respiratory syncytial virus (BRSV) between herds, knowledge of indirect transmission by personnel and fomites is fundamental. The aims of the study were to determine the duration of viral RNA carriage and the infectivity of viral particles on fomites and human nasal mucosa after exposure to BCoV and BRSV. During two animal infection experiments, swabs were collected from personnel (nasal mucosa) and their clothes, boots and equipment after contact with calves shedding either virus. Viral RNA was quantified by RT-qPCR or droplet digital RT-PCR (RT-ddPCR), and selected samples with high levels of viral RNA were tested by cell culture for infectivity.

Results: For BCoV, 46% ($n = 80$) of the swabs from human nasal mucosa collected 30 min after exposure were positive by RT-qPCR. After two, four and six hours, 15%, 5% and 0% of the swabs were positive, respectively. Infective virions were not detected in mucosal swabs ($n = 2$). A high viral RNA load was detected on 97% ($n = 44$) of the fomites 24 h after exposure, and infective virions were detected in two of three swabs. For BRSV, 35% ($n = 26$) of the human nasal mucosa swabs collected 30 min after exposure, were positive by RT-ddPCR, but none were positive for infective virions. Of the fomites, 89% ($n = 38$) were positive for BRSV RNA 24 h after exposure, but all were negative for infective viruses.

Conclusions: The results indicate that human nasal mucosa can carry both BCoV and BRSV RNA after exposure to virus shedding calves, but the carriage seems short-lived and the transmission potential is likely limited. High viral loads on contaminates fomites 24 h after exposure to infected animals, and detection of infective BCoV, indicate that contaminated fomites represent a significant risk for indirect transmission between herds.

Keywords: Indirect transmission, Virus infectivity, Biosecurity, Bovine respiratory disease, Human nasal mucosa, Cattle

Background

Bovine coronavirus (BCoV) and bovine respiratory syncytial virus (BRSV) are contagious pathogens detrimentally affecting production and animal welfare in the cattle industry. The viruses are part of the bovine respiratory disease complex and are endemic worldwide. BRSV and BCoV can cause epidemics of respiratory disease and additionally BCoV cause diarrhea in calves and adult cattle (winter dysentery) [1–4]. The traditional way of handling and preventing these diseases is through metaphylactic antibiotic treatment, use of vaccines, or changes in management to improve calf health in herds [5]. The within-herd prevalence and morbidity of BCoV and BRSV infections are high [6, 7] and once the virus enters a herd, circulation is difficult to mitigate. An additional preventive strategy is therefore to reduce inter-herd transmission of virus. Movement of live animals between herds is an important transmission route [8]. If this risk is under control,

* Correspondence: Veslemoy.Oma@nmbu.no
[1]Department of Production Animal Clinical Sciences, Norwegian University of Life Sciences, P.O. Box 8146 Dep, 0033 Oslo, Norway

the next question concerns the contribution of indirect spread of virus between herds. Indirect spread can occur via e.g. personnel travelling between herds, their clothes or equipment.

Important risk factors for indirect spread are the level of virus contamination of relevant surfaces and the infectivity of the viruses. Enveloped respiratory viruses like BCoV and BRSV are generally fragile outside the host [9]. However, as related viruses like human respiratory syncytial virus (HRSV) and human coronavirus 229E remain infective for several hours on contaminated surfaces like countertops and surgical gloves [10, 11], there is a potential for indirect transmission. Epidemiological studies also point out the importance of indirect transmission; Ohlson et al. found that lack of boot provision for visitors was a risk factor for infections with both viruses [12] and Toftaker et al. found that a herd's BCoV and BRSV antibody status was influenced by the status of its neighboring herds [8].

Human nasal mucosa might also be a vector for interherd virus transmission, as traffic of personnel between herds is common. Carriage of BCoV and BRSV in human nostrils has not been studied. Generally, there are few studies on indirect transmission of these viruses, and no experimental studies have been performed. Molecular methods and virus isolation in cell culture can be used to study the level of virus carriage and infectivity, which are determinants for virus transmission. Combined, these methods provide sensitive quantification of viral genomes and assessment of virus infectivity.

Consequently, the aim of the present study was to investigate whether personnel (nostrils) and fomites carry viral RNA and infective viruses after exposure to BCoV or BRSV infected animals.

Methods

Study design and animal experiments

The present study was performed during two animal experiments, one with BCoV and one with BRSV, and during a field outbreak of winter dysentery. Swabs were rubbed in the nostrils of personnel and on their coats, boots, wristwatches and stethoscopes at different time points after animal contact, and examined for viral RNA and infective viruses.

The BCoV experiment was conducted in 2014 at the Swedish University of Agricultural Sciences (SLU) as described by Oma et al. [13]. A total of ten bull calves between six and twelve weeks of age were included, six were Swedish red and white, three were Swedish Holstein and one Swedish mountain breed. Briefly, four calves at SLU were exposed to a group of six calves brought in from SLU's research farm that experienced an outbreak of winter dysentery. The field outbreak was confirmed by RT-qPCR and serology to have been caused by BCoV. After comingling for 24 h, the calves were housed in the isolation unit within their original groups of four and six animals. As six of the calves were naturally exposed to BCoV in the field, the dates of infection were unknown. The presented contamination study was conducted within a three week period while the calves showed signs of disease and shed virus as detected by RT-qPCR. The number of BCoV RNA copies in nasal swabs from the ten calves varied between $\log_{10}$ 2.9 and 10.4 (mean of $\log_{10}$ 6.9) during the study period.

The BRSV experiment took place at the Norwegian Veterinary Institute in 2015 (to be published). A total of eight Norwegian Red calves between two and four months of age were included, six bulls and two heifers. Briefly, six of the calves were infected after contact with two calves inoculated with a field isolate of BRSV, O4-4B/N-11 [14]. The calves were housed in isolation units in groups of four including one inoculated calf. The contamination study was conducted on three different days within one week while the calves showed signs of respiratory disease and shed virus. The number of BRSV RNA copies in nasal swabs from the calves varied between $\log_{10}$ 2.7 and 8.1 (mean of $\log_{10}$ 5.6) during the study period.

Both experiments were conducted in line with the ARRIVE guidelines for planning and reporting in vivo experiments and the concept of the 3R's (Reduction, Replacement and Refinement) [15, 16]. In both experiments, efforts were made to minimize the stress and discomfort for the animals. The animals were closely monitored and medical treatment were administered in line with national Norwegian and Swedish recommendations for treatment of pneumonia and diarrhea in calves.

Exposure procedure and sampling schemes

Table 1 presents an overview of exposed personnel and fomites. During ten minutes, the personnel handled and examined animals that showed clinical signs and shed either BCoV or BRSV. In the BCoV experiment, swabs were collected from human nostrils prior to and 0.5, 2, 4 and 6 h after exposure to the animals. The BRSV experiment included only a single time point (0.5 h), as viral RNA was not detected in nasal swabs collected during a BRSV pilot study.

Clean boots, coats, wristwatches and stethoscopes were used. After exposure to the animals, boots were rinsed in lukewarm water until visually clean and left to dry. All fomites were stored at 16–18 °C, in a room separate from the animals.

Sampling procedure

A detailed protocol was developed for collection of material from fomites and human nostrils. The same person collected all the material from fomites in each

Table 1 Overview of personnel and fomites that were sampled after exposure to virus shedding animals

	BCoV[a] experiment	BRSV[b] experiment	Winter dysentery outbreak in dairy herd
No. of animals	10	8	300
No. of persons	16	12	19
Personnel			
No. of challenges	86	26	19
Hours between exposure and sampling	−0.5, 0.5, 2, 4 and 6	0.5	0.5, 2, 4
Fomites			
No. of challenges	44	38	–
No. and types	12 rubber coats, 16 rubber boots, 8 stethoscopes, 8 wrist watches	19 rubber coats, 19 rubber boots	–
Hours between exposure and sampling	0, 2 and 24	2 and 24	–

Sample collection was performed during two animal experiments and one outbreak of winter dysentery (caused by BCoV). [a]BCoV – bovine coronavirus, [b]BRSV - bovine respiratory syncytial virus

experiment, and instructed the personnel that took part in the human mucosa trial. Specimens were collected with ESwab™ (Copan, Brescia, Italy) and stored in 1 ml of Liquid Amies medium. Gloves were used throughout the experiments.

Specimens from human nasal mucosa were collected by rotating a swab inside one nostril for a couple of seconds. When a person was sampled more than once, the left and right nostrils were sampled alternately. Sampling of fomites was performed by moistening the tip of the swab with Amies medium before lightly rubbing a defined area (5 cm × 10 cm of coats and boots) without visible contamination. For wristwatches and stethoscopes, the area was approximately 3 cm × 3 cm and 2 cm × 5 cm, respectively. At later time points, new areas were sampled. After sample collection, swabs were stored at 4 °C for no more than two hours and thereafter at −70 °C until use.

RNA extraction and quantification of viral genomes

BCoV

RNA was extracted from 140 μl of the Amies medium by the QIAamp Viral RNA Mini QIAcube kit (Qiagen, Hilden, Germany) according to the manufacturer's instructions, eluted in 50 μl buffer and stored at −80 °C. RT-qPCR was performed, in duplicates for nasal swabs, using RNA UltraSense™ One-Step Quantitative RT-PCR System (Invitrogen, MA, USA) and the target was an 85 bp fragment of the M protein gene [17]. Two μl of RNA was used in a total volume of 20 ul containing 200 nM each of forward and reverse primer and 250 nM TaqMan probe. The thermal profile included an RT step at 55 °C for 30 min followed by 95 °C for 2 min and thereafter 40 cycles of 15 s at 95 °C and 1 min at 60 °C. RT-qPCR was run in a Stratagene Mx3005p™ (Agilent Technologies, CA, USA) and each run included a positive (RNA from the nostril of a BCoV positive trial calf) and negative control (water).

Positive swabs from human nasal mucosa were subjected to Kaplan-Meier survival analysis in Stata (Stata SE/14, Stata Corp., College Station, TX, USA). The function shows the cumulative survival, i.e. carriage of BCoV RNA over time, which descends as personnel turns BCoV RNA negative. As the exact time-point a person turned negative was unknown, the mid-point between the last positive and the first negative sample was used in the analysis [18].

In order to estimate the number of BCoV RNA genome copies (GC), a standard curve was prepared using tenfold dilutions of a plasmid containing the BCoV target sequence. The BCoV RNA positive control was aliquoted and included in every RT-qPCR plate as a calibrator to adjust for inter-plate variation. The number of GC in clinical samples was calculated using the formula from Livak and Schmittgen [19]:

$$Q_s = Q_c{}^*(1+E)^{-(Ct_s - Ct_c)}$$

Where Q_s = *sample RNA copy number,* Q_c = *calibrator RNA copy number,* Ct_s = *sample Ct value,* Ct_c = *calibrator Ct value* and E = *efficiency of target amplification.*

The standard curve covered the range from 10.8 to 1.08 × 10^{10} copies. The curve showed a strong linear relationship with a high coefficient of determination (R^2 = 0.996) and a high amplification efficiency (E = 0.965). The limit of quantification represented $\log_{10}$ 3.6 BCoV GC per swab from human nasal mucosa and fomites.

BRSV

RNA was extracted from 200 μl of Amies medium, using the automated NucliSens easyMAG protocol (Biomérieux, Marcy l'Etoile, France), according to the manufacturer's instructions. Quantification of BRSV genomes was conducted in duplicate, with Bio-Rad's QX200 ddPCR System (droplet digital PCR). Each run included a positive (RNA from the nostril of a BRSV positive trial calf) and negative control (water). Droplet generation and transfer of droplets were as described by the manufacturer. The One-Step RT-ddPCR Advanced Kit for Probes (BioRad, CA, USA) and 2 μl RNA were used. The sequence of primers and probe (5'FAM and BHQ1 as quencher) was as described [20], targeting a 123 bp region of the BRSV N gene. Primers and probe concentrations were as recommended by the kit manufacturer and with the following cycling conditions; 50 °C for 60 min, 95 °C for 10 min and 40 cycles of 95 °C for 30 s and 60 °C for 1 min. The ramp rate was set to 2 °C/s. Data processing and absolute quantification of BRSV genomes per μl RNA was performed with QuantaSoft Version 1.7 (BioRad).

Half-life calculation for BRSV RNA carriage was unattainable due to single sampling.

Testing of virus infectivity

BCoV

Virus infectivity was tested in five samples with the highest level of BCoV RNA, using integrated cell culture RT-qPCR; swabs from a wristwatch, a stethoscope and a coat collected 24 h after exposure, and from two human nostrils, collected 30 min after exposure. The swab medium was diluted 1:10 in Dulbecco's Modified Eagle Medium (DMEM, Thermo Fisher Scientific, DE, USA) and added, in duplicate wells, to monolayers of 4-days-old human rectal tumor cells (HRT-18, ATCC CRL-11663) in a 24-well plate. Positive control (cultivated BCoV from a calf in the experiment), positive control in Amies medium and negative controls (cells only) were included. After incubation at 37 °C for one hour, the inoculum was removed, the cells washed and DMEM with 1% fetal calf serum (FCS) and antibiotics (5000 IU penicillin and 5 mg streptomycin/ml) was added. Simultaneously, cells were harvested from one parallel well of each sample as a time zero replication control. After three days incubation at 37 °C in 5% CO_2, cells were harvested from the remaining wells and RNA extracted with Qiazol (Qiagen), chloroform phase separation (mixed 1:1 with 70% ethanol), and RNeasy Mini Kit column (Qiagen). The amount of RNA was measured using Nanodrop (Thermo Fisher Scientific) and equal amounts analyzed by BCoV RT-qPCR run in duplicates as described. Relative quantification of target RNA from incubated and time zero replication control cells was performed using the standard curve.

BRSV

BRSV infectivity was tested in ten swabs showing the highest level of viral RNA by RT-ddPCR; eight swabs from coats and two from human nostrils, collected 24 h and 30 min after exposure, respectively. Fetal bovine turbinate cells (courtesy of Swedish Veterinary Institute) propagated in Eagle's minimal essential medium (BioWhittaker, Belgium) in 96 well plates were incubated with 50 μl filtered swab samples for 30 min. Medium with 2% FCS was added, the plates were incubated at 37 °C in 5% CO_2 and the supernatant passaged after seven days. Samples were cultivated in duplicates with positive (cultivated BRSV from a calf in the experiment) and negative controls (cells only). The cells were observed for cytopathic effect (CPE) and infection visualized by direct immunofluorescence test using FITC Moab a-BRSV (Bio-X Diagnostics, Rochefort, Belgium). Culture supernatants were harvested and tested by the BRSV RT-ddPCR as described.

Results

Viral RNA in human nasal mucosa

The positive controls were consistently positive throughout the analyses, and all negative controls were negative. No viral RNA was detected in human nasal mucosa that was sampled prior to exposure to animals, however, positives were found among samples collected after exposure (Table 2). The number of BCoV GC per swab is shown in Fig. 1. Estimated half-life of BCoV RNA carriage was less than 90 min and the estimated longest persistence was five hours (Fig. 2). The positive BRSV swabs contained between $\log_{10}$ 1.3 to 3.3 genome copies.

Viral RNA on fomites

The positive controls were consistently positive throughout the analyses, and all negative controls were negative. BCoV RNA was detected on all boots, coats and stethoscopes, and on seven out of eight wristwatches, 24 h after exposure. The eighth watch was positive for BCoV RNA 15 min and two hours after exposure. The copy numbers of BCoV RNA 24 h after exposure are presented in Fig. 3. BRSV RNA was detected on 18 out of 19 boots sampled after two hours and 16 out of 19 boots after 24 h. For the coats, 17 out of 19 were positive two hours after exposure, and 18 out of 19 were positive after 24 h. There were minor differences in BRSV RNA copy numbers between samples collected 2 and 24 h after exposure and no tendency of reduction in copy numbers (Fig. 4).

Virus infectivity

BCoV

RT-qPCR results from cells inoculated with swab material from a wristwatch or a stethoscope indicated a 1000-fold increase in the number of RNA-copies after three days of incubation. Cells inoculated with swab material from human nostrils and from a rubber coat showed no increase in viral RNA during incubation. Positive virus controls were positive, and the Amies medium showed no inhibition of virus replication. No BCoV RNA was detected in negative control wells.

Table 2 BCoV and BRSV RNA in human nasal swabs

	BCoV[a] experiment		BRSV[b] experiment		Winter dysentery outbreak	
Hours between exposure and sampling	Total no. of swabs	No. of positive swabs (%)	Total no. of swabs	No. of positive swabs (%)	Total no. of swabs	No. of positive swabs (%)
−0.5	67	0	ND[c]	ND	ND	ND
0.5	80	37 (46%)	26	9 (35%)	7	1 (14%)
2	68	10 (15%)	ND	ND	1	0
4	38	2 (5%)	ND	ND	12	0
6	28	0	ND	ND	ND	ND
24	11	0	ND	ND	ND	ND
Total	292	49 (17%)	26	9 (35%)	20	1 (5%)

RT-qPCR and droplet digital RT-PCR results in swabs from the nasal cavity of personnel before and after exposure to BCoV or BRSV infected calves
[a]BCoV – bovine coronavirus
[b]BRSV – bovine respiratory syncytial virus
[c]ND = Not done

BRSV

No CPE was seen in cells incubated with swab material or with passaged material and RT-ddPCR results did not indicate any virus replication after two passages in the cells. Positive control wells were positive, and negative control wells were negative.

Discussion

This is the first time BCoV and BRSV contamination of personnel and fomites has been described. The PCR results indicate that fomites (like clothes, boots, wristwatches and stethoscopes) exposed to virus pose an infection risk to cattle. For BCoV in particular, fomites seem to represent a high risk, as virus isolation detected infective viruses after 24 h. Consequently, measures to prevent inter-herd transmission should include actions against indirect spread of virus.

As high copy numbers of viral RNA on fomites indicated a transmission potential, further investigations were performed in order to assess whether the detected RNA could represent infective viruses. Although infectivity ideally should be studied in live animals, cell culture was used due to practical, ethical and economic reasons. Virus isolation in cell culture may have a low sensitivity [21], but the use of integrated cell culture RT-qPCR increases the possibility of detecting infective viruses [22]. Using this method, we showed that visually clean surfaces of fomites can carry infective BCoV for at least 24 h after exposure to infected animals.

As reviewed by La Rosa et al., related coronaviruses and HRSV can be transmitted by fomites in addition to direct transmission through droplets and aerosols [23]. It is therefore plausible that BCoV and BRSV could be transmitted between farms via personnel and fomites. Even if protective clothing is used and changed between herds, personnel might constitute a risk of virus transmission as human nasal mucosa could be a potential hideaway for infective viruses. In addition, BCoV has been isolated from a diarrheic child and is most likely

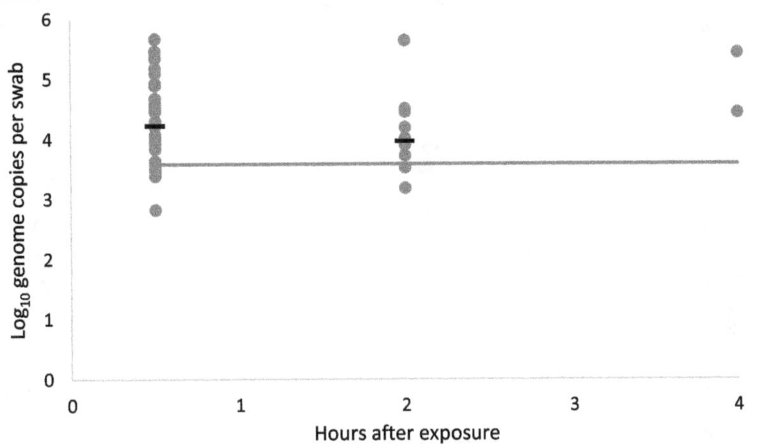

Fig. 1 Bovine coronavirus (BCoV) contaminated human nasal mucosa. Log_{10} genome copies of BCoV per positive swab. Personnel had close contact for at least ten minutes with calves shedding BCoV. Swabs were taken from human nostrils at different time points after exposure to the calves. The grey line shows the limit of quantification and the black short lines indicate median genome copies per positive swab

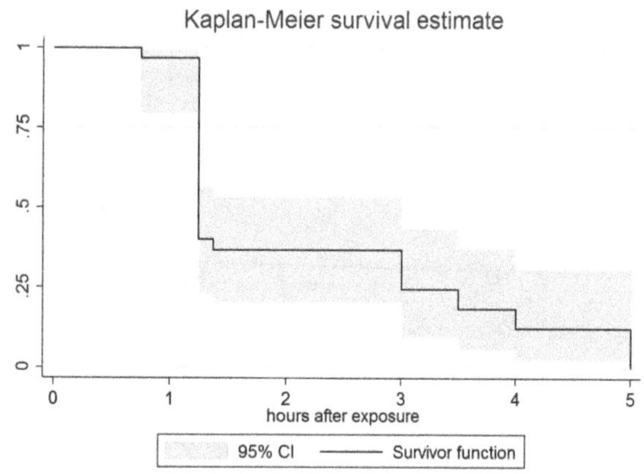

Fig. 2 Kaplan-Meier survival function for BCoV RNA carriage in human nasal mucosa

the ancestor of a related human pathogen [24–26], thus the ability to replicate in cells in the human nasal mucosa cannot be excluded. Human nasal mucosa was therefore studied in place of skin, oral mucosa or hair that could also act as passive vectors for the viruses.

Sellers et al. have shown that human nasal mucosa is a possible vehicle for foot-and-mouth-disease virus even when a high level of biosecurity is implemented [27]. This was refuted by Amass et al. who found a low risk of virus transmission by personnel after hand wash and change of outerwear [28]. Wright et al. found a low risk of prolonged human nasal carriage of the virus [29]. In the present study, we aimed to study whether human nasal mucosa is a possible vehicle for transmission of BCoV and BRSV. Based on our results, this is a possibility, but the low level of viral RNA and the failure to detect infective virus after a few hours, indicate a low risk of virus transmission from human nasal mucosa.

In the present study, personnel was sampled during an experimental setting and during an outbreak of winter dysentery in the field. The results showed that nasal carriage of BCoV in humans was less common in the outbreak situation than during the animal experiment. Factors that could have influenced the amount of virus in the two settings were differences in virus exposure, degree of contact between animals and personnel and environmental conditions. Other factors could be repeated swabbing of the same nostril, nose touching and nose blowing.

The finding that neither BCoV nor BRSV could be cultivated from human nasal swabs resembles the rapid inactivation on skin for respiratory syncytial virus [11] and human coronavirus 229E [30]. This could be due to substances or microorganisms in the mucosa that neutralize or inactivate the virus. Although there is a chance of underestimating the risk, due to e.g. freezing and thawing, dilution and filtering of the samples, the virus transmission potential of mucosa is probably low. There were no sign of BCoV replicating in human nasal mucosa, as the amount of BCoV RNA found were low and declining over time.

Despite the general view that enveloped viruses are fragile outside the host, several coronaviruses remain infective after drying on surfaces for more than 24 h as reviewed by Otter et al. [31]. The present study indicates that BCoV has a similar property. Infective BRSV, on the other hand, was not detected in any of the samples, which was similar to HRSV after drying on surfaces for seven hours [11]. Studies of HRSV survival in cell culture medium and aerosols also showed a higher inactivation rate compared to coronaviruses [32–35]. This suggests that BRSV is more susceptible than BCoV to degradation by environmental factors, and that the importance of indirect BRSV transmission after 24 h, is probably low. As demonstrated by Mullis et al. [36], viral infectivity is more rapidly lost than viral RNA.

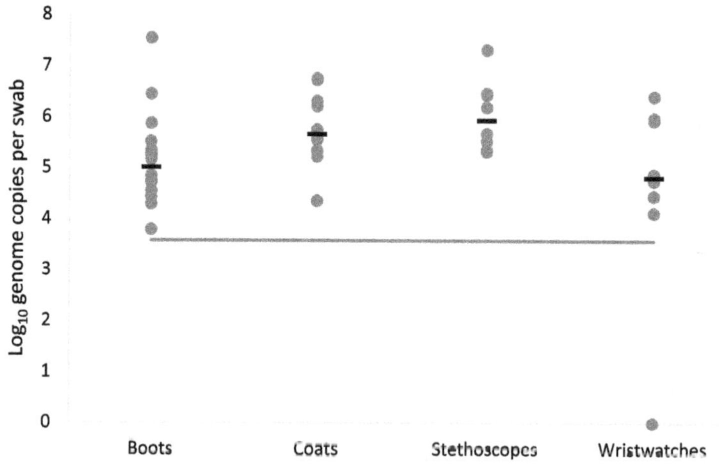

Fig. 3 Bovine coronavirus (BCoV) contaminated fomites. Log_{10} numbers of genome copies per swab taken 24 h after exposure to BCoV-infected calves. The grey line shows the limit of quantification and the black short lines indicate median genome copies per swab

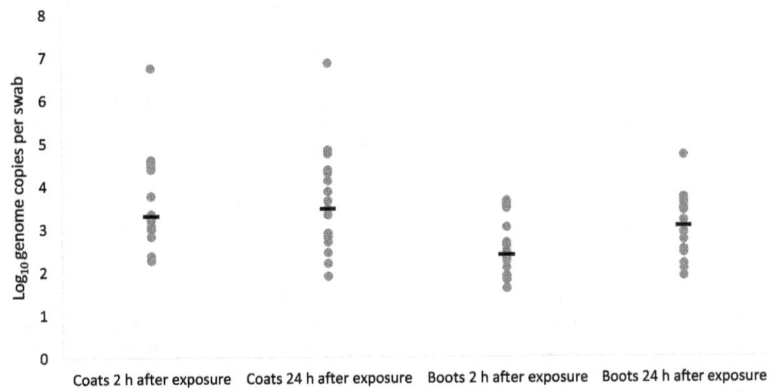

Fig. 4 Bovine respiratory syncytial virus (BRSV) contaminated fomites. Log_{10} numbers of genome copies in positive swabs taken after exposure to BRSV-infected calves. The black short lines indicate median genome copies per swab

For both viruses, the viral RNA level recovered from boots was lower than from coats, possibly due to the rinsing with water. However, as high genome copy numbers remained, rinsing might not be sufficient to prevent virus transmission. This is supported by epidemiologic data that show an increased risk of seropositivity for BRSV and BCoV in herds that do not provide boots to visitors [37]. The present BCoV experiment indicated that also stethoscopes and wristwatches could serve as vehicles. These items are often brought between farms without cleaning/disinfection, and can carry infective virus particles for at least 24 h after exposure to infected cattle.

Conclusions

Personnel pose a risk in inter-herd transmission of BRSV and BCoV when bringing fomites between herds. In order to control the spread of these viruses, biosecurity measures should be implemented, including herd-specific clothing and equipment and washing/disinfection of fomites. Although personnel may carry the viruses intra-nasally for shorter periods of time, the relative importance of contaminated mucosa for indirect transmission is less than that of contaminated fomites.

Abbreviations

BCoV: Bovine coronavirus; BRSV: Bovine respiratory syncytial virus; CPE: Cytopathic effect; Ct: Cycle threshold; DMEM: Dulbecco's Modified Eagle Medium; FCS: Fetal calf serum; GC: Genome copies; HRSV: Human respiratory syncytial virus; HRT: Human rectal tumor cells; RT-ddPCR: Reverse transcription droplet digital polymerase chain reaction; RT-qPCR: Quantitative reverse transcriptase polymerase chain reaction; SLU: Swedish University of Agricultural Sciences; µl: Microliter

Acknowledgements

The authors would like to acknowledge Mamata Khatri, Ingvild Berg Nyman and Elisabeth Furuseth Hansen at the Department of Food Safety and Infection Biology, Norwegian University of Life Sciences for skillful technical assistance.

Funding

This study was supported by the Research Council of Norway (NFR-project No 224771/E40), the Norwegian Research Funding for Agriculture and Food Industry and TINE Norwegian Dairies BA. The funders had no role in the design of the study, data collection, analyses, decision to publish or preparation of the manuscript.

Authors' contributions

VSO, MT, SA, MM and MS designed and performed the BCoV experiment. TK, BG, MM and MS designed and performed the BRSV experiment. All authors wrote, read and approved the manuscript.

Ethics approval and consent to participate

The animal experiments were conducted in agreement with national and international guidelines for the care and use of animals. Approval for the BCoV experiment was given by the Ethics Committee for Animal Experiments, Uppsala, Sweden (protocol no. C45/14) and by the Norwegian Animal Research Authority (approval no. 7468) for the BRSV experiment. Verbal consent was obtained from the personnel participating in the study. Written consent was deemed not necessary, as the aim of the study was purely animal health related, and the Southeastern Regional Committee for Medical and Health Research Ethics in Norway exempted the study from application (reference no. 2016/2012 A).

Consent for publication

Not applicable.

Competing interests

The authors declare that they have no competing interests.

Author details

[1]Department of Production Animal Clinical Sciences, Norwegian University of Life Sciences, P.O. Box 8146 Dep, 0033 Oslo, Norway. [2]Department of Clinical Sciences, Swedish University of Agricultural Sciences, 75007 Uppsala, Sweden. [3]Norwegian Veterinary Institute, P.O. Box 750 Sentrum, 0106 Oslo, Norway. [4]Department of Food Safety and Infection Biology, Norwegian University of Life Sciences, P.O. Box 8146 Dep, 0033 Oslo, Norway.

References

1. Tråvén M, Näslund K, Linde N, Linde B, Silván A, Fossum C, Hedlund KO, Larsson B. Experimental reproduction of winter dysentery in lactating cows using BCV – comparison with BCV infection in milk-fed calves. Vet Microbiol. 2001;81:127–51.
2. Brodersen BW. Bovine respiratory syncytial virus. Vet Clin North Am Food Anim Pract. 2010;26:323–33.
3. Boileau MJ, Kapil S. Bovine coronavirus associated syndromes. Vet Clin North Am Food Anim Pract. 2010;26:123–46.
4. Su S, Wong G, Shi W, Liu J, Lai AC, Zhou J, Liu W, Bi Y, Gao GF. Epidemiology, genetic recombination, and pathogenesis of coronaviruses. Trends Microbiol. 2016;24:490–502.
5. Murray GM, O'Neill RG, More SJ, McElroy MC, Earley B, Cassidy JP. Evolving views on bovine respiratory disease: an appraisal of selected control measures - part 2. Vet J. 2016;217:78–82.
6. Klem TB, Gulliksen SM, Lie KI, Løken T, Østerås O, Stokstad M. Bovine respiratory syncytial virus: infection dynamics within and between herds. Vet Rec. 2013;173:476.
7. Hägglund S, Svensson C, Emanuelson U, Valarcher JF, Alenius S. Dynamics of virus infections involved in the bovine respiratory disease complex in Swedish dairy herds. Vet J. 2006;172:320–8.
8. Toftaker I, Sanchez J, Stokstad M, Nødtvedt A. Bovine respiratory syncytial virus and bovine coronavirus antibodies in bulk tank milk - risk factors and spatial analysis. Prev Vet Med. 2016;133:73–83.
9. Boone SA, Gerba CP. Significance of fomites in the spread of respiratory and enteric viral disease. Appl Environ Microbiol. 2007;73:1687–96.
10. Sizun J, MW Y, Talbot PJ. Survival of human coronaviruses 229E and OC43 in suspension and after drying on surfaces: a possible source of hospital-acquired infections. J Hosp Infect. 2000;46:55–60.
11. Hall CB, Douglas RG, Geiman JM. Possible transmission by fomites of respiratory syncytial virus. J Infect Dis. 1980;141:98–102.
12. Beaudeau F, Ohlson A, Emanuelson U. Associations between bovine coronavirus and bovine respiratory syncytial virus infections and animal performance in Swedish dairy herds. J Dairy Sci. 2010;93:1523–33.
13. Oma VS, Tråvén M, Alenius S, Myrmel M, Stokstad M. Bovine coronavirus in naturally and experimentally exposed calves; viral shedding and the potential for transmission. Virol J. 2016;13:100.
14. Klem TB, Rimstad E, Stokstad M. Occurrence and phylogenetic analysis of bovine respiratory syncytial virus in outbreaks of respiratory disease in Norway. BMC Vet Res. 2014;10:15.
15. Kilkenny C, Browne WJ, Cuthill IC, Emerson M, Altman DG. Improving bioscience research reporting: the ARRIVE guidelines for reporting animal research. PLoS Biol. 2010;8:e1000412.
16. Russell WMS. The increase of humanity in experimentation: replacement, reduction and refinement. Laboratory Animals Bureau, Collected Papers. 1957;6:23–5.
17. Decaro N, Elia G, Campolo M, Desario C, Mari V, Radogna A, Colaianni ML, Cirone F, Tempesta M, Buonavoglia C. Detection of bovine coronavirus using a TaqMan-based real-time RT-PCR assay. J Virol Methods. 2008;151:167–71.
18. Dohoo I, Martin W, Stryhn H. Modelling survival data. In: Veterinary epidemiologic research. 2nd ed. Charlottetown, Prince Edward Island, Canada: VER Inc.; 2009. p. 467–522.
19. Livak KJ, Schmittgen TD. Analysis of relative gene expression data using real-time quantitative PCR and the $2^{-\Delta\Delta CT}$ method. Methods. 2001;25:402–8.
20. Boxus M, Letellier C, Kerkhofs P. Real time RT-PCR for the detection and quantitation of bovine respiratory syncytial virus. J Virol Methods. 2005;125: 125–30.
21. Kapil S, Richardson KL, Radi C, Chard-Bergstrom C. Factors affecting isolation and propagation of bovine coronavirus in human rectal tumor-18 cell line. J Vet Diagn Investig. 1996;8:96–9.
22. Reynolds KA, Gerba CP, Pepper IL. Detection of infectious enteroviruses by an integrated cell culture-PCR procedure. Appl Environ Microbiol. 1996;62:1424–7.
23. La Rosa G, Fratini M, Della Libera S, Iaconelli M, Muscillo M. Viral infections acquired indoors through airborne, droplet or contact transmission. Ann Ist Super Sanita. 2013;49:124–32.
24. Zhang XM, Herbst W, Kousoulas KG, Storz J. Biological and genetic characterization of a hemagglutinating coronavirus isolated from a diarrhoeic child. J Med Virol. 1994;44:152–61.
25. Han MG, Cheon DS, Zhang X, Saif LJ. Cross-protection against a human enteric coronavirus and a virulent bovine enteric coronavirus in gnotobiotic calves. J Virol. 2006;80:12350–6.
26. Vijgen L, Keyaerts E, Moes E, Thoelen I, Wollants E, Lemey P, Vandamme AM, Van Ranst M. Complete genomic sequence of human coronavirus OC43: molecular clock analysis suggests a relatively recent zoonotic coronavirus transmission event. J Virol. 2005;79:1595–604.
27. Sellers RF, Herniman KAJ, Mann JA. Transfer of foot-and-mouth disease virus in the nose of man from infected to non-infected animals. Vet Rec. 1971;89:447–9.
28. Amass SF, Mason PW, Pacheco JM, Miller CA, Ramirez A, Clark LK, Ragland D, Schneider JL, Kenyon SJ. Procedures for preventing transmission of foot-and-mouth disease virus (O/TAW/97) by people. Vet Microbiol. 2004;103:143–9.
29. Wright CF, Gloster J, Mazelet L, Paton DJ, Ryan ED. Short-lived carriage of foot-and-mouth disease virus in human nasal cavities after exposure to infected animals. Vet Rec. 2010;167:928–31.
30. Wolff MH, Sattar SA, Adegbunrin O, Tetro J. Environmental survival and microbicide inactivation of coronaviruses. In: Schmidt A, Wolff MH, Weber O, editors. Coronaviruses with special emphasis on first insights concerning SARS. Basel: Birkhäuser Verlag; 2005. p. 201–12.
31. Otter JA, Donskey C, Yezli S, Douthwaite S, Goldenberg SD, Weber DJ. Transmission of SARS and MERS coronaviruses and influenza virus in healthcare settings: the possible role of dry surface contamination. J Hosp Infect. 2016;92:235–50.
32. Hambling MH. Survival of the respiratory syncytial virus during storage under various conditions. Br J Exp Pathol. 1964;45:647–55.
33. Rechsteiner J, Winkler KC. Inactivation of respiratory syncytial virus in aerosol. J Gen Virol. 1969;5:405–10.
34. Ijaz MK, Brunner AH, Sattar SA, Nair RC, Johnson-Lussenburg CM. Survival characteristics of airborne human coronavirus 229E. J Gen Virol. 1985;66:2743–8.
35. Rabenau HF, Cinatl J, Morgenstern B, Bauer G, Preiser W, Doerr HW. Stability and inactivation of SARS coronavirus. Med Microbiol Immunol. 2005;194:1–6.
36. Mullis L, Saif LJ, Zhang Y, Zhang X, Azevedo MS. Stability of bovine coronavirus on lettuce surfaces under household refrigeration conditions. Food Microbiol. 2012;30:180–6.
37. Ohlson A, Emanuelson U, Tråvén M, Alenius S. The relationship between antibody status to bovine corona virus and bovine respiratory syncytial virus and disease incidence, reproduction and herd characteristics in dairy herds. Acta Vet Scand. 2010;52:37.

13

Influence of border disease virus (BDV) on serological surveillance within the bovine virus diarrhea (BVD) eradication program in Switzerland

V. Kaiser[1], L. Nebel[1], G. Schüpbach-Regula[2], R. G. Zanoni[1*] and M. Schweizer[1*]

Abstract

Background: In 2008, a program to eradicate bovine virus diarrhea (BVD) in cattle in Switzerland was initiated. After targeted elimination of persistently infected animals that represent the main virus reservoir, the absence of BVD is surveilled serologically since 2012. In view of steadily decreasing pestivirus seroprevalence in the cattle population, the susceptibility for (re-) infection by border disease (BD) virus mainly from small ruminants increases. Due to serological cross-reactivity of pestiviruses, serological surveillance of BVD by ELISA does not distinguish between BVD and BD virus as source of infection.

Results: In this work the cross-serum neutralisation test (SNT) procedure was adapted to the epidemiological situation in Switzerland by the use of three pestiviruses, i.e., strains representing the subgenotype BVDV-1a, BVDV-1h and BDSwiss-a, for adequate differentiation between BVDV and BDV. Thereby the BDV-seroprevalence in seropositive cattle in Switzerland was determined for the first time. Out of 1,555 seropositive blood samples taken from cattle in the frame of the surveillance program, a total of 104 samples (6.7%) reacted with significantly higher titers against BDV than BVDV. These samples originated from 65 farms and encompassed 15 different cantons with the highest BDV-seroprevalence found in Central Switzerland. On the base of epidemiological information collected by questionnaire in case- and control farms, common housing of cattle and sheep was identified as the most significant risk factor for BDV infection in cattle by logistic regression.

Conclusion: This indicates that pestiviruses from sheep should be considered as a source of infection of domestic cattle and might well impede serological BVD surveillance.

Keywords: Bovine viral diarrhea virus (BVDV), Border disease virus (BDV), Pestivirus, Serum neutralisation test (SNT), Seroprevalence, Small ruminants, Cross-neutralisation, Eradication, Risk factor

Background

The genus *Pestivirus* in the family *Flaviviridae* comprises the four established species border disease virus (BDV), bovine viral diarrhea virus type-1 (BVDV-1), bovine viral diarrhea virus type-2 (BVDV-2) and classical swine fever virus (CSFV). Additional putative pestivirus species were isolated from giraffe ("Giraffe-1 pestivirus"), cattle ("atypical pestiviruses"), antelopes ("Pronghorn antelope pestivirus ") und piglets ("Bungowannah virus") [1–3]. Recently, an additional new strain termed "atypical porcine pestivirus" was isolated from pigs and piglets with congenital tremor [4, 5]. The ruminant pestiviruses BVDV and BDV are important pathogens with a worldwide distribution [6] causing substantial economic losses in farm animal husbandry [7, 8].

Acute, transient infections of seronegative, immunocompetent animals with ruminant pestiviruses are frequently asymptomatic or are accompanied by mild respiratory or enteric symptoms [9, 10]. By contrast,

* Correspondence: reto.zanoni@vetsuisse.unibe.ch; matthias.schweizer@vetsuisse.unibe.ch
[1]Institute of Virology and Immunology, Federal Food Safety and Veterinary Office (FSVO) and Vetsuisse Faculty, University of Bern, Laenggass-Strasse 122, POB, CH-3001 Bern, Switzerland

acute infection of pregnant cattle between approx. day 40 to 120 of gestation may cause transplacental transmission of non-cytopathogenic (ncp) biotypes to the fetus leading to the birth of persistently infected (PI) calves. These animals shed virus life-long and, thereby, comprise the primary pestivirus reservoir [11–13]. Similarly, lambs persistently infected with border disease virus caused by transplacental transmission display alterations in their fleece and show tremor (hence there are also called 'hairy shakers'), and they may succumb by a syndrome resembling Mucosal disease in cattle [14–17].

BDV in small ruminants occurs worldwide but with very variable seroprevalence depending, e.g., on the geographic location and the type of animal husbandry [14, 17–20]. It was for the first time isolated in Switzerland in a flock of sheep that gave birth to lambs with generalized tremors and excessively hairy fleece in 2001 [21]. In a study published in 1995, seroprevalence in registered sheep flocks of breeding associations and in large flocks was around 20 and 65%, respectively [22]. More recent data pointed to a slightly lower seroprevalence of 13.5% [23] or 16.1% [18] in sheep and 25.4% in goats [18]. In the latter study, it was demonstrated by means of cross-serum neutralisation tests (cross-SNT) that 9% of the sheep and 6% of the goats were infected with BDV. However, 31% and 66% of the seropositive sheep and goats, respectively, could not be assigned to BVDV or BDV leaving the source of infection unidentified.

Thus, even though broad serological cross-reactivity occurs among pestiviruses, considerable quantitative differences in neutralisation efficiency can be measured by SNT between different species (also called genotype) [24–26] and even subgenotypes [26–33]. To date, up to 21 (1a to 1u) and three (2a to 2c) subgenotypes were described within the pestivirus species BVDV-1 and BVDV-2, respectively [34, 35], mostly based on comparison of the 5′-UTR (untranslated region) or N^{pro} (N-terminal protease) region of the pestiviral genomes. In Switzerland, BVDV-1e, -1h, -1k, and -1b are the most prevalent subgenotypes identified in cattle, while BVDV-2 was never detected [30, 36, 37]. Similarly, BD viruses exhibit a large heterogeneity of strains [38] with 7 main subgenotypes and several atypical BDV strains described [39]. An additional phylogenetic group was detected in our institute exclusively in Switzerland and provisionally named BD Switzerland or BDSwiss [40, 41]. Lately, a similar isolate was identified in Italy and labeled as BDV-8 [42].

As ruminant pestiviruses are not strictly species specific, they are able to infect a variety of even-toed ungulates (Artiodactyla) [3, 13, 20]. Virus transmission was described between cattle and both, sheep and goats ([19, 43, 44], and references therein). Natural infections of cattle with BDV were reported in England and Wales [45, 46], Austria [47, 48], Italy [49] and New Zealand [50]. Common housing of cattle with persistently infected sheep was the most important cause for seroconversions, and resulted in reduced fertility and abortions in pregnant cows [19, 44, 46]. BDV-specific seroconversion in cattle was reported in Switzerland after pasturing them with BDV-positive sheep on common alpine meadows [51].

Based on the economic impact of BVDV infections in livestock, several European countries therefore initiated programs to eradicate BVDV in cattle [52, 53]. In Switzerland, such a program was started in 2008 that particularly targeted on a nationwide identification and elimination of PI animals [54, 55]. Initially, 1.4% of all newborn calves were persistently infected with BVDV, which dropped to less than 0.02% by the end of 2012. From that time on, BVD control is based predominantly on risk-based serological surveillance of bulk milk and blood samples [56], with continuously decreasing seroprevalence. As serology performed by ELISA does not distinguish between an infection with BVDV from one with BDV, the impact of infection with BDV on serological surveillance on BVDV is not known. Thus, the aim of this study was to determine the frequency of BDV infections in cattle by using an optimized cross-neutralisation SNT, to identify potential risk factors for interspecies transmission with special emphasis on small ruminants, and to assess their possible influence on the serological surveillance of BVD in bovines.

Methods

Sera

Sera used in this study were from the period 2012 to 2014 and were initially rated as "indeterminate" or "positive" to antibodies (Ab) against pestivirus by ELISA performed by a primary laboratory and later confirmed as positive by the national BVD reference laboratory (Institute of Veterinary Virology/Institute of Virology and Immunology, Bern, Switzerland). A serum was confirmed as positive when it was either positive in the institutes "in house"-ELISA [57] or when it was rated positive in a serum neutralisation test (SNT) using BVDV-1a (Table 1) as challenge virus. Sera were only included if they were obtained from animals that were at least 6 months old at the time of sampling and that were born later than Sept. 30, 2009, i.e., after phase 2 of the Swiss BVD eradication program [58]. If several samples from the same animal were obtained, only the one that was analysed first in the reference laboratory was used for the analysis. All sera were stored at -20 °C prior to use. Overall, 1,568 sera fulfilled the criteria mentioned

Table 1 Ruminant pestivirus isolates selected for SNT

Pestivirus	Subgenotype	Isolate	Species	Source[d]	Reference
BDV	Swiss-a	R9336/11	Cattle	IVV/IVI BE	[40]
BDV	Swiss-b	R4785/06/CH-BD4	Sheep	IVV/IVI BE	[44]
BDV	3	R1343/01/CH-BD1	Sheep	IVV/IVI BE	[36]
BDV	1a	Moredun	Sheep	P. Nettleton[a]	[72]
BVDV-1	1h	CH-04-01b	Cattle	IVV/IVI BE	[30]
BVDV-1	1e	CH-Maria	Cattle	IVV/IVI BE	[30]
BVDV-1	1k	CH-Suwa (ncp)[c]	Cattle	IVV/IVI BE	[30]
BVDV-1	1b	CH-04-05	Cattle	IVV/IVI BE	[30]
BVDV-1	1a	R1935/72 (cp)[c]	Cattle	IVV/IVI BE	[59]
BVDV-2	2a	890	Cattle	J. F. Ridpath[b]	[73]

[a]Moredun Research Institute, Edinburgh, Scotland
[b]National Animal Disease Center, Ames IA, USA
[c]*cp* cytopathogenic, *ncp* non-cytopathogenic
[d]IVV/IVI BE = Institute of Veterinary Virology/Institute of Virology and Immunology, Bern

above, with 506, 536, and 526 sera obtained in the years 2012, 2013, and 2014, respectively.

Cells

Bovine turbinate (BT) cells were prepared at the Institute of Veterinary Virology (University of Bern, Switzerland) from bovine fetuses obtained from a local abattoir and were maintained in Earle's minimal essential medium (E-MEM; Biochrom GmbH, Berlin, Germany) supplemented with 15% fetal calf serum (FCS) (2% during experiments), 100 U/ml penicillin, and 100 μg/ml streptomycin at 37 °C in a humidified 5% CO_2 atmosphere. FCS was free of pestivirus and antibodies to BVDV/BDV as tested by virus isolation and SNT, respectively. BT cells were found to be free of pestivirus by immunoperoxidase staining, and they were used to produce virus stocks, and to perform SNTs and virus (back-)titrations.

Production of challenge viruses used for SNT

Overall, 10 strains of ruminant pestiviruses from different subgenotypes were selected as challenge virus for the SNT (4× BDV, 5× BVDV-1 and 1× BVDV-2) (Table 1). All major subgenotypes hitherto isolated in Switzerland were represented by one isolate (BDSwiss-a, BDSwiss-b, BDV-3 and BVDV-1h, -1e, -1k, -1b) [37, 40]. With the exception of the North American strain Oregon C24 (R1935/72, BVDV-1a, [59]), all isolates were of the non-cytopathogenic (ncp) biotype. Each isolate was propagated in 150 cm^2 cell culture flasks (TPP AG, Trasadingen, Switzerland) seeded with 3 × 10^6 BT cells in 50 ml E-MEM with 15% FCS. One day post-seeding, the cells were infected for 1 h at a multiplicity of infection (moi) of 0.01 in 10 ml E-MEM with 7% FCS followed by the addition of 40 ml of E-MEM with 7% FCS. Cell infected with an ncp biotype of pestivirus were further incubated for 5 days at 37 °C and 5% CO_2, whereas the cells infected with the cytopathogenic (cp) BVDV-1a strain were harvested at the time when 80% of the cells showed signs of cytopathic effect (CPE) as judged by light microscopy. Virus stocks were obtained by freeze-thawing the cells at -20 °C and removal of cell debris by centrifugation for 15 min at 10,000g (HiCen® 21C, Hemotec GmbH, Gelterkinden, Switzerland). Aliquots of 0.5 ml were stored at -80 °C until use.

Homologous sera

For each of the 10 subgenotypes, a serum as homologous as possible was chosen. Eight out of 10 sera were Swiss field sera, whereas the immune sera to the genotypes BVDV-1a (strain R1935/72) and BVDV-2 were obtained from vaccine trials in Switzerland and Germany, respectively. The latter was raised against the BVDV-2 strain CS8644 [60] and was kindly provided by G. Wolf (LMU, Munich, Germany). For the subgenotype BDV-1a that was never detected in Switzerland to date, no homologous serum was available and, thus, we used a field serum from an antibody-positive heifer with unknown source of infection that displayed a rather high titer against BDV-1a. All sera were stored at -20 °C.

Serum neutralisation test

For the detection and quantification of pestivirus-specific neutralising antibodies in cattle sera, a serum neutralisation test (SNT), which is considered the gold standard in BVDV serology [61, 62], was developed and optimized for the current situation in Switzerland. Basically, the SNT was done according to the directions of the OIE [63]. Briefly, sera to be tested were pre-diluted tenfold in E-MEM with 2% FCS and inactivated for 30 min at 56 °C. In the cases were only an insufficient amount of serum was available for all experiments, the serum was pre-diluted 20- ($n = 46$) or 40-fold ($n = 6$).

Thereafter, sera were further diluted seven times in twofold steps up to a dilution of 1 in 1,280. Dilutions were directly done in 96-well plates with 4 wells per dilution and in a volume of 50 µl of serum per well. Afterwards, 100 tissue culture infectious dose 50 ($TCID_{50}$) of challenge virus in 50 µl E-MEM with 2% FCS per well were added and the plates were incubated for one hour at 37 °C in a humidified 5% CO_2 atmosphere. Subsequently, a cell suspension in E-MEM with 2% FCS with 20,000 cells in a volume of 100 µl was added to each well and further incubated for 4 to 5 days. For each experiment, a positive (α-BVDV-1b) and a negative control serum was included, and to control for possible serum cytotoxicity, 50 µl of serum of the first pre-dilution was added to another well in the absence of challenge virus. For evaluation of the ncp strains, immunoperoxidase staining was performed, whereas the neutralisation of the cp strain was directly quantified by analysis of the CPE by light microscopy. The neutralisation titer was calculated according to Spearman-Kaerber and expressed as reciprocal value of the dilution required for 50% of the wells exhibiting neutralisation of the challenge virus. Samples with a titer greater than 8 were rated as positive.

In order to differentiate the source of infection, we performed cross-neutralisation tests using different strains of ruminant pestiviruses as challenge virus in parallel SNTs. To identify the strains with the best discriminatory power, we screened various combinations of BVDV and BDV as challenge virus, whereby an at least fourfold difference of their SNT titers were regarded as significant [63]. Taking the samples with low volume (20-fold pre-dilution) into account, sera with BVDV and BDV titers lower than 15 were regarded as negative. A ratio of the BVDV- and BDV-titer of lower than 4 was considered as "indeterminate".

Virus titration

Virus titrations were performed according to the direction of the OIE [63]. Briefly, the virus suspensions were diluted seven times in tenfold steps in E-MEM with 2% FCS directly in 96-well plates with 6 wells per dilution using a volume of 50 µl per well. Six wells were used as cell control with the simple addition of medium. After addition of the cell suspension (20,000 cells per well in 100 µl E-MEM with 2% FCS), the plates were incubated for 4 to 5 days at 37 °C in a humidified 5% CO_2 atmosphere. Virus titers were calculated according to Spearman-Kaerber and presented as tissue culture infectious dose 50 ($TCID_{50}$) per ml. In every SNT, the amount of challenge virus applied (100 $TCID_{50}$) was controlled by back-titration in parallel to the SNT, and a variation of half a log level was considered as acceptable ($10^{1.5}$–$10^{2.5}$ $TCID_{50} = 32$ to 316) [63]. SNTs with the back-titration being outside of this range were repeated.

Immunoperoxidase staining

An immunoperoxidase staining was applied for cells infected with an ncp biotype of pestivirus. Cells were washed with PBS and subsequently thoroughly dried for at least one hour in the air flow of a safety cabinet. Thereafter, cells were fixed and permeabilised by incubation for two hours at 80 °C. After the plates regained room temperature, the primary antibody (polyclonal swine-α-BVDV hyperimmune serum prepared at the Institute of Veterinary Virology, University of Bern) was applied at a dilution of 1 to 750 in PBS with 5% Tween-20 (PBS-T) and incubated for 90 min at ambient temperature. After washing the cells three times with PBS-T, the secondary antibody (monoclonal peroxidase-labeled goat-α-swine IgG; KPL, Gaithersburg, MD, USA) at a dilution of 1 to 1,000 in PBS-T with 5% low-fat powdered milk was applied for 90 min at room temperature. Subsequently, cells were washed twice with PBS-T and once with distilled water, followed by the addition of the substrate solution (0.4 mg/ml 3-Amino-9-ethylcarbazole (AEC), 6% dimethylformamide, 0.3‰ H_2O_2 in 0.05 M sodium acetate at pH 5.0) and incubation for approx. 30 min until an adequate staining was observed. Staining was stopped by washing the cells with distilled water.

Case-control-study

In order to detect risk factors for infection with border disease virus in livestock, we performed a retrospective case-control-study in farms whose young stock was surveyed serologically in the BVD eradication program in Switzerland and the Principality of Liechtenstein [64]. Young stocks are used for sero-surveillance in non-dairy herds. Each year, a random sample of one third of all farms that did not have a PI animal within the last 24 months, and all herds that either had a PI animal or a conspicuous result in bulk milk testing within this time period, are tested serologically. A group of young stock is tested that consists of 5 calves within a herd of an age of at least 6 months, which were born later than September 30, 2009 and more than one month after the removal of the last known PI animal from the farm. In addition, these calves should have stayed on a given farm for at least 6 months and not previously had a known contact with a PI animal. Based on these criteria, 54 farms with at least two seropositive animals in the time interval from February 2014 to June 2015 were selected as potential case farms. Sera from such potential case farms were tested in the cross-SNT, and the farm was considered a definite case farm if the herd contained at least one single seropositive animal that was rated as infected with BDV. Potential case farms which proved negative for BDV in the cross-SNT were excluded. Control farms were selected out of a pool of 10,753 farms (data collection by the Swiss Veterinary Service) that were never tested seropositive for BVDV since the start of the eradication program in 2008 and that had tested at least 5 animals as seronegative since the beginning of 2012. Control farms were selected using the Swiss animal movement database [56]. Because only 16 definitive case farms that also returned the questionnaire (see below) could be identified, the number of controls was increased to reach a relation of at least three controls per case.

Data acquisition

To identify potential risk factors for the occurrence of BDV infections in cattle, variables such as presence of small ruminants, herd size, use and distribution of breed, animal movement (summer pasturing), common housing and pasture management, and external contact to (wild) ruminants, were collected by a standardized, bipartite questionnaire for cattle farmers. The first part of the questionnaire contained a total of 55 questions regarding the farm itself, contacts to animals on neighboring farms, or purchase and sale (animal movement) with special emphasis on possible direct and indirect exposure to small ruminants. The second part mainly aimed at receiving information on the origin of seropositive and seronegative animals in case and control farms, respectively. Additionally, information on the breeds of the animals was obtained from the animal movement database. The questionnaires were issued in German, French and Italian, and were distributed to the Cantonal Veterinary Services starting in the middle of August 2014. Official veterinarians or the farm veterinarians were instructed by the Cantonal Veterinary Services to perform the interviews that were conducted on-site or by phone.

Statistics

Data of the SNTs and the questionnaires were entered into the spreadsheet Excel® 2010 (Microsoft Corporation). The frequency distribution of the different potential risk factors in the 37 case and 280 control animals from 16 case and 56 control farms were statistically compared using the Chi-Square test for categorical variables and with the two-sample *t*-Test for normally distributed continuous variables, whereas the Mann-Whitney-*U*-Test was used for not normally distributed variables. After the univariable analysis of each individual risk factor according to the questionnaire, a multivariable logistic regression model with the potential risk factors ($p < 0.1$) was developed. In the model, clustering on the level of the farm was corrected with the GEE (generalized estimating equation) method [65]. None of the 13 identified potential risk factors were highly correlated with each other (correlation coefficient <0.7), therefore all of them were offered to the model. The final model was established by stepwise forward selection of the risk factors, until only significant factors ($p < 0.05$) and confounders remained in the model. Risk factors were considered as confounders and kept in the model if they altered the regression coefficient of another risk factor by more than 25%. The univariable and multivariable statistical analysis of the data was performed with NCSS (version 9; NCSS LLC, Kaysville UT) and with PROC GENMOD in SAS (version 9.4; SAS Institute Inc., Cary NC), respectively. Model fit was assessed with QIC and QICu fit criteria, as well as with visual assessment of the residuals.

Results

Distribution of field sera

Around three quarters (74.8%) of the 1,568 confirmed antibody-positive sera (see Methods section) were from cattle that were born prior to 2012 with 44% ($n = 512$) thereof in the year 2010. Animals at the age of 6 to 11 months at the time of examination represented the largest age group ($n = 347$, 22.1%) which is in accordance with the specifications of the «spot test» [55]. Overall, samples from 24 out of the 26 Swiss (half-) cantons and of the Principality of Liechtenstein were at our disposition, and the sample distribution was in good agreement (Fig. 1) with the cattle density per km^2 per locality (based on postal code) in Switzerland.

Cross-neutralisation of the selected BDV and BVD strains

For the selection of the most appropriate BD and BVD virus strains to be used in the cross-SNT, we tested all possible combinations of the 10 isolates of ruminant pestiviruses (BDV, BVDV-1, BVDV-2; Table 1). For each subgenotype, we selected a serum as homologous as possible and titrated them in the SNT with all the pestivirus strains (see Additional file 1: Table S1). With only few exceptions, the highest titers were detected with the homologues pairs of virus and serum. The highest titer of 5,120 was observed with the anti-BVDV-1e serum and the corresponding virus strain CH-Maria. The only exceptions were the sera raised against BDSwiss-a, BDV-1a, and BVDV-1k, which reacted equally or even stronger in response to other virus isolates than their homologous virus strain. As expected [26, 66], the anti-BVDV-2a serum exhibited a low (average titer = 48) and a medium (average titer = 135)

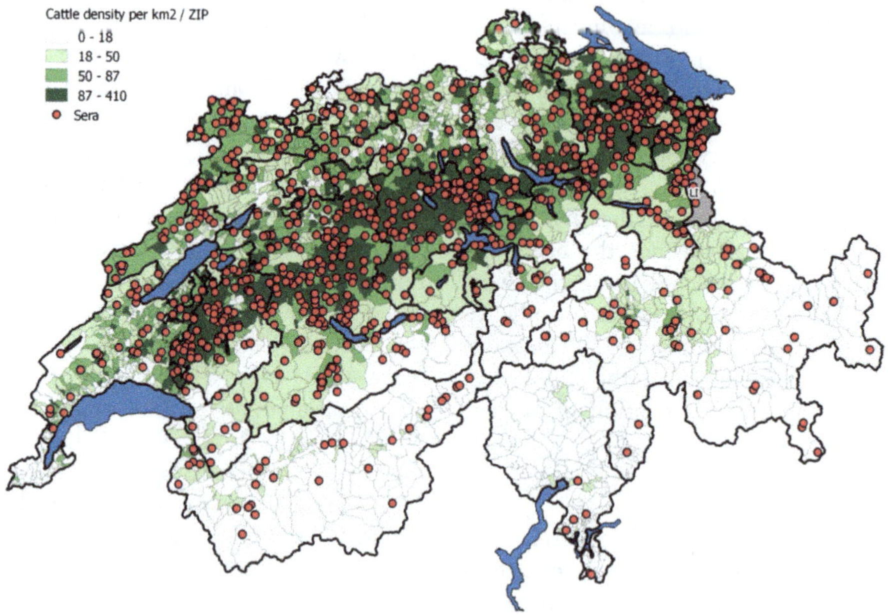

Fig. 1 Geographic distribution of field sera and density of the cattle population in Switzerland and of the Principality of Liechtenstein (LI). The map was created with the geographic information system QGIS 2.8.1-Wien. The number of cattle per ZIP (postal code; data 2014) were obtained from the Swiss Federal Statistical Office (https://www.pxweb.bfs.admin.ch); n (Farms providing sera) = 898

neutralisation reactivity against BDV and BVDV-1, respectively. Conversely, the anti-BDV sera poorly (mean titer = 19) and moderately (mean titer = 164) neutralized the BVDV-2a strain 890 and the BVDV-1 strains, respectively. All BVDV-1 sera except the one directed against BVDV-1h only weakly neutralized BD viruses (mean titer = 47), whereas their cross-neutralisation was rather high against heterologous BVDV-1 strains (mean titer = 523). Finally, sera directed against BDV showed a low and medium neutralisation activity towards BVDV-1 strains (mean titer = 57) and heterologous BD viruses (mean titer = 192), respectively.

Coefficients of antigenic similarity (R)

The antigenic relatedness of a pair of viruses was determined by calculating the coefficient of antigenic similarity (R) [67] according to the following formula:

$$R = 100 \times \sqrt{\frac{\text{titer virus isolate A with serum B} \times \text{titer virus isolate B with serum A}}{\text{titer virus isolate A with serum A} \times \text{titer virus isolate B with serum B}}}$$

A value of $R \leq 25$ is considered to denote a significant antigenic difference between the virus strains [24, 30]. All possible combinations of BD and BVD viruses displayed an *R*-value lower than 25, which is indicative of a significant antigenic difference between two strains. The largest differences with *R*-values lower than two were observed between BDV and BVDV-1 strains, whereas the four strains of BDV displayed the highest *R*-values among each other (Table 2) with no significant difference between any two strains.

Selection of the pestivirus strains used in cross-neutralisation of field sera

The quotients of titers of the cross-neutralisation assays of the 10 sera with the four BD and the six BVD viruses were calculated for all possible pairwise combinations (Table 3), with the absolute numbers of the SN titers listed in supplementary Additional file 1: Table S1. Values of the quotient below 4 were considered as indeterminate. All antisera against BDV were correctly assigned with 4 different combinations of virus pairs (BVDV-1a and BDSwiss-a; BVDV-2a and BDSwiss-a or BDSwiss-b; BVDV-1k and BDSwiss-a). The six BVDV antisera were correctly classified in three pairwise combinations of the subgenotype BVDV-1h, e.g., with the strains BDSwiss-a, BDSwiss-b and BDV-3. Overall, the virus pair BVDV-1a and BDSwiss-a displayed the best differentiation of all sera, with correct classification of 8 out of 10 sera in the cross-neutralisation assays. By adding the results of the BVDV-1h strain CH-04-01b in combination with the BDV strain Swiss-a, both remaining BVD-antisera (anti-BVDV-1h and anti-BVDV-2) could be properly assigned. Therefore, the pestivirus strains BVDV-1a, BVDV-1h and BDSwiss-a were selected as being optimally suited for the differentiation between BVDV and BDV as source of infection in the current epidemiological situation in Switzerland.

Table 2 Coefficients of antigenic similarity *(R)* between BDV and BVDV isolates

BDV & BVDV isolates	BDV Swiss-a	BDV Swiss-b	BDV–3	BDV-1a	BVDV–1h	BVDV–1e	BVDV–1k	BVDV–1b	BVDV–1a	BVDV–2a
BDSwiss-a	**100**	77.1	40.3	71.0	*19.2*	*≤ 2.3*	*≤ 13.6*	*6.0*	*≤ 2.2*	*4.8*
BDSwiss-b		**100**	29.8	73.8	*10.1*	*1.8*	*19.2*	*5.5*	*1.6*	*3.0*
BDV-3			**100**	59.5	*14.8*	*3.7*	*15.6*	*5.5*	*3.0*	*3.4*
BDV-1a				**100**	*20.1*	*4.8*	*≤ 23.8*	*12.0*	*2.5*	*6.7*
BVDV-1h					**100**	*13.6*	52.3	32.4	*12.5*	*11.5*
BVDV-1e						**100**	*22.9*	*13.1*	*7.1*	*6.5*
BVDV-1k							**100**	52.3	28.5	*11.5*
BVDV-1b								**100**	*21.0*	*5.0*
BVDV-1a									**100**	*3.6*
BVDV-2a										**100**

R-Values ≤ 25 indicate significant antigenic differences between two isolates (in italics). For nondescript titers (≤ 14), the value of 14 was taken for calculation and, accordingly, the *R*-values are marked with "≤"

Table 3 Ratios of the cross-SNT titers of 10 sera tested with all combinations of BDV and BVDV isolates

Sera	Virus pair											
	BVDV-1h				BVDV-1e				BVDV-1k			
	BD Swiss-a	BD Swiss-b	BDV-3	BDV-1a	BD Swiss-a	BD Swiss-b	BDV-3	BDV-1a	BD Swiss-a	BD Swiss-b	BDV-3	BDV-1a
α-BDSwiss-a	4.4	4.4	**1.9**	**2.0**	5.1	5.1	**2.2**	**2.4**	8.8	8.8	**3.7**	4.1
α-BDSwiss-b	**4.0** [a]	6.7	**1.8**	**2.2**	5.6	9.5	**2.6**	**3.1**	8.8	14.7	4.0	4.8
α-BDV-3	**1.1**	**0.9**	**2.8**	**0.9**	**1.5**	**1.3**	4.0	**1.3**	4.4	**3.7**	11.4	**3.7**
α-BDV-1a	**1.7**	**2.6**	**1.7**	**1.5**	**3.1**	4.8	**3.1**	**2.8**	11.4	17.6	11.4	10.5
α-BVDV-1h	6.2	14.7	16.0	16.0	**0.6**	**1.5**	**1.7**	**1.7**	**0.5**	**1.1**	**1.2**	**1.2**
α-BVDV-1e	64.6	60.3	32.3	26.6	365.7	341.3	182.9	150.6	11.4	10.7	5.7	4.7
α-BVDV-1k	22.7	6.8	13.3	6.2	10.4	**3.1**	6.1	**2.8**	6.2	**1.8**	**3.6**	**1.7**
α-BVDV-1b	22.7	19.0	26.7	16.0	10.4	8.7	12.2	7.3	4.8	4.0	5.6	**3.4**
α-BVDV-1a	38.1	10.4	15.9	24.5	17.5	4.8	7.3	11.3	19.1	5.2	7.9	12.3
α-BVDV-2a	8.0	4.3	8.0	**3.4**	7.3	**4.0**[a]	7.3	**3.1**	**1.7**	**0.9**	**1.7**	**0.7**

Sera	Virus pair											
	BVDV-1b				BVDV-1a				BVDV-2a			
	BD Swiss-a	BD Swiss-b	BDV-3	BDV-1a	BD Swiss-a	BD Swiss-b	BDV-3	BDV-1a	BD Swiss-a	BD Swiss-b	BDV-3	BDV-1a
α-BDSwiss-a	5.1	5.1	**2.2**	**2.4**	8.8	8.8	**3.7**	4.1	8.2	8.2	**3.5**	**3.8**
α-BDSwiss-b	4.3	7.3	**2.0**	**2.4**	34.5	58.1	15.8	18.8	22.6	38.0	10.3	12.3
α-BDV-3	**2.0**	**1.7**	5.2	**1.7**	4.4	**3.7**	11.4	**3.7**	6.2	5.2	16.0	5.2
α-BDV-1a	**2.0**	**3.1**	**2.0**	**1.8**	11.4	17.6	11.4	10.5	10.7	16.5	10.7	9.8
α-BVDV-1h	**1.5**	**3.7**	4.0	4.0	**0.6**	**1.4**	**1.5**	**1.5**	**0.5**	**1.3**	**1.4**	**1.4**
α-BVDV-1e	32.4	30.2	16.2	13.3	24.9	23.3	12.5	10.3	11.4	10.7	5.7	4.7
α-BVDV-1k	19.1	5.7	11.2	5.2	6.2	**1.8**	**3.6**	**1.7**	**2.6**	**0.8**	**1.5**	**0.7**
α-BVDV-1b	54.0	45.3	63.5	38.1	8.0	6.7	9.4	5.7	**3.7**	**3.1**	4.3	**2.6**
α-BVDV-1a	69.8	19.1	29.1	45.0	234.8	64.3	97.8	151.5	5.7	**1.5**	**2.4**	**3.6**
α-BVDV-2a	**2.0**	**1.1**	**2.0**	**0.8**	**2.8**	**1.5**	**2.8**	**1.2**	53.5	29.1	53.5	22.7

For the calculation of the ratios of the α-BD sera, the titer of BDV was taken as numerator and the titer of BVDV as denominator. For the calculation of the ratios of the α-BVD sera the reverse ratio was used. Ratios < 4 are highlighted in bold

[a]The exact value was slightly below 4.0

Neutralisation and cross-neutralisation of field sera

In order to differentiate BVDV and BDV as source of infection, we employed the cross-SNT with the three challenge strains BVDV-1a, BVDV-1h, and BDSwiss-a as described above. From 1,568 samples that were initially confirmed to be positive for pestivirus antibodies and that originated from 898 farms, we were able to analyze 1,555 in the cross-SNT. This includes the six results to the BVDV-1a strain from the initial SNT that was applied by the reference laboratory for samples where not sufficient material was available to perform full cross-neutralisation.

Both, the average titer (geometric mean titer (GMT) = 346.3; median = 495.5) of the field sera and the neutralisation titer of the majority of the individual samples (1,394 out of 1,549 samples) were higher towards the BVDV-1h isolate than the ones towards the BVDV-1a strain (GMT = 125.5; median = 135). The average (GMT = 74.8; median = 80) as well as most of the single titers to BDSwiss-a were clearly lower than the values to the BVD viruses (Fig. 2a). The highest titer (≥1,810) was reached with 153, 14, and one sera against BVDV-1h, BVDV-1a, and BDV, respectively. Most sera with negative neutralisation titers (≤ 8) were observed using BDV in the SNT ($n = 257$).

Identification of the source of infection by cross-SNT

By combining the results of the SNTs with the virus strains of the subgenotype BVDV-1a and BDSwiss-a, we were able to differentiate the source of infection between BVDV and BDV in 550 out of 1,555 samples (35.4%). By adding the BVD/BD-quotient with the BVDV-1h strain to the remaining samples (i.e., 870 samples that could not be assigned using only the BVDV-1a strain, and to 135 samples that were rated as negative), further 666 sera could be sourced to an infection with BVDV (Table 4). In 93% of all BVD-positive cases, the sera had a higher titer towards BVDV-1h than BVDV-1a. Conversely, for samples that were assigned to an infection with BDV, their low reactivity towards the BVDV-1a strain was pivotal for their differentiation from BVDV (Fig. 2b). In no instance a contradictory combination of the results was observed. Overall, the majority of samples were assigned to BVD ($n = 1{,}112$, 71.5%; CI 95%: 69.2–73.7%), and only 104 cattle sera (6.7%; CI 95%: 5.5–8.0%) were attributed to an infection with BDV. In 28 samples that could initially not be differentiated because they still exhibit full neutralisation even at the highest dilution, repetition of the experiments with more dilutions steps allowed their differentiation to an infection with BVDV ($n = 27$) or BDV ($n = 1$). The remaining sera could either not be differentiated ($n = 286$, 18.4%; CI 95%: 16.5–20.4%) or were regarded as negative ($n = 53$, 3.4%; CI 95% 2.6–4.4%) based on their low titers (titers ≤ 14).

BDV as source of infection: distribution of samples

The 104 cattle sera that were assigned to BDV as source of infection originated from 65 farms within 15 cantons (Fig. 3). A large part of the samples came from Central Switzerland ($n = 36$), in accordance with the observation that the two cantons with the highest prevalence of

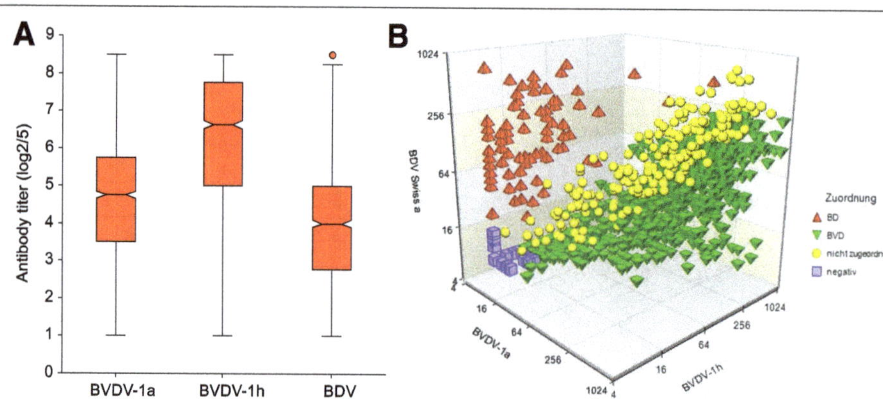

Fig. 2 Distribution of SNT titers of sera against BVDV-1a, BVDV-1h and BDV. **a** The box plots of titers of positive sera were created with the statistical software NCSS 9. Only positive titers (> 8) were included ($n_{BVDV-1a} = 1{,}400$; $n_{BVDV-1h} = 1{,}496$; $n_{BDV} = 1{,}292$). For titers with a value of ≤ 14 (20-fold pre-dilution), ≤ 28 (40-fold pre-dilution) and ≥ 1,810, the values 14, 28 and 1,810 were taken for calculation, respectively. The SNT titers are represented in the y-axis as logarithm to the base 2 including the standard 5-fold pre-dilution, i.e., multiplication with 5 yields the final titer. **b** The scatter plot of the sera samples ($n = 1{,}555$) based on their titers against BDSwiss-a, BVDV-1a and BVDV-1h was created with the statistical software NCSS 9. The black dotted line represents the threefold rotational axis on which each of the three titers per sample would have the same value and, therefore, no assignment to "BVD" or "BD" is possible (*yellow*). BDV-specific sera with low reaction against BVDV-1a are located in the upper left section of the cube (*red*), whereas BVDV-specific sera with high reaction against BVDV-1h are located in the lower right section of the cube (*green*). Negative sera are in the lower left corner (*purple*)

Table 4 Combinations of results from the cross-SNT tested with 3 isolates

Evaluation with BVDV-1a & BDV		Evaluation with BVDV-1h & BDV	Assignment[a]	$n =$	Percentage of combination per assignment [%]	Proportion overall [%]
BVD	&	BVD	BVD	428	38.5	71.5
BVD	&	Negative		4	0.4	
BVD	&	Indeterminate		13	1.2	
Indeterminate	&	BVD		594	53.5	
Negative	&	BVD		72	6.5	
BVD		n.d.[b]		1	0.1	
BD	&	BD	BD	68	65.4	6.7
BD	&	Indeterminate		32	30.8	
BD		n.d.[b]		1	3.8	
Indeterminate	&	Indeterminate	Indeterminate	276	96.5	18.4
Negative	&	Indeterminate		10	3.5	
Negative	&	Negative	Negative	52	98.1	3.4
Negative		n.d.[b]		1	1.9	
Total				1,555		100

[a]Sera with a ratio of ≥ 4 were assigned as "BVD" or "BD", sera with a ratio < 4 were "indeterminate", and sera with both titers < 15 are "negative"
[b]*n.d* not done, due to insufficient amount of material

BDV-reactive sera (64.7 and 32.4%) were from this area. Notably, in three of these farms in Central Switzerland, calves that were persistently infected with BDV were detected within the scope of the Swiss BVD eradication program (H.P. Stalder, personal communication and [37]). In 9 cantons and in the Principality of Liechtenstein, we could not detect any BDV positive sample within the 206 sera that we analyzed.

In 21 out of 29 farms where a BDV-specific sample and more than one positive serum in the cross-SNT was obtained, all of the samples were specific for BDV as source of infection. From the remaining farms, four had samples that could not be differentiated or were negative, whereas in the other four, BVDV could also be identified as source, sometimes in combination with indeterminate samples.

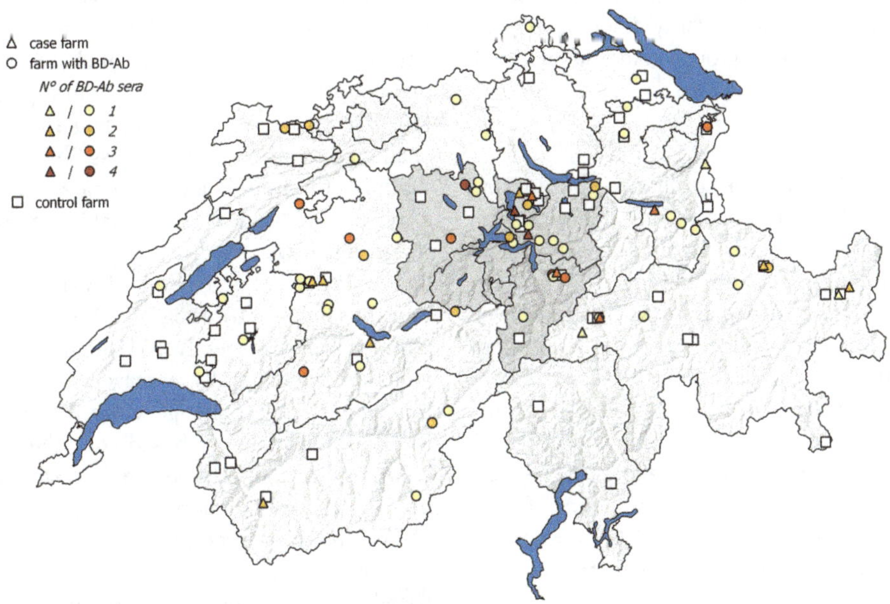

Fig. 3 Farms with BDV-positive sera and case and control farms in Switzerland and the Principality of Liechtenstein (LI). The location of farms with BDV-positive sera (*circles*) including the definitive case farms (*triangle*) and the control farms (*squares*) are presented. The number of BDV-specific sera from one to four is displayed by a color gradient from yellow to red as indicated in the figure

According to information retrieved from the Swiss animal movement database [58], small ruminants were present on 44 farms (68%) where BDV was identified as source of infection in at least one sample. Thereof, 27 (61%) of these farms housed sheep, 4 (9%) goats, and 13 (30%) both, sheep and goats. By contrast, only a third of the farms without BDV-positive samples ($n = 259$, 31%) kept small ruminants according to information obtained from the animal movement database.

Selection of farms for the case-control-study

We obtained detailed data on 54 potential case farms that were recruited between February 2014 and June 2015 based on the serological screening results using a standardized questionnaire. BDV was later identified by means of cross-SNT as the source of infection in 37 samples from 16 farms that were distributed across 7 cantons and the Principality of Liechtenstein (Fig. 3). Additional 3 samples from these definitive case farms were indeterminate and one was negative. In no case farm could we detect any BVDV-specific serum.

Descriptive, univariable statistics

By means of Pearsons's chi square test, we identified significant differences ($p < 0.1$) in 9 out of 50 categorical variables and, using Mann-Whitney-*U*-test or *t*-test, in 4 out of 5 continuous variables (Additional file 2: Table S2a and S2b). The variables sheep farming, sheep breed, and origin of cattle display the lowest *p* values ($p < 0.0001$). Within the continuous variables, the number of cattle was considerably ($p = 0.0006$) and the number of sheep, goats, and loss of lambs was fairly different ($p = 0.1$) between the case and control groups (Additional file 3: Table S3).

Logistic regression

Variables that were significant according to the univariable analysis and additional variables of epidemiological relevance regarding animal movement (purchase, provenance) were introduced into the model of logistic regression by stepwise forward selection. At the farm level, the risk factors "same stable" (sheep and cattle are kept in the same stable; OR (odds ratio) = 167.23, $CI_{95\%}$: 15.37–1,819.29, $p < 0.0001$) and "cattle purchase" (purchase of cattle within the last 12 months; OR = 9.57, $CI_{95\%}$: 1.08–84.95; $p = 0.0426$), and at the level of the individual animal, the risk factor "cattle provenance" (cattle was purchased; OR = 4.16, $CI_{95\%}$: 1.64–10.60; $p = 0.0028$) could be confirmed as significant in the final model.

Discussion

In this study, we investigated the frequency of BDV infections in cattle and evaluated their possible influence on the serological surveillance in the Swiss BVD eradication program. Thus, we selected one strain out of each major subgenotype of ruminant pestiviruses detected in Switzerland so far [37] in addition to the two strains previously used for differentiation in routine diagnostic, i.e., R1935/72 (BVDV-1a) and Moredun (BDV-1a). In order to choose the most appropriate challenge viruses in the cross-SNT to differentiate BVDV from BDV infections, all possible pairwise combinations out of the 6 BVDV and 4 BDV strains were tested (Table 1) together with sera that were as homologous as possible to the corresponding virus isolates. The combination of BVDV-1a and BDSwiss-a turned out to be an optimal combination that correctly assigned 8 out of 10 test sera (Table 3), including all BDV-specific samples, and that yielded the highest ratios of the neutralisation titers (mean = 33.6). In addition, by analyzing the effect of varying the dose of challenge virus employed in the SNT (10^2 $TCID_{50} \pm$ 0.5 log as acceptable range for the challenge viruses) on the SN titers essentially confirmed the usefulness of a quotient of four used in classic serology as a threshold for significance (not shown). And notably, the use of cytopathogenic strains such as the strain R1935 (BVDV-1a) was advantageous as it allowed for direct microscopical evaluation of the results. In field situations, however, only non-cytopathogenic strains are available and, thus, the need for fixation and immunostaining of the cells cannot be avoided.

In contrast to the test sera, the combination of BVDV-1a and BDSwiss-a was not able to differentiate between BVDV and BDV as the source of infection in more than half of the field sera (55.9%) (Table 4). In addition to the BVDV-1a subgenotype, which was never found to circulate in Switzerland, we thus included a strain of the in our country most commonly found genotype BVDV-1h [30, 37] into the analysis. Using these three strains as challenge viruses in the cross-SNT, we were able to diminish the rate of indeterminate samples to 18.4%. As a result, we could considerably improve the former triage which was used in the laboratory that applied only two strains (BVDV-1a and BDV-1a Moredun) and that was thus unable to assign 31.0 and 66.4% of the sheep and goat sera, respectively [18]. In the case of a BVDV infection, we generally observed higher neutralisation titers against the challenge virus BVDV-1h than towards BVDV-1a, which is based on the higher antigenic homology of the field sera to a strain circulating in the Swiss cattle population. But the exclusive use of the BVDV-1h strain in combination with BDSwiss-a was not favorable as it only poorly identified BDV infections. Based on the quotients of the neutralisation titers observed (Table 3), further reduction of the number of indeterminate sera proved to be difficult. In case of acute infections in the field, direct virus isolation and partial sequencing of the viral genome would be required [37]. Nonetheless, the adaptation of the serological triage system to the current epidemiological situation proved to be an important prerequisite for optimal selectivity to differentiate between ruminant pestivirus infections.

A small number of samples (3.4%) that were initially confirmed to be antibody positive by ELISA could not be confirmed in the cross-SNT (Table 4). In spite of the significant correlation between the SNT and the ELISA results (not shown), the two assays are inherently different. Thus, the epitopes important for the neutralisation assay are mainly part of the E2 envelope glycoprotein in its native conformation, whereas the "in-house" ELISA detects primarily antibodies directed to the conserved non-structural protein NS2-3 [57, 62, 68].

The results of the cross-SNTs comprising the three challenge viruses BVDV-1a, BVDV-1h, and BDSwiss-a, clearly demonstrated that the majority of pestivirus infections in cattle discovered 4 to 7 years after the start of the eradication can be ascribed to BVDV (71.5%), whereas only 6.7% were caused by infections with BD viruses. Nevertheless, it has to be considered that at least some of the indeterminate sera (18.4%) might also have BDV as source of infection. The BDV seroprevalence within the pestivirus-antibody positive animals slightly increased between 2012 and 2014 from 4.2 to 8.1% (not shown), whereas the overall seroprevalence in the Swiss cattle population continuously decreases [69]. These results point to a potential interference by small ruminants, in particular sheep, with the eradication of BVDV from cattle. Interestingly, small ruminants were kept – according to the entries in the animal movement database – on all case farms and on the majority of farms where BDV infections were detected by serology, with sheep being more prominent than goats. Notably, sheep flocks with more than 100 animals were exclusively found within the case farms. The highest portion of BDV-specific sera ($n = 36$; 34.6%) as well as the highest BDV seroprevalence (19% of all seropositive samples) were obtained from Central Switzerland (Fig. 3) while the number of animals and of samples in this region are within average of all greater areas of Switzerland (Fig. 1).

By means of a logistic regression model to determine the influence of potential risk factors for an infection with BDV, we provided evidence for the impact of small ruminants. Common indoor housing of sheep and cattle was identified as the risk factor with the highest odds ratio in the final model (OR = 167.23). Further variables with significant but smaller influence were the purchase of cattle on the level of the farm (purchase within the last 12 month) (OR = 9.57) and on the level of the individual animal (OR = 4.16). As the sample size of seropositive farms was limited, the power of the study was not sufficient to identify risk factors with weak association. Nonetheless, the final model confirms previous reports that pointed to the relevance of common housing of sheep and cattle as a main risk factor [19, 23, 44, 70]. In particular, sheep persistently infected with BDV pose a considerable risk when housed together with BVDV-free livestock. In the course of a BVD eradication scheme, the importance of interspecies transmission might increase with the continuous decrease in antibody seroprevalence in cattle and their ensuing increase in susceptibility to (re-)infection with pestiviruses [19].

Conclusion

Collectively, our study proposes that farmers with common housing of cattle and sheep should be aware of interspecies virus transmission, especially during lambing, where a high infection pressure exists [71]. In situations where contact between cattle and sheep cannot be avoided or minimized, surveillance of pestiviruses in sheep might be considered [46]. The Swiss eradication program encompasses only bovines, but not sheep and goats. Thus, the mean BDV seroprevalence in pestivirus-antibody positive cattle of at least 6.7% with an increasing trend between 2012 and 2014 indicates that the serological surveillance by ELISA, which does not differentiate BVDV from BDV infections, might be critical. Even though discrimination by cross-SNT as described in this study is laborious, it adds to classical epidemiological investigations and allows the identification of possible sources of infection, which is of particular importance in the late phase of an eradication program [37, 48]. In summary, we determined for the first time the prevalence of BDV in pestivirus-positive cattle in Switzerland, and we provide strong evidence that common housing of cattle and sheep is the most significant risk factor for the interspecies transmission of BD virus from small ruminants to cattle.

Additional files

Additional file 1: Antibody titer of 10 sera against homologous and heterologous BDV and BVDV isolates. Cross-neutralisation titers of 10 sera against five, four, and one BVDV-I, BDV, and BVDV-II strain, respectively.

Additional file 2: Categorical risk factors with significant differences. Categorical risk factors with significant differences between case (BDV-seropositive) and control (seronegative) farms (Additional file 2: Table S2a) and, on the animal level, between case (BDV-seropositive) and control (seronegative) farms (Additional file 2 Table S2b).

Additional file 3: Continuous risk factors with significant differences between case (BDV-seropositive) and control (seronegative) farms. Within the continuous variables, differences between the number of cattle, sheep, goats, and loss of lambs between the case and control groups.

Abbreviations

Ab: Antibody; AEC: 3-amino-9-ethylcarbazole; BDV: Border disease virus; BT cells: Bovine turbinate cells; BVDV: Bovine virus diarrhea virus; CI: Confidence interval; cp: Cytopathogenic; CSFV: Classical swine fever virus; ELISA: Enzyme-linked immunosorbent assay; E-MEM: Earle's minimal essential medium; FCS: Fetal calf serum; moi: Multiplicity of infection; n.d: Not done; ncp: Non-cytopathogenic; N^{pro}: N-terminal protease; OIE: Office International des Epizooties (World Organisation for Animal Health); OR: Odds ratio; PBS: Phosphate buffered saline; PBS-T: PBS with 5% Tween-20; PI: Persistently infected; SNT: Serum neutralisation test; $TCID_{50}$: Tissue culture infectious dose 50; UTR: Untranslated region

Acknowledgements
Special thanks go to Lupe Camina and Antoinette Golomingi for the preparation of the bovine turbinate cell suspension, to Claudia Bachofen for advice to compile the questionnaire and the help in finding appropriate challenge virus strains and corresponding sera, to Georg Wolf for providing the α-BVDV-2 serum, and to Peter Nettleton and Julia Ridpath for providing BDV-1a and BVDV-2 isolates, respectively. The support by Elena Di Labio and Patrick Schaller of the Federal Food Safety and Veterinary Office in providing data on the Swiss BVD eradication program and in helping to establish the list of potential control farms, and by the persons in charge of BVD of the Cantonal Veterinary Services and the Veterinary Service of the Principality of Liechtenstein, and the assigned (official) veterinarians in the collection of data on the case and control farms is highly appreciated. Last but not least, we thank Eveline Kindler for critically reading the manuscript.

Funding
This research was supported by the Federal Food Safety and Veterinary Office (grant 1.14.09 to RGZ, MS and VK).

Authors' contributions
RGZ and MS designed the study; VK with the aid of LN performed the experiments; GSR helped to design the questionnaire and to perform statistical analyses; VK, RGZ, and MS analyzed the data and wrote the paper. All authors critically read the manuscript and approved the final version.

Competing interests
The authors declare that they have no competing interests.

Consent for publication
Not applicable.

Author details
[1]Institute of Virology and Immunology, Federal Food Safety and Veterinary Office (FSVO) and Vetsuisse Faculty, University of Bern, Laenggass-Strasse 122, POB, CH-3001 Bern, Switzerland. [2]Veterinary Public Health Institute, Vetsuisse Faculty, University of Bern, Schwarzenburgstrasse 155, CH-3097 Liebefeld, Switzerland.

References

1. Liu L, Xia H, Wahlberg N, Belák S, Baule C. Phylogeny, classification and evolutionary insights into pestiviruses. Virology. 2009;385:351–7.
2. Becher P, Thiel H-J. Genus *Pestivirus (Flaviviridae)*. In: Tidona CA, Darai G, editors. The springer index of viruses. 2nd ed. Heidelberg: Springer Verlag; 2011. p. 483–8.
3. Ridpath JF. Emerging pestiviruses infecting domestic and wildlife hosts. Anim Health Res Rev. 2015;16:55–9.
4. Hause BM, Collin EA, Peddireddi L, Yuan FF, Chen ZH, Hesse RA, et al. Discovery of a novel putative atypical porcine pestivirus in pigs in the USA. J Gen Virol. 2015;96:2994–8.
5. Arruda BL, Arruda PH, Magstadt DR, Schwartz KJ, Dohlman T, Schleining JA, et al. Identification of a divergent lineage porcine pestivirus in nursing piglets with congenital tremors and reproduction of disease following experimental inoculation. PLoS ONE. 2016;11:e0150104.
6. Nettleton PF, Entrican G. Ruminant pestiviruses. Br Vet J. 1995;151:615–42.
7. Houe H. Epidemiological features and economical importance of bovine virus diarrhoea virus (BVDV) infections. Vet Microbiol. 1999;64:89–107.
8. Moennig V, Houe H, Lindberg A. BVD control in Europe: current status and perspectives. Anim Health Res Rev. 2005;6:63–74.
9. Müller-Doblies D, Arquint A, Schaller P, Heegaard PMH, Hilbe M, Albini S, et al. Innate immune responses of calves during transient infection with a noncytopathic strain of bovine viral diarrhea virus. Clin Diagn Lab Immunol. 2004;11:302–12.
10. Brodersen BW. Bovine viral diarrhea virus infections: manifestations of infection and recent advances in understanding pathogenesis and control. Vet Pathol. 2014;51:453–64.
11. Peterhans E, Schweizer M. BVDV: a pestivirus inducing tolerance of the innate immune response. Biologicals. 2013;41:39–51.
12. Lanyon SR, Hill FI, Reichel MP, Brownlie J. Bovine viral diarrhoea: pathogenesis and diagnosis. Vet J. 2014;199:201–9.
13. Schweizer M, Peterhans E. Pestiviruses. Annu Rev Anim Biosci. 2014;2:141–63.
14. Nettleton PF, Gilray JA, Russo P, Dlissi E. Border disease of sheep and goats. Vet Res. 1998;29:327–40.
15. García-Pérez AL, Minguijon E, Estevez L, Barandika JF, Aduriz G, Juste RA, et al. Clinical and laboratorial findings in pregnant ewes and their progeny infected with border disease virus (BDV-4 genotype). Res Vet Sci. 2009;86:345–52.
16. Hilbe M, Camenisch U, Braun U, Peterhans E, Stalder HP, Zlinszky K, et al. Mucosal lesions in a sheep infected with the Border Disease Virus (BDV). Schweiz Arch Tierheilkd. 2009;151:391–6.
17. Oğuzoğlu TC. A review of border disease virus infection in ruminants: molecular characterization, pathogenesis, diagnosis and control. Animal Health Product Hyg. 2012;1:1–9.
18. Danuser R, Vogt HR, Kaufmann T, Peterhans E, Zanoni R. Seroprevalence and characterization of pestivirus infections in small ruminants and new world camelids in Switzerland. Schweiz Arch Tierheilkd. 2009;151:109–17.
19. Krametter-Froetscher R, Duenser M, Preyler B, Theiner A, Benetka V, Moestl K, et al. Pestivirus infection in sheep and goats in West Austria. Vet J. 2010;186:342–6.
20. Passler T, Walz PH. Bovine viral diarrhea virus infections in heterologous species. Anim Health Res Rev. 2010;11:191–205.
21. Braun U, Hilbe M, Ehrensperger F, Salis F, Alther P, Strasser M, et al. Border disease in a flock of sheep. Schweiz Arch Tierheilkd. 2002;144:419–26.
22. Schaller P, Vogt HR, Strasser M, Nettleton PF, Peterhans E, Zanoni R. Seroprevalence of maedi-visna and border disease in Switzerland. Schweiz Arch Tierheilkd. 2000;142:145–53.
23. Braun U, Bachofen C, Schenk B, Hässig M, Peterhans E. Investigation of border disease and bovine virus diarrhoea in sheep from 76 mixed cattle and sheep farms in eastern Switzerland. Schweiz Arch Tierheilkd. 2013;155:293–8.
24. Becher P, Avalos-Ramirez R, Orlich M, Rosales SC, König M, Schweizer M, et al. Genetic and antigenic characterization of novel pestivirus genotypes: implications for classification. Virology. 2003;311:96–104.
25. Thabti F, Letellier C, Hammami S, Pepin M, Ribiere M, Mesplede A, et al. Detection of a novel border disease virus subgroup in Tunisian sheep. Arch Virol. 2005;150:215–29.
26. Ridpath JF, Fulton RW, Kirkland PD, Neill JD. Prevalence and antigenic differences observed between bovine viral diarrhea virus subgenotypes isolated from cattle in Australia and feedlots in the southwestern United States. J Vet Diagn Invest. 2010;22:184–91.
27. Nagai M, Ito T, Sugita S, Genno A, Takeuchi K, Ozawa T, et al. Genomic and serological diversity of bovine viral diarrhea virus in Japan. Arch Virol. 2001;146:685–96.
28. Couvreur B, Letellier C, Collard A, Quenon P, Dehan P, Hamers C, et al. Genetic and antigenic variability in bovine viral diarrhea virus (BVDV) isolates from Belgium. Virus Res. 2002;85:17–28.
29. Pizarro-Lucero J, Celedón MO, Aguilera M, De Calisto A. Molecular characterization of pestiviruses isolated from bovines in Chile. Vet Microbiol. 2006;115:208–17.
30. Bachofen C, Stalder HP, Braun U, Hilbe M, Ehrensperger F, Peterhans E. Co-existence of genetically and antigenically diverse bovine viral diarrhoea viruses in an endemic situation. Vet Microbiol. 2008;131:93–102.
31. Nagai M, Hayashi M, Itou M, Fukutomi T, Akashi H, Kida H, et al. Identification of new genetic subtypes of bovine viral diarrhea virus genotype 1 isolated in Japan. Virus Genes. 2008;36:135–9.

32. Minami F, Nagai M, Ito M, Matsuda T, Takai H, Jinkawa Y, et al. Reactivity and prevalence of neutralizing antibodies against Japanese strains of bovine viral diarrhea virus subgenotypes. Comp Immunol Microbiol Infect Dis. 2011;34:35–9.
33. Alpay G, Yeşilbağ K. Serological relationships among subgroups in bovine viral diarrhea virus genotype 1 (BVDV-1). Vet Microbiol. 2015;175:1–6.
34. Giammarioli M, Ceglie L, Rossi E, Bazzucchi M, Casciari C, Petrini S, et al. Increased genetic diversity of BVDV-1: recent findings and implications thereof. Virus Genes. 2015;50:147–51.
35. Decaro N, Lucente MS, Lanave G, Gargano P, Larocca V, Losurdo M et al. Evidence for circulation of bovine viral diarrhoea virus type 2c in ruminants in Southern Italy. Transbound Emerg Dis. 2016. In press. doi:10.1111/tbed.12592.
36. Stalder HP, Meier P, Pfaffen G, Wageck-Canal C, Rüfenacht J, Schaller P, et al. Genetic heterogeneity of pestiviruses of ruminants in Switzerland. Prev Vet Med. 2005;72:37–41.
37. Stalder HP, Hug C, Zanoni R, Vogt HR, Peterhans E, Schweizer M, et al. A nationwide database linking information on the hosts with sequence data of their virus strains: a useful tool for the eradication of bovine viral diarrhea (BVD) in Switzerland. Virus Res. 2016;218:49–56.
38. Dubois E, Russo P, Prigent M, Thiéry R. Genetic characterization of ovine pestiviruses isolated in France, between 1985 and 2006. Vet Microbiol. 2008;130:69–79.
39. Giammarioli M, La Rocca SA, Steinbach F, Casciari C, De Mia GM. Genetic and antigenic typing of border disease virus (BDV) isolates from Italy reveals the existence of a novel BDV group. Vet Microbiol. 2011;147:231–6.
40. Peterhans E, Bachofen C, Stalder HP, Schweizer M. Cytopathic bovine viral diarrhea viruses (BVDV): emerging pestiviruses doomed to extinction. Vet Res. 2010;41:44.
41. Braun U, Hilbe M, Janett F, Hässig M, Zanoni R, Frei S, et al. Transmission of border disease virus from a persistently infected calf to seronegative heifers in early pregnancy. BMC Vet Res. 2015;11:43.
42. Peletto S, Caruso C, Cerutti F, Modesto P, Zoppi S, Dondo A, et al. A new genotype of border disease virus with implications for molecular diagnostics. Arch Virol. 2016;161:471–7.
43. Bachofen C, Vogt HR, Stalder H, Mathys T, Zanoni R, Hilbe M, et al. Persistent infections after natural transmission of bovine viral diarrhoea virus from cattle to goats and among goats. Vet Res. 2013;44:32.
44. Braun U, Reichle SF, Reichert C, Hässig M, Stalder HP, Bachofen C, et al. Sheep persistently infected with Border disease readily transmit virus to calves seronegative to BVD virus. Vet Microbiol. 2014;168:98–104.
45. Cranwell MP, Otter A, Errington J, Hogg RA, Wakeley P, Sandvik T. Detection of border disease virus in cattle. Vet Rec. 2007;161:211–2.
46. Strong R, La Rocca SA, Ibata G, Sandvik T. Antigenic and genetic characterisation of border disease viruses isolated from UK cattle. Vet Microbiol. 2010;141:208–15.
47. Hornberg A, Fernández SR, Vogl C, Vilček Š, Matt M, Fink M, et al. Genetic diversity of pestivirus isolates in cattle from Western Austria. Vet Microbiol. 2009;135:205–13.
48. Krametter-Froetscher R, Benetka V, Rasser K, Tockner F, Moesslacher G, Moestl K, et al. BVDV control program in Austria - is a monitoring of the BDV status in sheep in Austria necessary? Vet Med (Praha). 2009;54:517–24.
49. Schirrmeier H, Strebelow G, Tavella A, Stifter E: Border disease virus infection in cattle - epidemiological and diagnostic impact. In: 7th ESVV Pestivirus Symposium. Uppsala (Sweden); 2008: 172.
50. McFadden AMJ, Tisdall DJ, Hill FI, Otterson P, Pulford D, Peake J, et al. The first case of a bull persistently infected with Border disease virus in New Zealand. N Z Vet J. 2012;60:290–6.
51. Braun U, Bachofen C, Büchi R, Hässig M, Peterhans E. Infection of cattle with Border disease virus by sheep on communal alpine pastures. Schweiz Arch Tierheilkd. 2013;155:123–8.
52. Greiser-Wilke I, Grummer B, Moennig V. Bovine viral diarrhoea eradication and control programmes in Europe. Biologicals. 2003;31:113–8.
53. Ståhl K, Alenius S. BVDV control and eradication in Europe - an update. Jpn J Vet Res. 2012;60:S31–9.
54. Presi P, Heim D. BVD eradication in Switzerland-a new approach. Vet Microbiol. 2010;142:137–42.
55. Bachofen C, Stalder HP, Vogt HR, Wegmüller M, Schweizer M, Zanoni R, et al. Bovine Virusdiarrhöe (BVD): von der Biologie zur Bekämpfung. Berl Munch Tierarztl Wochenschr. 2013;126:452–61.
56. Schärrer S, Widgren S, Schwermer H, Lindberg A, Vidondo B, Zinsstag J, et al. Evaluation of farm-level parameters derived from animal movements for use in risk-based surveillance programmes of cattle in Switzerland. BMC Vet Res. 2015;11:149.
57. Canal CW, Strasser M, Hertig C, Masuda A, Peterhans E. Detection of antibodies to bovine viral diarrhoea virus (BVDV) and characterization of genomes of BVDV from Brazil. Vet Microbiol. 1998;63:85–97.
58. Schwermer H, Bernet D, Presi P, Schaller P, Stern M, Heim D. Data management systems for the bovine viral diarrhoea eradication programme in Switzerland. Rev Sci Tech. 2013;32:741–50.
59. Steck F, Lazary S, Fey H, Wandeler A, Huggler C, Oppliger G, et al. Immune responsiveness in cattle fatally affected by bovine virus diarrhea-mucosal disease. Zentralbl Veterinarmed B. 1980;27:429–45.
60. Wolfmeyer A, Wolf G, Beer M, Strube W, Hehnen HR, Schmeer N, et al. Genomic (5'UTR) and serological differences among German BVDV field isolates. Arch Virol. 1997;142:2049–57.
61. Edwards S. The diagnosis of bovine virus diarrhoea-mucosal disease in cattle. Rev Sci Tech. 1990;9:115–30.
62. Sandvik T. Laboratory diagnostic investigations for bovine viral diarrhoea virus infections in cattle. Vet Microbiol. 1999;64:123–34.
63. Kirkland PD, Lung O, Drew T. Bovine viral diarrhoea. chapter 2.4.8. In: Manual of standards for diagnostic tests and vaccines for terrestrial animals office international des epizooties (OIE). 2015. p. 690–703.
64. Presi P, Struchen R, Knight-Jones T, Scholl S, Heim D. Bovine viral diarrhea (BVD) eradication in Switzerland-experiences of the first two years. Prev Vet Med. 2011;99:112–21.
65. Liang KY, Zeger SL. Longitudinal data analysis using generalized linear models. Biometrika. 1986;73:13–22.
66. Avalos-Ramirez R, Orlich M, Thiel H-J, Becher P. Evidence for the presence of two novel pestivirus species. Virology. 2001;286:456–65.
67. Archetti I, Horsfall FL. Persistent antigenic variation of influenza A viruses after incomplete neutralization *in ovo* with heterologous immune serum. J Exp Med. 1950;92:441–62.
68. Bachofen C, Bollinger B, Peterhans E, Stalder HP, Schweizer M. Diagnostic gap in Bovine viral diarrhea virus serology during the periparturient period in cattle. J Vet Diagn Invest. 2013;25:655–61.
69. Di Labio E. Stand der BVD-Ausrottung in der Schweiz. In: 9 Stendaler Symposium: Tierseuchenbekämpfung, Tierschutz und Tierarzneimittel bei Rindern. Stendal: 2015.
70. Carlsson U, Belák K. Border disease virus transmitted to sheep and cattle by a persistently infected ewe: epidemiology and control. Acta Vet Scand. 1994;35:79–88.
71. Lindberg A, Stokstad M, Løken T, Alenius S, Niskanen R. Indirect transmission of bovine viral diarrhoea virus at calving and during the postparturient period. Vet Rec. 2004;154:463–7.
72. Barlow RM. Experiments in Border disease: IV. Pathological changes in ewes. J Comp Pathol. 1972;82:151–7.
73. Bolin SR, Ridpath JF. Differences in virulence between two noncytopathic bovine viral diarrhea viruses in calves. Am J Vet Res. 1992;53:2157–63.

14

Primary surveys on molecular epidemiology of bovine viral diarrhea virus 1 infecting goats in Jiangsu province, China

Li Mao[1], Wenliang Li[1], Leilei Yang[1], Jianhui Wang[2], Suping Cheng[3], Yong Wei[2], Qiusheng Wang[3], Wenwen Zhang[1], Fei Hao[1], Yonglong Ding[3], Yinhua Sun[3] and Jieyuan Jiang[1*]

Abstract

Background: Bovine viral diarrhea virus (BVDV) is a pathogen of domestic and wildlife animals worldwide and is associated with several diseases. In China, there are many reports about genotyping of BVDV strains originated from cattle and pigs, and some of them focused on the geographical distributions of BVDV. Currently, the goat industry in Jiangsu province of China is under going a rapid expansion. Most of these goat farms are backyard enterprises and in close proximity to pig and cattle farms. However, there was very limited information about BVDV infections in goats. The objective of this study was to assess the frequency of BVDV infections of goats, the relationship of these infections to clinical signs and determine what BVDV genotypes are circulating in Jiangsu province.

Results: From 236 goat sera collected from six regions in Jiangsu province between 2011 and 2013, BVDV-1 was identified in 29 samples from the five regions by RT-PCR. The BVDV-1 infections occurred with/without clinical signs. Eight different BVDV-1 strains were identified from these positive samples based on the 5′-untranslated region (5′-UTR) sequences, and further clustered into four BVDV-1 subtypes on the phylogenetic analysis. Three were BVDV-1b, two BVDV-1m, two BVDV-1o, and one BVDV-1p, respectively.

Conclusions: To our knowledge, this is the first report to investigate the occurrence of BVDV and the genotypes of BVDV infecting goats in China. The results indicated that BVDV-1 infections were indeed present and the viruses were with genetic variations in Chinese goat herds. The information would be very useful for prevention and control of BVDV-1 infections in China.

Keywords: BVDV-1, Goats, Genotyping

Background

Bovine viral diarrhea viruses (BVDVs) are members of the genus *Pestivirus* (family *Flaviviridae*), and cause considerable economic losses in cattle industry worldwide [1–3]. Traditionally, two different BVDV species have been recognized as BVDV-1 and BVDV-2, and co-circulated in cattle, sheep, goats and pigs [4–7]. In recent years, several atypical bovine pestivirus strains named HoBi-like pestivirus were detected from contaminated fetal bovine sera and infected animals [8]. Phylogenetic analysis of these HoBi-like viruses showed a closer relationship to the recognized BVDV-1 and BVDV-2, and these viruses have been proposed as a new pestivirus species, BVDV-3 or HoBi-like pestivirus [8–11]. BVDV-1 and -2 infections are associated with several cattle diseases, including subclinical infections, immunosuppression, acute diarrhea, respiratory diseases, reproductive failures, and mucosal diseases in persistently infected calves [2]. Reproductive disorders caused by BVDVs appear with various clinical signs including infertility, abortion, stillbirth and persistently infected calves depending on the stage of pregnancy a cow is infected at [7, 12].

Genetic typing of the pestiviruses is based on different genomic regions, such as 5′-UTR, N^{pro}, and E2

* Correspondence: 1776556843@qq.com
[1]Institute of Veterinary Medicine, Jiangsu Academy of Agricultural Sciences; Key Laboratory of Veterinary Biological Engineering and Technology, Ministry of Agriculture, National Center for Engineering Research of Veterinary Bio-products, Nanjing 210014, China

genes [13–15]. The 5′-UTR is the most frequently analyzed portion of the genome [16], and provides meaningful phylogenetic inferences as this region has the highest degree of sequence conservation and efficient amplification by RT-PCR [13, 17]. Based on the 5′-UTR sequences, BVDV-1 strains have further been divided into 21 subtypes (1a-1u) [13, 17–20], BVDV-2 into four subtypes (2a-2d) [21, 22], and BVDV-3 species has also been divided into two genotypes of Brazilian and Thai origin [10].

Investigating on the molecular epidemiology of BVDV can provide invaluable information about the diversity of viruses in a population and, in turn, inform control programs, drive vaccine development and determine likely infection sources. So far, epidemiologic surveys on BVDV infections have been carried out in cattle, pig and yak populations in several provinces of China [19, 23–26]. These findings highlighted that corresponding measures should be taken for control and prevention against BVDV infections.

Recently, the goat industry has developed increasingly in Jiangsu province, China. There were over five million goats reared in the province during 2013 according to the official statistics. However, there were different sizes and kinds of the goat farms because the farmers had various economic conditions, most of them were backyard reared with various sizes and numbers, and some goat flocks were adjoined to pigs and cattle. The risk of cross infection might occur between cattle, pigs and goats, so we investigated the BVDV infections in the goat herds and further determined the subtypes of BVDVs on the basis of 5′-UTR genomic region amplification and sequencing.

Methods

Sampling

A total of 236 goat serum samples in the laboratory were from 31 farms in six regions of Jiangsu province, China during 2011 to 2013. Of the samples examined in this study, 71 were from goats with clinical signs of diarrhea or abortion, while the remaining 165 samples from the animals with no overt clinical sign. No BVDV vaccines have been used for animal immunization in Jiangsu province yet.

RNA extraction and RT-PCR detection

Total RNA was extracted from the serum samples using TRIzol Reagent (Invitrogen), and suspended in 20 μL of ultrapure water, following the manufacturer's recommendations. RT-PCR was carried out in a 50 μL reaction mixture containing 1× RT-PCR buffer (TAKARA, Bio, Inc.), 20 pM of Panpesti generic primers targeting 5′-UTR, 2 U of one-step Enzyme Mix (TAKARA, Bio, Inc.) and 4 μL of RNA for the expected product sizes of 290 bp [27], the reaction was run in a thermocycler (Mjmini, BIO-RAD) according to the following program: reverse transcription at 50 °C for 30 min, denaturation at 95 °C for 5 min, 35 cycles at 94 °C for 30 s, 54 °C for 30 s and 72 °C for 45 s, and termination with a final extension of 10 min at 72 °C. The amplification products were electrophoresed in 2 % agarose gels. Part of the interesting N^{pro} genes were amplified as the same method and the primer sequences were as followed: N1F: ATG GAGTTGATTTCAAATGAACT; N504R: GCAGCTT GAAACCCATAGAG, the expected product was 504 bp.

Sequencing

The expected bands were excised and recovered from the agarose gels using the Axygen gel extraction kit (Axygen, Hangzhou, China), and cloned into the pMD-18-T vectors (TAKARA, Bio, Inc.). The recombinant plasmids of the three positive bacteria were sequenced. The resultant gene sequences were compared with the sequences in the GenBank database using the Basic Local Alignment Search Tool (BLAST) and submitted to GenBank for confirmation of their identities.

Phylogenetic analysis

Sequence data were assembled and analyzed using the DNAStar version 7.0 package (DNAStar Inc., USA). Multiple sequences were aligned with the corresponding regions of BVDV-1, BVDV-2 and atypical BVDV reference sequences retrieved from GenBank by the ClustalW program. Phylogenetic reconstructions for genotyping were compiled using the 224 bp fragment of the 5′-UTR region and 504 bp of N^{pro} gene, phylogenetic trees were constructed by MEGA version 4.1. The reliability of the neighbor-joining tree was estimated by boostrap analysis with 1000 replicates.

Results

Sample detection

RNA extraction was performed directly from sera in order to exclude possible pestivirus contamination during virus isolation in cell culture. The 5′-UTR fragments, with expected size of 289–291 bp were positive detected out in 78 of the 236 samples (32.6 %, Table 1). The amplified fragments were purified, cloned and sequenced. The pestiviruses were further identified and the prevalence of BVDV was 12.3 % (29/236) (Table 1). The remaining 48 pestiviruses were identified as border disease viruses (BDV) and reported before [28]. One pestivirus strain to be further testified was not included now. The 29 positive BVDV sera were from most regions, including Nanjing, Xuzhou, Nantong, Zhenjiang and Huai'an, and no BVDV was detected in Suqian (Table 1). Fifteen BVDV positive goats had different clinical signs as three from Zhenjiang (2012) and six from

Table 1 The prevalence of pestivirus in Jiangsu province

Year of collection	Origin	Samples	Host	BVDV Positive	BVDV prevalence (%)	BDV positive	BDV prevalence (%)
2011	Nanjing	3	goat	2	66.7	0	0
2012	Xuzhou	55	goat	6	10.9	1	1.8
2012	Suqian	59	goat	0	0	1	1.7
2012	Nantong	105	goat	14	13.3	46	43.8
2012	Zhenjiang	7	goat	3	42.9	0	0
2013	Huai'an	5	goat	2	40	0	0
2013	Zhenjiang	2	goat	2	100	0	0
Total		236		29	12.3	48	20.3

Nantong with diarrhea, two from Nanjing and two from Zhenjiang (2013) suffering with abortion, and two weak lambs from Huai'an, the remaining 14 were clinically health (Table 2).

Sequence analysis

Based on the 5′-UTR sequences, 29 BVDV samples were classified as eight BVDV-1 strains (Table 2): the strains JS12/02 and JR1-2 were identified from two abortion goats, BH789 from animals with diarrhea, HA2-12 from six diarrhea and four healthy goats, XY-3 from two weakness kids, HA6-5 from four healthy goats, the strains XZ5-9 and XZ5-8 from healthy goats, respectively. The 5′-UTR sequences of these strains were deposited in GenBank under following accession number: KP749794- KP749800, KP749802 (Table 2), the interested N^{pro} sequences of HA2-12, JS12-02 and JR1-2 were as KX218370-218372.

The 5′-UTR sequence analysis of eight BVDV-1 strains as Table 3 showed 83.4–99.7 % homology with each other, and 83.1–88.9 % and 83.4–100 % identifies with the reference strains NADL (BVDV-1a) and VEDEVAC (BVDV-1b), respectively. The strains JR1-2, XZ5-8 and XZ5-9 from two regions (Zhenjiang and Xuzhou) shared very high identity of 100, 96.5 and 96.2 % with the BVDV vaccine strain VEDEVAC, respectively; BH789 (Zhenjiang) and HA6-5 (Nantong) showed a high identity of 95.5 and 94.8 % with the representative BVDV-1m strain ZM-95 originated from pigs in 5′-UTR, respectively. Compared with the reference strain IS25CP/01 (BVDV-1o), the strains JS12/02 (Nanjing) and HA2-12 (Nantong) shared nucleotide homology as low as 85.9 and 88.4 % homology respectively (Table 3). Based on the N^{pro} sequence, the strain JR1-2 was 98.2 % homology with the VEDEVAC (BVDV-1b), JS12/02 and HA2-12 had 97.6 and 98.2 % identity with the strain IS25CP/01, respectively.

Phylogenetic analysis

The 5′-UTR phylogenetic tree was created using their nucleotide sequences and 44 BVDV reference strains covering all subtypes retrieved from GenBank, and the eight detected BVDV-1 strains were further subdivided into four subtypes (Fig. 2): BVDV-1b (XZ5-9, XZ5-8, and JR1-2), BVDV-1p (XY-3), BVDV-1m (BH789 and HA6-5) and BVDV-1o (JS12/02 and HA2-12), respectively. The same result showed in N^{pro} phylogenetic tree including the strains HA2-12, JS12-02 and JR1-2 (Fig. 3).

BVDV-1b, BVDV-1m and BVDV-1o were detected in two regions, respectively; while BVDV-1p was only found from two weak goats in Huai'an (Table 2 and Figs. 1 and 2). Two strains XZ5-8 and XZ5-9 clustered into BVDV-1b were detected from Xuzhou. In Nantong, there were two circulating strains HA6-5 and HA2-12 that were classified into two different subtypes BVDV-1m and 1o; and two subtypes, BVDV-1m (BH789) and

Table 2 Details of the BVDV positive samples identified

Year of collection	Strains	Origin	Clinical signs (Quantity)	Subtypes	Accession numbers
2011	JS12/02	Nanjing	Abortion (2)	1o	KP749794
2012	XZ5-9	Xuzhou	None (2)	1b	KP749800
	XZ5-8	Xuzhou	None (4)	1b	KP749795
2012	BH789	Zhenjiang	Diarrhea (3)	1m	KP749799
2012	HA6-5	Nantong	None (4)	1m	KP749798
	HA2-12	Nantong	Diarrhea (6)/None (4)	1o	KP749802
2013	XY-3	Huai'an	Weakness (2)	1p	KP749796
2013	JR1-2	Zhenjiang	Abortion (2)	1b	KP749797

Table 3 Identity of 5'-UTR between the new strains and reference strains

Reference strain	1b (VEDEVAC, 289 bp)			1m (ZM-95, 290 bp)		1o (IS25CP/01, 256 bp [a])		1p (BJ0701, 245 bp [a])
Strain	JR1-2	XZ5-8	XZ5-9	BH789	HA6-5	JS12/02	HA2-12	XY-3
Identity (%)	100	96.5	96.2	95.5	94.8	85.9	88.4	95.9

[a]The sequence information of reference strain was limited

1b (JR1-2) were detected from Zhenjiang as well (Figs. 1, 2 and Table 2).

Discussion

Cattle are generally considered as the main host of BVDV infections, however, BVDV has also been found to infect an extensive range of animals including sheep, goats, swine, yaks, deer and members of the *Camelidae* family [4–7, 25, 29]. In China, BVDV infections were first detected in cattle in 1980 while the pig origin BVDV strain ZM-95 was isolated in 1995 [30]. BVDV infections in cattle, pigs, camels and Sika deer (Cervus nippon) have continually been reported in different regions of China [6, 18, 19, 23–26, 31–34]. The BVDV prevalence of pig population has significantly increased recent years [6, 26, 35]. Until now, the BVDV infections of goats in China were not examined, and the viral epidemic situation of BVDV in the goat flocks was not clear either. In this study, the serum samples from 31 goat herds of six regions in Jiangsu were detected and the results showed that BVDV infections were indeed present as 12.3 % of the prevalence in the samples. Eight different BVDV-1 strains were identified from five regions, and the infections might occur in the goat herds no matter the animals were with clinical signs or normal situations.

Currently, ten BVDV subtypes have been detected to be circulating among multiple hosts in China: BVDV-1a, BVDV-1b, BVDV-1c, BVDV-1d, BVDV-1m, BVDV-1o, BVDV-1p, BVDV-1q, BVDV-1u and BVDV-2a [18, 19, 23, 24, 26, 31, 33]. The four BVDV-1 subtypes determined in goats of Jiangsu were consistent with the previous subtypes reported from pigs and cattle on BVDV phylogenetic groups. At the same time, the different subtypes infecting goat herds might indicate that the epidemic situation of BVDV infections in the animal population of China was very complicated. The same subtype viruses found in cattle, pigs and goats meant the possibility of interspecies transmission of the viruses.

BVDV-1b was first isolated in Jilin province, China in 1980, and continually detected in Hebei, Xinjiang, Heilongjiang, Tianjin and Qinghai [23, 24], it is considered to be the major predominant subtype in Chinese cattle [23, 24]. The BVDV-1b strains JR1-2, XZ5-8 and XZ5-9 from two different regions were 96.2–99.7 % of identity with each other, and had very high identity (>96 %) with BVDV vaccine strain VEDEVAC [24],

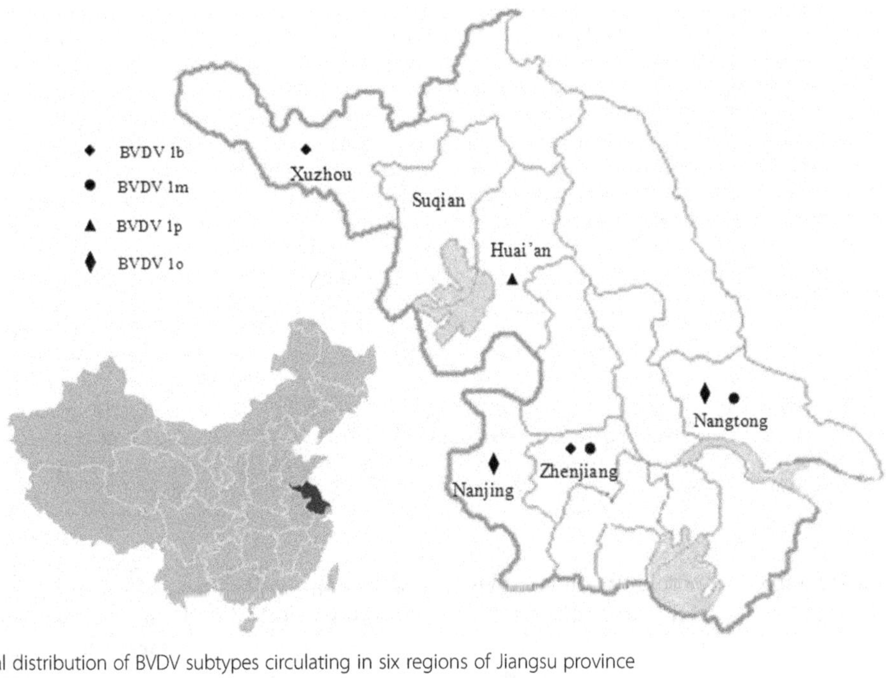

Fig. 1 Geographical distribution of BVDV subtypes circulating in six regions of Jiangsu province

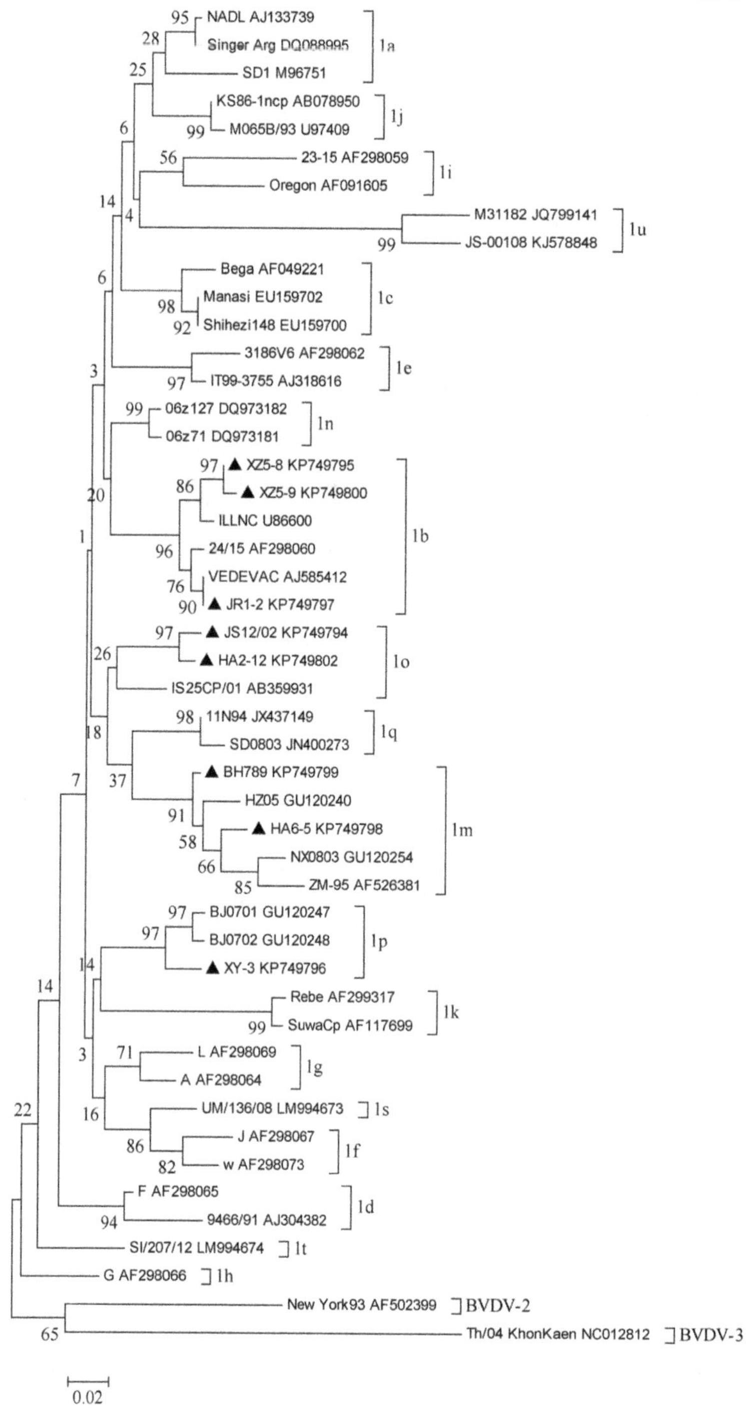

Fig. 2 Unrooted phylogeneitc tree based on the 5′-UTR sequences. Phylogenetic tree of the 5′-UTR from eight BVDV positive samples was constructed by the neighbor-joining (NJ) method with the sequences published in GenBank. The nucleotide length of the 5′-UTR used for the analysis was 224 bp. The *numbers* at the phylogenetic branches indicated the bootstrap values (1000 replicates) in percentage supporting each group. The *bar* represented a genetic distance

especially the JR1-2 had the 100 % identity in 5′-UTR and 98.2 % in N^{pro} gene sequences with the strain VEDEVAC, respectively, and these viruses also showed closed relationship based on the evolutionary trees. VEDEVAC was the main strain in batches of BVDV vaccine Oregon C24V in Hungary, however, no BVDV vaccines have been licensed for cattle and goats in China yet. Additionally, the current 1b strains shared 95.5–100.0 %

identity with BVDV-1b isolates YL07 and TJ0802 originated from cattle in 5′-UTR gene, which showed that cross infections of the BVDV-1b viruses might occur between cattle and goats.

Chinese BVDV-1 strain ZM-95 of pig origin was first isolated in 1995 and classified into subtype BVDV-1m [30, 35]. There were no further reports until 2010, new BVDV-1m strains in Chinese cattle were detected [24]. According to the surveys, BVDV-1m infections appeared to be more popular in China recently [18]. As Table 3, two BVDV-1m strains BH789 and HA6-5 showed as high as 95.5 and 94.8 % identity with ZM-95, respectively. BVDV-1m infections in goats might be closely related with its prevalence in pigs since there were many pig farms in the regions. Some goats infected with BVDV-1m had diarrhea, however, the pathogenicity of BVDV-1m infections in goats was not clear yet.

The BVDV strains JS12/02 and HA2-12 from goats were classified as BVDV1o, but they shared very low nucleotide identity (85.9 and 88.4 %) with the reference strain IS25CP/01 in 5′-UTR. The subtype viruses have been detected from bovine, pigs and camels in China [26, 32, 33], and the low prevalence of this subtype was observed worldwide. Interestingly, two BVDV-1o strains from goats shared high homology with each other, and significantly differed from other 1o strains based on and the Neighbor-joining unrooted trees of the 5′-UTR. Furthermore, these two viruses shared high homology

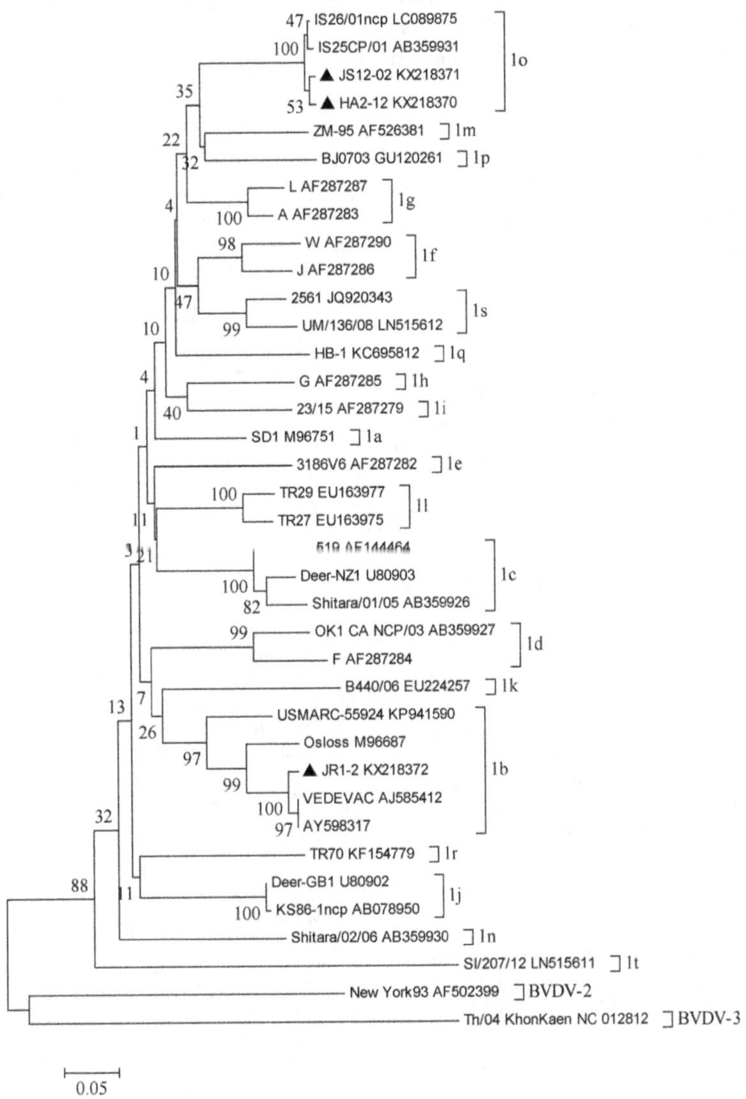

Fig. 3 Unrooted phylogeneitc tree based on the N^{pro} sequences. Phylogenetic tree of the N^{pro} from three BVDV positive samples was constructed by the neighbor-joining (NJ) method with the sequences published in GenBank. The nucleotide length of the N^{pro} sequence used for the analysis was 504 bp. The *numbers* at the phylogenetic branches indicated the bootstrap values (1000 replicates) in percentage supporting each group. The *bar* represented a genetic distance

(97.6 and 98.2 %) with IS25CP/01 strain based on the N^{pro} sequence. According to the phylogenetic trees (Figs. 2 and 3), the results showed that the present viruses might form one different branch with the 1o representative strains and the genetic divergence might exist within this subtype.

It should be noticed that four isolates detected from cattle in Beijing and Tianjing of China in 2010, formed a separate branch as BVDV-1p in the 5′-UTR phylogenetic tree, three of them were detected in the aborted bovine fetuses [24]. Now, the strain XY-3 detected from weak kids was in the same cluster with the BVDV-1p strains. It might indicate that the subtype viruses could infect animals to cause productive disorders.

This is the first report that BVDV-1 infections occurred in goats in Jiangsu province, China, and the genetic diversity of the viruses was determined as four BVDV-1 subtypes. Although numbers of the infected goats with clinical signs were limited, these infections might cause economic loss to the developing goat industry. More epidemiological investigations should be performed to better understand BVDV infections in goats as well as other farm animals, and design and develop effective control strategies.

Conclusions

The study showed that BVDV were circulating in Jiangsu, and present as 12.3 % of the prevalence in the goats, eight different BVDV-1 strains were identified from the samples, which were further divided into four subtypes: 1b, 1p, 1m, 1o. BVDV cross infections might be occur between goats, cattle and pigs, for control and prevention against BVDV infections, epidemiologic surveys should be carried out in these animals.

Acknowledgements
The Authors acknowledged the precious collaboration of all the farmers involved during the study. We thanked the technical staff at institute of agricultural ecomomics and information, Jiangsu academy of agricultural sciences for their support on the mapping. We thanked Dr. Kevin Coombs (University of Manitoba in Canada) read and revised the manuscript kindly and carefully.

Funding
This work was supported by the Special Fund for Independent innovation of Agricultural Science and Technology in Jiangsu province (CX (14)2090) and Jiangsu Provincial Natural Science Foundation of China (BK20130729).

Authors' contributions
LM and JJ designed and initiated the study. LM, WZ and LY carried out RT-PCR work; LM, WL, and FH collected samples and analyzed the sequence information; JW, SC, YW, QW, YD and YS joined in collecting the goat serum samples at the different regions; ML drafted the manuscript, JJ revised and all authors read and approved the final manuscript.

Competing interests
The authors declare that they have no competing interests.

Consent for publication
Not applicable.

Author details
[1]Institute of Veterinary Medicine, Jiangsu Academy of Agricultural Sciences; Key Laboratory of Veterinary Biological Engineering and Technology, Ministry of Agriculture, National Center for Engineering Research of Veterinary Bio-products, Nanjing 210014, China. [2]Suining Animal Husbandry and Veterinary station, Suining 221200, China. [3]Hai'an Animal Husbandry and Veterinary Station, Hai'an 226600, China.

References

1. Peterhans E, Schweizer M. BVDV: a pestivirus inducing tolerance of the innate immune response. Biologicals. 2013;41(1):39–51.
2. Baker JC. Bovine viral diarrhea virus: a review. J Am Vet Med Assoc. 1987; 190(11):1449–58.
3. Houe H. Epidemiological features and economical importance of bovine virus diarrhoea virus (BVDV) infections. Vet Microbiol. 1999;64(2–3):89–107.
4. Julia S, Craig MI, Jimenez LS, Pinto GB, Weber EL. First report of BVDV circulation in sheep in Argentina. Prev Vet Med. 2009;90(3–4):274–7.
5. Kim IJ, Hyun BH, Shin JH, Lee KK, Lee KW, Cho KO, Kang MI. Identification of bovine viral diarrhea virus type 2 in Korean native goat (Capra hircus). Virus Res. 2006;121(1):103–6.
6. Tao J, Liao J, Wang Y, Zhang X, Wang J, Zhu G. Bovine viral diarrhea virus (BVDV) infections in pigs. Vet Microbiol. 2013;165(3–4):185–9.
7. Walz PH, Grooms DL, Passler T, Ridpath JF, Tremblay R, Step DL, Callan RJ, Givens MD. Control of bovine viral diarrhea virus in ruminants. J Vet Intern Med. 2010;24(3):476–86.
8. Schirrmeier H, Strebelow G, Depner K, Hoffmann B, Beer M. Genetic and antigenic characterization of an atypical pestivirus isolate, a putative member of a novel pestivirus species. J Gen Virol. 2004;85(Pt 12):3647–52.
9. Decaro N, Lucente MS, Mari V, Sciarretta R, Pinto P, Buonavoglia D, Martella V, Buonavoglia C. Hobi-like pestivirus in aborted bovine fetuses. J Clin Microbiol. 2012;50(2):509–12.
10. Liu L, Xia H, Wahlberg N, Belak S, Baule C. Phylogeny, classification and evolutionary insights into pestiviruses. Virology. 2009;385(2):351–7.
11. Stahl K, Beer M, Schirrmeier H, Hoffmann B, Belak S, Alenius S. Atypical 'HoBi'-like pestiviruses–recent findings and implications thereof. Vet Microbiol. 2010;142(1–2):90–3.
12. Brigstocke T. Cattle Health Certification Standards and BVDV control. Vet Rec. 2010;167(21):834–5.
13. Vilcek S, Paton DJ, Durkovic B, Strojny L, Ibata G, Moussa A, Loitsch A, Rossmanith W, Vega S, Scicluna MT, et al. Bovine viral diarrhoea virus genotype 1 can be separated into at least eleven genetic groups. Arch Virol. 2001;146(1):99–115.
14. Vijayaraghavan B, Xia H, Harimoorthy R, Liu L, Belak S. Evaluation of envelope glycoprotein E(rns) of an atypical bovine pestivirus as antigen in a microsphere immunoassay for the detection of antibodies against bovine viral diarrhea virus 1 and atypical bovine pestivirus. J Virol Methods. 2012; 185(2):193–8.
15. Ridpath JF, Bolin SR, Dubovi EJ. Segregation of bovine viral diarrhea virus into genotypes. Virology. 1994;205(1):66–74.
16. Oem JK, Hyun BH, Cha SH, Lee KK, Kim SH, Kim HR, Park CK, Joo YS. Phylogenetic analysis and characterization of Korean bovine viral diarrhea viruses. Vet Microbiol. 2009;139(3–4):356–60.
17. Yamamoto T, Kozasa T, Aoki H, Sekiguchi H, Morino S, Nakamura S. Genomic analyses of bovine viral diarrhea viruses isolated from cattle imported into Japan between 1991 and 2005. Vet Microbiol. 2008;127(3–4):386–91.
18. Deng M, Ji S, Fei W, Raza S, He C, Chen Y, Chen H, Guo A. Prevalence study and genetic typing of bovine viral diarrhea virus (BVDV) in four bovine species in China. PLoS One. 2015;10(4):e0121718.

19. Gong X, Cao X, Zheng F, Chen Q, Zhou J, Yin H, Liu L, Cai X. Identification and characterization of a novel subgenotype of bovine viral diarrhea virus isolated from dairy cattle in Northwestern China. Virus Genes. 2013;46(2): 375–6.
20. Tajima M, Frey HR, Yamato O, Maede Y, Moennig V, Scholz H, Greiser-Wilke I. Prevalence of genotypes 1 and 2 of bovine viral diarrhea virus in Lower Saxony, Germany. Virus Res. 2001;76(1):31–42.
21. Flores EF, Ridpath JF, Weiblen R, Vogel FS, Gil LH. Phylogenetic analysis of Brazilian bovine viral diarrhea virus type 2 (BVDV-2) isolates: evidence for a subgenotype within BVDV-2. Virus Res. 2002;87(1):51–60.
22. Barros SC, Ramos F, Pauperio S, Thompson G, Fevereiro M. Phylogenetic analysis of Portuguese bovine viral diarrhoea virus. Virus Res. 2006;118(1–2): 192–5.
23. Zhong F, Li N, Huang X, Guo Y, Chen H, Wang X, Shi C, Zhang X. Genetic typing and epidemiologic observation of bovine viral diarrhea virus in Western China. Virus Genes. 2011;42(2):204–7.
24. Xue F, Zhu YM, Li J, Zhu LC, Ren XG, Feng JK, Shi HF, Gao YR. Genotyping of bovine viral diarrhea viruses from cattle in China between 2005 and 2008. Vet Microbiol. 2010;143(2–4):379–83.
25. Gong X, Liu L, Zheng F, Chen Q, Li Z, Cao X, Yin H, Zhou J, Cai X. Molecular investigation of bovine viral diarrhea virus infection in yaks (Bos gruniens) from Qinghai, China. Virol J. 2014;11:29.
26. Deng Y, Sun CQ, Cao SJ, Lin T, Yuan SS, Zhang HB, Zhai SL, Huang L, Shan TL, Zheng H, et al. High prevalence of bovine viral diarrhea virus 1 in Chinese swine herds. Vet Microbiol. 2012;159(3–4):490–3.
27. Vilcek S, Herring AJ, Herring JA, Nettleton PF, Lowings JP, Paton DJ. Pestiviruses isolated from pigs, cattle and sheep can be allocated into at least three genogroups using polymerase chain reaction and restriction endonuclease analysis. Arch Virol. 1994;136(3–4):309–23.
28. Li W, Mao L, Zhao Y, Sun Y, He K, Jiang J. Detection of border disease virus (BDV) in goat herds suffering diarrhea in eastern China. Virol J. 2013;10:80.
29. Rodriguez-Prieto V, Kukielka D, Rivera-Arroyo B, Martinez-Lopez B, de las Heras AI, Sanchez-Vizcaino JM, Vicente J. Evidence of shared bovine viral diarrhea infections between red deer and extensively raised cattle in south-central Spain. BMC Vet Res. 2016;12:11.
30. Wang X, Tu C, Li H, Jin K, Xuan H, Chang G, Sun H, Zhu W, Fei E, Yin Z. Detection and isolation of bovine viral diarrhea virus from classical swine fever suspected pigs. Chin J Vet Sci. 1996;16(4):341–5.
31. Zhu LQ, Lin YQ, Ding XY, Ren M, Tao J, Wang JY, Zhang GP, Zhu GQ. Genomic sequencing and characterization of a Chinese isolate of Bovine viral diarrhea virus 2. Acta Virol. 2009;53(3):197–202.
32. Lang Y, Gao S, Du J, Shao J, Cong G, Lin T, Zhao F, Liu L, Chang H. Polymorphic genetic characterization of E2 gene of bovine viral diarrhea virus in China. Vet Microbiol. 2014;174(3–4):554–9.
33. Gao S, Luo J, Du J, Lang Y, Cong G, Shao J, Lin T, Zhao F, Belák S, Liu L, et al. Serological and molecular evidence for natural infection of Bactrian camels with multiple subgenotypes of bovine viral diarrhea virus in Western China. Vet Microbiol. 2013;163(1–2):172–6.
34. Gao Y, Wang S, Du R, Wang Q, Sun C, Wang N, Zhang P, Zhang L. Isolation and identification of a bovine viral diarrhea virus from sika deer in china. Virol J. 2011;8:83.
35. Xu X, Zhang Q, Yu X, Liang L, Xiao C, Xiang H, Tu C. Sequencing and comparative analysis of a pig bovine viral diarrhea virus genome. Virus Res. 2006;122(1–2):164–70.

15

Evaluation of a multiplex real-time PCR for detection of four bacterial agents commonly associated with bovine respiratory disease in bronchoalveolar lavage fluid

Henk J. Wisselink*, Jan B.W.J. Cornelissen, Fimme J. van der Wal, Engbert A. Kooi, Miriam G. Koene, Alex Bossers, Bregtje Smid, Freddy M. de Bree and Adriaan F.G. Antonis

Abstract

Background: *Pasteurella multocida*, *Mannheimia haemolytica*, *Histophilus somni* and *Trueperella pyogenes* are four bacterial agents commonly associated with bovine respiratory disease (BRD). In this study a bacterial multiplex real-time PCR (the RespoCheck PCR) was evaluated for the detection in bronchoalveolar lavage fluid (BALF) of these four bacterial agents.

Results: The analytical sensitivity of the multiplex real-time PCR assay determined on purified DNA and on bacterial cells of the four target pathogens was one to ten fg DNA/assay and 4×10^{-1} to 2×10^{0} CFU/assay. The analytical specificity of the test was, as evaluated on a collection of 118 bacterial isolates, 98.3% for *M. haemolytica* and 100% for the other three target bacteria. A set of 160 BALF samples of calves originating from ten different herds with health problems related to BRD was examined with bacteriological methods and with the RespoCheck PCR. Using bacteriological examination as the gold standard, the diagnostic sensitivities and specificities of the four bacterial agents were respectively between 0.72 and 1.00 and between 0.70 and 0.99. Kappa values for agreement between results of bacteriological examination and PCRs were low for *H. somni* (0.17), moderate for *P. multocida* (0.52) and *M. haemolytica* (0.57), and good for *T. pyogenes* (0.79). The low and moderate kappa values seemed to be related to limitations of the bacteriological examination, this was especially the case for *H. somni*.

Conclusion: It was concluded that the RespoCheck PCR assay is a valuable diagnostic tool for the simultaneous detection of the four bacterial agents in BALF of calves.

Keywords: Bovine respiratory disease, PCR, *Histophilus somni*, *Mannheimia haemolytica*, *Pasteurella multocida*, *Trueperella pyogenes*

* Correspondence: henk.wisselink@wur.nl
Wageningen Bioveterinary Research, P.O. Box 65, 8200, AB, Lelystad, The Netherlands

Background

In veal production, bovine respiratory disease (BRD) is the most common and economically important disease [1–3]. BRD is a multifactorial disease, involving multiple potentially pathogenic microorganisms that causes economic losses due to morbidity, mortality, medication costs, increased time on feeding and associated labour costs [4]. In the Dutch veal industry, calves at an age between ten days and five weeks are sold by dairy farmers to traders who transport calves to the calf collection centres, from where they are transported to fattening farms where they are mingled with other young calves, reared and fattened until slaughter at an age of 6–7 months. The young age of the calves, their inadequate immunity and several stress factors including transportation and mingling with other young calves makes that the calves are vulnerable for infections, a scenario which is recognised as resulting in the frequent occurrence of disease problems including BRD [5]. To protect calves, at the end of the twentieth century veal farmers commonly administered antibiotics prophylactically immediately after arrival on the farm. Growing concerns about the increase in antibiotic resistance including methicillin resistant *Staphylococcus aureus* (MRSA) and extended-spectrum βeta-lactamase producing *Escherichia coli* in livestock resulted in a policy aimed to substantially reduce the use of antibiotics in The Netherlands [6]. Part of this policy was a ban of all prophylactic use of antimicrobials in the Netherlands in 2012 (Besluit Diergeneesmiddelen, 2 November 2012; http://wetten.overheid.nl/BWBR0032386/2015-01-01).
To implement and regulate this reduction in the use of antibiotics in the Netherlands, in recent years a collaboration was set up between stakeholders in livestock production and the national government [6, 7]. This lead to registration of all antibiotics administered to livestock at farm level [8] and to mandatory treatment and health plans to be developed for each farm to enforce that antimicrobial treatment is solely based on proper diagnosis [6]. However, the availability of rapid and reliable diagnostic tools is limited.

In BRD, several viral and bacterial pathogens may be involved [9–11]. For detection of these potential pathogens a variety of diagnostic tests have been described, including culture and molecular methods, see review by Fulton and Confer [12]. These methods all have their benefits and limitations in terms of speed of analysis, sensitivity and specificity [12]. In case of BRD, interpretation of results is sometimes challenging and diagnostic laboratories usually offer diagnostic tests for a limited number of potential pathogens which can be involved in BRD and which are sometimes also time-consuming. These are drawbacks that can hamper veterinarians and farmers for sending samples to diagnostic laboratory for analysis of BRD associated organisms. To offer veterinarians a fast and complete laboratory result, a set of three multiplex real-time PCRs was developed for detection of i) viral, ii) bacterial and iii) mycoplasma BRD-associated pathogens in bronchoalveolar lavage fluid (BALF) of calves, under the name RespoCheck (WBVR, Lelystad, The Netherlands). In this study the RespoCheck bacterial multiplex real-time PCR (abbreviated to RespoCheck PCR) was evaluated with commonly used culture methods as reference. This PCR has been developed for the detection of four bacterial pathogens commonly associated with BRD [9, 13], i.e. *Pasteurella multocida*, *Mannheimia haemolytica*, *Histophilus somni* and *Trueperella pyogenes*. The results showed that the RespoCheck PCR assay is a sensitive and valuable diagnostic tool suitable for the simultaneous detection of four BRD associated bacterial targets in BALF of calves.

Methods

Bacterial strains and growth conditions

A total of 118 bacterial isolates, representing 46 different species, were used for evaluation of specificity of the RespoCheck PCR. At least five representatives of each of the four target species, as well as bacterial strains from non-target species, consisting of isolates of phylogenetically closely related non-target species, and isolates of species from cattle with diseases other than BRD were included. Reference strains were obtained from the American Type Culture Collection (ATCC) and the Culture Collection of the University of Gothenburg (Sweden) (CCUB) and field strains were isolated from lungs of calves in a Dutch field study on BRD. An overview of the strains is provided in Table 1. Bacteria other than *H. somni* were grown overnight at 37 °C on heart infusion agar (ACU 7269C, Acumedia Manufacturers Inc. Lansing, MI) supplemented with 5% defibrinated sheep blood. *H. somni* was grown overnight at 37 °C in air with 5% CO_2 on chocolate blood agar, using 7% defibrinated horse blood and Columbia blood agar (CM 331, Oxoid, Badhoevedorp, The Netherlands).

Field samples

In a Dutch field study on BRD from October 2013 till March 2014, veal calf farms (n = 10) were selected. On each farm, the study period started shortly after arrival of the calves (D0) and ended 84 days (D84) later. At the start (D0) and end (D84) of the study period and during outbreaks of BRD, bronchoalveolar lavage (BAL) samples were taken from calves without and with clinical problems of BRD (as defined by severe upper respiratory tract and/or lower respiratory tract disease).

BAL samples were collected as previously described [14]. Between 35 and 75 mL BAL was obtained from each

Table 1 List of 118 bacterial isolates and results of the RespoCheck bacterial multiplex real-time PCR

Identification[a] (number of isolates tested)	CCUG identification[b]	Source	RespoCheck bacterial multiplex real-time PCR(Ct values)			
			P. multocida	*M. haemolytica*	*H. somni*	*T. pyogenes*
Acidovorax spp. (*n* = 3)	NA[c]	WBVR collection[d]	-[g]	-	-	-
Actinomyces spp	NA	WBVR collection[d]	-	-	-	-
Aerococcus viridans	NA	WBVR collection[e]	-	-	-	-
Biberstenia trehalosi	NA	WBVR collection[d]	-	-	-	-
Brucella abortus	NA	WBVR collection[e]	-	-	-	-
Comamonas kerstersii	NA	WBVR collection[d]	-	-	-	-
Corynebacterium bovis (*n* = 2)	NA	WBVR collection[e]	-	-	-	-
Corynebacterium pseudotuberculosis	NA	WBVR collection[e]	-	-	-	-
Escherichia coli	NA	WBVR collection[e]	-	-	-	-
Gallibacterium anatis (*n* = 5)	NA	WBVR collection[d]	-	-	-	-
Hafnia alvei	NA	WBVR collection[d]	-	-	-	-
Histophilus somni	NA	ATCC 22132[f]	-	-	15.2	-
Histophilus somni (*n* = 4)	NA	WBVR collection[d]	-	-	12.3–14.1	-
Klebsiella oxytoca	NA	WBVR collection[e]	-	-	-	-
Klebsiella pneumoniae	NA	WBVR collection[e]	-	-	-	-
Lactobacillus mucosae	NA	WBVR collection[d]	-	-	-	-
Lactococcus garvieae	NA	WBVR collection[e]	-	-	-	-
Lactococcus lactis	NA	WBVR collection[e]	-	-	-	-
Listeria monocytogenes	NA	WBVR collection[e]	-	-	-	-
Mannheimia haemolytica	NA	ATCC 14003[f]	-	16.7	-	-
Mannheimia haemolytica (*n* = 5)	NA	WBVR collection[d]	-	13.0–14.0	-	-
Mannheimia haemolytica	*Mannheimia ruminalis*	CCUG 38470-T	-	15.5	-	-
Mannheimia haemolytica	*Mannheimia glucosida*	CCUG 38457-T	-	16.0	-	-
Mannheimia granulomatis	*Mannheimia granulomatis*	CCUG 45422-T	-	-	-	-
Mannheimia varigena	*Mannheimia varigena*	CCUG 38462-T	-	-	-	-
Micrococcus luteus	NA	WBVR collection[e]	-	-	-	-
Moraxella bovis	NA	WBVR collection[e]	-	-	-	-
Moraxelle lacunata (*n* = 2)	NA	WBVR collection[d]	-	-	-	-
Pantoea agglomerans (*n* = 13)	NA	WBVR collection[d]	-	-	-	-
Pasteurella multocida	NA	ATCC 15743[f]	21.5	-	-	-
Pasteurella multocida (*n* = 23)	NA	WBVR collection[d]	15–17	-	-	-
Pasteurella multocida	Bisgaard Taxon 13	CCUG 16497	17.9	-	-	-
Pasteurella multocida	Bisgaard Taxon 13	CCUG 16498	17.6	-	-	-
Pasteurella multocida	*Pasteurella multocida* subsp. *gallicida*	CCUG 17978-T	18.0	-	-	-
Pasteurella multocida	*Pasteurella multocida* subsp. *septica*	CCUG 17977-T	17.0	-	-	-
Not typable	*Pasteurella aerogenes*	CCUG 27905	-	-	-	-
Proteus mirabillis	NA	WBVR collection[d]	-	-	-	-
Pseudomonas aeruginosa	NA	WBVR collection[e]	-	-	-	-
Psychrobacter spp.	NA	WBVR collection[d]	-	-	-	-
Salmonella enterica subsp. *enterica* serovar Dublin	NA	WBVR collection[e]	-	-	-	-
Salmonella enterica subsp. *enterica* serovar Typhimurium	NA	WBVR collection[e]	-	-	-	-
Serratia marcescans	NA	WBVR collection[e]	-	-	-	-
Staphylococcus aureus	NA	WBVR collection[e]	-	-	-	-

Table 1 List of 118 bacterial isolates and results of the RespoCheck bacterial multiplex real-time PCR *(Continued)*

Staphylococcus epidermidis	NA	WBVR collection[e]	-	-	-	-
Streptococcus agalactiae	NA	WBVR collection[e]	-	-	-	-
Streptococcus bovis (*n* = 5)	NA	WBVR collection[d]	-	-	-	-
Streptococcus dysgalactiae	NA	WBVR collection[e]	-	-	-	-
Streptococcus faecalis	NA	WBVR collection[e]	-	-	-	-
Streptococcus pluranimalium (*n* = 5)	NA	WBVR collection[d]	-	-	-	-
Streptococcus pneumoniae	NA	WBVR collection[e]	-	-	-	-
Streptococcus spp. (*n* = 3)	NA	WBVR collection[d]	-	-	-	-
Streptococcus uberis	NA	WBVR collection[e]	-	-	-	-
Trueperella pyogenes	NA	ATCC 9731[f]	-	-	-	15.8
Trueperella pyogenes (*n* = 5)	NA	WBVR collection[d]	-	-	-	12.8–16.0
Yersinia enterolytica	NA	WBVR collection[e]	-	-	-	-

[a]Isolates were identified with MALDI-TOF mass spectrometry (as described in Materials and Methods)
[b]CCUG: Culture Collection University of Gothenburg (Sweden)
[c]NA: Not applicable
[d]Isolated by Wageningen Bioveterinary Research (WBVR) from lungs of calves in the course of an earlier field study to BRD
[e]Cattle isolates from tissue other than lung
[f]ATCC: American Type Culture Collection (USA)
[g]-: Ct value above 40

calf after instillation of 100 mL PBS with 10% fetal calf serum. The BAL samples were transported within 24 h under cooled conditions (ice packs) to the laboratory. At the laboratory, clots of mucus were removed and 10 mL BAL was spun down for 10 min at 4600×g at 4 °C. The BAL pellet was used for direct bacteriological examination and the remaining BAL material was stored in 250 µL aliquots at −80 °C in the presence of 15% glycerol.

Bacteriological examination was performed using media and growth conditions as described above. Single colonies were subcultured twice and identified with MALDI-TOF mass spectrometry (Bruker MALDI Biotyper Microflex, version 3.1 with reference database V4.0 (5627 MSPs) Bruker Daltonics GmbH, Germany).

For evaluation of the RespoCheck PCR, BALF samples were selected from calves which were positive in bacteriological examination for one of the four BRD associated targets. In total 160 BALF samples were selected from the ten herds (with 12, 7, 7, 11, 9, 9, 5, 4, 6 and 90 samples being collected from these herds).

RespoCheck PCR

Starting material for the DNA extraction was 200 µL of BAL pellet or a loopful of colonies from a pure culture on agar plates. DNA extraction was performed on the MagNA Pure LC Instrument (Roche Life Science) and carried out with the MagNA Pure LC Total NA Isolation Kit (Roche-Diagnostics, Almere, The Netherlands) according to the manufacturer's instructions and using the "Total NA External lysis" protocol in the accompanying software (MagNA Pure LC Software Version 2.11) of the MagNA Pure LC Instrument.

The RespoCheck PCR reaction was on a 7500 Fast Real Time PCR system (Applied Biosystems, Bleiswijk, The Netherlands) using the QuantiFast Multiplex Kit RT-PCR kit (Qiagen, Venlo, The Netherlands).The PCR assay was run in a 20 µL reaction mix containing 5 µL of the nucleic acid sample, 250 nM of each primer, 100 nM of each MGB probe, 1 × QuantiFast Multiplex RT-PCR Master Mix (with ROX dye) and sterile deionised water. An initial denaturation/activation for 60 s at 95 °C was followed by 50 cycles of 10 s at 95 °C and 30 s at 60 °C.

The threshold cycle (Ct) value was determined for each sample by singleplex and multiplex real-time PCR with a threshold of 50% of the Delta Rn value (log). The threshold was manually set at 0.04 in the linear phase of the amplification plot, whereby the slope and correlation coefficient values were approximately 3.2 and 99.9% respectively. The PCR results were scored negative (−) when the generated Ct value was 40 or more and positive (+) for Ct values below 40.

A positive control of each of the four bacterial pathogens was included in every DNA isolation run and in each multiplex PCR reaction. Positive controls were prepared using the ATCC supplied strains of the four bacterial species (Table 1). *P. multocida*, *M. haemolytica* and *T. pyogenes* were cultured in brain heart infusion broth (BHI) broth (CM 225; Oxoid) and *H. somni* was cultured in BHI supplemented with 5 µg/mL β-nicotinamide adenine dinucleotide (NAD, Calbiochem, La Jolla, CA). To obtain early stationary phase cultures, the overnight cultures were diluted (1:100) and incubated at 37 °C. Incubation was stopped after approximately 4 h at an optical density of 0.5 at 600 nm. Cultures containing approximately

1×10^9 CFU/mL were then centrifuged at 4000 x g for 15 min. And stored at −80 °C in Nutrient broth No. 2 with 15% glycerol. In every DNA isolation run a "reagent blank" control (sterile water) was included and in each multiplex PCR run a "no-template" control (also sterile water).

Sensitivity and specificity

The analytical sensitivity of the RespoCheck PCR was defined as the ability of the assay to detect the lowest concentration of DNA/assay and CFU/assay [15] of the four target bacteria in PBS and BALF. For determination of the analytical sensitivity on DNA, a mixture was prepared of 10 ng DNA per assay of each of the four target bacteria. Ten-fold serial dilutions (n = 7) of this mixture were prepared in PBS resulting in a range from 10 ng to 1 fg DNA per assay of each of the four target bacteria. The Ct was determined for each sample by the singleplex and multiplex real-time PCR for each of the four target bacteria. For determination of the analytical sensitivity on bacterial cells in BALF, a mixture of cells of the four target bacteria was prepared in BALF of specific pathogen free (SPF) calves of 3–4 weeks old. Ten-fold serial dilutions (n = 7) of this mixture were prepared in BALF and DNA was isolated from these mixtures as described above. This resulted in a range from a 1×10^6 to 1×10^{-1}*P. multocida* CFU/assay, 6×10^5 to 6×10^{-2} *M. haemolytica* CFU/assay, 4×10^6 to 4×10^{-1}*H. somni* CFU/assay and 6×10^5 to 6×10^{-2} *T. pyogenes* CFU/assay. The Ct was determined for each sample by the singleplex and multiplex real-time PCR for each of the four target bacteria.

The analytical specificity was defined as the ability of the assay to distinguish the target organisms from non-target organisms [15]. For determining the analytical specificity of the RespoCheck PCR a panel of 118 bacterial isolates was used (Table 1).

The diagnostic sensitivity was defined as the proportion of samples from known infected reference animals that tested positive in an assay and the diagnostic specificity as the proportion of samples from known uninfected reference animals that tested negative in an assay [15]. The diagnostic sensitivity and specificity with 95% confidence intervals (95% CI) of the RespoCheck PCR was calculated for each target species using the result of the bacteriological examination of the BALF samples of the calves as reference method. Cohen's kappa coefficient with 95% CI was calculated to describe to what extent the RespoCheck PCR agrees with the bacteriological examination. For calculations the DAG Stat spreadsheet was used [16]. Kappa values were interpreted as follows: Kappa = 0.00–0.20, poor agreement; Kappa = 0.21–0.40, fair agreement; Kappa = 0.41–0.60, moderate agreement; Kappa = 0.61–0.80, good agreement; Kappa = 0.81–1.00, near-perfect agreement [17].

Results

Analytical sensitivity and specificity

The analytical sensitivity of the RespoCheck PCR was determined on mixtures of 10-fold serial dilutions of DNA of the four target bacteria and compared with the sensitivity of the singleplex PCRs. Positive multiplex real-time PCR results (Ct < 40) were obtained on at least one fg of chromosomal DNA of *H. somni* and ten fg chromosomal DNA of *P. multocida*, *M. haemolytica* and *T. pyogenes* (Fig. 1). Compared to the multiplex real-time PCR, the singleplex real-time PCR for *T. pyogenes* was less sensitive (one step in the serial dilution) (Fig. 1). For the other three target bacteria the sensitivity of the singleplex was similar to the multiplex real-time PCR (Fig. 1). The coefficient of correlation (R^2) between the Ct values and the amount of DNA/assay was higher than 0.99 for all assays. The slopes varied between 3.036 (multiplex real-time PCR for *M. haemolytica*) and 3.500 (singleplex real-time PCR for *T. pyogenes*) which corresponds to an efficiency (E) of 113.5% and 93.1% respectively (Fig. 1).

The analytical sensitivity of the RespoCheck bacterial multiplex real-time PCR was also determined in BALF of SPF calves, spiked with ten-fold serial dilutions of mixtures of cells of the four target bacteria (Fig. 2). Positive PCR results (Ct < 40) were obtained from the highest dilution corresponding to 1×10^{-1}*P. multocida* CFU/assay, 6×10^{-2} *M. haemolytica* CFU/assay, 4×10^{-1}*H. somni* CFU/assay and 6×10^{-2} *T. pyogenes* CFU/assay, both in single- and in multiplex real-time PCRs (Fig. 2). A good linear correlation ($R^2 > 0.96$) between the Ct values and the number of CFU/assay was found for all assays. The slopes ranged from 2.711 (singleplex real-time PCR for *M. haemolytica*) to 3.290 (singleplex real-time PCR for *T. pyogenes*) which corresponds to an efficiency (E) of 133.8% and 101.3% respectively (Fig. 1).

The RespoCheck PCR was evaluated on a panel of 118 isolates including the four target species (Table 1). The results of the RespoCheck PCR showed that for isolates of the four target species Ct-values were obtained below 40 (Table 1). On the reference strains of *M. ruminalis* and *M. glucosida* also positive test results were obtained in the *M. haemolytica* PCR, indicating a cross reaction. The other bacteria tested all negative (Table 1).

Diagnostic sensitivity and specificity

A set of 160 BALF samples of calves, originating from herds with health problems related to BRD, was examined using bacteriological culture as well as the RespoCheck PCR. Diagnostic sensitivity and specificity, and Cohen's Kappa Coefficient for the four target species in

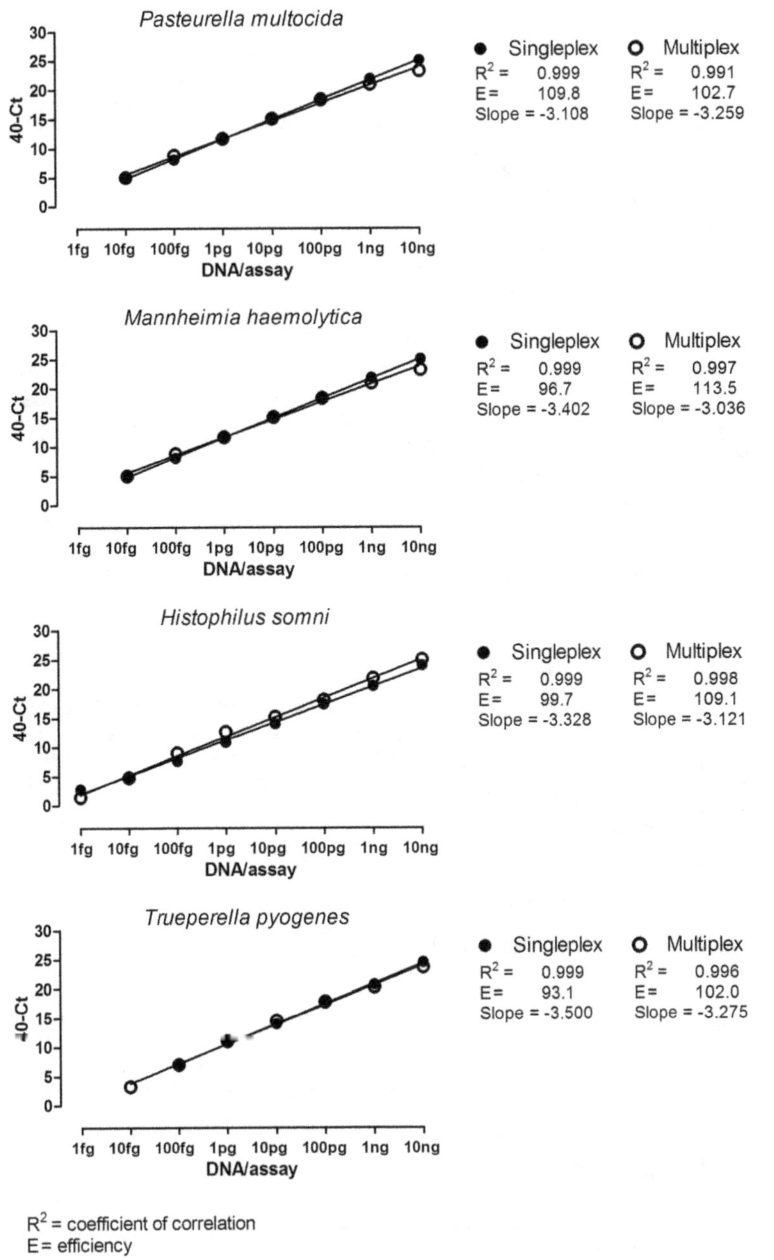

Fig. 1 Analytical sensitivity of the RespoCheck bacterial real-time PCR determined in singleplex and multiplex assays on mixtures of DNA of *P. multocida, M. haemolytica, H. somni* and *T. pyogenes* isolates

the multiplex real-time PCR, were calculated with the results of the bacteriological culture as reference. For three of the four target pathogens, a high number of the RespoCheck PCR-positive samples tested bacteriologically positive; 53 of 82 PCR-positives for *P. multocida*, 46 of 76 for *M. haemolytica* and 21 of 22 for *T. pyogenes*). A low number of RespoCheck PCR -positive samples tested bacteriologically positive for *H. somni* (5 of 42). For all four target pathogens, a high number of RespoCheck PCR -negative samples tested bacteriologically negative (68 of 78 for *P. multocida*, 80 of 84 for *M. haemolytica*, 118 of 118 for *H. somni* and 125 of 133 for *T. pyogenes*) (Table 2), resulting in moderate to high diagnostic sensitivities (resp. 0.84, 0.92, 1.00, and 0.72) and in moderate to high diagnostic specificities (resp. 0.70, 0.73, 0.76, 0.99) (Table 2). Kappa values for agreement between results of bacteriological examination and RespoCheck PCRs were low for *H. somni* (0.17), moderate for *P. multocida* (0.52) and *M. haemolytica* (0.57), and good for *T. pyogenes* (0.79) (Table 2).

To corroborate the results of the bacteriological examination for *H. somni*, ten BALF samples that were

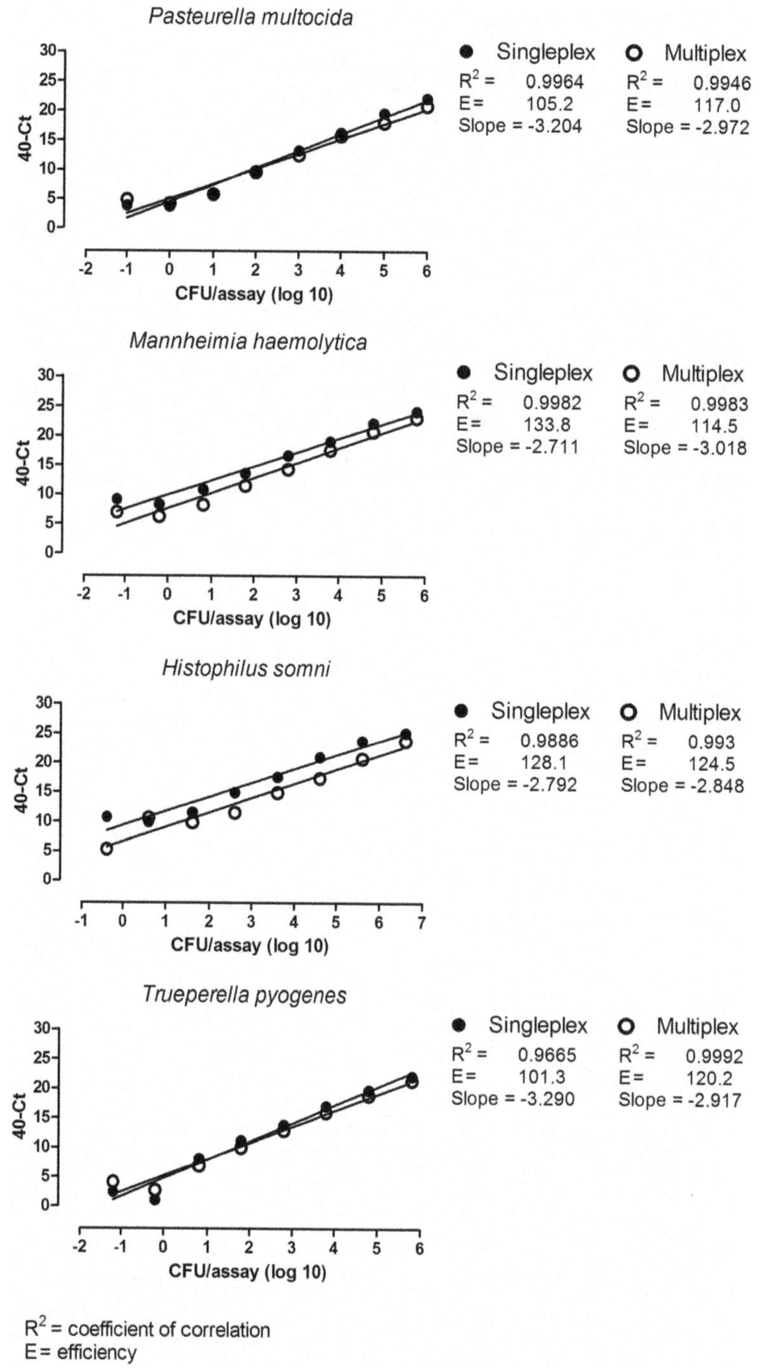

Fig. 2 Sensitivity of the RespoCheck bacterial real-time PCR determined in singleplex and multiplex assays on BALF spiked with mixtures of cells of *P. multocida*, *M. haemolytica*, *H. somni* and *T. pyogenes* isolates

bacteriologically negative and PCR-positive for *H. somni* were cultured again for *H. somni*. No *H. somni* bacteria were detected in these ten samples.

Discussion

To improve the detection and identification of BRD-associated pathogens in BALF samples, real-time PCR tests were developed under the name RespoCheck (WBVR, Lelystad, The Netherlands). In this study, the analytical and diagnostic sensitivity and specificity was evaluated of the RespoCheck PCR targeting four bacterial agents associated with BRD, i.e. *P. multocida*, *M. haemolytica*, *H. somni* and *T. pyogenes*. The results showed that the RespoCheck PCR assay is specific and sensitive for the

Table 2 Results of bacteriological examination (BE) and of the RespoCheck bacterial multiplex real-time PCR of BALF samples of calves originating from herds with BRD associated health problems

Agent	No (%) of BALF samples					Diagnostic Sensitivity	Diagnostic Specificity	Cohen's Kappa Coefficient
	BE + PCR +	BE + PCR-	BE PCR +	BE - PCR -	Total			
P. multocida	53 (33)	10 (6)	29 (18)	68 (43)	160	0.84 (0.73–0.92)[a]	0.70 (0.60–0.79)	0.52 (0.39–0.64)
M. haemolytica	46 (28)	4 (3)	30 (19)	80 (50)	160	0.92 (0.81–0.98)	0.73 (0.63–0.81)	0.57 (0.45–0.69)
H. somni	5 (3)	0 (0)	37 (23)	118 (74)	160	1.00 (NA)[b]	0.76 (0.69–0.83)	0.17 (0.04–0.30)
T. pyogenes	21 (13)	8 (5)	1 (1)	125 (81)	155	0.72 (0.53–0.87)	0.99 (0.96–0.99)	0.79 (0.65–0.92)

[a]95% CI
[b]NA: Not applicable

detection of these four bacterial agents. The RespoCheck PCR is easy to perform and allows large-scale application as 96 samples can be run simultaneously. Compared to standard bacteriological assays, the RespoCheck PCR is much more rapid to perform. Therefore, this RespoCheck PCR may be an important diagnostic tool for bacterial pathogens associated with BRD in calves. The results can be used by veterinarians and farmers for selection of specific intervention measures, including appropriate biosecurity, vaccination and possible antimicrobial treatments [12, 18].

Recently, we evaluated a triplex real-time PCR for detection of in BALF of *Mycoplasma (M.) dispar, M. bovis* and *M. bovirhinis* [19]. These three Mycoplasma's are associated with BRD [20] and application of the multiplex bacterial real-time PCR and the triplex Mycoplasma real-time PCR together improves detection of BRD-associated pathogens further.

In terms of labour and costs, a multiplex PCR approach is more favourable approach than several singleplex assays run in parallel. However, in a multiplex assay the sensitivity of the test can be affected for example due to competition for the reagents in the assay. Therefore we determined the analytical sensitivity in singleplex and in multiplex real-time PCR assays. The results showed that the analytical sensitivity was equally high for the singleplex and multiplex real-time PCR (Figs. 1 and 2) demonstrating that the sensitivity of the individual PCRs was not reduced when multiplexing the PCR assays.

The RespoCheck PCR showed a high analytical sensitivity as the number of CFU/assay detected ranged from 0.6×10^{-1} to 4×10^{-1}. Primers and probes of the RespoCheck PCR are based on the V3 region of the 16S rDNA of the four bacterial target pathogens. Bacterial cells can contain multiple copies of the 16S rDNA gene [21] and this number influences the analytical sensitivity of the multiplex real-time PCR. *P. multocida, M. haemolytica, H. somni* and *T. pyogenes* contain six, six, five and two 16S rDNA copies per cell respectively (https://rrndb.umms.med.umich.edu). The analytical sensitivity is for *P. multocida, M. haemolytica, H. somni* and *T. pyogenes* is, in theory, 6×10^{-1}, 4×10^{-1}, 2×10^{0} and 1×10^{-1} CFU/assay respectively. In summary, the RespoCheck PCR can detect 4×10^{-1} to 2×10^{0} CFU/assay of each of the four target pathogens. The sensitivity as determined on purified DNA, was found to be one to ten fg DNA/assay and assuming 4.6 fg of DNA per bacterial cell [22] the sensitivity is 2×10^{-1} to 2×10^{0} CFU/assay for the RespoCheck PCR. It seems that sensitivity of the RespoCheck PCR as determined on DNA and on bacterial cells in BALF is similar. Apparently, BALF does not influence the analytical sensitivity of the RespoCheck PCR.

In the RespoCheck PCR, the reference strains of the four target bacteria resulted in positive PCR results (Ct-values <40) (Table 1). Two reclassified *P. multocida* isolates, Bisgaard Taxon 13 strains CCUG 16497 and CCUG 16498, were correctly identified as *P. multocida* in the multiplex real-time PCR (Table 1). As these strains originally were classified as *P. avium* biovar 2 [23], these results underline the high analytical specificity of the RespoCheck PCR.

For determining the analytical specificity, 118 bacterial isolates representing 46 different species including the target species were examined in the RespoCheck PCR (Table 1). This panel consisted of strains that were isolated in the course of an earlier study to BRD in calves (Table 1; Source WBVR collection[d]) including the target species but also nontarget species. For *P. multocida, H. somni* and *T. pyogenes* only the target species were positive in the RespoCheck PCR, and these results confirm the analytical specificity of the RespoCheck PCR. The reference strains for *M. glucosida* and *M. ruminalis* were positive for *M. haemolytica* in the RespoCheck PCR whereas negative results were expected according to the CCUG identification (Table 1) and the phylogenetic analysis by 165 rRNA comparison of these strains [24]. Interestingly, MALDI-TOF MS identified *M. ruminalis* as "probable" an *M. haemolytica* (score 1.789) and *M. glucosida* as "highly probable" an *M. haemolytica* (score 2.257). Thus the RespoCheck PCR results for *M. haemolytica* seemed to be in agreement with the MALDI-TOF MS results. An explanation could be that in the reference database of the Bruker MALDI Biotyper (V4.0) there are no *M. ruminalis* isolates and only one *M. glucosidase* isolate (DSM 19638 T). The results by MALDI-TOF MS suggest that there is a lack of differentiation using this method between *M. haemolytica*, *M. ruminalis* and *M. glucosidase*. This lack of differentiation was also observed using 16S rRNA gene sequencing and a tRNA-intergenic spacer PCR [24, 25]. On the other hand, a clear separation between the species of the genus *Mannheimia* is possible by real-time PCR or multiplex PCR on other targets [26, 27]. The consequences of the lack in differentiation for interpretation of the results of the RespoCheck PCR between these *Mannheimia* strains seems to be limited since *M. ruminalis* and *M. glucosida* are mainly associated with sheep [24].

Comparison of the results of the RespoCheck PCR with bacterial culture on BALF samples showed that the PCR assay was more frequently positive than the bacteriological examination. Similar observations concerning differences between PCR and culture of these pathogens in lung specimens have been reported by others [9, 28]. Low levels of bacterial cells (live or dead) and overgrowth by contaminating microflora in the BALF specimens may explain the differences in results obtained with the two methods. Suboptimal transport conditions or presence of antibiotics could also influence the viability of the bacterial cells, both, as previously suggested [29] leading to false negative results.

The difference in number of *H. somni* culture positive samples and *H. somni* RespoCheck PCR positive samples highlights the difficulties in isolating this organism. In other studies on *H. somni*, differences were found between PCR and culture of lung specimens [28–30]. The explanation for this difference is that *H. somni* is a slow growing organism with small colonies that can be easily overgrown by other organisms [30]. These findings explain also the low kappa values for agreement between results of bacteriological examination and multiplex real-time PCRs for *H. somni* (0.17). This low kappa value seems relate to a low sensitivity of the bacteriological examination of BALF for *H. somni* and for this reason the use of the multiplex real-time PCR may be preferable over bacteriological examination.

The evaluation of the RespoCheck PCR on BALF samples from naturally infected calves showed that, for *P. multocida*, *M. haemolytica* and *T. pyogenes*, there were a number of samples (respectively 10, 4 and 8) which were positive by cultivation and negative by RespoCheck PCR. Analysis of these results showed that the majority of these culture positive samples was on the basis of the presence of one single colony only (results not shown). Thus although PCR on BALF samples is far more sensitive than bacteriological examination (as described above), the PCR can also be false negative.

Conclusion

The RespoCheck bacterial multiplex real-time PCR-test has been shown to represent a sensitive and reliable test for the simultaneous detection of *P. multocida*, *M. haemolytica*, *H. somni* and *T. pyogenes* in BALF samples of calves.

Abbreviations

ATCC: American Type Culture Collection; BALF: Bronchoalveolar lavage fluid; BRD: Bovine respiratory disease; CCUB: Culture Collection of the University of Gothenburg (Sweden); PCR/DGGE: PCR with denaturing gradient gel electrophoresis fingerprinting

Acknowledgements

This research was part of the project "Development and application of diagnostics" within the public-private partnership "One Health for Food" in The Netherlands. The authors thank Prof. Dr. D.J. Mevius (Wageningen Bioveterinary Research, Lelystad, The Netherlands; Utrecht University, Faculty of Veterinary Medicine, Utrecht, The Netherlands) for editing the Introduction about details of policy for reduction of antimicrobial use in The Netherlands.

Funding

The research was funded by the Dutch Ministry of Economic Affairs, Productschap Vee en Vlees / Stichting Brancheorganisatie Kalversector (SBK) and supported by the Van Drie Group, MSD Animal Health and Denkavit.

Authors' contributions

HJW, JBWJC, FJW, EAK, MGK, AB, FMB and AFGA designed the study. HJW, JBWJC, BS and AFGA acquired the data. HJW, JBWJC and AFGA analysed the data. HJW, JBWJC and FJW interpreted the data. HJW drafted the manuscript. All authors revised and approved the final version of the manuscript.

Consent for publication

Not applicable.

Competing interests

All authors are employee of Wageningen Bioveterinary Research, which commercialised the RespoCheck PCR.

References

1. Pardon B, Hostens M, Duchateau L, Dewulf J, De Bleecker K, Deprez P. Impact of respiratory disease, diarrhea, otitis and arthritis on mortality and carcass traits in white veal calves. BMC Vet Res. 2013;9:79.
2. Sans P, De Fontguyon G. Veal calf industry economics. Revue Méd Vét. 2009,160.420–4.
3. Snowder GD, Van Vleck LD, Cundiff LV, Bennett GL, Koohmaraie M, Dikeman ME. Bovine respiratory disease in feedlot cattle: phenotypic, environmental, and genetic correlations with growth, carcass, and longissimus muscle palatability traits. J Anim Sci. 2007;85:1885–92.

4. Gorden PJ, Plummer P. Control, management, and prevention of bovine respiratory disease in dairy calves and cows. Vet Clin North Am Food Anim Pract. 2010;26:243–59.
5. Cusack PMV, McMeniman N, Lean IJ. The medicine and epidemiology of bovine respiratory disease in feedlots. Austr Vet J. 2003;81:480–7.
6. Speksnijder DC, Mevius DJ, Bruschke CJM, Wagenaar JA. Reduction of veterinary antimicrobial use in the Netherlands. The Dutch success model. Zoonoses Public Health. 2015;62:79–87.
7. Mevius D, Heederik D. Reduction of antibiotic use in animals "let's go Dutch". J Verbr Lebensm. 2014;9:177–81.
8. Bos MEH, Mevius DJ, Wagenaar JA, van Geijlswijk IM, Mouton JW, Heederik DJJ. Antimicrobial prescription patterns of veterinarians: introduction of a benchmarking approach. J Antimicrob Chemother. 2015;70:2423–5.
9. Angen O, Thomsen J, Larsen LE, Larsen J, Kokotovic B, Heegaard PM, Enemark JM. Respiratory disease in calves: microbiological investigations on trans-tracheally aspirated bronchoalveolar fluid and acute phase protein response. Vet Microbiol. 2009;137:165–71.
10. Griffin D, Chengappa MM, Kuszak J, McVey DS. Bacterial pathogens of the bovine respiratory disease complex. Vet Clin North Am Food Anim Pract. 2010;26:381–94.
11. Moore SJ, O'Dea MA, Perkins N, O'Hara AJ. Estimation of nasal shedding and seroprevalence of organisms known to be associated with bovine respiratory disease in Australian live export cattle. J Vet Diagn Investig. 2015;27:6–17.
12. Fulton RW, Confer AW. Laboratory test descriptions for bovine respiratory disease diagnosis and their strengths and weaknesses: gold standards for diagnosis, do they exist? Can Vet J. 2012;53:754–61.
13. Tegtmeier C, Uttenthal A, Friis NF, Jensen NE, Jensen HE. Pathological and microbiological studies on pneumonic lungs from Danish calves. Zentralblatt fur Veterinarmedizin Reihe B Journal of veterinary medicine Series B. 1999;46:693–700.
14. Fogarty U, Quinn PJ, Hannan J. Bronchopulmonary lavage in the calf - a new technique. Irish Vet J. 1983;37:35–8.
15. Office International des Epizooties. Terrestrial animal health code. Paris: Office International des Epizooties 2013. .
16. Mackinnon A. A spreadsheet for the calculation of comprehensive statistics for the assessment of diagnostic tests and inter-rater agreement. Comput Biol Med. 2000;30:127–34.
17. Landis JR, Koch GG. The measurement of observer agreement for categorical data. Biometrics. 1977;33:159–74.
18. Speksnijder DC, Jaarsma DAC, Verheij TJM, Wagenaar JA. Attitudes and perceptions of Dutch veterinarians on their role in the reduction of antimicrobial use in farm animals. Prev Vet Med. 2015;121:365–73.
19. Cornelissen JBWJ, de Bree FM, van der Wal FJ, Kooi EA, Koene MG, Bossers A, Smid B, Antonis AF, Wisselink HJ. Mycoplasma detection by triplex real-time PCR in bronchoalveolar lavage fluid from bovine respiratory disease complex cases. BMC Vet Res. 2017;13:97.
20. Nicholas R, Ayling R, McAuliffe L. Bovine respiratory disease. In: In: Mycoplasma diseases of ruminants CAB International. United Kingdom: Wallingford; 2008. p. 132–68.
21. Větrovský T, Baldrian P. The variability of the 16S rRNA gene in bacterial genomes and its consequences for bacterial community analyses. PLoS One. 2013;8:e57923.
22. Lichtensteiger CA, Steenbergen SM, Lee RM, Polson DD, Vimr ER. Direct PCR analysis for toxigenic *Pasteurella multocida*. J Clin Microbiol. 1996;34:3035–9.
23. Christensen H, Angen O, Olsen JE, Bisgaard M. Revised description and classification of atypical isolates of *Pasteurella multocida* from bovine lungs based on genotypic characterization to include variants previously classified as biovar 2 of *Pasteurella canis* and *Pasteurella avium*. Microbiology. 2004;150:1757–67.
24. Angen O, Ahrens P, Bisgaard M. Phenotypic and genotypic characterization of *Mannheimia (Pasteurella) haemolytica*-like strains isolated from diseased animals in Denmark. Vet Microbiol. 2002;84:103–14.
25. Catry B, Baele M, Opsomer G, de Kruif A, Decostere A, Haesebrouck F. tRNA-intergenic spacer PCR for the identification of *Pasteurella* and *Mannheimia* spp. Vet Microbiol. 2004;98:251–60.
26. Guenther S, Schierack P, Grobbel M, Lubke-Becker A, Wieler LH, Ewers C. Real-time PCR assay for the detection of species of the genus *Mannheimia*. J Microbiol Methods. 2008;75:75–80.
27. Alexander TW, Cook SR, Yanke LJ, Booker CW, Morley PS, Read RR, Gow SP, McAllister TA. A multiplex polymerase chain reaction assay for the identification of *Mannheimia haemolytica*, *Mannheimia glucosida* and *Mannheimia ruminalis*. Vet Microbiol. 2008;130:165–75.
28. Bell CJ, Blackburn P, Elliott M, Patterson TI, Ellison S, Lahuerta-Marin A, Ball HJ. Investigation of polymerase chain reaction assays to improve detection of bacterial involvement in bovine respiratory disease. J Vet Diagn Investig. 2014;26:631–4.
29. Tegtmeier C, Angen O, Ahrens P. Comparison of bacterial cultivation, PCR, in situ hybridization and immunohistochemistry as tools for diagnosis of *Haemophilus somnus* pneumonia in cattle. Vet Microbiol. 2000;76:385–94.
30. Angen O, Ahrens P, Tegtmeier C. Development of a PCR test for identification of *Haemophilus somnus* in pure and mixed cultures. Vet Microbiol. 1998;63:39–48.

16

Incidence of bovine neonatal pancytopenia in 243 farms in Germany

Frederike Reichmann[1*], Annette Pfitzner[1], Guenter Rademacher[1], Elke Schwedinger[2], Klaus Cussler[2] and Carola M. Sauter-Louis[1,3]

Abstract

Background: Several research groups from different European countries have worked on the aetiopathogenesis of bovine neonatal pancytopenia (BNP) and an association between the use of the vaccine PregSure BVD (Pfizer, Germany) and the development of this haemorrhagic disease was confirmed. Because BNP is not a notifiable disease, it is difficult to obtain information on its incidence. Based on pharmacovigilance (PhV) data, which are the only officially available data at the national level, the incidence of BNP is considered low. However, voluntary reporting of the disease can lead to underreporting.

To gain more insight into the incidence of BNP among the affected herds, an epidemiological study was performed, which focused on 243 farms in Germany with cases of BNP. Farmers were asked to report the occurrence of BNP, including the number of cases, which allowed calculation of incidence in the affected herds. Matching such data with the registered cases in the National PhV System (NPhVS) gave us an opportunity to assess the extent of BNP underreporting.

Results: On 243 farms, a total of 1195 calves younger than 4 weeks with haemorrhagic diathesis were registered. In 58 % of the reports, a diagnosis of BNP was confirmed by blood analysis and or by necropsy. The number of cases observed on individual farms ranged from 1 to 80. Based on these results, the incidence of BNP on affected farms ranged from 0.3 to 15.2 % (median 2.9 %). The maximal incidence in the year with the highest number of BNP calves ranged between 0.4 and 18.6 % (median 3.3 %). Comparing the number of cases registered in the NPhVS to the numbers found in this study revealed considerable underreporting to the national database: only 44 % of the farms and 41 % of the BNP calves included in the study were registered in the NPhVS.

Conclusions: In spite of the opportunity to report BNP calves to the Paul-Ehrlich-Institut (Langen, Germany), the estimated number of undetected BNP cases is remarkably high. However, even if the revealed substantial underreporting is taken into account, the incidence of BNP is low. Nevertheless, the incidence on some affected farms is very high, resulting in considerable financial losses that should not be underestimated. Although the exact pathomechanism of BNP at the molecular level is still not known, its incidence is clearly declining following withdrawal of PregSure BVD from the market.

Keywords: BNP, Calf, Epidemiology, Haemorrhagic diathesis

* Correspondence: frederike.reichmann@lmu.de

[1]Clinic for Ruminants with Ambulatory and Herd Health Services at the Centre for Clinical Veterinary Medicine, LMU Munich, Sonnenstrasse 16, 85764 Oberschleissheim, Germany

Full list of author information is available at the end of the article

Background

First cases of bovine neonatal pancytopenia (BNP) were reported in Germany in 2006 and then in several other European countries and New Zealand [1]. The work of several research groups in various European countries has yielded important information about the aetiology and pathogenesis:

The consumption of colostrum-containing alloreactive antibodies plays a crucial role in the development of BNP [2, 3]. Vaccination against bovine viral diarrhoea virus (BVDV) was suspected to be correlated with the pathogenesis of BNP [4]. Intensive investigations identified the inactivated vaccine PregSure BVD (Pfizer, Germany) as a major risk factor for developing BNP [5–7]. In some cows, the PregSure BVD vaccine induces alloreactive antibodies that target antigens derived from bovine cells used for vaccine production [8]. Owing to the consumption of colostrum from these cows, some calves and foster calves develop thrombocytopenia, leukocytopenia, and panmyelophthisis [8–10]. As a consequence, affected calves show bleeding symptoms and most often, immunosuppression [2, 11, 12]. Owing to the absence of a promising therapy for BNP, only about 10 % of affected calves survive [4, 10, 11]. A subclinical course of the disease has been described [13]. The most important differential diagnoses for the clinical signs of BNP are the haemorrhagic form of BVD (type 2), a severe septicaemia, and other haemorrhagic diatheses. The haemorrhagic form of BVD is very uncommon in Germany; in 95 % of BVDV infections, the virus is type 1 [14]. Furthermore, bleeding disorders of cattle are relatively rare. If these disorders occur, they generally happen not only in calves within the first 4 weeks of age, but at all ages. No other haemorrhagic diatheses show similar processes to BNP [15]. Classic septicaemia affects newborn calves between 2 and 6 days of age, and is attributed to other diseases. Bleeding signs in septicaemic calves are possibly due to capillary fragility or thrombocytopenia, but they are less typical signs [16].

There is a remarkable difference in the incidence of BNP among different regions within a country [17]. In addition to this geographical difference, the number of cases per herd also varies [2, 4, 12]. BNP is not a notifiable disease and, therefore, exact figures about its occurrence are not available. Because the occurrence of BNP is believed to be related to the use of a vaccine, each case should be reported as a possible adverse event within the pharmacovigilance (PhV) scheme. Veterinary surgeons and farmers have the opportunity to report BNP cases to their national competent authority, which for Germany, is the Paul-Ehrlich-Institut in Langen. The agency recorded 4623 cases in Europe (date of download: 28.02.2011), including 3040 cases from Germany..

However, there is currently no information available as to what degree these numbers truly reflect the BNP incidence. Furthermore, little information exists about the incidence within an individual herd; most often, the information is based on investigations on single farms. According to reported results, within-herd incidence are no higher than 10 % [8].

The aim of our study was to provide an overview of the occurrence of BNP based on 243 affected farms in Germany. The incidence of BNP was calculated and the data of this study were compared with cases registered in the National PhV System (NPhVS) to identify to what extent BNP cases were reported within the PhV scheme.

Methods

Participating farms

Since the first case of BNP, affected calves have been registered in a database at the Clinic for Ruminants with Ambulatory and Herd Health Services, LMU Munich (Oberschleissheim, Germany). Calls to encourage registration from all over Germany were published in *Deutsches Tieraerzteblatt*, a journal of the German Veterinary Association. The database included 310 farms on January 3, 2011. Farms were eligible as case farms if they had one or more confirmed cases of BNP by necropsy (panmyelophthisis) and or by blood analysis (thrombocytopenia [less than 200×10^9 platelets (PLT)/L] and leukocytopenia [less than 4×10^9 white blood cells (WBC)/L]). Blood analyses were performed by the laboratory of the Clinic for Ruminants with Ambulatory and Herd Health Services, LMU Munich. Necropsies were performed by one of the following institutions: Bavarian Authority for Health and Food Safety in Oberschleißheim or Erlangen, Bavarian Animal Health Service in Poing, Institute of Veterinary Pathology, Center for Clinical veterinary Medicine, LMU Munich or Animal Health service in Fellbach, Baden-Wuerttemberg.

Twenty-seven farms were excluded, because the blood analysis of affected calves with haemorrhagic diathesis revealed only thrombocytopenia or leukocytopenia. Further, farms which only reported on BNP calves without any confirmed case, were excluded ($n = 28$). In total, 255 farms with at least one confirmed case of BNP were selected from the database.

Interview

Farmers were interviewed by telephone between January 3 and March 31, 2011. The interviews were conducted in a uniform fashion by one veterinarian (FR) at the Clinic for Ruminants. The veterinarian was trained prior to the interviews and instructed to avoid leading questions. The following data were collected for the years from 2005 to 2011:

- Number of cows per year
- Number of calves born per year
- Breed
- Type of farm (dairy farm or beef farm)
- Number of observed calves younger than 4 weeks of age with haemorrhagic diathesis each year
- Proportion of confirmed cases of calves with BNP among the observed calves with haemorrhagic diathesis
- Use of PregSure BVD.

One year was defined as the 12 month period from the first of January until the 31st of December. During the interview, the answers were recorded on paper. In case that the herd manager needed more time to look up the information, the farm was contacted a second time. If the farmer did not have any information about the vaccine that was used to vaccinate against BVDV, the veterinarian was contacted. If the farm had changed the veterinarian during the considered timeframe, the former veterinarian was called to get the information needed. Only farms that used PregSure BVD were included in the analysis. The questionnaire is available as Additional file 1.

Incidence of BNP at the herd level

The total incidence on each farm was calculated as:

$$\text{Total incidence }(\%) = \frac{\text{Number of BNP calves}}{\text{Average number of calves born per year} \times \text{Years with observed BNP cases}} \times 100$$

The considered time period for each farm was not equal due to the fact that BNP occurred on some farms over several years and on other farms only within 1 year. The aim of the study was to evaluate the incidence of BNP on farms while the disease was prevalent on the farm. Using the calculation above, allowed the comparison of incidences across farms. In addition, the maximal incidence was calculated, which was related to the year with the highest number of calves with BNP on the farm:

$$\text{Maximal incidence }(\%) = \frac{\text{Highest number of calves with BNP in one specific year}}{\text{Number calves born in that year}} \times 100$$

Data matching between the NPhVS and the clinic for ruminants

Data matching was used to compare the number of farms and cases of BNP registered in the national PhV database at the Paul-Ehrlich-Institut against the numbers calculated in our study. Therefore, the first digits of the postal zip codes and the initials of the farmers were compared. Farms and cases not listed in the NPhVS were counted.

Descriptive and statistical analyses

The geographical distribution of the farms affected by BNP was illustrated in a map created in ArcGIS software (ESRI, version 9.3). A descriptive statistical analysis was conducted using Excel version 13.0 (Microsoft Inc., USA) and SPSS for Windows version 19 (SPSS Inc., USA). Binomial confidence intervals were calculated in Excel using the method of Agresti and Coull [18].

Results

Participating farms

Two of the 255 contacted farmers were not willing to give an interview and one farmer had stopped farming, thus they were excluded from the analysis. Nine farms were excluded, because the herd was either not vaccinated against BVDV or PregSure BVD was not used. Finally, 243 farms could be included and farmers were available and willing to participate in the study. The veterinarians were always available to complete the data about the occurrence of BNP calves and the vaccination against BVDV, in case they were needed. Most of the farms ($n = 241$) were dairy farms with 168 having German Fleckvieh (70 %), 47 farms with Holstein-Frisian (19 %), 12 farms with Brown Swiss (5 %), nine farms with Red Holstein (4 %) and five farms with German Fleckvieh crossbred (2 %). Two farmers had a cow-calf operation with Limousin and German Fleckvieh, respectively. The geographical distribution of the participating farms is shown in Fig. 1. Of the 243 farms, 76 % were located in Bavaria and 24 % were in eight other federal states of Germany. The herd size ranged from 14 to 950 cows (average 67).

Total number of calves with haemorrhagic diathesis (four weeks old or younger)

During the interviews, the 243 farmers reported 1195 newborn calves with a bleeding disorder occurring between January 1, 2005, and March 31, 2011 (Fig. 2). The first four cases were observed in 2005 on three different farms. However, a remarkable increase in the number of calves being observed with BNP was not noted until 2008, when 253 cases were observed on 88 farms. The year 2009 had the highest number of observed cases: 68 farms that had previously reported BNP cases and 120 new farms were affected and recorded. A total of 539 calves had clinical signs of haemorrhagic diathesis during their first month of life. After 2009, the number of cases declined on all participating farms.

Number of cases observed per farm

The number of newborn calves with haemorrhagic diathesis on each farm is illustrated in Fig. 3, excluding two farms with 50 and 80 cases, because they are outliers.

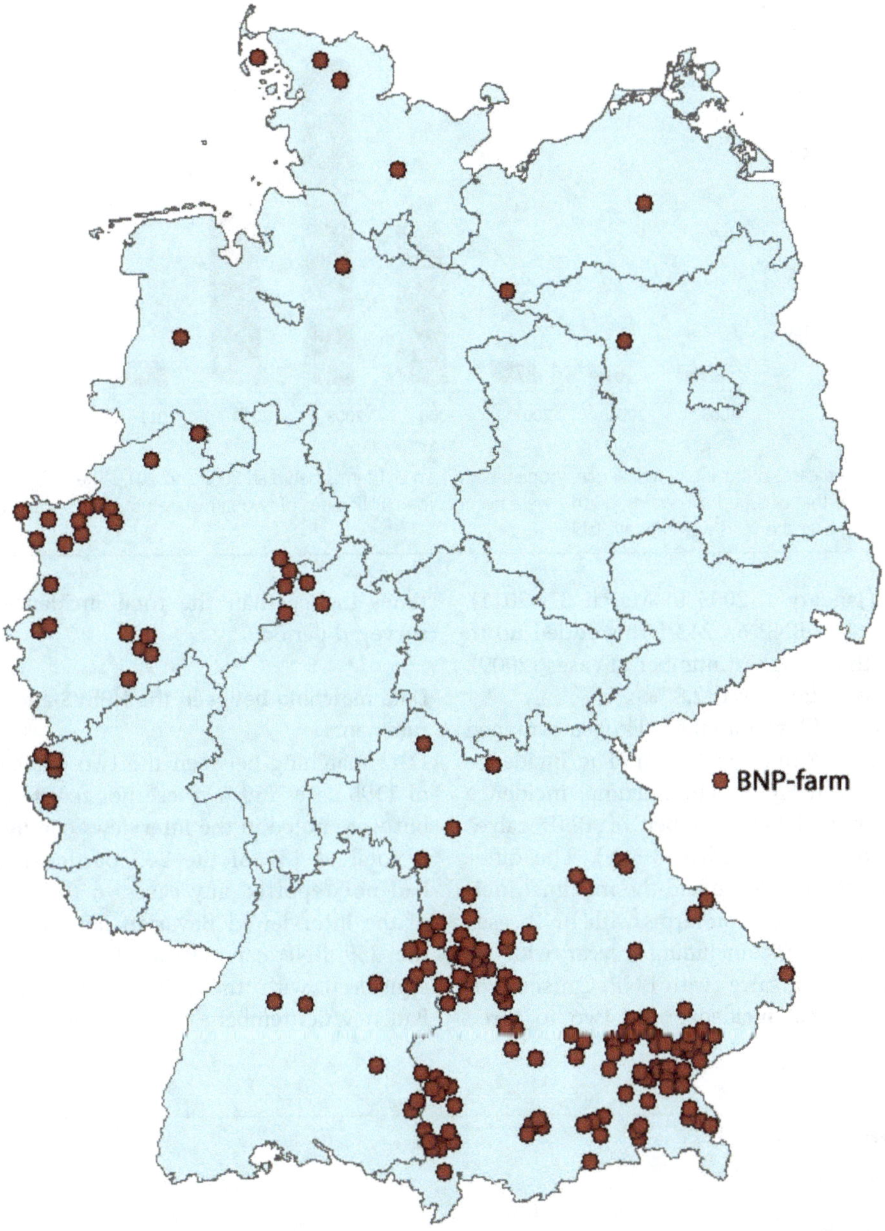

Fig. 1 Geographical distribution of participating farms with bovine neonatal pancytopenia (BNP) ($n = 243$) The farms were listed in a database of the Clinic for Ruminants with Ambulatory and Herd Health Services at the Centre for Clinical Veterinary Medicine, LMU Munich (Oberschleissheim, Germany). At least one calf younger than four weeks of age was registered on each farm, with haemorrhagic diathesis that was confirmed as BNP by blood analysis and or necropsy

BNP cases were registered during only 1 year on nearly one-half of the farms (44 %). Thirty-five percent of the 241 farmers noted calves with BNP during two of the years, 36 farmers during three of the year, and another 12 farmers during four of the years. One interviewed owner reported cases during 5 years and another one during 6 years. Some farms had a year without having calves with BNP between the years with cases of BNP.

In 58 % of the registered cases, BNP was confirmed either by blood analysis and or by necropsy. In 42 % of the cases, farmers did not initiate an examination to clarify the aetiology of the bleeding disorder.

Overall incidence and incidence of BNP at the herd level

The incidence could not be calculated for two farms, because the farmers did not remember the exact distribution of BNP calves over the years. Regarding the

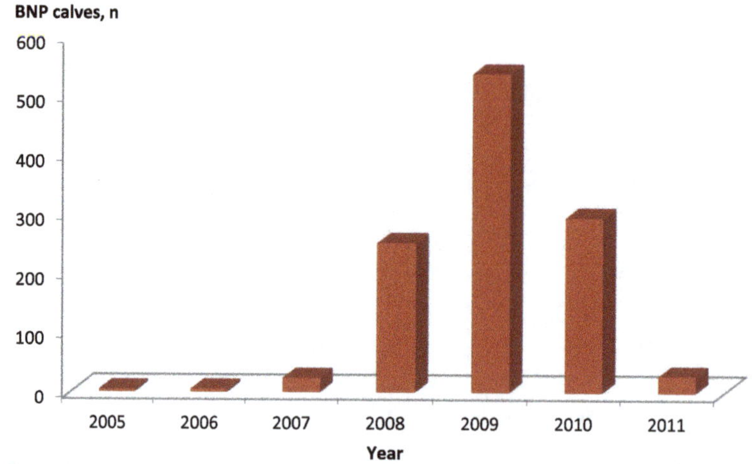

Fig. 2 Number of observed cases of bovine neonatal pancytopenia (BNP) on 241 farms between 2005 and 2011. The study was finished on March 31, 2011. BNP cases that occurred afterwards in 2011 were not recorded. BNP cases of two farms are not shown, because the exact distribution of BNP calves over the years were not available

whole timeframe (January 1, 2005 to March 31, 2011), the overall incidence of BNP on 243 farms added up to 1 %. In the year with the highest number of cases (2009), the incidence across all farms was 2.8 %.

The incidence of BNP within the 241 farms ranged between 0.3 and 15.2 % (median 2.9 %). The incidence distribution is shown in Fig. 4. The maximal incidence in the year with the highest number of BNP calves ranged from 0.4 to 18.6 % (median 3.3 %). The difference between the total incidence and the maximal incidence was particularly marked on farms with BNP cases scattered over several years, including 1 year with an irregularly high number of calves with BNP. On some of those farms, the maximal incidence was two to three times higher than the total incidence throughout the surveyed period.

Data matching between the NPhVS and the clinic for ruminants

Data matching between the two databases revealed 701 of 1195 cases (59 %) were not registered in the NPhVS, but were noted in the interviews conducted in our study. A total of 137 of the 243 participating farms (56 %) had not reported any cases to the NPhVS. Only 41 % of the interviewed Bavarian farmers and only 33 % of the 759 BNP calves from Bavaria in our study were registered with the NPhVS. Ten of 12 farms from Baden-Wuerttemberg reported their cases to the NPhVS,

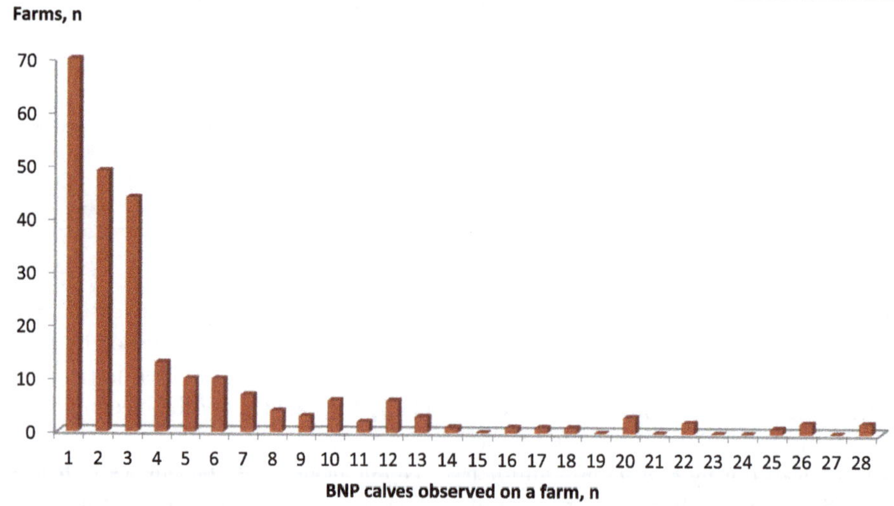

Fig. 3 Number of farms that have observed a particular number of bovine neonatal pancytopenia (BNP) cases between 2005 and March 2011. One farm with 50 cases and another one with 80 are not included in the figure, because they are outliers

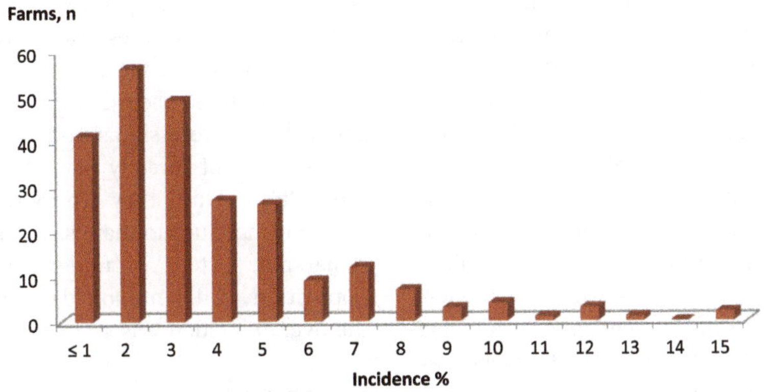

Fig. 4 Within-herd incidence of bovine neonatal pancytopenia (BNP). The incidence was calculated using the number of BNP calves on 241 affected farms from January 1, 2005, to March 31, 2011. The rate ranged from 0.3 to 15.2 % (median 2.9 %). The incidence could not be calculated for two farms, because the exact distribution of BNP cases over the years was not available

but only 24 of 47 interviewed farmers from the other federal states of Germany reported their cases. The data matching results are shown in Table 1.

Discussion

Incidence of BNP at the herd level

One of the main objectives of this study was to identify the incidence of BNP at the herd level. For the calculation of the incidence the average number of calves born on the farm was available. This could have introduced a bias, as stillborn calves would not have been included in the denominator of the calculation. Using the average number of calves born on the farm also could have introduced a bias, which we cannot exclude. However, all farmers included in the study had reported that their herd size had been stable over the considered time period, therefore we are confident that the number of calves born over several years did not vary considerably. BNP affects calves during their first 4 weeks of life. Male Holstein-Frisian and Red Holstein calves are usually sold at an age of two weeks, thus it is possible that some cases of BNP were missed when BNP occurred after the calves were sold to another farm. However, most of the participating farms (77 %) had German Fleckvieh, crossbred or Brown Swiss. Male calves from those farms are typically not sold before they have reached a certain age or weight and do not leave the farm during the time when BNP may occur.

The highest incidence reported in a herd from our study (19 %) was substantially higher than the incidence rates reported in other studies, which was up to only 10 % [8]. Although the BNP incidence was low on most of the affected farms, some herds were obviously more severely affected. The incidence of BNP varies markedly among different countries and also among the federal states of Germany [4, 6]. The farms enrolled in our study were from nine federal states of Germany. However, most of the farms with cases of BNP were localized in

Table 1 Results of data matching between the Clinic for Ruminants with Ambulatory and Herd Health Services at the Centre for Clinical Veterinary Medicine, LMU Munich (Oberschleissheim, Germany), and the National Pharmacovigilance System (NPhVS): Comparison of bovine neonatal pancytopenia (BNP) farms and the number of BNP cases on those farms

Federal State	Farms reported to the Clinic for Ruminants	Farms from the study registered at the NPhVS	Proportion of NPhVS-registered farms in % (Confidence Interval)	Total BNP cases on the farms of the LMU study	BNP cases on the farms of the LMU study registered at the NPhVS	Proportion of NPhVS-registered BNP cases in % (Confidence Interval)
Bavaria	184	75	40.76 (33.92–47.98)	759	248	32.67 (29.43–36.09)
Baden-Wuerttemberg	12	10	83.33 (55.20–95.30)	67	54	80.6 (69.58–88.29)
Brandenburg	1	1	100.00 (20.65–100.00)	6	6	100 (60.97–100.00)
Hesse	5	4	80.00 (37.55–96.38)	47	7	14.89 (7.14–27.69)
Lower Saxony	1	0	0.00 (0.00–79.35)	2	0	0.00 (0.00–65.76)
North Rhine-Westphalia	26	14	53.85 (35.46–71.24)	242	133	54.96 (48.66–61.1)
Mecklenburg-Vorpommern	2	2	100.00 (34.24–100.00)	39	32	82.05 (67.33–91.02)
Schleswig-Holstein	6	3	50.00 (18.76–81.24)	19	9	47.37 (27.33–68.29)
Rhineland-Palatinate	6	0	0.00 (0.00–39.03)	14	0	0.00 (0.00–21.53)

Bavaria. Kasonta et al. [17] investigated the influence of different patterns of PregSure BVD vaccinations on the incidence of BNP. Their data showed that different vaccination regimens resulted in differences in alloreactive antibody concentrations. Therefore, the differences in incidence rates among federal states can potentially be explained by the different vaccination protocols that exist. Apart from the varying incidence among the federal states of Germany, we found considerable differences among the participating farms within the federal states. Presumably, the use of different vaccination regimens at the herd level is the reason for that observation, but other factors (e.g. colostrum management) could have had an influence on the incidence as well.

The present study shows a remarkable increase of BNP cases in 2008, reaching a peak in 2009. Although PregSure BVD was licensed in 2004, an increased BNP incidence was not noted until 2007. The same phenomenon was observed in New Zealand: the first cases of BNP occurred there in 2011, which was 3 years after the introduction of PregSure BVD [19]. Kasonta et al. [17] interpreted the temporal delay as being associated with the repeated immunizations that are required for creating a titre of alloreactive antibodies high enough to induce clinical BNP in the calves.

Various reasons exist for the decline of BNP cases on farms in our study since 2010. Owing to increased knowledge concerning aetiopathogenesis, farmers adapted their management practices, especially colostrum and vaccination management [20]. The safest way to prevent further BNP cases is to dispose of colostrum coming from known 'BNP dams' and to feed calves colostrum from cows that have not been vaccinated with PregSure BVD. Before PregSure BVD was withdrawn from the German market in April 2010, many farmers had already switched to a different BVD vaccine or discontinued vaccination against the BVDV. Nevertheless, calves with BNP are still observed on some farms, but in much smaller numbers [21]. Therefore, the antibody titre seems to remain high for a long time, even though PregSure BVD is no longer used. It is uncertain how much longer further cases of BNP will occur, but they will most likely stop when the last PregSure BVD-vaccinated cows are removed.

Data matching between the two databases

Another objective of our study was to investigate whether the number of calves with BNP registered at the NPhVS reflects the real number of cases in Germany. More than one-half of the cases and farms had not been listed in the NPhVS database of the Paul-Ehrlich-Institut. Three-hundred cases of BNP were reported in our study that occurred before 2009, which was the year when BNP was recognized as a potential serious adverse effect of PregSure BVD. A retrospective report of these cases was technically possible, but seems unlikely to happen, so most of these cases are most likely not in the database. Although veterinarians are requested to report adverse events in accordance with the veterinary medical association's professional code of conduct, they and the farmers do not directly benefit from reporting cases to the NPhVS. Thus, they are increasingly reluctant to fill in reporting questionnaires. Although we discovered a remarkable extent of underreporting, considering the population of 4.3 million dairy cows in Germany [22], the overall incidence is still very low.

Economic relevance

The severity of economic losses in affected herds due to medical treatments of BNP calves, costs for the veterinarian, and especially case fatalities, is difficult to estimate. However, an appreciable pecuniary loss can be assumed for farms with high incidence. In addition, less obvious losses could be caused by subclinical BNP cases in which calves could be more susceptible to other diseases, especially infections due to immunosuppression. Thus, in an epidemiological study, 2.5 % of control calves had thrombocytopenia and leucocytopenia without clinical signs of BNP [15], and another research group induced subclinical BNP in 20 % of calves after feeding with colostrum from 'BNP dams' [23].

Confirmation of BNP via blood analysis or necropsy

To reduce the bias of misclassification and, thus, maximize the probability of including farms with actual cases, only farms with at least one confirmed case of BNP were included in this study. Therefore, we excluded some farms from the study because even though the farmers had observed calves with signs consistent with BNP within their herds, they did not initiate an examination to verify the aetiology.

As reflected by the high proportion (58 %) of cases that were confirmed by haematology and or necropsy, veterinary practitioners demonstrated a great commitment to elucidate the previously unknown aetiology of this disease.

The spectacular clinical signs that are most often seen in affected calves, and the severe process of BNP are very similar among cases. Therefore, it can be assumed that affected farmers and their veterinarians are aware of the disease pattern and that most of the calves included in this study without disease confirmation likely did have BNP. Some cases of BNP were probably missed at the very beginning of the occurrence when farmers and veterinarians were not aware of the disease or in case of sudden death. Furthermore it is not possible to get any information about subclinical cases of BNP within a herd without blood sampling the calves, thus subclinical BNP could not be taken into account.

Because there are no advantages (e.g. compensation or specific treatments and better prognoses for the calves after disease diagnosis) for farmers and veterinarians in confirming additional cases, the proportion of investigated cases will decline and reports on the occurrence of BNP to official agencies will become less common over the years.

Conclusions

The number of calves with BNP in Germany is higher than officially assumed. Many cases have not been registered as PhV reports at the NPhVS, and we revealed a remarkable underreporting of the disease.

The incidence of BNP varies considerably, not only among federal states, but also among affected herds within one area. Considerable financial losses can be assumed, especially on farms with a high incidence.

Considering the whole bovine population of Germany, the incidence of BNP is low, even when accounting for the fact that many affected calves have not been registered.

Due to the well-advanced clarification of the aetiopathogenesis of BNP and the suspension of the license for PregSure BVD, the peak incidence for this disease has passed. However, a few cases of BNP still occur. It is likely that cases of BNP will occur until the last dam vaccinated with PregSure BVD has left the farms.

Abbreviations

BNP: Bovine Neonatal Pancytopenia; BVD: Bovine Viral Diarrhoea; BVDV: Bovine Viral Diarrhoea Virus; NPhVS: National Pharmacovigilance System; PLT: Platelets; WBC: White Blood Cells

Acknowledgement

We are grateful to all farmers and veterinarians who participated in the project.

Funding

The study is part of the project 2809HS0205 funded by the Federal Ministry of Food, Agriculture and Consumer Protection (BMELV).

Authors' contributions

FR conducted the study, analysed the data, and wrote the manuscript. AP and CSL participated in the design of the study and revised the manuscript. GR helped design the study. ES and KC matched the data from the study with the Paul-Ehrlich-Institut database. CSL substantially contributed to statistical analysis of the data and editing of the paper. All authors have read and approved the final manuscript.

Competing interests

The authors declare that they have no competing interests.

Consent for publication

Not applicable.

Author details

[1]Clinic for Ruminants with Ambulatory and Herd Health Services at the Centre for Clinical Veterinary Medicine, LMU Munich, Sonnenstrasse 16, 85764 Oberschleissheim, Germany. [2]Paul-Ehrlich-Institut, Paul-Ehrlich-Straße 51-59, 63225 Langen, Germany. [3]Institute of Epidemiology, Friedrich-Loeffler-Institute, Suedufer 10, 17493 Greifswald, Isle of Riems, Germany.

References

1. Stöber M. Infektionsbedingte Krankheiten des Blutes. In: Innere Medizin und Chirurgie des Rindes. 5th ed. Stuttgart: Parey; 2006. p. 214–21.
2. Friedrich A, Buttner M, Rademacher G, Klee W, Weber BK, Muller M, Carlin A, Assad A, Hafner-Marx A, Sauter-Louis CM. Ingestion of colostrum from specific cows induces Bovine Neonatal Pancytopenia (BNP) in some calves. BMC Vet Res. 2011;7:10.
3. Bridger PS, Bauerfeind R, Wenzel L, Bauer N, Menge C, Thiel HJ, Reinacher M, Doll K. Detection of colostrum-derived alloantibodies in calves with bovine neonatal pancytopenia. Vet Immunol Immunopathol. 2011;141(1–2):1–10.
4. Friedrich A, Rademacher G, Weber BK, Kappe E, Carlin A, Assad A, Sauter-Louis C, Hafner-Marx A, Büttner M, Böttcher J, Klee W. Gehäuftes Auftreten von hämorrhagischer Diathese infolge Knochenmarksschädigung bei jungen Kälbern. Tierärztl Umschau. 2009;64:423–31.
5. Update on possible vaccine link to bleeding calf syndrome. Vet Rec. 2011; 168(19):497. http://veterinaryrecord.bmj.com/.
6. Weiss DJ, Evanson OA, Sykes J. A retrospective study of canine pancytopenia. Veterinary clinical pathology/American Society for Veterinary Clinical Pathology. 1999;28(3):83–8.
7. Corapi WV, French TW, Dubovi EJ. Severe thrombocytopenia in young calves experimentally infected with noncytopathic bovine viral diarrhea virus. J Virol. 1989;63(9):3934–43.
8. Bastian M, Holsteg M, Hanke-Robinson H, Duchow K, Cussler K. Bovine Neonatal Pancytopenia: is this alloimmune syndrome caused by vaccine-induced alloreactive antibodies? Vaccine. 2011;29(32):5267–75.
9. Deutskens F, Lamp B, Riedel CM, Wentz E, Lochnit G, Doll K, Thiel HJ, Rumenapf T. Vaccine-induced antibodies linked to bovine neonatal pancytopenia (BNP) recognize cattle major histocompatibility complex class I (MHC I). Vet Res. 2011;42(1):97.
10. Pardon B, Stuyven E, Stuyvaert S, Hostens M, Dewulf J, Goddeeris BM, Cox E, Deprez P. Sera from dams of calves with bovine neonatal pancytopenia contain alloimmune antibodies directed against calf leukocytes. Vet Immunol Immunopathol. 2011;141(3–4):293–300.
11. Kappe EC, Halami MY, Schade B, Alex M, Hoffmann D, Gangl A, Meyer K, Dekant W, Schwarz BA, Johne R, et al. Bone marrow depletion with haemorrhagic diathesis in calves in Germany: characterization of the disease and preliminary investigations on its aetiology. Berl Munch Tierarztl Wochenschr. 2010;123(1–2):31–41.
12. Pardon B, Steukers L, Dierick J, Ducatelle R, Saey V, Maes S, Vercauteren G, De Clercq K, Callens J, De Bleecker K, et al. Haemorrhagic diathesis in neonatal calves: an emerging syndrome in Europe. Transbound Emerg Dis. 2010;57(3):135–46.
13. Bell CR, Kerr MG, Scott PR, Morrison WI, Brown H. Evidence of a high incidence of subclinically affected calves in a herd of cattle with fatal cases of Bovine Neonatal Pancytopenia (BNP). BMC Vet Res. 2014;10:245.
14. Beer M, Wolf G. [Selection of BVDV genotype II isolates using a monoclonal antibody and FACS analysis]. Berl Munch Tierarztl Wochenschr. 1999;112(9):345–50.
15. Holliman A, Willoughby K, Scholes S, Colloff A, Caldow G, Howie F, Smith R, Lambton S, Bell C. Bovine Neonatal Pancytopenia: results of a 2 year study into this emerging disease in the United Kingdom. In: AAVLD Annual Conference: 2011; Buffalo. 2011. p. 97.
16. Fecteau G, Smith BP, George LW. Septicemia and meningitis in the newborn calf. The Veterinary clinics of North America Food animal practice. 2009;25(1):195–208. vii-viii.

17. Mizuno NS, Perman V, Bates FW, Sautter JH, Schultze MO. Life span of thrombocytes and erythrocytes in normal and thrombocytopenic calves. Blood. 1959;14(6):708–19.
18. Agresti A, Coull B. Aproximate is better than exact for interval estimation of binominal proportions. Am Stat. 1998;52(2):119–26.
19. Recall of BVD vaccine in New Zealand [www.foodsafety.govt.nz/elibrary/industry/bvd-vaccine-nz.htm]
20. Reichmann F. Case–control-study of Bovine Neonatal Pancytopenia in farms with high and low incidence. Ludwig-Maximilians University Munich. 2012.
21. Cussler K, Schwedinger E, Sauter-Louis C, Kappe E. Bovine Neonatale Panzytopenie. Deutlicher Rückgang der Fälle, aber noch keine Entwarnung für betroffene Betriebe. In: Deutsches Tierärzteblatt. vol. 6. 2015. p. 841–3.
22. Battilani M, Balboni A, Ustulin M, Giunti M, Scagliarini A, Prosperi S. Genetic complexity and multiple infections with more Parvovirus species in naturally infected cats. Vet Res. 2011;42(1):43.
23. Steficek BA, Thomas JS, Baker JC, Bell TG. Hemorrhagic diathesis associated with a hereditary platelet disorder in Simmental cattle. J Vet Diagn Invest. 1993;5(2):202–7.

17

Risk factors associated with exposure to bovine respiratory disease pathogens during the peri-weaning period in dairy bull calves

Gerard M. Murray[1*], Simon J. More[2], Tracy A. Clegg[2], Bernadette Earley[3], Rónan G. O'Neill[4], Dayle Johnston[3], John Gilmore[5], Mikhail Nosov[5], Máire C. McElroy[4], Thomas J. Inzana[6] and Joseph P. Cassidy[7]

Abstract

Background: Bovine respiratory disease (BRD) remains among the leading causes of death of cattle internationally. The objective of this study was to identify risk factors associated with exposure to BRD pathogens during the peri-weaning period (day (d)-14 to d 14 relative to weaning at 0) in dairy bull calves using serological responses to these pathogens as surrogate markers of exposure.

Clinically normal Holstein-Friesian and Jersey breed bull calves ($n = 72$) were group housed in 4 pens using a factorial design with calves of different breeds and planes of nutrition in each pen. Intrinsic, management and clinical data were collected during the pre-weaning (d − 56 to d − 14) period. Calves were gradually weaned over 14 days (d − 14 to d 0). Serological analysis for antibodies against key BRD pathogens (BRSV, BPI3V, BHV-1, BHV-4, BCoV, BVDV and *H. somni)* was undertaken at d − 14 and d 14. Linear regression models (for BVDV, BPI3V, BHV-1, BHV-4, BCoV and *H. somni)* and a single mixed effect random variable model (for BRSV) were used to identify risk factors for changes in antibody levels to these pathogens.

Results: BRSV was the only pathogen which demonstrated clustering by pen. Jersey calves experienced significantly lower changes in BVDV S/P than Holstein-Friesian calves. Animals with a high maximum respiratory score (≥8) recorded significant increases in *H. somni* S/P during the peri-weaning period when compared to those with respiratory scores of ≤3.

Haptoglobin levels of between 1.32 and 1.60 mg/ml at d − 14 were significantly associated with decreases in BHV-1 S/N during the peri-weaning period. Higher BVDV S/P ratios at d − 14 were significantly correlated with increased changes in serological responses to BHV-4 over the peri-weaning period.

Conclusions: Haptoglobin may have potential as a predictor of exposure to BHV-1. BRSV would appear to play a more significant role at the 'group' rather than 'individual animal' level. The significant associations between the pre-weaning levels of antibodies to certain BRD pathogens and changes in the levels of antibodies to the various pathogens during the peri-weaning period may reflect a cohort of possibly genetically linked 'better responders' among the study population.

Keywords: Calves, Antibodies, Bovine respiratory disease, Dairy, Exposure

* Correspondence: gerard.murray@agriculture.gov.ie
[1]Sligo Regional Veterinary Laboratory, Department of Agriculture, Food and Marine, Doonally, Sligo, Ireland
Full list of author information is available at the end of the article

Background

Bovine respiratory disease (BRD) remains among the leading causes of death of dairy and beef cattle of all ages in Ireland [1] and internationally [2]. Dairy calf pneumonia (enzootic calf pneumonia) represents an epidemiologically distinct component of the BRD complex, typically affecting 2 to 6 month old calves; shipping fever of feedlot cattle and atypical interstitial pneumonia are other recognised syndromes [3].

While indoor or outdoor individual housing of calves has been recognised as beneficial to dairy calf health [4] legislative changes in animal welfare in Ireland and Europe (EU Directive 91/629/EC and EU Decision 97/182/EC) have encouraged group housing of calves, which also facilitates more efficient use of labour and space. However, this shift has presented the dairy industry with new challenges in disease control [5] and increased the risk of BRD among young dairy calves [6–8].

Stress is an important co-factor in the pathogenesis of BRD [9, 10]. Clinically affected calves shedding large numbers of pathogens into the environment act as an important source of exposure for other calves. In addition, BRD pathogens can also be carried and shed by apparently healthy animals [11, 12]. Weaning has been traditionally recognised as a stressful time for calves with both nutritional and non-nutritional factors contributing to weaning distress [13]. Non-nutritional factors such as the breaking of the maternal bond and the rearrangement of the social group tend to be critical factors for beef suckled calves in particular. As dairy calves are generally separated from the dam shortly after birth, weaning distress in the dairy calf typically arises predominantly from nutritional factors [14]. When coupled with the increased interaction and contact of calves in group housing, weaning distress can facilitate the efficient spread of BRD pathogens leading to exposure and, potentially, disease. Inevitably there are interactions and cross-effects between pathogens which together lead to variable and complex immunological and pathological effects in each calf.

Many studies have examined the effects of intrinsic [15], management [16, 17] or clinical [18] risk factors for BRD; however, reports as to the various potential risks these variables pose in exposing calves to BRD pathogens are lacking. The objective of this study was to identify risk factors associated with exposure to BRD pathogens during the peri-weaning period in dairy bull calves using serological responses to these pathogens as surrogate markers of exposure. These responses were determined by calculating changes in sample to positive ratios (S/P), percentage positivity (PP) or sample to negative ratios (S/N), to a wide range of recognised BRD pathogens during the peri-weaning period (day (d) -14 to d 14 relative to weaning at d 0).

Methods

Animal management

This study was conducted as part of a larger study designed to examine changes in haematological profiles and gene expression in response to gradual weaning. Ethical approval for this study was sought, and received, from the Teagasc Animal Ethics Committee. Animal management, sample collection and haematological analysis have been outlined previously by Johnston et al. [14]. Briefly, 72 clinically normal bull calves of Jersey or Holstein-Friesian breed were sourced from 2 preferential supplier farms at a mean age of 19 (S.D. 8) days and were group housed indoors in 4 sawdust-floored pens (16, 18, 18 and 20 per pen) from d – 56 (relative to weaning (d 0)) to d 28 of the study. A sample size calculation for linear regression based on a power level of 0.8, significance level of 0.05 and a range of anticipated effects and numbers of predictors suggested a minimum sample range of 59 to 65 animals. Calves were immunised on arrival against bovine herpesvirus 1 (BHV-1; Rispoval IBR Marker Live administered intramuscularly, Zoetis), and, using a combined vaccine (Bovilis Bovipast RSP, MSD, inactivated vaccine administered subcutaneously), they were also immunised against bovine parainfluenza 3 virus (BPI3V), bovine respiratory syncytial virus (BRSV), and *Mannheimia haemolytica* serotypes A1 and A6. A single combined booster dose against BRSV, BPI3V and *Mannheimia haemolytica* was administered 4 weeks later. Vaccination against *Salmonella* Dublin and *Salmonella* Typhimurium (Bovivac S, inactivated vaccine administered subcutaneously) was also administered on arrival. The vaccination status of the dams was not available. The study was structured as a factorial design with two breeds (Holstein-Friesian and Jersey), and three planes of nutrition (high (H), medium (M) and low (L)) within each breed and calves were stratified to a nutrition treatment within each breed, on the basis of live-weight, age at the first day of the study (d – 56) and sire [14]. Each pen contained calves of each breed and each plane of nutrition and calves were fed using automatic milk (Vario Powder; Förster-Technik GmbH, Engen, Germany) and concentrate (KFA3-MA3; Förster-Technik GmbH) feeders. All calves were offered approximately 400 g straw daily, from a rack within the group pen during the peri-weaning period. During the pre-weaning period (d-56 to d-14), Holstein-Friesian calves on the H, M and L planes of nutrition were offered 1.2 kg milk replacer (8 l at 150 g/l) with ad libitum concentrate, 0.8 kg milk replacer (6 l at 133.33 g/l) with a maximum of 1.5 kg concentrate and 0.5 kg milk replacer (4 l at 125 g/l) with a maximum of 1 kg concentrate, daily, respectively. The Jersey calves on the H, M and L planes of nutrition were offered 0.8 kg milk replacer (6 l at 133.33 g/l) with ad libitum concentrate, 0.5 kg milk replacer (4 l at 125 g/l) with a maximum of 1.5 kg concentrate and 0.35 kg milk replacer (3.5 l at 100 g/l) with a maximum of 1 kg concentrate, daily, respectively. During the weaning phase (d-14 to d0), daily milk replacer was gradually reduced and by d − 1, all calves had been consuming at least 1 kg of concentrate per day for 3 consecutive days. On d 0, milk replacer was eliminated from the diet of all calves.

Animals were maintained on different planes of nutrition which were devised for each breed using National Research Council guidelines [19] to achieve a target growth rate of ⩾1.0, 0.7 and < 0.5 kg/day, for Holstein-Friesian breed calves on the H, M and L planes of nutrition and a target growth rate of 0.7, 0.5 and ⩽0.3 kg/day, for Jersey breed calves on the H, M and L planes of nutrition, respectively [14].

Clinical assessment

Clinical assessments were carried out on all calves twice weekly during the pre-weaning and weaning periods (d − 56 to d 0). A modified version of the Wisconsin health scoring criteria was used to score clinical criteria [20]. A cumulative respiratory score (0–12) was devised from nasal discharge (0–3), eye (0–3) or ear (0–3) score (whichever was greatest), cough index (0–3) and rectal temperature based on the method described by Lago et al. [21]; the maximum respiratory score achieved by each calf between d − 56 and d − 14 was recorded.

Sampling

Blood samples were collected from all calves on arrival, and on d − 14 and d 14 relative to weaning (d 0), via jugular venipuncture. Serum was harvested and samples were stored at − 20 °C pending analysis. Blood samples were also collected in 6 ml K_3 Ethylenediaminetetraacetic acid (K_3EDTA) tubes (Vacuette; Cruinn Diagnostics, Dublin, Ireland) and in 9 ml Lithium Heparin (LH) tubes (Vacuette; Cruinn Diagnostics) on d − 14 and d 14 relative to weaning (d 0) for haematological analysis and to determine the concentration of the acute phase protein, haptoglobin respectively. LH blood samples were centrifuged at 4 °C (1600×g for 15 min) and the plasma was harvested and stored at − 20 °C until analysed.

Serological testing

The K_3EDTA blood samples were analysed immediately after collection using an ADVIA 2120 analyser (AV ADVIA 2120; Bayer Healthcare, Siemens, UK), containing software necessary for the analysis of bovine blood as described by Johnston et al. [14]. Zinc sulphate turbidity (ZST) test analysis was carried out at 520 nm using a spectrophotometer [22]. The haptoglobin concentration was measured using an automatic analyser (Olympus AU 400 Analyser; Beckman Coulter, Inc., Clare, Ireland) and a commercial assay kit (Tridelta Development Ltd., Maynooth, Ireland).

Serum samples were analysed for the presence of immunoglobulin (Ig) G against BRSV, BPI3V, BCoV (SVANOVA Biotech, Uppsala, Sweden) BHV-4 (Bio − X Diagnostics S.A., Rochefort, Belgium), BVDV (IDEXX Laboratories Inc., Maine, USA) and BHV-1 glycoprotein E (IDEXX Laboratories Inc., Maine, USA) by commercially available enzyme-linked immunosorbent assays (ELISA) each of which were performed in accordance with the manufacturer's instructions. *H. somni* serology was performed by an in-house ELISA to detect IgM/IgG against *H. somni* exopolysaccharide (EPS) as described by Pan et al. [23]. Details on the calculations used in the ELISA analyses, cut-off values employed and the correlation between ELISA results and changes in antibody status are presented in Table 1.

Table 1 Details on the calculations used in the ELISA analyses, the cut-off values employed and the correlation between the ELISA results and antibody levels

ELISA	Optical density (OD) measured at	Calculation used	Positive results	Correlation between ELISA value and antibody level
BRSV	450 nm	PP = $Sample_{COD}$ /P_{COD} X100	PP ≥ 10	Positive
BPI3V	450 nm		PP ≥ 10	Positive
BCoV	450 nm		PP ≥ 10	Positive
BHV-1 gE	650 nm	S/N ratio = $Sample_{OD}/N_2$	S/N ≤ 0.6	Negative
BHV-4	450 nm	S/P ratio = ($Sample_{OD}$ − N_1)/(P_1-N_1)	S/P ≥ 0.3	Positive
BVD	450 nm	S/P ratio = ($Sample_{OD}$ − N_2)/(P_2-N_2)	S/P ≥ 0.3	Positive
H. somni	450 nm	S/P ratio = ($Sample_{OD}$ − N_3)/(P_3-N_3)	S/P > 0.6	Positive

Note that the subsequent computation applied to the raw Optical Density (OD) result can differ between ELISA kits and has been designed to give optimal diagnostic performance. Therefore the different comparison parameters (e.g. S/N, S/P or PP values) used here are inherent to the test rather than to the pathogen
COD = corrected optical density (calculated by subtraction of the mean OD value for the negative control)
Positive control (P_1), Mean of 2 (P_2) or 3 (P_3) positive controls
Negative control (N_1), Mean of 2 (N_2) or 3 (N_3) negative controls

Statistical methods

Separate multivariable linear regression models (for BHV-1, BHV-4, BoCV, BPI3V, BVDV, BRSV and *H. somni*) were developed to quantify the effects of the various risk factors on the exposure of the calves to BRD pathogens between d – 14 and d 14. A plot of the residuals of a simple model for each dependent variable was created against the predicted values of that model and the normality of the residuals was assessed to determine if a transformation of the dependent variable was warranted; the optimal transformation being determined by Tukey's ladder of powers [24]. As the diseases studied are transmissible, the potential for clustering within pens (i.e. group) was accounted for by including pen as a random effect for each of the outcome variables examined and its significance was tested using a likelihood ratio test. For outcome variables in which clustering by pen was not significant ($p < 0.05$), a linear regression model was preferred. The change in antibody status, as reflected by changes in PP values (BPI3V, BCoV and BRSV), S/P ratios (BHV-4, BVDV and *H. somni*) or S/N (BHV-1) ratios on ELISA analyses, between d – 14 and d 14 was the dependent variable in each analysis. The S/P ratios, S/N ratios or PP values at d – 14 of each BRD pathogen was considered as a risk factor for changes in the antibody levels against other BRD pathogens in the analysis during the peri-weaning period (d – 14 to d 14) but were not included in the analysis for that specific pathogen (i.e. BHV-1 S/N at d-14 was not considered as a risk factor for BHV-1 S/N change during the peri-weaning period). Univariable analyses were performed individually to determine the association of each potential risk factor with the dependent variable using Stata v11. The risk factors included in the univariable analysis are listed in Table 2.

Risk factors identified as significant in the univariable analyses ($p < 0.20$) were included in the multivariable analysis. Independent variables were plotted against the dependent variable and, if a linear relationship did not exist between both and a transformation of the

Table 2 The potential risk factors for exposure to BRD pathogens (as reflected by S/P ratio (for BHV-4, BVDV, *H. somni*), PP value (for BRSV, BPI3V, BCoV) and S/N ratio (BHV-1)) which were included in the univariable analysis

Risk factor type	Details
Intrinsic risk factors	Calf breed (Jersey or Friesian)
	ZST score on arrival
	Arrival age
	Arrival weight
Management risk factors	Pen to which the animal was assigned ($n = 4$)
	Nutrition plane (high, medium or low)
	UFL (*unité* fourragère *lait*) intake in the pre-weaning period (d-56 to d-14)
	Percentage of UFL allowance consumed in pre-weaning period (d-56 to d-14)
	Average daily gain in pre-weaning period (d-56 to d-14)
Clinical risk factors	Maximum temperature recorded during the pre-weaning period (d-56 to d-14)
	Mean temperature recorded during the pre-weaning period (d-56 to d-14)
	Maximum respiratory score in the pre-weaning period (d-56 to d-14)
	Number of times treated for BRD in pre-weaning period (d-56 to d-14)
	Number of sickness bouts recorded in the pre-weaning period (d-56 to d-14)
	Total white blood cell count at d – 14
	Lymphocyte count at d – 14
	Haptoglobin level at d – 14
	H. somni S/P at d – 14
	BHV-4 S/P at d – 14
	BVDV S/P at d – 14
	BHV-1 S/N at d – 14
	BPI3V PP at d – 14
	BRSV PP at d – 14
	BCoV PP at d – 14

The pre-weaning period was from d – 56 to d – 14 relative to weaning at d 0. Note that BRD pathogen S/P ratios, S/N ratios or PP values at d – 14 were not considered as risk factors for changes in antibody levels for that specific pathogen during the peri-weaning period (d – 14 to d 14)

independent variable was not appropriate due to a non-linear relationship between the transformed independent variable and the dependent variable, continuous independent variables were then included as categorical variables based on the quartiles of the continuous independent variable. Plausible interactions were considered in the full model during model construction and two-way interactions identified as significant (i.e. between arrival age, arrival weight, ZST and breed) were included in the multivariable model during development. The variance inflation factor was used to detect multicollinearity. A backward stepwise selection procedure based on a likelihood ratio test ($p > 0.05$) was used to eliminate terms from the model. Regression diagnostics (heteroskedasticity, skewness and kurtosis) and plots of the residuals versus the predicted values were used to assess the presence of outliers and to check the fit of the final model.

Results

Of the 72 calves in the study, 43 were Holstein-Friesian and 29 were Jersey calves. In total, 24 animals were on the high plane of nutrition with 25 and 23 on the medium and low planes of nutrition respectively. The mean age and weight of calves on arrival were 19 (S.D. 8) days and 41.4 (S.D. 8.4) kgs respectively and the mean ZST result on arrival was 17.95 units (SD 3.46 units). A bout of sickness was defined as a concurrent recorded temperature > 39.5 °C and respiratory score of ≥5 in a calf. Sickness was not recorded during the pre-weaning period (d − 56 to d − 14) in 36 calves, was recorded once in 26 calves, twice in 7 calves, 3 times in 2 calves and 4 times in 1 calf. The mean values of S/P, PP or S/N for each of the respective BRD pathogens of interest and the change in mean values from d − 14 to d 14 (relative to weaning at day 0) are presented in Fig. 1.

A schematic diagram of significant linkages identified in the seven regression models is presented in Fig. 2. The model for the change in BRSV PP during the peri-weaning period included the pen the animal occupied as a random effect. Transformed outcome variables were used in the models of BVDV S/P change (square of the outcome variable), BHV-1 S/N change (cubed root of the outcome variable) and *H. somni* S/P change (cubed root of the outcome variable). The variables which were present in each model are displayed in Table 3.

Breed ($p < 0.001$) and higher BPI3V P *P* values at d − 14 ($p \leq 0.006$) were significantly associated with changes in the square of BVDV S/P ratios during the peri-weaning period (d − 14 to d 14 relative to weaning at d 0). Jersey breed calves experienced a significantly reduced change in squared BVDV S/P value compared to Holstein-Friesian calves as did calves with higher BPI3V PP values (between 76.5 and 130) at d − 14 compared to those with lower BPI3V PP values (between 39 and 76.5). A mean rectal temperature of 38.7 °C during the pre-weaning period was also significantly associated ($p = 0.014$) with changes in the square of BVDV S/P (Table 3).

Arrival weight was a significant risk factor ($p \leq 0.001$) for BHV-4 S/P change during the same period with calves between 34.5 kg and 59 kg all experiencing greater increases in BHV-4 S/P when compared to lighter calves (between 28.5 and 34.5 kg). BRSV PP values between 46.5 and 74 on d − 14 were also significantly associated ($p \leq 0.026$) with increases in BHV-4 S/P ratios during the peri-weaning period when compared to those with BRSV PP values between 18 and 46.5. Likewise, higher BVDV S/P ratios (between 0.547 and 1.617) at d − 14 were significantly associated ($p \leq 0.004$) with increases in BHV-4 S/P ratios during the peri-weaning period when compared to calves with lower BHV-4 S/P ratios (≤0.300).

A smaller BPI3V PP change during the peri-weaning period was significantly associated ($p \leq 0.031$) with a heavier arrival weight (34.5–49 kgs) among calves entering the calf house when compared to lighter calves. The reduced change in BPI3V PP value was also associated ($p = 0.001$) with a high respiratory score (≥8) during the pre-weaning period (d-56 to d-14) when compared to calves with the lowest respiratory scores (≤3). Increases in BPI3V PP values were associated ($p = 0.033$) with BHV-1 S/N ratios at d − 14 of > 0.434 and ≤1.046 when compared to lower S/N ratios between 0.074 and 0.140. Lymphocyte counts at the beginning of weaning (d − 14) of > 5.13 $X10^9$/l to ≤5.88 $X10^9$/l were also significantly associated with an increase in the change of BPI3V PP value ($p = 0.035$) during the peri-weaning period, when compared to those with a lymphocyte count of ≤5.13 $X10^9$/l.

BHV-1 S/N ratios decreased ($p = 0.038$) during the peri-weaning period in calves that arrived at 16–20 days of age when compared to those aged 5–12 days on arrival. This change was reversed in calves aged 27 to 41 days of age on arrival ($p = 0.026$) with an increase in BHV-1 S/N ratios when compared to the youngest arriving calves (aged 5–12 days). Decreases in BHV-1 S/N ratios during the peri-weaning period were significantly associated with haptoglobin levels of between 1.32 and 1.60 mg/ml ($p = 0.007$) when compared to haptoglobin levels between 0.75 and 1.08 mg/ml at the beginning of weaning (d − 14). Lymphocyte counts at the beginning of weaning (d − 14) between 5.13 $X10^9$/l and 5.88 $X10^9$/l were significantly associated with a decrease in BHV-1 S/N value ($p = 0.002$) during the peri-weaning period when compared to those with lower lymphocyte counts (≤ 5.13 $X10^9$/l).

Changes in the BCoV PP values during the peri-weaning period were significantly less ($p = 0.004$) in the

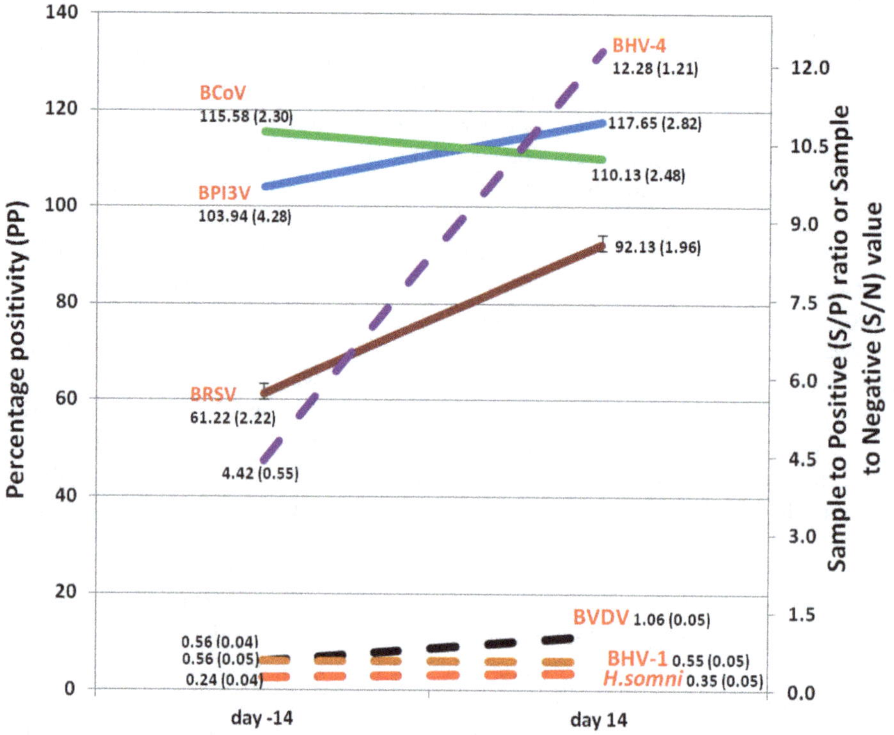

Fig. 1 The mean values of the antibody levels (as reflected by S/P. PP or S/N values) to each of the BRD pathogens of interest and the standard error of the mean (in parantheses) at d − 14 and d 14 relative to weaning at d 0. Note that solid lines (BCoV PP, BPI3V PP, BRSV PP) are plotted against the left hand y-axis while the broken lines (BVDV S/P, BHV-4 S/P, BHV-1 S/N and *H. somni* S/P) are plotted against the right hand y-axis

oldest calves (aged 27–41 days) when compared to the youngest (aged 5–12 days) calves on arrival.

A maximum respiratory score of ≥8 prior to the commencement of gradual weaning was significantly associated ($p = 0.010$) with an increase in the change of *H. somni* S/P value in the peri-weaning period. Changes in BRSV PP value during the same period were associated with the maximum rectal temperature recorded during the pre-weaning period and were also significantly clustered ($p < 0.001$) by the pen the calves occupied prior to the commencement of gradual weaning (Table 4).

Discussion

The objective of this study was to identify risk factors associated with exposure to BRD pathogens during the peri-weaning period in dairy bull calves using serological responses to these pathogens as proxy indices of exposure. These responses were determined by calculating changes in sample to positive ratios (S/P), percentage positivity (PP) or sample to negative ratios (S/N), to a wide range of recognised BRD pathogens during the peri-weaning period. The calves in this study were housed in a situation of minimum stress and management advantage to reduce adverse effects from those sources on our findings. By focusing less on disease but rather on good health, and factors influencing the calf immune response, our study identified intrinsic (breed, arrival age and arrival weight), management (pen) and clinical (rectal temperature and respiratory score) factors which were significantly associated with changes in BRD pathogen serological status.

Arrival age at housing was significantly associated with a reduction in BCoV PP ($p = 0.004$) and an increase in BHV-1 S/N ($p = 0.026$) in the oldest cohort (27–41 days) when compared to the youngest cohort (5–12 days) of calves. As both of these results are consistent with a decrease in antibody levels, it is likely that they reflect the natural decline of maternally derived antibody (MDA) with age. BCoV has been identified as a prevalent pathogen in BRD in Ireland [25] and internationally [26, 27]. Recently, Toftaker et al. [28] in a bulk tank serology cross-sectional study in Norway, identified herd size, purchase of livestock, geographic location, proximity to neighbours and seropositivity to BRSV as risk factors for seropositivity to BCoV in dairy herds. As acknowledged by Pardon et al. [29], BCoV seroconversion does not distinguish between respiratory or enteric infections, although concurrent faecal and nasal shedding is a relatively frequent event in calves [30]. Seventeen calves in the study population were recorded with a faecal

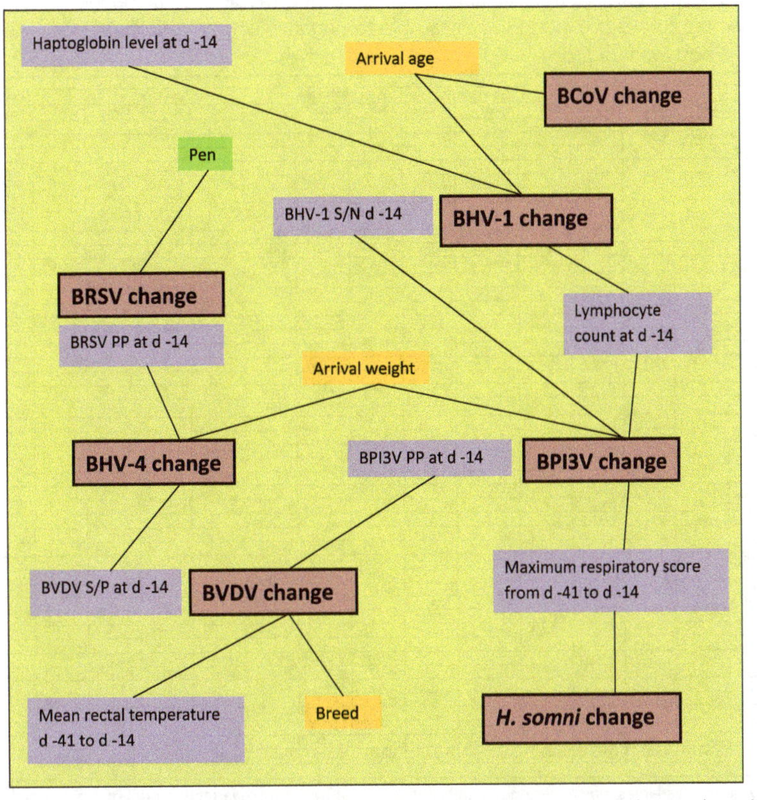

Fig. 2 A schematic diagram of the significant linkages identified in the seven models between intrinsic (orange boxes), management (green box) and clinical (purple boxes) independent variables and the dependent (pink boxes) variables. The dependent variables were antibody level changes during the peri-weaning period (d – 14 to d14 relative to weaning at d 0) as reflected by changes in S/N ratio (for BHV-1), PP value (for BRSV, BCoV and BPI3V) or S/P ratio (for BHV-4, BVDV and *H. somni*)

score of '2' (loose and watery faeces) suggestive of enteric infection during the pre-weaning period. As BCoV infection typically causes diarrhoea in calves between 3 and 21 days of age [31], it is quite likely that younger calves in this study housed at the peak age of enteric infection were more likely to acquire and spread BCoV infection than older animals arriving after the peak of enteric infection had passed.

In this study we recorded lower unit increases in BPI-3 V PP and higher unit increases in BHV-4 S/P during the peri-weaning period among heavier calves on arrival into the house when compared to the lightest group. Previously, low bodyweight at housing has been identified as a risk factor for the development of BRD at the individual [32, 33] but not at the group [34] level. While the reported BHV-4 S/P increase would appear to contradict these findings, it is important to recall that BRD pathogen antibodies reflect exposure history rather than disease. One possibility is that lower weight calves were exposed and seroconverted to BHV-4 at an earlier stage than heavier calves and the change in antibody level during the peri-weaning period reflects delayed exposure in heavier calves. This may be, in part, due to more prolonged systemic effects of BHV-4 infection as a result of its capability for latency or inherent host animal traits such as the respiratory rate or the metabolism of heavier calves may also play a role. Alternatively, lower bodyweight calves may not be capable of mounting an equivalent antibody response to that of heavier animals [35]. Further focused research examining the mechanism of the effect of bodyweight on calf immune response to BRD pathogens is warranted.

The finding of a clustering effect of exposure to BRD pathogens would seem intuitive. Brscic et al. [34] reported a higher prevalence of respiratory distress in young calves when the number of animals per pen increased. Miller et al. [36] reported that BRD outbreaks tend to cluster within calf housing systems; however, of the BRD pathogens examined in the present study, only changes in BRSV PP were significantly clustered by pen. This may indicate that BRSV is significant at the group rather than the individual level. MDA can suppress responses to BRSV resulting in a minority of BRSV seronegative calves seroconverting. Such an event may have impacted the detection of associations between risk factors and changes in BRSV PP in the current study [29, 37–39].

During the pre-weaning period, a high maximum respiratory score (≥8), reflecting typical clinical signs of

Table 3 Variables significantly associated in a multivariable linear regression model with a change in respiratory pathogen serology percentage positivity (PP), sample to positive ratio (S/P) or sample to negative control ratio (S/N) from the beginning of gradual weaning (day – 14) to day 14 post weaning (day+ 14) in housed dairy calves

		BVDV (S/P)[i]			BHV-4 (S/P)			BPI3V (PP)			BHV-1 (S/N)[ii]			BCoV (PP)			*H.somni* (S/P)[iii]		
		Coeff	Std err	p-value	Coeff	Std err	p-value	Coeff	Std err	p-value	Coeff	Std err	p-value	Coeff	Std err	p-value	Coeff	Std err	p-value
Breed	Holstein-Friesian																		
	Jersey	-0.172	0.043	<0.001															
Arrival age (days)	5-12																		
	13-15										0.182	0.143	0.210	6.869	8.176	0.404			
	16-20										-0.259	0.122	0.038	-1.709	6.919	0.806			
	21-26										-0.033	0.134	0.804	-11.217	7.229	0.125			
	27-41										0.307	0.135	0.026	-21.294	7.229	0.004			
Arrival weight (kg)	28.5 to ≤34.5																		
	>34.5 to ≤41.0				5.161	1.469	0.001	-19.812	7.356	0.009									
	>41.0 to ≤49.0				10.031	1.537	<0.001	-19.656	8.922	0.031									
	>49.0 to ≤59.0				7.283	1.634	<0.001	2.022	8.923	0.821									
Nutrition plane	Low																		
	Medium	0.060	0.045	0.183															
	High	-0.051	0.043	0.240															
Mean rectal temperature in the pre- weaning period (d -56 to d-14) (◦C)	38.2-38.5																		
	38.6-38.6	-0.069	0.051	0.181															
	38.7-38.7	0.133	0.053	0.014															
	38.8-39.4	-0.063	0.047	0.183															
Maximum respiratory score in pre- weaning period (d -56 to d -14)	1-3																		
	4-7							-9.953	7.044	0.163							0.287	0.189	0.133
	8-12							-27.675	7.918	0.001							0.566	0.213	0.010
BRSV PP at the beginning of gradual weaning (d -14)	18 to ≤46.5																		
	>46.5 to ≤60.5				5.028	1.577	0.002												
	>60.5 to ≤74				3.352	1.473	0.026												
	>74 to ≤110				2.849	1.573	0.075												
BPI3V PP at the beginning of gradual weaning (d -14)	39 to ≤76.5																		
	>76.5 to ≤100	-0.142	0.050	0.006															
	>100 to ≤130	-0.167	0.055	0.004															
	>130 to ≤200	-0.096	0.061	0.117															
BHV-1 S/N ratio at the beginning of gradual weaning (d -14)	0.074 to ≤0.140																		
	>0.140 to ≤0.434							-7.862	6.386	0.223									
	>0.434 to ≤1.046							15.309	7.010	0.033									
	>1.046 to ≤1.152							14.436	8.788	0.106									
BVDV S/P ratio at the beginning of gradual weaning (d-14)	0.003 to ≤0.300																		
	>0.300 to ≤0.547				2.490	1.538	0.111												
	>0.547 to ≤0.763				4.614	1.521	0.004												
	>0.763 to ≤1.617				5.002	1.573	0.002												
Lymphocyte count at the beginning of gradual weaning (d-14) (X10⁹/litre)	3.53 to ≤5.13																		
	>5.13 to ≤5.88							13.594	6.312	0.035	-0.391	0.124	0.002						
	>5.88 to ≤6.64							-4.133	6.568	0.532	-0.195	0.121	0.113						
	>6.64 to ≤9.82							-0.865	6.390	0.893	-0.225	0.131	0.090						
Haptoglobin level at the beginning of gradual weaning (d-14) (mg/ml)	0.75 to ≤1.08																		
	>1.08 to ≤1.32										-0.178	0.128	0.170						
	>1.32 to ≤1.60										-0.331	0.118	0.007						
	>1.60 to ≤2.80										0.005	0.133	0.967						
Intercept		0.478	0.062	<0.001	-3.504	1.557	0.028	28.649	11.299	0.014	0.129	0.123	0.299	0.091	4.406	0.984	-0.175	0.172	0.314

Note that boxes coloured in grey denote variables which were not significant ($p > 0.05$) in the multivariable model of a given pathogen
[a]The square of BVDV percentage positive change was used as the dependent variable
[b]The cubed root of the change in sample to negative control ratio was used as the dependent variable
[c]The cubed root of the change in sample to positive control percentage was used as the dependent variable

BRD, was significantly associated ($p = 0.001$) with a decrease in BPI3V PP during the peri-weaning period. While this result may appear counter-intuitive, a similar finding was reported by Tuncer and Yesilbag [40]. In that study, despite an increase in the proportion of BPI3V seropositive calves between sampling points at 1 and 2 months of age, the mean antibody levels of all calves decreased; a feature they attributed to a low BPI3V antibody response combined with a high rate of maternally-derived BPI3V antibody catabolisation. It is possible that the decrease in peri-weaning BPI3V PP in the present study, in association with a high pre-weaning maximum respiratory score, may be due to a similar phenomenon.

A high maximum pre-weaning respiratory score was also the only risk factor for *H. somni* S/P increase during the peri-weaning period. The *H. somni* ELISA employed in this study detects antibody against EPS, a major component of the *H. somni* biofilm matrix, which is produced in much greater quantities by pathogenic strains

Table 4 Linear mixed regression model of the variables associated with a change in BRSV percentage positivity (PP) from the beginning of gradual weaning (day -14) to 2 weeks post weaning (day +14)

Variable	Categories	Coefficient	Standard error	*P* value
Intercept		31.577	5.292	<0.001
Maximum rectal temperature (°C) in the pre-weaning period (d-56 to d-14)	38.9–39.4	Baseline		
	39.5–39.6	7.956	4.955	0.108
	39.7–40.0	-7.442	4.513	0.099
	40.1–41.3	-3.732	4.474	0.404
Random effects		Variance estimate	Standard error	
Pen		78.259	73.088	<0.001
Residual		194.018	34.026	

during the disease process compared to commensal strains, helping to differentiate between diseased calves and those with antibodies derived from harbouring commensal organisms [23]. Increases in *H. somni* S/P over the peri-weaning period reflect exposure to pathogenic strains of *H. somni* in the preceding weeks. Pan et al. [23] in a study of calves experimentally exposed to *H. somni*, reported antibody peaks to *H. somni* EPS between 3 and 8 weeks post-exposure. The temporal association between respiratory signs (as reflected by a high maximum respiratory score ≥ 8) during the 42 days prior to weaning and the increases in *H. somni* S/P in the peri-weaning period (a range of 4 to 10 weeks after respiratory signs were recorded) concur with the observations of Pan et al. [23] and support their conclusions that assays measuring the antibody response to EPS can identify cattle with disease due to *H. somni* infection.

Levels of antibodies to some BRD pathogens at the beginning of the peri-weaning period were significantly associated with changes in BHV-4 S/P during that period. Higher levels of BRSV and BVDV antibodies (as reflected by BRSV PP and BVDV S/P respectively) at the beginning of weaning were significantly associated ($p < 0.026$) with increases in BHV-4 S/P during the peri-weaning period. Glass et al. [41] have shown that animal genetics influence both the responses to vaccination and infection in cattle. It is possible that higher levels to BVDV and BRSV at the beginning of weaning may reflect a cohort of genetically determined 'better responders' in the study population; larger increases in BHV-4 levels in response to exposure during weaning would also be expected among these calves when compared to the poorer responders. It is notable that of the 30 highest responders to BHV-4 during the peri-weaning period, all were Holstein-Friesian from the same farm with the exception of two Jersey calves from a different farm. It is also feasible that these results reflect exposure of the calves to multiple pathogens with humoral responses to BHV-4 elicited at a slower rate when compared to those triggered by exposure to BRSV and BVDV. Although Thiry et al. [42] reported that the immune response of cattle after BHV-4 infection is characterised by low levels of neutralising antibodies, to the authors' knowledge, studies examining the rate of serological response to BRD pathogens in calves have not been undertaken.

Haptoglobin, a protein produced primarily by hepatocytes following stimulation by proinflammatory cytokines, is part of the acute-phase response to inflammation [43]. While haptoglobin has been identified as a potentially sensitive indicator of BRD in calves owing to its prolonged response [44], this response can be quite variable [45]. Orro et al. [46] concluded that the haptoglobin response requires more profound tissue damage, which is more likely to occur in bacterial rather than viral infections. Young et al. [47], following the screening of cattle at feedlot entry and at 40 and 65 days later, reported that haptoglobin had a poor predictive ability for clinical respiratory disease. In this present study, haptoglobin was only identified as a significant predictor ($p = 0.007$) of the reduction of BHV-1 S/N ratios during the peri-weaning period. Lower S/N ratios reflect higher levels of BHV-1 antibodies, and haptoglobin levels between 1.32 and 1.60 mg/ml were associated with reduced BHV-1 S/N ratios when compared to lower haptoglobin levels (≤ 1.08 mg/ml). This finding supports a potential role for haptoglobin in identifying exposure of calves to BHV-1.

Data analysis in this study presented some challenges in terms of determining antibody level changes that were significant and in considering the potential effect of MDA. We decided to use all the available information on the change in serological antibody levels (i.e. the change in BRD-specific antibody levels as reflected by the S/P, PP or S/N) rather than assuming that a particular antibody level or antibody level change should be classed as biologically significant. There were a number of reasons for the adoption of this approach. Martin and Bohac [48] reported that such an approach (using ANOVA in their study) was more powerful and more likely to identify significant associations. Windeyer et al. [49] also concluded that the detection of seroconversion was difficult in calves with high levels of MDA as these calves must produce more endogenous antibodies than calves with lower levels of MDA to achieve the four-fold increase in antibodies required for seroconversion. The range of BRD pathogen-specific MDA will vary considerably among calves following the consumption of colostrum [50] but, while BRD pathogen-specific MDA begins to decline from their second month in dairy calves [40], it was estimated by Fulton et al. [51], in a study of non-vaccinated beef calves, that calves may be approximately 6 months of age before MDA against these pathogens is no longer detectable. Calves were enrolled into the present study at a mean age of 19 days and had received colostrum, as reflected by ZST results.

Calf vaccination was necessary in this study, firstly to minimise disease in a group of calves assembled and housed as part of a larger feeding trial and, secondly, as it also reflects the reality of practices on farms where such measures are typically in place. Nevertheless it is difficult to quantify the effect of calf vaccination in this study. MDA inhibits the production of endogenous antibody in response to vaccination [52] and the half-lives of vaccinal antibodies in dairy calves [48] appear to be shorter than those in beef calves [51]. Although calves were vaccinated on arrival, of the BRD pathogen levels examined, only BPI3V and BRSV antibody levels should have been affected by MDA as the BHV-1 ELISA employed does not detect antibodies against the BHV-1 vaccine used. Studies of the serological responses of calves with MDA to inactivated BPI-3 V [53] and inactivated BRSV [54] vaccines have shown that antibody levels continue to decline following initial vaccination and that a second vaccine dose is required to increase levels (BPI3V) or slow the rate of maternal antibody decay (BRSV). The authors thus acknowledge that while MDA and vaccination may have influenced the level of some BRD pathogen-specific antibodies recorded at the beginning of gradual weaning (42 days post-vaccination), they do not consider that these interventions significantly distorted the measured changes in antibody status between the sampling points at the beginning and end of the peri-weaning period. Therefore, this study remains valid in what it reveals about BRD pathogen exposure within housed calves. The authors concede that the control of other aspects of the study (e.g. nutritional status) would suggest that the findings may not be generalisable to the wider population.

Conclusions

Serological analysis and linear regression modelling has identified associations between a number of intrinsic, management and clinical risk factors such as age at housing, the sharing of pens, pre-weaning respiratory clinical score and serum haptoglobin concentration with exposure to BRD-pathogens among group-housed dairy bull calves. The identification of such factors will contribute to the ongoing evidence-based approach to the prevention of BRD at housing.

Abbreviations

BCoV: Bovine coronavirus; BHV-1: Bovine herpesvirus 1; BHV-4: Bovine herpesvirus 4; BPI3V: Bovine parainfluenza 3 virus; BRD: Bovine respiratory disease; BRSV: Bovine respiratory syncytial virus; BVDV: Bovine viral diarrhoea virus; EDTA: Ethylenediaminetetraacetic acid; ELISA: Enzyme linked immunosorbent assay; EPS: Exopolysaccharide; Ig: Immunoglobulin; LH: Lithium heparin; PP: Percentage positivity; S/N: Sample to negative value; S/P: Sample to positive ratio; SD: Standard deviation; ZST: Zinc sulphate turbidity

Acknowledgements

We thank Anna Champion for assistance in preparation and purification of *H. somni* EPS.

Funding

No funding was obtained for this study.

Authors' contributions

GM, JC, BE, RON conceived this project. BE and DJ coordinated field sampling. TI, JG, MN, RON developed/performed the diagnostic tests. GM, TC analyzed the data. GM, JC, SM, TC, MMcE, RON, BE, DJ wrote and edited the manuscript. All authors read and approved the final manuscript.

Ethics approval and consent to participate

An experimental licence was issued by the Irish Department of Health and Children (licence number B100/2869) for all animal procedures performed in this study and procedures were compliant with the Cruelty to Animals Act (Ireland 1876, as amended by European Communities regulations 2002 and 2005) and the European Community Directive 86/609/EC. Ethical approval for this study was sought, and received, from the Teagasc Animal Ethics Committee, Teagasc, Grange, Ireland.

Consent for publication

Not applicable.

Competing interests

The authors declare that they have no competing interests.

Author details

[1]Sligo Regional Veterinary Laboratory, Department of Agriculture, Food and Marine, Doonally, Sligo, Ireland. [2]Centre for Veterinary Epidemiology and Risk Analysis, UCD School of Veterinary Medicine, University College Dublin, Belfield Dublin 4, Ireland. [3]Animal and Bioscience Research Department, Animal & Grassland Research and Innovation Centre, Teagasc, Grange, Dunsany, Co. Meath, Ireland. [4]Central Veterinary Research Laboratory, Department of Agriculture, Food and Marine, Backweston, Celbridge, Co. Kildare, Ireland. [5]Farmlab Diagnostics, Emlagh Lodge, Elphin, Co. Roscommon, Ireland. [6]Department of Biomedical Sciences & Pathobiology, Virginia-Maryland College of Veterinary Medicine, Virginia Tech, Blacksburg, VA, USA. [7]School of Veterinary Medicine, University College Dublin, Belfield Dublin 4, Ireland.

References

1. Anon. The all-island animal disease surveillance report 2015, A joint AFBI/DAFM publication. 2016. http://www.agriculture.gov.ie/rvlreport/. Accessed 8 Sept 2017.
2. Gulliksen SM, Lie KI, Løken T, Osterås O. Calf mortality in Norwegian dairy herds. J Dairy Sci. 2009;92(6):2782–95.
3. Lillie LE. The bovine respiratory disease complex. Can Vet J. 1974;15:233–42.
4. Svensson C, Emanuelsson U, Pettersson K. Health status of dairy calves kept in individual pens or in group pens with or without automatic milk feeder. Proc Xth Int Congr Anim Hyg, Maastricht, the Netherlands. 2000; 426–430.
5. Windeyer MC, Timsit E, Barkema H. Bovine respiratory disease in pre-weaned dairy calves: are current preventative strategies good enough? Vet J. 2017;224:16–7.
6. Maatje K, Verhoeff J, Kremer WD, Cruijsen AL, van den Ingh TS. Automated feeding of milk replacer and health control of group-housed veal calves. Vet Rec. 1993;133(11):266–70.
7. Svensson C, Lundborg K, Emanuelson U, Olsson SO. Morbidity in Swedish dairy calves from birth to 90 days of age and individual calf-level risk factors for infectious diseases. Prev Vet Med. 2003;58:179–97.
8. Svensson C, Linder A, Olsson SO. Mortality in Swedish dairy calves and replacement heifers. J Dairy Sci. 2006;89(12):4769–77.
9. Dabo SM, Taylor JD, Confer AW. *Pasteurella multocida* and bovine respiratory disease. Anim Health Res Rev. 2007;8(2):129–50.
10. Ackermann MR, Derscheid R, Roth JA. Innate immunology of bovine respiratory disease. Vet Clin North Am Food Anim Pract. 2010;26(2):215–28.
11. Hodgins DC, Conlon JA, Shewen PE. Respiratory Viruses and Bacteria in Cattle. In: Brogden KA, Guthmiller JM, editors. Polymicrobial Diseases. Washington (DC): ASM Press; 2002. p. 213–30.
12. Murray GM, More SJ, Sammin D, Casey MJ, McElroy MC, O'Neill RG, Byrne WJ, Earley B, Clegg TA, Ball H, Bell CJ, Cassidy JP. Pathogens, patterns of pneumonia, and epidemiologic risk factors associated with respiratory disease in recently weaned cattle in Ireland. J Vet Diagn Investig. 2017;29(1):20–34.

13. Jasper J, Budzynska M, Weary DM. Weaning distress in dairy calves: acute behavioural responses by limit-fed calves. Appl Anim Behav Sci. 2008;110:136–43.
14. Johnston D, Kenny DA, Kelly AK, McCabe MS, McGee M, Waters SM, Earley B. Characterisation of haematological profiles and whole blood relative gene expression levels in Holstein-Friesian and Jersey bull calves undergoing gradual weaning. Animal. 2016;10(9):1547–56.
15. Babcock AH, White BJ, Renter DG, Dubnicka SR, Scott HM. Predicting cumulative risk of bovine respiratory disease complex (BRDC) using feedlot arrival data and daily morbidity and mortality counts. Can J Vet Res. 2013;77(1):33–44.
16. Quimby WF, Sowell BF, Bowman JG, Branine ME, Hubbert ME, Sherwood HW. Application of feeding behaviour to predict morbidity of newly received calves in a commercial feedlot. Can J Anim Sci. 2001;81:315–20.
17. Wolfger B, Schwartzkopf-Genswein KS, Barkema HW, Pajor EA, Levy M, Orsel K. Feeding behavior as an early predictor of bovine respiratory disease in North American feedlot systems. J Anim Sci. 2015;93:377–85.
18. Aich P, Babiuk LA, Potter AA, Griebel P. Biomarkers for prediction of bovine respiratory disease outcome. OMICS. 2009;13:199–209.
19. National Research Council. Nutrient Requirements of Dairy Cattle. 7th ed. Washington, DC: The National Academies Press; 2001.
20. McGuirk S, Peek S. Timely diagnosis of dairy calf respiratory disease using a standardized scoring system. Anim Health Res Rev. 2014;15(2):145–7.
21. Lago A, McGuirk SM, Bennett TB, Cook NB, Nordlund KV. Calf respiratory disease and pen microenvironments in naturally ventilated calf barns in winter. J Dairy Sci. 2006;89:4014–25.
22. McEwan AD, Fisher EW, Selman IE, Penhale WJ. A turbidity test for the estimation of immune globulin levels in neonatal calf serum. Clin Chim Acta. 1970;27:155–63.
23. Pan Y, Fisher T, Olk C, Inzana TJ. Detection of antibodies to the biofilm exopolysaccharide of *Histophilus somni* following infection in cattle by enzyme-linked immunosorbent assay. Clin Vaccine Immunol. 2014;21(10):1463–7.
24. Tukey JW. Relation among powers and logs (optional). In: Turkey JW, editor. Exploratory Data Analysis. Reading: Addison-Wesley; 1977. p. 89.
25. O'Neill R, Mooney J, Connaghan E, Furphy C, Graham DA. Patterns of detection of respiratory viruses in nasal swabs from calves in Ireland: a retrospective study. Vet Rec. 2014;175(14):351–6.
26. Storz J, Stine L, Liem A, Anderson GA. Coronavirus isolation from nasal swab samples in cattle with signs of respiratory tract disease after shipping. J Am Vet Med Assoc. 1996;208:1452–5.
27. Moore SJ, O'Dea MA, Perkins N, Barnes A, O'Hara AJ. Mortality of live export cattle on long-haul voyages: pathologic changes and pathogens. J Vet Diagn Investig. 2014;26:252–65.
28. Toftaker I, Sanchez J, Stokstad M, Nødtvedt A. Bovine respiratory syncytial virus and bovine coronavirus antibodies in bulk tank milk - risk factors and spatial analysis. Prev Vet Med. 2016;133:73–83.
29. Pardon B, De Bleecker K, Dewulf J, Callens J, Boyen F, Catry B, Deprez P. Prevalence of respiratory pathogens in diseased, non-vaccinated, routinely medicated veal calves. Vet Rec. 2011;169:278–83.
30. Hasoksuz M, Hoet AE, Loerch SC, Wittum TE, Nielsen PR, Saif LJ. Detection of respiratory and enteric shedding of bovine coronaviruses in cattle in an Ohio feedlot. J Vet Diagn Investig. 2002;14(4):308–13.
31. Saif LJ, Heckert RA. Enteric coronaviruses. In: Saif LJ, Theil KW, editors. Viral diarrheas of man and animals. Boca Raton: CRC Press; 1990. p. 185–252.
32. Martin SW, Bateman KG, Shewen PE, Rosendal S, Bohac JE. The frequency, distribution and effects of antibodies, to seven putative respiratory pathogens, on respiratory disease and weight gain in feedlot calves in Ontario. Can J Vet Res. 1989;53(3):355–62.
33. Gummow B, Mapham PH. A stochastic partial-budget analysis of an experimental *Pasteurella haemolytica* feedlot vaccine trial. Prev Vet Med. 2000;43(1):29–42.
34. Martin SW, Bateman KG, Shewen PE, Rosendal S, Bohac JG, Thorburn M. A group level analysis of the associations between antibodies to seven putative pathogens and respiratory disease and weight gain in Ontario feedlot calves. Can J Vet Res. 1990;54(3):337–42.
35. Brscic M, Leruste H, Heutinck LF, Bokkers EA, Wolthuis-Fillerup M, Stockhofe N, Gottardo F, Lensink BJ, Cozzi G, Van Reenen CG. Prevalence of respiratory disorders in veal calves and potential risk factors. J Dairy Sci. 2012;95(5):2753–64.
36. Miller WM, Harkness JW, Richards MS, Pritchard DG. Epidemiological studies of calf respiratory disease in a large commercial veal unit. Res Vet Sci. 1980;28:267–74.
37. Kimman TG, Westenbrink F, Schreuder BE, Straver PJ. Local and systemic antibody response to bovine respiratory syncytial virus infection and reinfection in calves with and without maternal antibodies. J Clin Microbiol. 1987;25:1097–106.
38. Kimman TG, Zimmer GM, Westenbrink F, Mars J, van Leeuwen E. Epidemiological study of bovine respiratory syncytial virus infections in calves: influence of maternal antibodies on the outcome of disease. Vet Rec. 1988;123:104–9.
39. Härtel H, Nikunen S, Neuvonen E, Tanskanen R, Kivelä SL, Aho R, Soveri T, Saloniemi H. Viral and bacterial pathogens in bovine respiratory disease in Finland. Acta Vet Scand. 2004;45(3–4):193–200.
40. Tuncer P, Yeşilbağ K. Serological detection of infection dynamics for respiratory viruses among dairy calves. Vet Microbiol. 2015;180(3–4):180–5.
41. Glass EJ, Baxter R, Leach RJ, Jann OC. Genes controlling vaccine responses and disease resistance to respiratory viral pathogens in cattle. Vet Immunol Immunopathol. 2012;148(1–2):90–9.
42. Thiry E, Dubuisson J, Bublot M, Van Bressem MF, Pastoret PP. The biology of bovine herpesvirus-4 infection of cattle. Dtsch Tierarztl Wochenschr. 1990;97(2):72–7.
43. Baumann H, Gauldie J. The acute phase response. Immunol Today. 1994;15:74–80.
44. Angen O, Thomsen J, Larsen LE, Larsen J, Kokotovic B, Heegaard PM, Enemark JM. Respiratory disease in calves: microbiological investigations on trans-tracheally aspirated bronchoalveolar fluid and acute phase protein response. Vet Microbiol. 2009;137(1–2):165–71.
45. Wittum TE, Young CR, Stanker LH, Griffin DD, Perino LJ, Littledike ET. Haptoglobin response to clinical respiratory tract disease in feedlot cattle. Am J Vet Res. 1996;57(5):646–9.
46. Orro T, Pohjanvirta T, Rikula U, Huovilainen A, Alasuutari S, Sihvonen L, Pelkonen S, Soveri T. Acute phase protein changes in calves during an outbreak of respiratory disease caused by bovine respiratory syncytial virus. Comp Immunol Microbiol Infect Dis. 2011;34(1):23–9.
47. Young CR, Wittum TE, Stanker LH, Perino LJ, Griffin DD, Littledike ET. Serum haptoglobin concentrations in a population of feedlot cattle. Am J Vet Res. 1996;57:138–41.
48. Martin SW, Bohac JG. The association between serological titers in infectious bovine rhinotracheitis virus, bovine virus diarrhea virus, parainfluenza-3 virus, respiratory syncytial virus and treatment for respiratory disease in Ontario feedlot calves. Can J Vet Res. 1986;50(3):351–8.
49. Windeyer MC, Leslie KE, Godden SM, Hodgins DC, Lissemore KD, LeBlanc SJ. Association of bovine respiratory disease or vaccination with serologic response in dairy heifer calves up to three months of age. Am J Vet Res. 2015;76:239–45.
50. Chamorro MF, Walz PH, Haines DM, Passler T, Earleywine T, Palomares RA, Riddell KP, Galik P, Zhang Y, Givens MD. Comparison of levels and duration of detection of antibodies to bovine viral diarrhea virus 1, bovine viral diarrhea virus 2, bovine respiratory syncytial virus, bovine herpesvirus 1, and bovine parainfluenza virus 3 in calves fed maternal colostrum or a colostrum-replacement product. Can J Vet Res. 2014;78(2):81–8.
51. Fulton RW, Briggs RE, Payton ME, Confer AW, Saliki JT, Ridpath JF, Burge LJ, Duff GC. Maternally derived humoral immunity to bovine viral diarrhea virus (BVDV) 1a, BVDV1b, BVDV2, bovine herpesvirus-1, parainfluenza-3 virus bovine respiratory syncytial virus, *Mannheimia haemolytica* and *Pasteurella multocida* in beef calves, antibody decline by half-life studies and effect on response to vaccination. Vaccine. 2004;22:643–9.
52. Nonnecke BJ, Waters WR, Goff JP, Foote MR. Adaptive immunity in the colostrum-deprived calf: response to early vaccination with Mycobacterium Bovis strain bacille Calmette Guerin and ovalbumin. J Dairy Sci. 2012;95(1):221–39.
53. Kaeberle ML, Sealock R, Honeyman M. Antibody responses of young calves to inactivated viral vaccines. In beef research report. 1997; https://lib.dr.iastate.edu/cgi/viewcontent.cgi?article=1028&context=beefreports_1997. Accessed 9 Aug 2017
54. Patel JR, Didlick SA. Evaluation of efficacy of an inactivated vaccine against bovine respiratory syncytial virus in calves with maternal antibodies. Am J Vet Res. 2004;65(4):417–21.

18

Perinatal immuno/inflammatory responses in the presence or absence of bovine fetal infection

Paulina Jawor[1*], John F. Mee[2] and Tadeusz Stefaniak[1]

Abstract

Background: It is known that the bovine fetus can mount an immune and inflammatory reaction to infection, but it is not known whether there is a contemporaneous maternal response. Nor is it known whether the response of calves which die perinatally, with or without infection, differs from that of live perinates. Hence, the objective of this study was to determine if acute phase reactant and immunoglobulin concentrations differed between calves (and their dams) in three groups: live calves (CC; $n = 21$) and dead calves with (PM INF+; $n = 22$) or without (PM INF-; $n = 89$) in utero infection. In calf plasma, serum amyloid A, haptoglobin, immunoglobulins M, G_1 and G_2 and interleukin-6 were measured. In dam serum, SAA and Hp was measured and in amniotic and abomasal fluid, IL-6 was measured.

Results: Live calves had higher plasma concentrations of SAA and IL-6 than dead calves with (PM INF+) or without (PM INF-) in utero infection. Calves in the PM INF-, but not PM INF+ group, had higher Hp concentrations than calves in the CC group. Calves in the PM INF+ group had higher IgG_1 concentrations than calves in the PM INF- and CC groups. Except for higher IgG_1 and IgG_2 concentrations, biomarker values did not differ significantly between dead calves with or without in utero infection. Live calves had higher IL-6 concentrations in abomasal fluid compared to PM INF- calves. There were no significant differences in blood biomarker concentrations between dams of the three groups of calves. Amniotic fluid IL-6 concentrations were higher from the dams of control calves than the dams of uninfected calves.

Conclusions: Differences in biomarkers (higher Hp and IgG_1; lower SAA and IL-6) between perinatal mortalities and live perinates probably reflect differences between these two groups in age at sampling (SAA and IL-6) and in utero infection (IgG_1). Out of the six analytes measured in calves, only IgG_1 and IgG_2 were biomarkers of (chronic) in utero infection.

Keywords: Acute phase proteins, Immunoglobulins, Interleukin-6, Stillborn, Infection in utero

Background

Bovine perinatal mortality (PM) may be defined as the death of a calf/calves prior to, during, or up to 48 h after birth, following full-term pregnancy (≥260 d) [1, 2]. Perinatal mortality rates internationally vary between 4 and 7% [3] and increasing stillbirth rates have been reported internationally [1, 4, 5]. While much of this loss has been attributed to dystocia [6], an increasing proportion of calf loss occurs at eutocia ('the unexplained stillbirth'). Some of this loss is caused by infections [7]. But, given the limitations in detecting infectious agents in calf carcasses, current estimates of the proportion of loss attributable to infections may underestimate the extent of their contribution. However, instead of attempting to detect infectious agents, it may be possible to use biomarkers of infection to explain some of this currently unexplained mortality.

The acute phase response is a non-specific, multifactorial response which develops after any disturbance of homeostasis, in particular during inflammation, designed to remove and replace damaged tissue. Damage to tissue activates cells which release acute phase cytokines including, in decreasing order of importance; interleukin-6 (IL-6), interleukin-1 beta (IL-1b) and tumor necrosis factor-alpha (TNF-a) [8]. The systemic reaction to proinflammatory

* Correspondence: paulina.jawor@upwr.edu.pl
[1]Department of Immunology, Pathophysiology and Veterinary Preventive Medicine, Wroclaw University of Environmental and Life Sciences, 31 C.K. Norwida, 50-375 Wrocław, Poland
Full list of author information is available at the end of the article

cytokines is production and secretion of acute phase proteins (APPs) by the liver [9]. In cattle immunoglobulins (Igs) from the dam do not cross the placenta and high fetal serum immunoglobulin (Ig) concentrations are strong indicators of fetal infection [10].

In human medicine the reaction of fetus to inflammatory stimuli is described as the fetal inflammatory response syndrome (FIRS) and is diagnosed by elevated IL-6 concentration in fetal blood. FIRS is typically diagnosed in pre-term fetuses, but can also occur in term fetuses [11, 12]. Because APPs are released early in the inflammatory response, testing for APPs allows early detection of inflammation. Hence, APPs have been widely used as biomarkers for detecting sick animals [13], monitoring of calf group health [14, 15], monitoring of response to treatment [16, 17], or to monitor the response to repeated liver biopsy [18, 19]. The most common APPs measured in cattle are haptoglobin (Hp) and serum amyloid A (SAA) [20].

Our recent research, showed, for the first time, that measurement of SAA and IgM concentrations may be used in the diagnosis of bacterial, but not of viral or parasitical infections in cases of bovine PM [21]. However, to date the APP and Igs concentrations in contemporaneous cases of PM, control live calves and their dams have not been examined. Thus, the hypotheses tested in cases of PM, control live calves and their dams in this study were (1) cases of PM without diagnosed infection in utero would exhibit a similar acute phase and immune response to control live calves, (2) cases of PM infected in utero would exhibit a greater acute phase and immune response than uninfected cases of PM or live control calves and (3) that the dams of these calf groups would not differ in acute phase response after calving. Hence, the objective of this study was to determine if APP and Ig concentrations differed between live calves and dead calves (and their respective dams) with or without diagnosed infection in utero.

Results

The plasma concentrations of SAA, Hp, Igs M, G_1, G_2 and IL-6 are shown in Table 1. Live calves (CC), had higher plasma concentrations of SAA and IL-6 than dead calves with (PM INF+, SAA $P < 0.05$; IL-6 $P < 0.01$) or without in utero infection (PM INF-, SAA and IL-6 $P < 0.01$). Calves in the PM INF-, but not PM INF+, group had higher Hp concentrations than calves in the CC group ($P < 0.05$). Calves in the PM INF+ group had higher IgG_1 median concentrations than calves in the PM INF- ($P < 0.05$) and CC ($P < 0.01$) groups and concentrations of IgG_1 tended to be higher in PM INF- compared to CC group ($P < 0.10$). Immunoglobulin M and IgG_2 values did not differ between control and dead calf groups. Except for higher IgG_1 and IgG_2 concentrations, biomarker values did not differ significantly between dead calves with or without in utero infection.

Table 1 Biomarkers in plasma of control calves and dead calves with and without in utero infection

Analyte	CC $n = 21$	PM INF+ $n = 22$	PM INF- $n = 89$
SAA (mg/L)	19.3^{Aa} (24.7 ± 15.9)	13.8^{b} (15.9 ± 17.3)	6.3^{B} (9.4 ± 9.9)
Hp (mg/L)	0.0^{a} (1.2 ± 3.6)	0.2 (3.6 ± 5.0)	0.0^{b} (6.0 ± 13.1)
IgM (mg/L)	21.4 (26.8 ± 25.6)	29.2 (42.2 ± 47.8)	16.6 (61.1 ± 153.5)
IgG_1 (mg/L)	4.2^{A}† (9.2 ± 17.8)	45.1^{Ba} (56.5 ± 79.1)	9.5^{b}† (59.3 ± 160.3)
IgG_2 (mg/L)	4.1 (4.8 ± 5.1)	10.8^{a} (31.1 ± 40.2)	2.8^{b} (8.6 ± 20.9)
IL-6 (pg/mL)	307.0^{A} (304.1 ± 17.2)	200.9^{B} (229.2 ± 70.4)	200.8^{B} (229.8 ± 69.0)

Median (mean ± SD) concentrations; SAA serum amyloid A; Hp haptoglobin; IgG_1 immunoglobulin G_1; IgM immunoglobulin M; IgG_2 immunoglobulin G_2; IL-6 interleukin-6; CC control calves; PM INF+ dead calves with infection in utero; PM INF- dead calves without in utero infection

A, B Values within a row with different uppercase superscripts differ ($P < 0.01$)

a, b Values within a row with different lowercase superscripts differ ($P < 0.05$)

† $P < 0.1$ and > 0.05

Significantly more calves in the CC group had high concentrations of SAA compared to the two PM calf groups (Table 2). The proportion of calves with high analyte concentrations for other biomarkers was not different between groups.

Live calves had higher IL-6 concentrations in abomasal fluid compared to PM INF- calves (mean ± SD 319.6 ± 41.4 pg/mL vs. 238.9 ± 7.6 pg/mL; $P < 0.05$) and tended to have higher concentrations than PM INF+ calves (mean ± SD 232.5 ± 92.5 pg/mL; $P < 0.10$).

There were no significant differences in blood biomarker concentrations between dams of the three groups of calves (Table 3). However, amniotic fluid IL-6 concentrations were higher from the dams of control calves than the dams of uninfected calves (Table 3).

Table 2 Percentage of calves within groups above reference thresholds for SAA, Hp, IgM and IgG values

Analyte	Threshold (mg/L)	CC $n = 21$	PM INF+ $n = 22$	PM INF- $n = 89$
SAA*	≥29.5	38.1	18.2	4.4
Hp	≥100	0	0	1.1
IgM	≥52.1	4.8	22.7	20.2
IgG	≥160	0	18.2	1.1

Reference thresholds for serum amyloid a and IgM from Jawor et al. [21], haptoglobin from Richter [22] and IgG from Sawyer [23]. SAA serum amyloid A; Hp haptoglobin; IgG sum of IgG_1 and IgG_2 classes; CC control calves; PM INF + dead calves with infection in utero; PM INF- dead calves without in utero infection

*$P \leq 0.0001$ Pearson's chi-square

Table 3 Plasma SAA, Hp and amniotic fluid IL-6 concentrations from dams of control and dead calves

Analyte	D CC $n = 21$	D INF+ $n = 21$	D INF- $n = 81$
SAA (mg/L)	58.6 (124.1 ± 129.1)	91.4 (98.2 ± 37.1)	89.9 (120.1 ± 105.1)
Hp (mg/L)	20.2 (362.5 ± 575.2)	45.8 (185.3 ± 288.1)	37.2 (236.6 ± 414.3)
IL-6 (pg/mL)	340.6[a] (331.1 ± 29.1)	NS	279.2[b] (274.6 ± 55.3)

NS not sampled; median (mean ± SD) concentrations; *SAA* serum amyloid A; *Hp* haptoglobin; *IL-6* interleukin-6; D CC dams of control calves; D INF+ dams of dead calves with in utero infection; D INF- dams of dead calves without in utero infection

[a,b]Values within a row with different lowercase superscripts differ ($P < 0.05$)

None of the amniotic fluid samples came from dams of PM INF+ calves.

Discussion

The results generated here present a unique opportunity to contemporaneously compare the acute phase and immune response in calves which survive with those who die at birth, whether infected or not, and acute phase proteins in their dams. Previous work only examined APPs in dead or live calves [24] or APPs and Igs in perinatal mortalities [21], but not both together and not in their dams.

The most rapidly responding markers of the acute phase reaction (IL-6, SAA) were significantly lower in the dead (irrespective of infection status) compared to the live calves, while the slower responding marker (Hp) was significantly higher only in the dead calves without infection in utero compared to the live calves. Among Igs, IgG_1 and IgG_2 concentrations were highest in infected dead calves while IgM concentrations did not differ between groups. Apart from amniotic IL-6, biomarker values did not differ between dams of live and dead calves.

Many of these are surprising findings which in some cases support, but in others, refute our a priori hypotheses that (1) calves which died perinatally without diagnosed infection in utero would exhibit a similar acute phase and immune response to calves which survived, (2) PM calves infected in utero would exhibit a greater acute phase and immune response than uninfected calves with perinatal mortality or control calves which survived and (3) that the dams of these calf groups would not differ in acute phase response after calving. The results of this study refute hypothesis one (for most biomarkers, except IgM and IgG_2) and two (for SAA, Hp, IgM, IL-6 and partly IgG_2), but support hypothesis three.

Regarding hypothesis one, the main reason why live and uninfected dead calves did not have similar SAA values may lie in their ages; the live calves were older (up to 9 h old) than the dead calves (99% < 1 h) when blood samples were collected. Immediately after birth calf SAA concentrations are low [24, 25], but rise rapidly, peaking at day 7 and decreasing after approximately 3 weeks [24]. However, in the present study the perinatal mortalities did not have the time to mount a postnatal SAA response, similar to that found in the live calves. They also did not mount a prenatal APP response whether they were infected in utero or not. This may be explained in the infected fetuses by the chronic nature of the infections detected here. Infected fetuses had time (days to weeks) to mount a pathogen-specific antibody response, thus the rise in SAA would have occurred sometime before birth and had subsided back to normal concentrations at birth. In a previous study on perinatal mortalities [21], SAA concentrations in calves without infection were also low, and all of these calves also died within 1 h of calving. The absence of infection [21, 26] or colostrum feeding [24] in the live calves in the present study, rule out the two other most likely causes of higher SAA concentrations in the live calves. This is the first study to report SAA concentrations in live healthy newborn calves which had not consumed colostrum up to 9 h after calving thus establishing a new reference range for such newborn calves. Samples in previous studies were either taken within a very short time after calving (10 min) [25] or after colostrum feeding [24].

Contrary to hypothesis one (that calves which died perinatally without diagnosed infection in utero would exhibit a similar acute phase response to calves which survived), Hp concentrations were higher in uninfected dead calves than in live calves. Haptoglobin is a slowly reacting APP [27]. Thus it might be expected that concentrations would be similar or higher in live compared to dead perinates as they had more time to mount a postnatal Hp response. However, the opposite was the case; Hp concentrations were higher in the dead perinates. This suggests that some prenatal factors may have influenced this response. But, Hp concentrations were not significantly higher in infected compared to uninfected dead calves. This suggests the antigenic stimulation of in utero infection was insufficient to maintain prolonged elevated Hp concentrations. Irrespective of calf group, Hp concentrations were very low suggesting that differences or lack of differences in concentrations between groups may have limited biological significance.

In support of hypothesis one, apart from IgG_1, Ig concentrations (M and G_2) did not differ between live and uninfected dead calves. The presence of (low) Ig concentrations in all three groups' pre-colostral samples is not surprising [23]. The bovine fetus has the ability to produce IgM and IgG after 90 and 111 days of pregnancy, respectively [28], and the older the fetus, the higher the concentrations of IgM and IgG [23]. Though numerically higher IgG_1 concentrations were found in uninfected dead calves compared to live calves, these concentrations were very low suggesting limited biological significance.

Contrary to hypothesis one (that calves which died perinatally without diagnosed infection in utero would exhibit a similar acute phase response to calves which survived), IL-6 concentrations (plasma and abomasal fluid) were higher in live compared to dead calves. While it is known that concentrations of IL-6 peak in cows before calving [29, 30] and that bidirectional transfer of IL-6 occurs between the human fetus and mother [31], neither adequately explain the discrepancy in IL-6 concentrations between the live and dead calves reported here. And while lower concentrations of IL-6 are secreted by human trophoblast cells under hypoxic compared to normoxic conditions [32], paradoxically, studies in human infants show higher plasma IL-6 concentrations in cases of severe hypoxia at birth [33, 34]. Thus, while one may speculate that the perinatal mortalities in the present study suffered more severe hypoxia (and ultimately, anoxia) than the live calves, the effects of hypoxia on IL-6 concentrations in bovine perinates have yet to be elucidated. Given this lack of basic information on bovine perinatal IL-6 profiles it is suggested that the higher concentrations detected here may again reflect the time lag required for perinates to mount a significant IL-6 response after birth. Interleukin-6 is the most potent inducer of APPs and it alone is able induce significant production of SAA, but not Hp, in bovine hepatocytes [35]. Thus, higher concentrations of IL-6 in control calves stimulated higher production of SAA in control calves.

In hypothesis two it was posited that if PM calves were infected in utero they would exhibit a greater acute phase and immune response than uninfected live or dead calves; this was not generally supported by the data generated here. The main reason for this may lie in the type of infections detected in the calves. The infections detected were chronic in nature rather than acute (fetuses had time to mount a pathogen-specific antibody response) and none of the live calves was infected. A previous study in dead perinates only has shown that differences in the APP response are detectable between dead calves with acute, but not chronic, infections [21]. Thus it is perhaps not surprising that SAA concentrations were not higher in infected dead calves as both the chronic nature of the in utero infections and the older age of the live calves at sampling contributed to higher SAA concentrations in the live rather than the dead calves, irrespective of infection status.

Infected (dead) calves had numerically higher and uninfected dead calves had significantly higher Hp concentrations than live (all uninfected) calves. Given the chronic nature of the infections and the slow rising Hp values (concentrations remain above baseline levels for 10–15 days) [36], the higher Hp concentrations in infected dead calves support hypothesis two (that PM calves infected in utero would exhibit a greater acute phase response than uninfected live or dead calves). However, the significantly higher Hp values in the uninfected group refute this hypothesis. The large within-group variability in Hp values (as discernable by the median and mean values) may indicate that biological differences may not have been detectable at these very low Hp concentrations.

In support of hypothesis two (that PM calves infected in utero would exhibit a greater immune response than uninfected live or dead calves), infected (dead) calves had significantly higher concentrations of IgG_1 and IgG_2 than uninfected (dead) calves and significantly higher concentrations of IgG_1 than live (uninfected) calves. These findings are in agreement with a previous report on infected and uninfected dead perinates [21]. High serum Ig concentrations, particularly IgG_1, the most abundant Ig, are a good marker of fetal infection [10]. In calves there is a lag phase between fetal infection and detectable Ig concentrations. For example, in experimental studies, IgM is detectable approximately 2 weeks and IgG_1 approximately 3 weeks after fetal infection with BVDv and as the interval between infection and cesarotomy increased, the concentration of IgG_1 increased more compared to IgM in the majority of fetuses [37]. Consistent with these findings, in the present study, the lag time between infection and sampling influenced the IgG_1 and IgG_2 classes more than IgM.

Perinatal mortality calves with diagnosed infection in utero were posited in hypothesis two to have higher concentration of IL-6, as human fetuses with diagnosed infection in utero showed elevated IL-6 concentrations [12]. However, the opposite was the case here; live calves had higher concentrations. As IL-6 values did not differ between infected and uninfected dead calves, it is suggested that the higher values in the live calves reflected older age at sampling and the lack of difference between dead calf groups was due to the chronic nature of the infections. While increased IL-6 production is a part of the acute phase reaction, the half-time of IL-6 elimination in humans lasts from minutes to hours [38].

The results generated here supported our third hypothesis that the dams of live and infected and uninfected dead calves would not differ in acute phase response after calving. While high concentrations of APPs (SAA, Hp) were detected in the three dam groups, these were expected in cows immediately after calving [39, 40]. These results are not surprising if it is accepted that the factors affecting perinate APP concentrations (e.g. age at sampling, infection, calving trauma, colostrum feeding) are not collinear with those affecting maternal APP concentrations. The only analyte which showed similar trends in dams and calves was IL-6. Differences in IL-6 concentrations between groups were consistent for amniotic fluid, abomasal and plasma samples which may be due to bidirectional transfer of IL-6 between the dam and her fetus. Apart from one study, in which amniotic fluid IL-6 concentration from a calving with a *Salmonella Stanley*-infected dead calf [41] was 6-times higher than in live calves in the present study, there are no other bovine perinatal IL-6 studies. Thus the data generated here represent baseline values for IL-6 concentrations in newborn calves and their dams.

Conclusions

Significant differences in biomarkers (higher Hp and IgG_1; lower SAA and IL-6) between perinatal mortalities and live perinates probably reflect differences between these two groups in age at sampling (SAA and IL-6) and in utero infection (IgG_1). Adjustment for age at sampling should be considered when interpreting acute phase protein and IL-6 concentrations in newborn calves. Additionally, the interval between infection and sampling affected some Ig classes more than others. Out of the six analytes measured in calves, only IgG_1 and IgG_2 were biomarkers of (chronic) in utero infection. While significantly higher IL-6 concentrations were found in samples from dams with live calves, the biological significance of these findings is unclear given the paucity of current knowledge on bovine perinatal IL-6 profiles.

Methods

Ethical authorisation

The study design was approved by the II Local Ethics Committee in the Wroclaw University of Environmental and Life Sciences (permission no. 23/2012, 58/2014, 60/2014).

Animals

This prospective study was carried out on calves which died in the perinatal period ($n = 121$) and their dams ($n = 110$; 11 cows had twins) and live control calves (CC; $n = 21$) and their dams ($n = 21$). All animals in this study came from Polish commercial dairy farms. Material was collected over a 20-month period (11.2013–06.2015). Perinatal mortalities were defined as calves (66 females and 55 males) born following a gestation of ≥260 days which died before, during or within 6 h after birth without receiving colostrum. A case definition of death within 24 h was planned, but only two calves survived the first hour and no calves died between 6 and 24 h after birth. These calves were either Holstein-Friesian ($n = 113$) or Holstein-Friesian crossbreds (Simmental, Jersey, Limousin, Brown Swiss sires; $n = 8$).

Control calf inclusion criteria were the same gestation length as perinatal mortalities, singleton calving, Holstein-Friesian breed and assisted but not difficult calving. None of the control calf had received colostrum before sampling (6 females and 15 males). From this group six male calves (selected on the basis that the farmers agreed to sell the newborn male calf) were euthanized between 1.30 and 8.30 h after birth (without receiving colostrum). They were first premedicated with 1 mL (20 mg) xylazine (Sedazin®, Biowet Pulawy) and 1 mL (100 mg) ketamine (Bioketan®, Vetoquinol) IV and then they were euthanized with a mixture of pentobarbital and pentobarbital sodium, 160 mg/mL (Morbital®, Biovet Pulawy) IV at a dose of 48–96 mg/kg bw.

Herds

The study was conducted in 30 herds located in the south-west of Poland. Herds were recruited on farms located within 2.5 h one-way driving distance from Wroclaw. Farm staff were provided with a mobile phone number to call the veterinary staff who collected the material as soon after calf death as possible, on all days and times (day and night). Details about farm management, herd records and the calving associated with the calf were collected at farm visits. The median herd size was 73 with a range of 1–1037 cows/herd and the mean (SD) inter-calving period and previous lactation milk yield was 415 (38) days and 8884 (1442) kg/cow/305 DIM, respectively. Fifteen herds were vaccinated against BVD alone ($n = 7$), BoHV-1 alone ($n = 4$) or against BVD and BoHV-1 ($n = 4$) in 2013–2015. Free-stall housing system was present on 14 farms, tie-stall housing system was present on 14 farms and both systems were present on two farms. Total mixed ration was fed on 18 farms and on 12 farms ingredients were fed separately.

PM cases and CC were collected from 29 and 5 herds, respectively. Between 1 and 35 cases of PM and between 2 and 7 CC were investigated per herd. The degree of calving assistance was recorded in seven categories (supplied to the farmers): unobserved, observed but not assisted, normal assisted calving without calving jack, normal assisted calving with calving jack, difficult calving with calving jack, difficult calving without calving jack (assistance by at least three people and/or a veterinarian) and caesarean section.

Samples and laboratory examinations

All carcasses (perinatal mortalities and euthanized control calves) were subjected to systematic external and internal gross examinations according to the same project-specific protocol. Necropsies of all PM calves were performed 8 ± 3 h after calving. Necropsies of the six CC were performed within 6 ± 3 h after calving and within 2 h ± 40 min after death. Abomasocentesis was performed with a sterile needle and syringe to collect abomasal fluid immediately after opening the abdominal cavity. Blood was aseptically collected from PM and CC by syringe from a jugular vein and immediately dispensed into 5 mL lithium heparinized tubes (MEUS Srl®, Piove di Sacco, Italy, 18,648) and centrifuged (14 min; 1860 x g). Plasma samples were aliquoted and frozen at − 80 °C until analysed. Farm staff collected amniotic fluid from the amniotic sac at six calvings of PM calves and from 20 CC calvings. Blood was collected from the dams of PM calves when the carcass was collected. Blood from the dams of CC and their calves was collected within 9 h of calving. Maternal samples were collected from the coccygeal vessels into 9 mL tubes with lithium-heparin (Sarstedt, Germany, cat no. 02.1065.001) and 10 mL serum separator tubes (MEUS Srl®, Piove di Sacco, Italy, cat no. 18203). Maternal samples (blood and amniotic fluid) were centrifuged within 4 h of collection (14 min at 1860 g at room temperature). Plasma and amniotic fluid were aliquoted and frozen at – 80 °C until analysis.

In calf plasma, SAA, Hp, IgM, IgG_1, IgG_2 and IL-6 were measured. In dam serum, SAA and Hp was measured and in amniotic and abomasal fluid, IL-6 was measured.

SAA was analyzed by a multispecies ELISA (TP 802, Tridelta Development, Maynooth, Ireland) with intra- and inter-assay coefficients of variation (CV) of 17.4 and 6.1%, respectively, and a detection limit of 4.6 mg/L. Haptoglobin was determined by ELISA (#2410–7 Life Diagnostics Inc., Knypersley, UK) with intra- and inter-assay CVs of 5.9% and 7.8%, respectively. The detection limit was 7.8 mg/L. Interleukin - 6 was analysed by ELISA (Qayee-Bio) with intra- and inter-assay CVs of 4.9 and 3.5%, respectively. The detection limit was 124.4 pg/mL. Immunoglobulin M, G_1 and G_2 concentration was determined by ELISA (E10–101, E10–116, E10–117 Bethyl Laboratories, Montgomery, TX). The intra- and inter-assay CVs for IgM was 4.3 and 1.2%, for IgG_1 12.8 and 2.0% and for IgG_2 14.4% and 2.1%, respectively. The limits of detection for IgM, IgG_1 and IgG_2 were 6.9, 13.2 and 4.1 mg/L, respectively.

In calf plasma, pathogen-specific antibodies were measured: *Neospora caninum*, glycoprotein B of BoHV-1 and BVDv using the IDEXX Neospora Ab Test, IDEXX IBR gB X3 Ab Test and IDEXX BVDV Total Ab Test (Hoofddorp, The Netherlands), SBV using the Serological ID Screen Schmallenberg Virus Competition Multi-Species ELISA (ID-vet, Grabels, France), *Leptospira hardjo* and *Leptospira pomona* using live strains of *L. hardjo* and *L. pomona* in the microscopic agglutination test at the veterinary diagnostic laboratory as described in Jawor et al. [7] and *Coxiella burnetii* in the IDEXX Q Fever Ab Test (Hoofddorp, The Netherlands) (Krol et al., unpublished).

Experimental groups

Calves and their dams were each divided into six mutually exclusive groups. Three calf groups were compared; two PM and one CC group. The first PM group (PM INF+; $n = 22$) included only calves in which pathogen-specific antibodies (BVDv, $n = 3$; *Coxiella burnetii*, $n = 6$; *Leptospira hardjo*, n = 3, *Leptospira hardjo* and *L. pomona*, $n = 1$; *Neospora caninum*, $n = 5$ or SBV, $n = 4$), but not antigens, were detected. Calves with infection in utero diagnosed by the detection of antigen/s [7] were excluded from the study as multiple infections and co-infections were detected and their number was too small for statistical analyses ($n = 10$). Diagnosis of infection was made only in the calves, not their dams. The second PM group (PM INF-; $n = 89$) included calves in which no infection (neither antibody nor antigen) was diagnosed. The third group was the control calves (CC; $n = 21$); infection in utero was not diagnosed in any of the CC. The majority of calvings in all three groups were assisted (CC 100%, INF+ 86% and INF- 92%). Three dam groups were compared; dams of PM INF+ calves (D INF+), dams of PM INF- calves (D INF-) and dams of control calves (D CC). Results from calves and dams were analysed separately.

Statistical analysis

For each test, the experimental unit was an individual animal. Data were checked for normality using the Lilliefors test. All parameters had abnormal distributions therefore logarithmic transformation (log10) was performed. After transformation only SAA values in cows were normally distributed. SAA values were tested using ANOVA and post hoc comparisons were made using Tukey's HSD test. Values without normal distributions were analyzed using the Kruskal-Wallis ANOVA with post hoc multiple comparisons of mean ranks in cases where three group were compared and the U-Mann-Whitney test was used in cases where two group were compared. The frequencies of cases above and below the normal threshold values for APPs and Igs were compared between groups by Pearson's chi-square analysis. Statistical analyses were conducted using Statistica 12.5 (StatSoft Inc., Tulsa, OK, USA).

Abbreviations

APP(s): acute phase protein(s); CC: control calves; D CC: dams of control calves; D INF-: dams of dead calves without in utero infection; D INF+: dams of dead calves with in utero infection; FIRS: fetal inflammatory response syndrome; Hp: haptoglobin; Ig(s): immunoglobulin(s); IL-6: interleukin 6; PM INF-: dead calves without in utero infection; PM INF+: dead calves with infection in utero.; PM: perinatal mortality; SAA: serum amyloid A

Acknowledgements

Not Applicable.

Funding

This study was funded by The National Centre for Research and Development [project NCBR no PBS2/A8/20/2013 "Diagnosis of intrauterine infection in weak and stillborn calves. Assessment of the usefulness of selected parameters as potential markers"].

Authors' contributions

PJ, TS designed and coordinated the study. PJ, TS collected the samples. PJ analyzed the samples and conducted the statistical analyses. PJ, JFM, TS performed the literature review. PJ, JFM drafted the manuscript. All authors read and approved the final manuscript.

Consent for publication
Not applicable.

Competing interests
The authors declare that they have no competing interest.

Author details
[1]Department of Immunology, Pathophysiology and Veterinary Preventive Medicine, Wroclaw University of Environmental and Life Sciences, 31 C.K. Norwida, 50-375 Wrocław, Poland. [2]Animal and Bioscience Research Department, Teagasc, Moorepark Research Centre, Fermoy, Co. Cork, Ireland.

References
1. Berglund B, Steinbock L, Elvander M. Causes of stillbirth and time of death in Swedish Holstein calves examined post mortem. Acta Vet Scand. 2003;44:111–20.
2. Mee JF, Sánchez-Miguel C, Doherty M. Influence of modifiable risk factors on the incidence of stillbirth/perinatal mortality in dairy cattle. Vet J. 2014; 199:19–23.
3. Mee JF, Berry DP, Cromie AR. Prevalence of, and risk factors associated with, perinatal calf mortality in pasture-based Holstein-Friesian cows. Animal. 2008;2:613–20.
4. Mee JF. Prevalence and risk factors for dystocia in dairy cattle: a review. Vet J. 2008;176:93–101.
5. Meyer CL, Berger PJ, Koehler KJ, Thompson JR, Sattler CG. Phenotypic trends in incidence of stillbirth for holsteins in the United States. J Dairy Sci. 2001;84:515–23.
6. Khodakaram-Tafti A, Ikede BO. A retrospective study of sporadic bovine abortions, stillbirths, and neonatal abnormalities in Atlantic Canada, from 1990 to 2001. Can Vet J. 2005;46:635–7.
7. Jawor P, Król D, Mee JF, Sołtysiak Z, Dzimira S, Larska M, et al. Infection exposure, detection and causes of death in perinatal mortalities in polish dairy herds. Theriogenology. 2017;103:130–6.
8. Heinrich PC, Castell JV, Andus T. Interleukin-6 and the acute phase response. Biochem J. 1990;265:621–36.
9. Eckersall PD, Conner JG. Bovine and canine acute phase proteins. Vet Res Commun. 1988;12:169–78.
10. Ohmann HB. Immunoglobulin levels in non-aborted and aborted fetuses from Danish herds of cattle. Acta Vet Scand. 1981;22:428–34.
11. Bashiri A, Burstein E, Mazor M. Cerebral palsy and fetal inflammatory response syndrome: a review. J Perinat Med. 2006;34:5–12.
12. Chaiworapongsa T, Romero R, Kim JC, Kim YM, Blackwell SC, Yoon BH, et al. Evidence for fetal involvement in the pathologic process of clinical chorioamnionitis. Am J Obstet Gynecol. 2002;186:1178–82.
13. Deignan T, Alwan A, Kelly J, McNair J, Warrens T, O'Farrelly C. Serum haptoglobin: an objective indicator of experimentally-induced Salmonella infection in calves. Res Vet Sci. 2000;69:153–8.
14. Furman-Fratczak K, Rzasa A, Stefaniak T. The influence of colostral immunoglobulin concentration in heifer calves' serum on their health and growth. J Dairy Sci. 2011;94:5536–43.
15. Gånheim C, Alenius S, Persson Waller K. Acute phase proteins as indicators of calf herd health. Vet J. 2007;173:645–51.
16. Smith BI, Donovan GA, Risco CA, Young CR, Stanker LH. Serum haptoglobin concentrations in Holstein dairy cattle with toxic puerperal metritis. Vet Rec. 1998;142:83–5.
17. Jawor P, Steiner S, Stefaniak T, Baumgartner W, Rzasa A. Determination of selected acute phase proteins during the treatment of limb diseases in dairy cows. Vet Med (Praha). 2008;53:173–83.
18. Vels L, Røntved CM, Bjerring M, Ingvartsen KL. Cytokine and acute phase protein gene expression in repeated liver biopsies of dairy cows with a lipopolysaccharide-induced mastitis. J Dairy Sci. 2009;92:922–34.
19. Jawor P, Brzozowska A, Słoniewski K, Kowalski ZM, Stefaniak T. Acute phase response in the primiparous dairy cows after repeated percutaneous liver biopsy during the transition period. Pol J Vet Sci. 2016;19:393–9.
20. Eckersall PD, Bell R. Acute phase proteins: biomarkers of infection and inflammation in veterinary medicine. Vet J. 2010;185:23–7.
21. Jawor P, Stefaniak T, Mee JF. Immune and inflammatory biomarkers in cases of bovine perinatal mortality with and without infection in utero. J Dairy Sci. 2017;100:1408–16.
22. Richter H. Haptoglobin bei haussäugtieren. III. Mitteilung: Der Haptoglobingehalt im blutplasma und serum von wiederkäuern und schweinen unter verschiedenen physiologischen bedingungen. Arch Exp Vet. 1974;28:505–19.
23. Sawyer M, Osburn BI, Knight HD, Kendrick J. A quantitative serologic assay for diagnosing congenital infections of cattle. Am J Vet Res. 1973;34:1281–4.
24. Orro T, Jacobsen S, LePage JP, Niewold T, Alasuutari S, Soveri T. Temporal changes in serum concentrations of acute phase proteins in newborn dairy calves. Vet J. 2008;176:182–7.
25. Alsemgeest SPM, Jonker FH, Taverne MAM, Kalsbeek HC, Wensing T, Gruys E. Serum amyloid-a (SAA) and haptoglobin (Hp) plasma concentrations in newborn calves. Theriogenology. 1995;43:381–7.
26. Erol E, Jackson C, Horohov D, Locke S, Smith J, Carter C. Elevated serum amyloid a levels in cases of aborted equine fetuses due to fetal and placental infections. Theriogenology. 2016;86:971–5.
27. Jacobsen S, Andersen PH, Toelboell T, Heegaard PMH. Dose dependency and individual variability of the lipopolysaccharide-induced bovine acute phase protein response. J Dairy Sci. 2004;87:3330–9.
28. Ellis WA, Logan EF, O'Brien JJ. Serum immunoglobulins in aborted and non-aborted bovine foetuses. Clin Exp Immunol. 1978;33:136–41.
29. Dervishi E, Zhang G, Hailemariam D, Dunn SM, Ametaj BN. Occurrence of retained placenta is preceded by an inflammatory state and alterations of energy metabolism in transition dairy cows. J Anim Sci Biotechnol. 2016;7:26.
30. Trevisi E, Jahan N, Bertoni G, Ferrari A, Minuti A. Pro-inflammatory cytokine profile in dairy cows: consequences for new lactation. Ital J Anim Sci. 2015;14:3862.
31. Zaretsky MV, Alexander JM, Byrd W, Bawdon RE. Transfer of inflammatory cytokines across the placenta. Obstet Gynecol. 2004;103:546–50.
32. Shirasuna K, Shimamura N, Seno K, Ohtsu A, Shiratsuki S, Ohkuchi A, et al. Moderate hypoxia down-regulates interleukin-6 secretion and TLR4 expression in human sw.71 placental cells. Cell Physiol Biochem. 2015;36:2149–60.
33. Šumanović-Glamuzina D, Čulo F, Čulo MI, Konjevoda P, Jerković-Raguž M. A comparison of blood and cerebrospinal fluid cytokines (IL-1β, IL-6, IL-18, TNF-α) in neonates with perinatal hypoxia. Bosn J Basic Med Sci. 2017;17:203–10.
34. Chaparro-Huerta V, Flores-Soto ME, Ernesto M, Sigala M, Barrera De León JC, De M, et al. Proinflammatory cytokines, enolase and S-100 as early biochemical indicators of hypoxic-ischemic encephalopathy following perinatal asphyxia in newborns. 2017;58:70–76.
35. Alsemgeest SP, van 't Klooster GA, van Miert AS, Hulskamp-Koch CK, Gruys E. Primary bovine hepatocytes in the study of cytokine induced acute-phase protein secretion in vitro. Vet Immunol Immunopathol. 1996;53:179–84.
36. Gånheim C, Hultén C, Carlsson U, Kindahl H, Niskanen R, Waller KP. The acute phase response in calves experimentally infected with bovine viral diarrhoea virus and/or Mannheimia haemolytica. J Vet Med B Infect Dis Vet Public Health. 2003;50:183–90.
37. Brown TT, Schultz RD, Duncan JR, Bistner SI. Serological response of the bovine fetus to bovine viral diarrhea virus. Infect Immun. 1979;25:93–7.
38. Castell JV, Geiger T, Gross V, Andus T, Walter E, Hirano T,et al. Plasma clearance, organ distribution and target cells of interleukin-6/hepatocyte-stimulating factor in the rat. Eur J Biochem. 1988;177:357-61.
39. Chan JP-W, Chang C-C, Hsu W-L, Liu W-B, Chen T-H. Association of increased serum acute-phase protein concentrations with reproductive performance in dairy cows with postpartum metritis. Vet Clin Pathol. 2010;39:72–8.
40. Uchida E, Katoh N, Takahashi K. Appearance of haptoglobin in serum from cows at parturition. J Vet Med Sci. 1993;55:893–4.
41. Jawor P, Stefaniak T, Sołtysiak Z, Dzimira S, Bednarski M. Salmonella enterica serovar Stanley intrauterine infection in a stillborn calf - case report. Acta Vet Brno. 2013;82:363–7.

19

Recombinant E2 protein enhances protective efficacy of inactivated bovine viral diarrhea virus 2 vaccine in a goat model

Yao-Chi Chung[1], Li-Ting Cheng[1], Jia-Yu Zhang[1], Yue-Jyun Wu[1], Shyh-Shyan Liu[2] and Chun-Yen Chu[1*]

Abstract

Background: Inactivated and subunit bovine viral diarrhea virus (BVDV) vaccines have shown limited protective efficacy. This study aimed to evaluate the effectiveness of a vaccine containing both inactivated BVDV (iBVDV) and baculovirus-expressed recombinant E2 (rE2), an important BVDV antigen with strongly neutralizing epitopes.

Results: Four groups of goats were immunized twice with one of four vaccine preparations: 1) iBVDV+rE2, 2) rE2, 3) iBVDV, and 4) saline, and challenged with BVDV. For goats vaccinated with the iBVDV+rE2 vaccine, no viremia was observed after challenge, and clinical signs, pyrexia, and leukopenia were reduced compared to the saline group. In contrast, for goats vaccinated with either iBVDV or rE2 alone, viremia was still detectable.

Conclusion: The combination of iBVDV and rE2 elicited stronger protective immune responses against BVDV than iBVDV or rE2 alone.

Keywords: BVDV, Inactivated vaccine, Subunit vaccine, E2

Background

Bovine viral diarrhea virus (BVDV) infects immune cells and causes short-term leukopenia, immunosuppression, pyrexia, and diarrhea in infected animals [1]. It is strongly associated with the bovine respiratory disease complex that can result in severe economic losses for the cattle industry [2]. Infection of pregnant cattle with BVDV can result in abortions and stillbirths. BVDV belongs to the *Pestivirus* genus of the *Flaviviridae* family and its single-stranded positive-sense RNA genome encodes one polyprotein that is cleaved into 11 or 12 viral proteins, including envelope proteins (E^{rns}, E1, and E2), the capsid protein (C), and nonstructural proteins. Based on the nucleotide sequences of the 5′ untranslated region, BVDV can be classified into two species, BVDV-1 and BVDV-2, each species containing a number of subgenotypes. The high genetic diversity of BVDV makes controlling of disease difficult [3]. In Taiwan, only BVDV-2 has been reported.

* Correspondence: cychu@mail.npust.edu.tw
[1]Graduate Institute of Animal Vaccine Technology, College of Veterinary Medicine, National Pingtung University of Science and Technology, 1, Shuehfu Road, Neipu, Pingtung 91201, Taiwan
Full list of author information is available at the end of the article

Both inactivated and subunit vaccines have been developed against BVDV, but neither offers complete protection. With inactivated BVDV (iBVDV) vaccines, protection comes mainly from humoral response directed at the E2 protein, and the duration and range (across different serotypes) of protection are limited [3]. When the E2 protein is used as a subunit vaccine, only partial protection is observed (lack of pyrexia and reduction in both leukopenia and nasal virus shedding) [4, 5].

In this study, we evaluated the protective efficacy of a vaccine that included both iBVDV and baculovirus-expressed, recombinant E2 (rE2) protein. Since data indicated that the E2 protein serves as the major antigen with neutralizing epitopes [3], we hypothesized that the effectiveness of iBVDV vaccines, which have production titers limited at around 10^8 $FAID_{50}/mL$ (50% fluorescent antibody infectious dose), can be enhanced by the addition of rE2. Furthermore, baculovirus expression of E2 protein in insect cells allows for post-translational modifications, such as protein folding and glycosylation. A water-in-oil-in-water (w/o/w) adjuvant was used for prolonged antigen presentation [6]. Immunization and challenge experiments were performed on goats since they are smaller and more economical for the evaluation of various BVDV vaccine formulations.

Methods

BVDV strains and virus titer determination

A BVDV-2 strain, BVDV/TW 2008, was obtained from the Animal Health Research Center, Council of Agriculture, Taiwan, and cultured for E2 gene cloning. For iBVDV vaccine production and to be used as the challenge strain, a more recent BVDV-2 field isolate, BVDV/TW 2014, was obtained from the Large Animal Hospital, NPUST, Taiwan. The virus strains were propagated in Madin-Darby bovine kidney (MDBK) cells (BCRC 60126; Bioresource Collection and Research Center, Taiwan) using Eagle's Minimum Essential Medium supplemented with 7% fetal bovine serum. Since the BVDV-2 strains used are non-cytopathic, 50% $FAID_{50}$ was employed to determine virus titer. In 96-well plates, virus sample (100 μL) and MDBK cells (100 μL of 1×10^5/mL) were co-cultured for 6 days at 37 °C, 5% CO_2. Buffered Formalde-Fresh (Thermo Fisher Scientific, MA, USA) was used for cell fixing and anti-BVDV fluorescein-conjugated polyclonal antiserum (VMRD, WA, USA) at 1:5 dilution was used for virus detection. Fluorescent wells were observed under a fluorescent microscope and BVDV $FAID_{50}$ was calculated using the Reed and Muench method [7].

Cloning and expression of BVDV E2

A partial segment (nucleotide 55 to 1026 of the full 1116 bps) of the E2 gene was cloned from the BVDV/TW2008 strain using reverse transcription-polymerase chain reaction (RT-PCR), with the following primers: forward: 5′- GCGGGATCCGGGTTATTGGGGCCA GAGAGT-3′, and reverse: 5′-ATAGCGGCCGCTAT GAACTCTGAAAAGTAATC-3′. Briefly, TRIzol® Reagent (Thermo Fisher Scientific, MA, USA) was used for viral RNA extraction according to the manufacturer's instructions. Applied Biosystems High Capacity cDNA Reverse Transcription Kits (Applied Biosystems, CA, USA) were used for cDNA production and PCR reaction was carried out using Ex Taq (Takara, Shiga, Japan) in the Thermocycler (Takara, Shiga, Japan). The E2 PCR products were inserted into the pGM-T vector using the pGM-T Cloning Kit (GeneMark, Taichung, Taiwan) and the sequence of the amplified E2 gene was determined. Using the Clustal W method in MegAlign of the DNAStar software (DNASTAR, WI, USA), the E2 sequence was compared to that of other BVDV strains from Taiwan, China, USA, and Japan.

To produce recombinant baculovirus expressing BVDV E2 protein, the BAC-TO-BAC® Baculovirus Expression System (Invitrogen, CA, USA) was employed following the manufacturer's instructions. Briefly, using *Bam*HI and *Not*I restriction sites, the E2 gene was subcloned from pGM-T into the pFastBac HT B vector, resulting in pFastBac HT B/E2. pFastBac HT B/E2 was then used to transform DH10Bac competent cells to produce recombinant bacmid DNA, which is the recombinant baculoviral genome. Using the Cellfectin II® Reagent (Invitrogen, CA, USA), the bacmid DNA was then used to transfect Sf9 insect cells for recombinant baculovirus production. The recombinant baculovirus stock was aliquoted in 2% fetal bovine serum (FBS)-Grace's medium and stored at − 80 °C. For rE2 protein expression, confluent Sf9 cells were infected with recombinant virus stock at multiplicity of infection (MOI) of 0.1 for 5 days at 27 °C and observed for cytopathic effect before cell collection.

For rE2 protein quantitation using sodium dodecyl sulfate-polyacrylamide gel electrophoresis (SDS-PAGE) analysis, 10 μL of protein sample was mixed with 2× sample buffer (100 mM Tri-HCl pH 6.8, 200 mM dithiothreitol, 4% sodium dodecyl sulfate, 0.2% bromophenol, 20% glycerol) in a 1.5 mL Eppendorf tube and boiled for 10 min in hot water. To establish a standard curve, serial dilution of bovine serum albumin (BSA, KPL, MD, USA) was prepared. Gel electrophoresis was performed on the samples and 0.1% Coomassie blue was used for staining for 12 h before decolorization. Image analysis of the resulting gel was performed using Uvitec to obtain a BSA standard curve. Protein concentration was deduced using regression analysis.

To evaluate the antigenicity of rE2 using Western blotting, after gel electrophoresis, protein samples were transferred onto polyvinylidene difluoride (PVDF) membranes (Amersham biosciences, Buckinghamshire, UK) using the TE22 Mini Tank Transfer Unit (GE Healthcare, WI, USA), with transfer buffer (25 mM Tris, 192 mM glycine, 20% methanol) run at 300 mA. After transfer, PVDF membranes were blocked with 1X BSA blocking buffer (KPL/SeraCare, MD, USA) and then washed three times using phosphate buffered saline-Tween 20 (PBST) (1xPBS, 0.5% Tween 20). For the primary antibody, 1:500 dilution of bovine BVDV convalescent serum was used for overnight incubation at 4 °C. After washing three times with PBST, the secondary antibody, 1:5000 dilution of goat anti-bovine IgG-HRP (KPL/SeraCare, MA, USA) was applied for 1 h at 37 °C before washing again with PBST three times. For color development, ECL plus Western Blot Detection Reagents (GE Healthcare, WI, USA) were used according to the manufacturer's instructions. Results were analyzed using the G Box Chemi-luminescence Imaging System (Syngene, MD, USA).

Vaccine preparation

Vaccines containing iBVDV and rE2 protein, emulsified in w/o/w adjuvant, were prepared. For rE2 protein preparation, baculovirus in Sf9 cells expressing rE2 proteins were inactivated using 0.3% binary ethylenimine (Sigma-Aldrich, MI, USA) at 37 °C for 10 h. Inactivation reaction was terminated by the addition of 1/10 volume $Na_2S_2O_3$ (Alfa Aesar, MA, USA). Sterility of the preparation was confirmed by streaking 1 mL of preparation on Tryptone soy agar plates for 14 days at 37 °C to make sure that no growth occurred. Recombinant protein preparation was spun at 1000×g for 10 min and the pellet was collected as the rE2 protein fraction. No protein purification was performed. For iBVDV preparation, 24 h of BEI inactivation was performed and sterility of the sample was also confirmed as above.

For single component vaccines, 1×10^8 $FAID_{50}$/mL iBVDV/TW2014 and 200 µg/mL rE2 protein were prepared. For vaccine containing both iBVDV and rE2, concentrated rE2 preparation was added to iBVDV to achieve the same final concentrations as the single component vaccines. Finally, antigen preparations were mixed with the adjuvant ISA 206 VG (SEPPIC, Puteaux, France) in a 1:1 (*v*/v) ratio using Top Three-way Stopcocks (Meditop, Persiaran Usahawan, Malaysia). The prepared vaccines were stored at 4 °C for animal experiments.

Immunization and challenge

Twelve 4–5-month-old, BVDV-negative goats (native to Taiwan, raised at NPUST farms), as confirmed using the LSIVet™ BVD/BD p80 Blocking One Step Kit (Thermo Fisher Scientific, MA, USA), were randomly divided into four groups of three animals each and immunized twice, 2 weeks apart, intramuscularly with one of the four vaccine preparations: 1) iBVDV+rE2, 2) rE2, 3) iBVDV, and 4) saline, at 2 mL per dose. On week 4 after primary immunization, the four groups were intranasally challenged with 2 mL of 1×10^8 $FAID_{50}$/mL BVDV/TW2014. For immune response evaluation, blood was collected on week 0, 2, and 4 post primary immunization for antibody and peripheral bone marrow cells (PBMCs) analyses. After challenge, blood samples were collected daily for analyses of leukopenia and viremia. Temperatures and clinical signs were also recorded daily after challenge. Goats were sacrificed 3 weeks after challenge by intravenous overdose of barbiturate. Animal experiments were carried out at the NPUST Barrier and Containment Facility, and all experimental protocols for animal trials were approved by the NPUST Animal Care and Use Committee.

Antibody analysis

Whole blood was collected and spun at 800×g, 4 °C, for 10 min to separate coagulated blood cells from the serum, which was collected for ELISA analysis. Flat-bottomed 96-well plates were coated with inactivated BVDV/TW2014 (1.25 µg/mL) in the coating buffer (15 mM Na_2CO_3, 35 mM $NaHCO_3$, 3 mM $NaNO_3$, pH = 9.6) at 4 °C overnight. The plates were washed with PBST (0.5% Tween) 3 times and blocked by 1% BSA in PBST at 37 °C for 1 h. After blocking, goat serum samples from the four experimental groups, diluted 1:4000 with PBS, were added into the wells (100 µL/well) and incubated at 37 °C for 1.5 h. Serum from the saline group acted as negative control. Thereafter, the plates were washed 3 times with PBST, and rabbit anti-goat IgG HRP at 1:10,000 dilution (KPL) was added (100 µl/well). The plates were incubated at 37 °C for 1 h, washed 3 times with PBST, and 100 µL TMB 2-component Microwell Peroxidase substrate (KPL) was added to each well. The reaction was stopped by adding 100 µL of TMB stop solution (KPL) after 5 min and the plates were read at 450 nm by a multi-well plate reader.

T-cell proliferation assay

To collect goat PBMCs for T-cell proliferation assay, whole blood was collected in ethylenediaminetetraacetic acid (EDTA, an anti-coagulant) tubes and spun at 800×g, 4 °C, for 10 min. The upper layer of serum was removed and the remaining blood cells were resuspended in equal volumes of PBS. The resuspended blood cells were then layered slowly on an equal volume Ficoll-Paque™ PLUS (GE Healthcare, Stgiles, Sweden) in a 50 mL tube and spun at 1500×g, 4 °C, for 40 min. The resulting middle PBMC layer was collected, rid of red blood cells using 1% acetic acid, washed with PBS, and resuspended in RPMI-1640 (10% FBS and 0.05 mM --mercaptoethanol). For T-cell proliferation assay, in 96-well plates, goat PBMCs (100 µL of 1×10^6 cell/mL) were seeded and stimulated with 10 µg/well of iBVDV for 72 h. Concanavalin A (ConA) (Sigma-Aldrich, MI, USA) at 1 mg/mL was used as positive control and non-stimulated cells acted as negative control. The proliferation extent of T-cells was then measured using the CellTiter 96® AQueous One Solution Cell Proliferation Assay (Promega, WI, USA), per the manufacturer's instructions. OD_{492nm} was read and stimulation index (SI) calculated, SI = (OD of treatment − OD of background)/(OD of the negative control − OD of background). Wells containing only RPMI-1640 and 20 mL of tetrazolium acted as the background level.

Clinical evaluation

Blinded clinical examinations were performed daily after BVDV challenge. Severity of clinical signs were judged by a veterinarian and assigned a numerical value according to Table 1 [8]. Daily clinical scores were calculated by summing up the numerical values of the five clinical parameters.

Table 1 Parameters used to calculate clinical scores in BVDV-infected goats

Clinical parameter	State description	Numerical value
General state	Normal	0
	Moves slowly, head down	1
	Lying down/staggers	2
	Recumbent	3
Appetite	Normal	0
	Reduced	1
	Absent	2
Rectal temperature	< 39.5 °C	0
	39.5 °C -39.9 °C	1
	≥ 40.0 °C	2
Nasal discharge	Normal	0
	Slight uni–/bilat. Serous	1
	Moderate bilat. Serous to purulent	2
	Copious bilat. Purulent	3
Feces state	Normal	0
	Celiac affection	2
	Mucohemorrhagic diarrhea	4

Hematology

To determine if leukopenia occurred, 0.5 mL of whole blood was collected daily post-challenge in EDTA tubes and 1% acetic acid was added to lyse red blood cells. White blood cells were counted under the microscope at 200X magnification.

Virus detection

Viremia was detected using quantitative RT-PCR (RT-qPCR). RNA was extracted from collected whole blood (in EDTA tubes) using the Total RNA Mini Kit (Geneaid Biotech, New Taipei City, Taiwan). Applied Biosystems High Capacity cDNA Reverse Transcription Kits (Applied Biosystems, Foster City, CA, USA) was used for cDNA production. qPCR detection of viral cDNA was performed using a continuous fluorescence detector with the SYBR1 Green Realtime PCR Master Mix (Takara, Shiga, Japan). Primers used are based on the 5′ untranslated region of BVDV, forward 5’-GCCA TGCCCTTAGTAGGACTAGC-3′, reverse: 5′- CAAC TCCATGTGCCATGTACAGC-3′ [9]. To control for mRNA levels, data were normalized against the values obtained for β-actin.

Statistical analysis

Statistical analysis was performed using SAS 9.0 general linear models and Duncan's Multiple Range Test was used for comparison between groups, with different letters indicating statistical significance, $p < 0.05$. Data are shown as mean ± standard deviation. To test for statistically significant difference in viremia for the vaccine groups, viremia results from 7 to 13 days post challenge were compared using the Pearson chi-square test in the software Statistica V.13.

Results

iBVDV and rE2 were prepared as vaccines

The E2 gene, excluding the c-terminal transmembrane domain, of BVDV/TW2008 was cloned for protein expression in a baculovirus expression system. The sequence of the cloned E2 gene was compared to that of other strains (Fig. 1a) to determine if the BVDV/TW2008 E2 gene would still be protective when expressed as an antigen. When compared to the E2 sequence of BVDV/TW2014, which was used as the challenge strain in this study, nucleotide sequences showed 99.8% identity and amino acid sequences showed 100% identity, suggesting that the cloned E2 gene will be protective against the challenge strain. Expressed rE2 moved at the 42 kDa position and quantitation was done using SDS-PAGE analysis (Fig. 1b). Antigenicity of rE2 was verified with convalescent bovine anti-sera (Fig. 1c). Four different formulations were prepared from iBVDV and rE2, with w/o/w adjuvant: 1) iBVDV+rE2, 2) rE2, 3) iBVDV, and 4) saline as negative control.

iBVDV+rE2 vaccine enhanced cell-mediated immunity

To evaluate the immune response elicited by the formulated vaccines, four groups of BVDV-naïve goats, three in each group, were immunized and boosted 2 weeks

a

Similarity matrix of nucleotide sequence

	1	2	3	4	5	6	7	8
1		99.8	99.2	99.1	93.7	93.1	91.7	84.2
2	100		99.0	98.9	93.5	92.9	91.5	84.0
3	98.5	98.5		99.7	93.3	92.8	91.4	83.7
4	98.8	98.8	99.7		93.2	92.7	91.5	83.6
5	95.1	95.1	94.8	95.1		97.0	90.4	83.6
6	93.5	93.5	92.9	93.2	95.7		90.0	82.7
7	91.7	91.7	91.0	91.4	91.0	90.7		81.4
8	85.5	85.5	84.6	84.9	84.9	84.0	82.4	

Similarity matrix of amino acid sequence

1. BVDV/TW 2014 (Taiwan)
2. BVDV/TW 2008 (Taiwan)
3. FJ527854.1 (China)
4. FJ795045.1 (China)
5. JN387139.1 (USA)
6. U18059.1 (USA)
7. AB359944.1 (Japan)
8. AF145968.1 (USA)

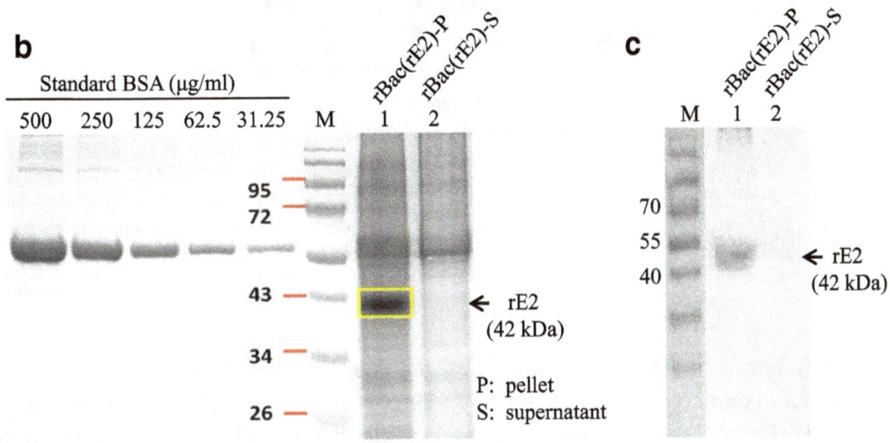

Fig. 1 Cloning and expression of the BVDV E2 gene. The E2 gene of BVDV/TW2008 was cloned for expression in a baculovirus expression system and the sequence was compared to that of other BVDV-2 strains (**a**). Expression and quantitation recombinant E2 (rE2) was done using SDS-PAGE analysis (**b**). Antigenicity was verified using bovine convalescent anti-serum (**c**)

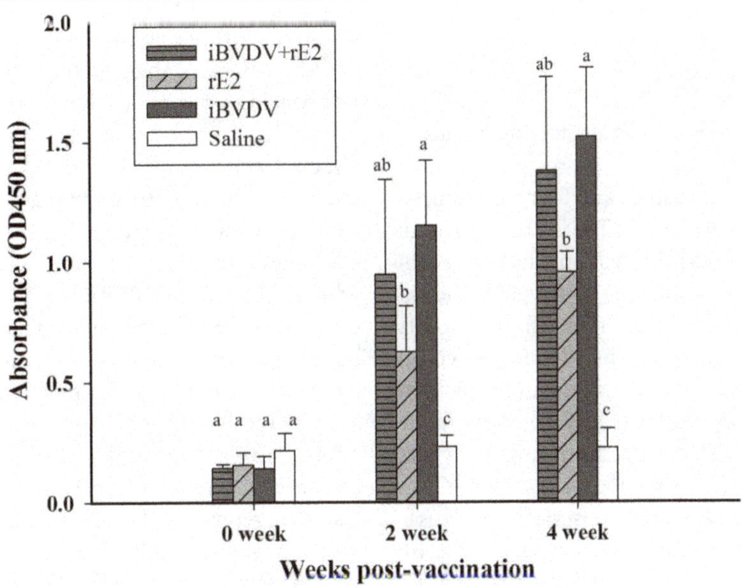

Fig. 2 Antibody response in goats immunized with BVDV vaccines. Goats were immunized twice intramuscularly with four different vaccine formulations: 1) iBVDV+rE2, 2) rE2, 3) iBVDV, and 4) saline. Total anti-BVDV IgG levels were analyzed by ELISA. Data represent means ± SD. Differences between groups were analyzed by Duncan's significance test (Different letters indicate significant difference at $p < 0.05$)

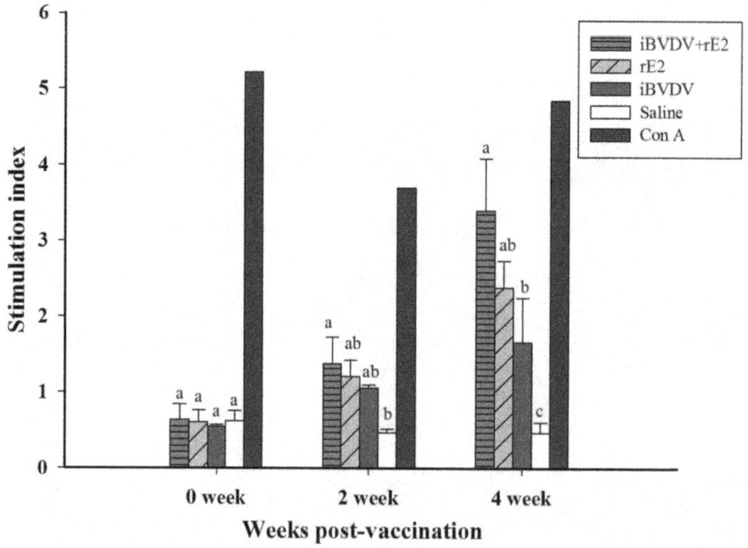

Fig. 3 T-cell proliferation assay of PBMC from goats immunized with BVDV vaccines. Goats were immunized twice intramuscularly with four different vaccine formulations: 1) iBVDV+rE2, 2) rE2, 3) iBVDV, and 4) saline. PBMC from blood were isolated and stimulated with inactivated BVDV or Con A (as a positive control). Proliferation extent was measured and stimulation index (SI) calculated. Data represent means ± SD. Differences between groups were analyzed by Duncan's significance test and different letters indicate significant difference at $p < 0.05$

apart with the four different formulations mentioned above. Analysis showed that, compared to that of the saline group, serum anti-BVDV antibody levels increased after vaccination for the vaccine groups (Fig. 2). However, vaccine containing both iBVDV and rE2 did not elicit higher antibody levels than vaccines with iBVDV or rE2 alone. In terms of cellular immunity, T-cell proliferation assay showed that the addition of rE2 enhanced T-cell response elicited by iBVDV 4 weeks post vaccination (Fig. 3).

iBVDV+rE2 vaccine showed improved protection against BVDV challenge

For protective efficacy evaluation, the four immunized groups were challenged with BVDV/TW 2014. In one of the first signs of BVDV infection, leukopenia, a 40% reduction in white blood cell count was observed in the saline group 3 days post challenge (Fig. 4). For goats receiving the iBVDV+rE2 vaccine, white blood cell count dropped by only about 20% initially, and then recovered at a faster pace and to a fuller extent than that of the saline group. In terms of pyrexia, rectal temperatures of goats vaccinated with iBVDV+rE2 remained below 39.5 °C, in contrast to the elevated temperature observed for the saline group between 7 and 14 days after challenge (Fig. 5). Observation of clinical signs showed that the iBVDV+rE2 vaccine achieved the most reduction in clinical scores between 7 and 14 days after challenge (Fig. 6). Overall, for the parameters of leukopenia, pyrexia, and clinical signs, the double-component iBVDV+rE2 vaccine showed similar efficacy as the single-component vaccines.

Differences in viremia were observed, however, between the vaccine groups. Using RT-qPCR to detect BVDV presence in immunized goats after challenge, results showed that the iBVDV+rE2 vaccine eliminated virus presence in all three goats. However, for the iBVDV group, viremia was detectable in two out of three goats 7–13 days after challenge. For the rE2 group, viremia was detectable in one out of three goats (Table 2). We conclude that the addition of rE2 enhances the efficacy of iBVDV vaccines by reducing virus replication.

Discussion

Our study sought to improve the protective efficacy of iBVDV vaccines by enhancing humoral immunity against the main antigen, E2, and results showed that the addition of rE2 to iBVDV vaccines abrogated virus presence in whole blood. The absence of BVDV viremia bodes well for potential sterilizing immunity. However, antigenic variation of E2 may lead to partial protection in a heterologous challenge. As shown in the E2 sequence comparison in Fig. 1a, up to 17.6% divergence in amino acid sequence can be seen among various strains. As divergence of the E2 sequence increases, effective vaccines might require the inclusion of E2 proteins of divergent strains, or other BVDV proteins.

Since inactivated pathogens and protein antigens are usually presented exogenously to the immune system, it can be expected that inactivated and subunit BVDV

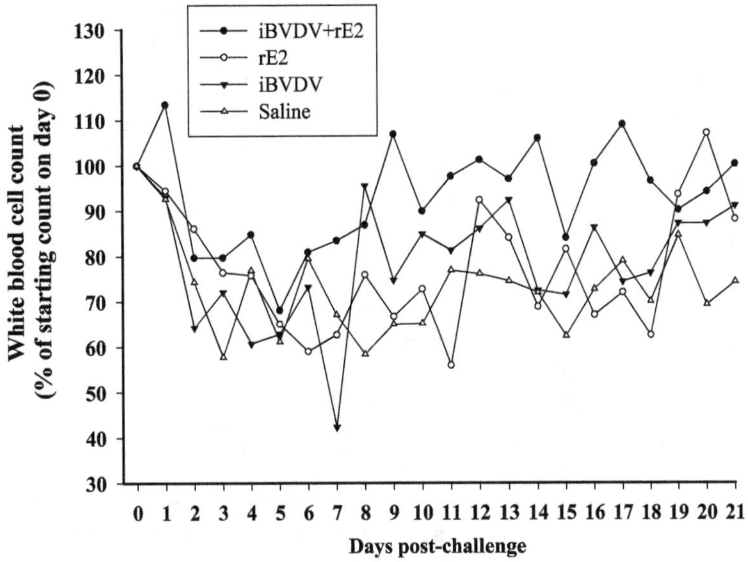

Fig. 4 Leukopenia of immunized goats following BVDV challenge. Goats were immunized twice intramuscularly with four different vaccine formulations: 1) iBVDV+rE2, 2) rE2, 3) iBVDV, and 4) saline, and challenged with 1×10^8 $FAID_{50}$ BVDV/TW 2014 4 weeks after primary immunization. Number of white blood cells in blood was counted. Means are presented

vaccines elicited very little cellular immunity. To further improve protective efficacy of these vaccines, the induction of stronger cellular immunity may be considered. Evidence showed that cellular immunity to BVDV can provide protection. In studies done on calves with high levels of maternal antibody, T cell responses were shown to be sufficient for protection in the absence of measureable humoral response [10–12]. Also, modified live BVDV vaccines, deemed more protective than inactivated or subunit vaccines, elicited antigen-specific activation of T lymphocyte subsets while a killed vaccine did not. Therefore, endogenous presentation (through DNA vaccination, for example) of important T cell epitopes within the capsid, nonstructural proteins NS2/3, or E^{rns} may contribute significantly to protection by eliciting cellular immunity against virus-infected cells.

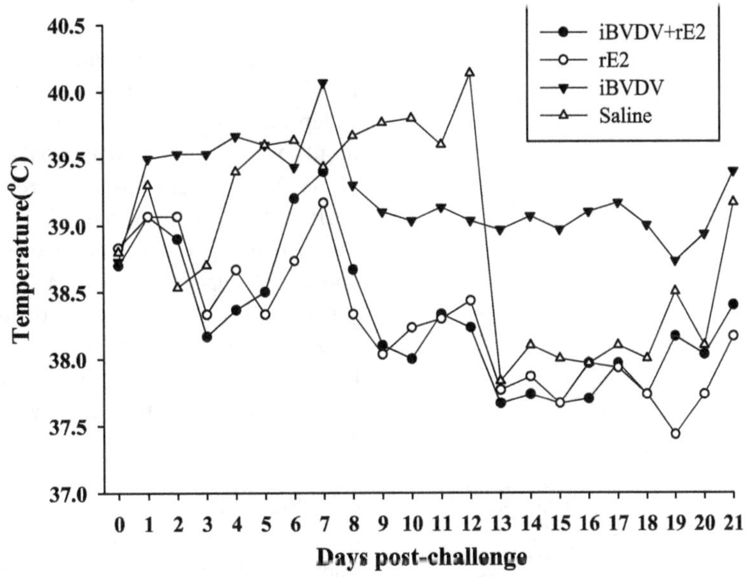

Fig. 5 Pyrexia of immunized goats following BVDV challenge. Goats were immunized twice intramuscularly with four different vaccine formulations: 1) iBVDV+rE2, 2) rE2, 3) iBVDV, and 4) saline, and challenged with 1×10^8 $FAID_{50}$ BVDV/TW 2014 4 weeks after primary immunization. Rectal temperatures after challenge were recorded. Means are presented

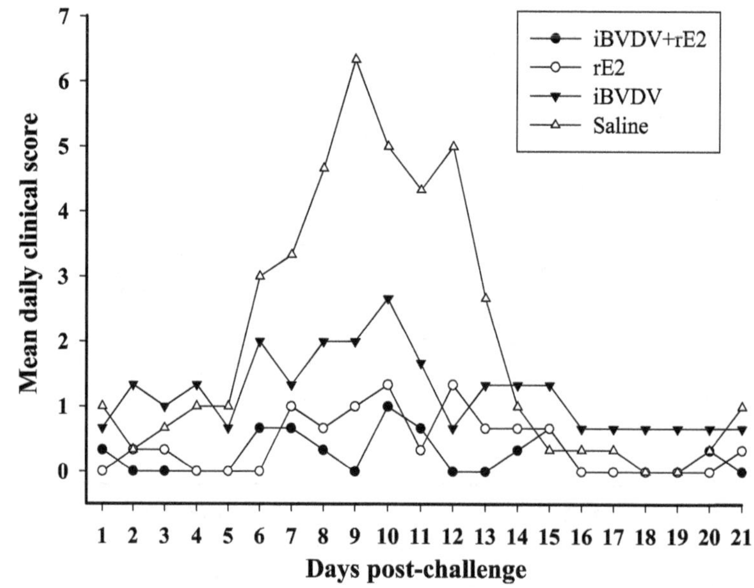

Fig. 6 Mean daily clinical scores of immunized goats following BVDV challenge. Goats were immunized twice intramuscularly with four different vaccine formulations: 1) iBVDV+rE2, 2) rE2, 3) iBVDV, and 4) saline, and challenged with 1×10^8 $FAID_{50}$ BVDV/TW 2014 4 weeks after primary immunization. Clinical signs were recorded and mean daily clinical scores calculated

Table 2 Viremia of immunized goats following BVDV challenge using RT-qPCR

Goat No.	[a]Days post-challenge														
	1	3	5	7	8	9	10	11	12	13	14	15	17	19	21
iBVDV+rE2															
241	–	–	–	–	–	–	–	–	–	–	–	–	–	–	–
250	–	–	–	–	–	–	–	–	–	–	–	–	–	–	–
546	–	–	–	–	–	–	–	–	–	–	–	–	–	–	–
rE2															
541	–	–	–	–	–	–	–	–	–	–	–	–	–	–	–
542	–	–	–	+	–	+	–	+	+	+[b]	–	–	–	+	–
543	–	–	–	–	–	–	–	–	–	–	–	–	–	–	–
iBVDV															
37	–	–	–	+	–	–	–	–	–	–	–	–	–	–	–
244	–	–	–	–	–	–	–	–	–	+	–	–	–	–	–
246	–	–	–	–	–	–	–	–	–	–	–	–	–	–	–
Saline															
538	–	–	–	–	+	+	+	+	+	+	–	–	–	–	–
539	–	–	–	+	+	+	+	+	+	+	–	–	–	–	–
533	–	–	–	+	+	+	+	+	+	+	–	–	–	–	–

[a]Goats were immunized twice intramuscularly with four different vaccine formulations: 1) inactivated BVDV+rE2, 2) rE2, 3) inactivated BVDV, and 4) saline, and challenged with 1×10^8 $FAID_{50}$ BVDV/TW 2014 4 weeks after primary immunization. Viremia was detected using RT-qPCR for 21 days after challenge for each animal. Samples that reacted with the RT-qPCR were designated positive (+) and non-reactive samples were designated negative (–)

[b]Using the Pearson chi-square test, for viremia between 7 and 13 days post challenge, the p value was 0.017 when comparing between iBVDV+rE2 and rE2, and 0.147 when comparing between iBVDV+rE2 and iBVDV

DNA vaccination using E2 [13] and NS3 [14] sequences was shown to induce protective T cell responses. By increasing cellular immunity, the range and duration of inactivated and subunit vaccines may be improved.

When studying BVDV biology, researchers have looked for alternative animal models since experimenting on pregnant cattle is cost-prohibitive. Because BVDV infection can result in abortions, it is important to study congenital BVDV-2 infection, which in turn necessitates more economical animal models. A study has demonstrated that the biology of BVDV-2 infection is essentially similar in pregnant sheep and cattle [15]. For our study, challenge experiments were performed in goats since natural infection of BVDV in goats has been reported [16, 17] and experimental infection has also been performed previously, causing reproductive diseases [18]. Because the course of disease of BVDV is essentially similar in cattle and goats, vaccine protective efficacy results obtained from a goat model offers reference value. Following our current study, vaccine efficacy studies can also be performed on pregnant goats.

Conclusions

Our study showed that the addition of rE2 protein may be an effective way to enhance the protective efficacy of inactivated BVDV vaccines, and this improved vaccine could prove useful in a BVDV control program.

Abbreviations

BSA: Bovine serum albumin; BVDV: Bovine viral diarrhea virus; EDTA: Ethylenediaminetetraacetic acid; $FAID_{50}$: 50% fluorescent antibody infectious dose; FBS: Fetal bovine serum; iBVDV: Inactivated bovine viral diarrhea virus; MDBK: Madin-Darby bovine kidney; MOI: Multiplicity of infection; PBMC: Peripheral bone marrow cells; PBST: Phosphate buffered saline-Tween 20; PVDF: Polyvinylidene difluoride; rE2: Recombinant E2; RT-PCR: Reverse transcription-polymerase chain reaction; RT-qPCR: Reverse transcription-quantitative polymerase chain reaction; SDS-PAGE: Sodium dodecyl sulfate-polyacrylamide gel electrophoresis; w/o/w: Water-in-oil-in-water

Acknowledgements

The authors thank Dr. Jai-wei Lee of the Department of Tropical Agriculture, NPUST, for helpful comments on the manuscript.

Funding

This research did not receive any specific grant from funding agencies in the public, commercial, or not-for-profit sectors.

Authors' contributions

YC performed data analysis and wrote the manuscript. LC was involved in data analysis and interpretation, and revised the manuscript for intellectual content. JZ produced the vaccines and executed the animal experiments. YW cloned and produced rE2 vaccine. SL was responsible for animal care and experimental protocols. CC conceived and designed the study, and was involved in the acquisition and interpretation of data. All authors read and approved the final manuscript.

Consent for publication

Not applicable.

Competing interests

The authors declare that they have no competing interests.

Author details

[1]Graduate Institute of Animal Vaccine Technology, College of Veterinary Medicine, National Pingtung University of Science and Technology, 1, Shuehfu Road, Neipu, Pingtung 91201, Taiwan. [2]Department of Veterinary Medicine, College of Veterinary Medicine, National Pingtung University of Science and Technology, 1, Shuehfu Road, Neipu, Pingtung 91201, Taiwan.

References

1. Lanyon SR, Hill FI, Reichel MP, Brownlie J. Bovine viral diarrhoea: pathogenesis and diagnosis. Vet J. 2014;199(2):201–9.
2. Richter V, Lebl K, Baumgartner W, Obritzhauser W, Kasbohrer A, Pinior B. A systematic worldwide review of the direct monetary losses in cattle due to bovine viral diarrhoea virus infection. Vet J. 2017;220:80–7.
3. Ridpath JF. Immunology of BVDV vaccines. Biologicals. 2013;41(1):14–9.
4. Thomas C, Young NJ, Heaney J, Collins ME, Brownlie J. Evaluation of efficacy of mammalian and baculovirus expressed E2 subunit vaccine candidates to bovine viral diarrhoea virus. Vaccine. 2009;27(17):2387–93.
5. Pecora A, Malacari DA, Perez Aguirreburualde MS, Bellido D, Nunez MC, Dus Santos MJ, Escribano JM, Wigdorovitz A. Development of an APC-targeted multivalent E2-based vaccine against bovine viral diarrhea virus types 1 and 2. Vaccine. 2015;33(39):5163–71.
6. Chu CY, Chen JJ, Chen H. Effect of water-in-oil-in-water (w/o/w) oil adjuvant on long-term antibody response of bovine ephemeral fever vaccine. Taiwan Vet J. 2007;33:211–6.
7. Reed LJ, Muench H. A simple method of estimating fifty percent endpoints. Am J Epidemiol. 1938;27:493–7.
8. Blodorn K, Hagglund S, Gavier-Widen D, Eleouet JF, Riffault S, Pringle J, Taylor G, Valarcher JF. A bovine respiratory syncytial virus model with high clinical expression in calves with specific passive immunity. BMC Vet Res. 2015;11:76.
9. Kosinova E, Psikal I, Robesova B, Kovarcik K. Real-time PCR for quantitation of bovine viral diarrhea virus RNA using SYBR green I fluorimetry. Veterinarni Medicina-Praha. 2007;52(6):253–61.
10. Endsley JJ, Roth JA, Ridpath J, Neill J. Maternal antibody blocks humoral but not T cell responses to BVDV. Biologicals. 2003;31(2):123–5.
11. Ridpath JE, Neill JD, Endsley J, Roth JA. Effect of passive immunity on the development of a protective immune response against bovine viral diarrhea virus in calves. Am J Vet Res. 2003;64(1):65–9.
12. Endsley JJ, Ridpath JF, Neill JD, Sandbulte MR, Roth JA. Induction of T lymphocytes specific for bovine viral diarrhea virus in calves with maternal antibody. Viral Immunol. 2004;17(1):13–23.
13. Nobiron I, Thompson I, Brownlie J, Collins ME. DNA vaccination against bovine viral diarrhoea virus induces humoral and cellular responses in cattle with evidence for protection against viral challenge. Vaccine. 2003;21(17–18):2082–92.
14. Young NJ, Thomas CJ, Thompson I, Collins ME, Brownlie J. Immune responses to non-structural protein 3 (NS3) of bovine viral diarrhoea virus (BVDV) in NS3 DNA vaccinated and naturally infected cattle. Prev Vet Med. 2005;72(1–2):115–20. discussion 215–119
15. Scherer CFC, Flores EF, Weiblen R, Caron L, Irigoyen LF, Neves JP, Maciel MN. Experimental infection of pregnant ewes with bovine viral diarrhea virus type-2 (BVDV-2): effects on the pregnancy and fetus. Vet Microbiol. 2001;79(4):285–99.
16. Kim IJ, Hyun BH, Shin JH, Lee KK, Lee KW, Cho KO, Kang MI. Identification of bovine viral diarrhea virus type 2 in Korean native goat (Capra hircus). Virus Res. 2006;121(1):103–6.
17. Mao L, Li W, Yang L, Wang J, Cheng S, Wei Y, Wang Q, Zhang W, Hao F, Ding Y, et al. Primary surveys on molecular epidemiology of bovine viral diarrhea virus 1 infecting goats in Jiangsu province, China. BMC Vet Res. 2016;12(1):181.
18. Passler T, Riddell KP, Edmondson MA, Chamorro MF, Neill JD, Brodersen BW, Walz HL, Galik PK, Zhang Y, Walz PH. Experimental infection of pregnant goats with bovine viral diarrhea virus (BVDV) 1 or 2. Vet Res. 2014;45:38.

Effect of different flooring types on pressure distribution under the bovine claw – an ex vivo study

Benjamin Oehme[1*], S. M. Geiger[1,2], S. Grund[1], K. Hainke[1], J. Munzel[1] and C. K. W. Mülling[1]

Abstract

Background: Mechanical interactions between hard floorings and the sole of bovine claws can be reasonable to cause traumatic claw lesions. In this ex vivo study, the direct kinetic impact of concrete and three types of rubber mats on the sole of dairy cattle claws was analyzed. In order to apply uniform loads, isolated distal hind limbs of adult Holstein Friesian dairy cows were functionally trimmed according to the Dutch method and attached to a load applicator. Kinetic data were recorded using a thin, foil-based pressure measurement system (Hoof™ System, Tekscan®).

Results: On concrete, the load distribution between the lateral and medial claw was less balanced than on the rubber floorings. The loaded area was significantly smaller on concrete (32.2 cm^2) compared to all rubber mats (48.3–58.0 cm^2). Average pressures (**P$_{av}$**) and maximum pressures (**P$_{max}$**) were significantly higher on concrete (P$_{av}$ 44.7 N/cm^2; P$_{max}$ 130.3 N/cm^2) compared to the rubber floorings (P$_{av}$ 24.9–29.7 N/cm^2; P$_{max}$ 71.9–87.2 N/cm^2). Pressure peaks occurred mainly in plantar and abaxial parts of the lateral claw and in apical and plantar regions of the medial claw. Load distribution displayed a widely unloaded slope region, but considering the pressure distribution under the claw, none of the zones showed a generally lower pressure exposure.

Conclusions: Altogether, rubber floorings lead to a significant mechanical relief of the sole compared to concrete. Furthermore, relevant differences between the tested rubber mats could be determined. Therefore the used system may be applied to compare further flooring types.

Keywords: Bovine claw, Dairy cattle, Pressure distribution, Concrete floor, Rubber mat

Background

Hard flooring systems such as concrete are the reason for a significantly higher risk of claw diseases and developing lameness in dairy cattle [1, 2]. Still, some studies proved a higher incidence of sole ulcers on rubber floorings than on concrete [3, 4], albeit with lower severity [5]. In contrast, Benz [6] found out that cows kept on rubber floorings showed a significantly lower incidence of claw diseases which increased again when the animals returned to concrete flooring. Besides, cows kept on rubber showed significantly more comfort behavior than those kept on concrete and walked longer distances, almost comparable to cows on pasture [7, 8].

* Correspondence: benjamin.oehme@vetmed.uni-leipzig.de
[1]Institute of Veterinary Anatomy, Histology and Embryology, Faculty of Veterinary Medicine, Leipzig University, Leipzig, Germany

To understand the underlying reasons of the mentioned differences in claw health and cow comfort, the effects of different floorings on the pressure load of bovine claws have been investigated in previous studies [9–11]. As most of the conducted surveys used stationary force or pressure measuring plates with hard surfaces, no direct interactions between flooring type and claw as well as the effects on the individual claw could be captured [12].

Franck et al. [12] applied thin, foil-based pressure sensitive sensors to investigate the direct impact of different concrete surface roughnesses on the pressure distribution under isolated bovine claws. With same type sensors, Nilsson et al. [13] analyzed the pressure distribution under isolated bovine distal limbs on slatted concrete flooring at different positions to the slats. In vivo measurements with large animals using these piezoresistive sensor foils have solely been performed in equine kinetic research so far [14–18].

An evaluation of these equine sensors (Tekscan® Hoof™System) found less accuracy in measuring vertical Ground Reaction Force (**vGRF**) compared to force plates, but overall results did not differ significantly [19]. Especially after proper calibration the piezoelectric sensors showed a reliable accuracy [20, 21]. The system was tested and recommended for scientific and clinical application by Lange et al. [17].

The aim of this study was to establish a laboratory setup for kinetic examinations of the bovine distal limb and to directly analyze the kinetic influence of different flooring types to the sole of dairy cattle's claws for the first time.

Methods

Material

Twelve left hind limbs of adult Holstein Friesian dairy cows (age 5.0 ± 1.6 years) from an abattoir were separated below the tarsometatarsal joint immediately after slaughter and transported to the lab. All claws were trimmed according to the Dutch method [22]. The limbs and claws showed no macroscopically visible pathologies. Metatarsi were cut with a band saw in the proximal third. The medullary cavity of each metatarsus was filled with Demotec®95 (Demotec Demel e.K., Nidderau, Germany) in order to fix a threaded bolt in the medullary canal to attach the limb to the load applicator.

Four different floorings were evaluated including concrete (**con**) and three types of rubber mats (Table 1): KARERA (**Kar**), KURA (**Kur**) and profiKURA (**proK**) (Gummiwerk KRAIBURG GmbH&Co. KG, Waldkraiburg, Germany).

Pressure measuring system

A foil-based piezoresistive pressure measurement system (M3200E, Hoof™System, Tekscan Inc., Boston, MA, USA) was used. The sensor foils were 0.23 mm thick with a sensor matrix of 167.6 × 167.6 mm and a resolution of 3.9 sensels/cm^2. A 1.5 mm adhesive protective foil covered the sensors on both sides.

Experimental setup

The distal limbs were attached to a load applicator and positioned centrally on the sensor with the metatarsus perpendicular to the tested floor. In order to simulate the in vivo situation as closely as possible the deep digital flexor tendon and the digital extensor tendons were pulled proximally with 500 and 50 N respectively. A securing strap was attached round the pastern joint to stabilize the phalangeal bone axis, which was monitored by latero-medial fluoroscopy. The limbs were loaded vertically with 240 kg resulting in a final load of 143 ± 9 kg under the claws, which was determined by a digital scale (Fig. 1). The lower resultant load was caused by the tendon traction in opposite direction and the horizontal Ground Reaction Forces, which were not captured by the sensors or the scale.

On every flooring a sequence of four repeated measurements was performed. The sensors were initially conditioned with loads up to 300 kg and subsequently calibrated before and after every sequence with 20 and 80% of the maximum applied load [23].

Data processing

In the subsequent data processing, the HoofSCAN Research software (version 6.85, Tekscan Inc., Boston, MA, USA) was used to analyze the force balance (**FB**), calculated from vertical Ground Reaction Forces (**vGRF**) of lateral and medial claw ($FB = \frac{vGRF_{lat}}{vGRF_{med}}$), the total loaded area (**A_{total}**), average pressure (**P_{av}**) and maximum pressure (**P_{max}**) on the four tested floorings under both claws. Afterwards, data were exported to RStudio [24] and by combining the pressure maps with computed tomography-based sole surface areas, a zonal subdivision was implemented according to Carvalho et al. [25] (Fig. 2). By this, the occurrence of P_{max} could be localized per claw. Parameters to compare the zones of each claw between the different floorings were the vGRF relative to the total vGRF per side (**F_{side}**), the loaded area per zone relative to the total zone area (**A_{zone}**) and the vGRF per loaded area per zone, which describes the pressure in each zone (**P_{zone}**).

Table 1 Overview and characteristics of tested rubber mats (Gummiwerk KRAIBURG GmbH&Co. KG)

	thickness [mm]	shore hardness	surface profile	bottom profile
KARERA	21	65	squares 	squares (height 2.5 mm) 
KURA	24	65	grip profile 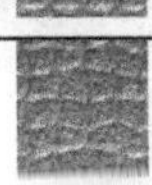	knobs (height 5.0 mm) 
profiKURA	24	65	star-shaped (with Corundum) 	knobs (height 5.0 mm) 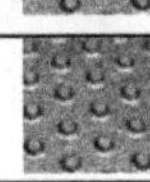

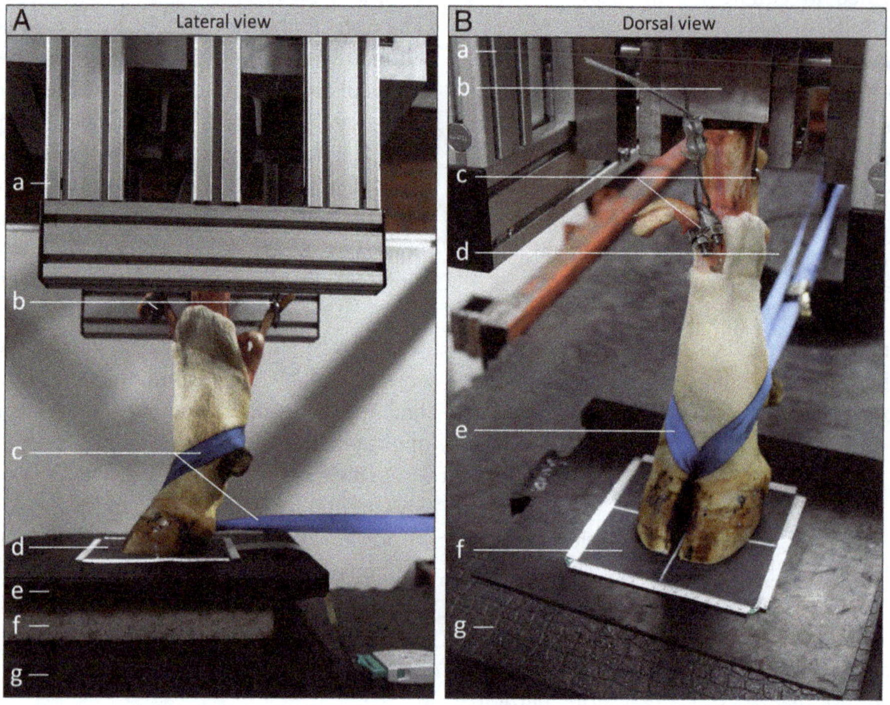

Fig. 1 Experimental setup. **A** Lateral view: a = load applicator, b = clamped tendons, c = securing strap, d = sensor foil, e = rubber mat, f = concrete, g = digital scale. **B** Dorsal view: a = load applicator, b = cam, c = clamped tendons, d = tendon weight, e = securing strap, f = sensor foil, g = rubber mat

Statistical analysis

Statistical analysis was conducted with the software RStudio [24]. Graphical assessment and the Shapiro-Wilk test were applied to determine normal distribution. As numerous data frames were non-parametric, statistical significances were tested using Wilcoxon signed-rank test. Post hoc correction was carried out with Bonferroni-Holm procedure. First of all, differences of overall parameter data (A_{total}, P_{av}, P_{max}) were analyzed between the four tested floors. Subsequently, parameters regarding the zones were examined separately for the lateral and medial claw. The parameters A_{zone} and P_{zone} were compared between the tested floorings for every zone. F_{side} was analyzed for differences between the floorings and between the five zones. Differences were considered to be significant at $p < 0.05$.

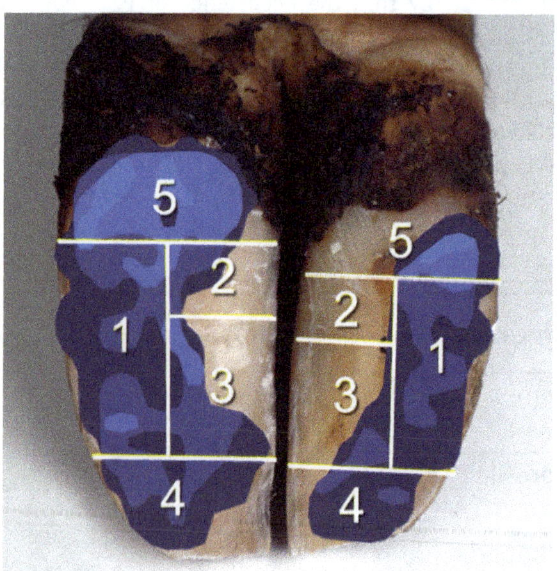

Fig. 2 Subdivision of the sole according to Carvalho et al. [25]

Results

For all tested floorings, the overall load was distributed unevenly between the lateral and medial claw on average. On concrete, the lateral claw was loaded 4.2 times higher than the medial claw. This imbalance decreased significantly on the tested rubber floorings, where the load on the lateral claw was 3.4 (Kar), 2.8 (Kur) and 2.7 (proK) times higher compared to the medial claw.

The total loaded area (Fig. 3a) was significantly smaller on concrete (33.2 ± 3.4 cm^2) than on all tested rubber floorings and differed significantly between all types of rubber mats (Kar 48.3 ± 3.5 cm^2, Kur 52.3 ± 4.8 cm^2, proK 58.0 ± 3.8 cm^2) (Fig. 4).

The lowest average pressure (Fig. 3b) was found on profiKURA (24.9 ± 2.6 N/cm^2). Average pressure differed significantly between all floorings and was highest on

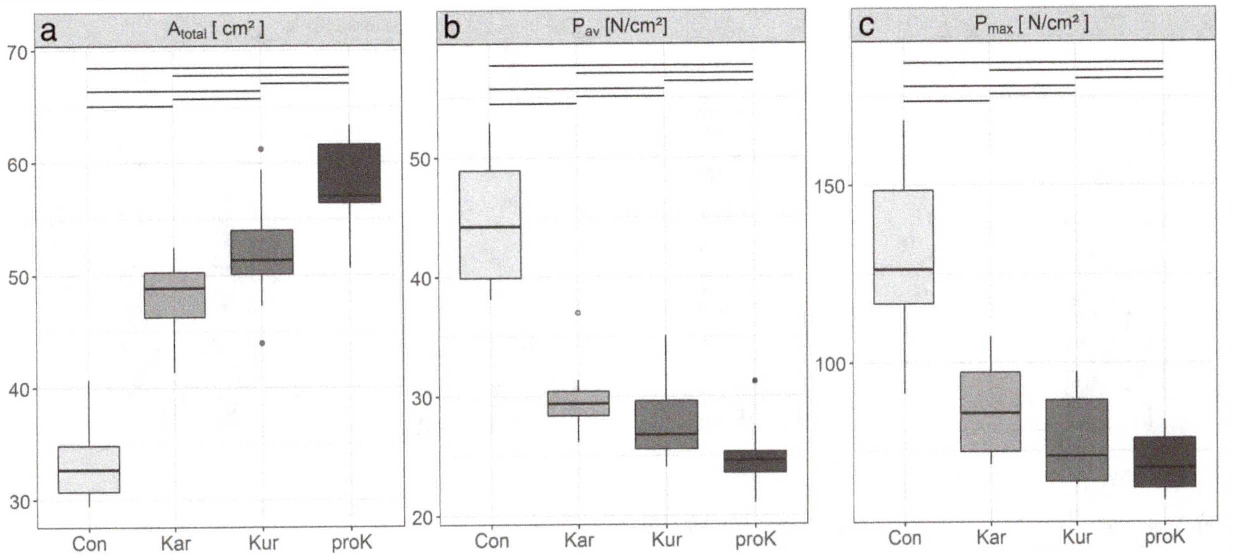

Fig. 3 **a** Contact area, **b** average pressure and **c** maximum pressure. Boxplots of values on the four tested floorings (concrete (con) and the three rubber mats KARERA (Kar), KURA (Kur) and profiKURA (proK)). Horizontal lines at the top indicate significant differences between the flooring types

concrete (Kur 27.6 ± 3.3 N/cm^2, Kar 29.7 ± 2.7 N/cm^2, con 44.7 ± 5.5 N/cm^2).

Also maximum pressure values (Fig. 3c) were significantly higher on concrete (130.3 ± 23.7 N/cm^2) compared to all rubber floorings. Significant differences occurred between the three types of rubber flooring as well (Kar 87.2 ± 13.4 cm^2, Kur 78.6 ± 12.6 cm^2, proK 71.9 ± 8.3 cm^2).

Analysis of kinetic characteristics in the different zones showed the following results: The majority of maximum pressure values in the lateral claw were measured in zones 1 and 5 and rarely in zone 4. Additionally, the occurrence of maximum pressures decreased in zone 1 and increased in zone 5 depending on type of flooring (Fig. 5). Considering the medial claw, the maximum pressures appeared mainly in zones 4 and 5 and to some extent in zone 1. In both claws, isolated pressure peaks occurred in zone 2 and none in zone 3.

There were no significant differences in F_{side} between the different floorings. Considering F_{side} averaged over all tested floors, zones 1 and 5 bore the highest vertical Ground Reaction Forces (vGRF) in the lateral claw (F_{side} mean: 40.1 and 32.1%). In the medial claw the highest amount of load appeared in zones 1, 4 and 5 (F_{side} mean: 24.8, 35.2 and 28.3%). Zones 2 and 3 carried the smallest loads on both sides (F_{side} between 3.4 and 8.6%) and were significantly different in terms of the vGRF to all other zones. Furthermore, significant differences could be determined between zone 1 and 4 as well as 4 and 5 in the lateral claw (Table 2).

Generally, the relative loaded area per zone (A_{zone}) increased on rubber mats compared to concrete (Fig. 6). Except for zone 2 the loaded area differed significantly between concrete and at least two rubber mats. There were also some significant differences in loaded areas between the tested rubber floorings (Table 3).

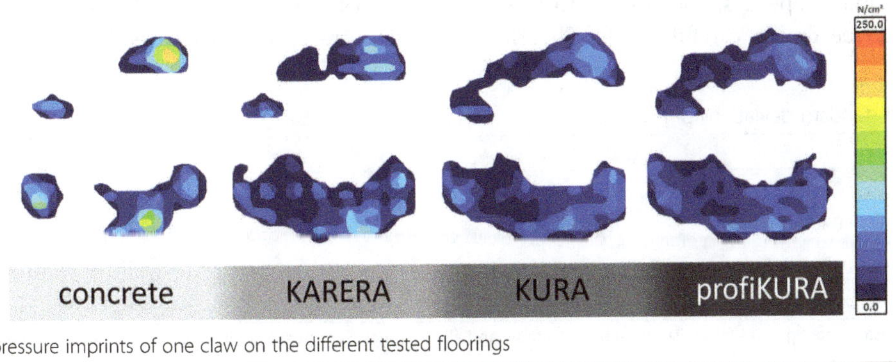

Fig. 4 Exemplary pressure imprints of one claw on the different tested floorings

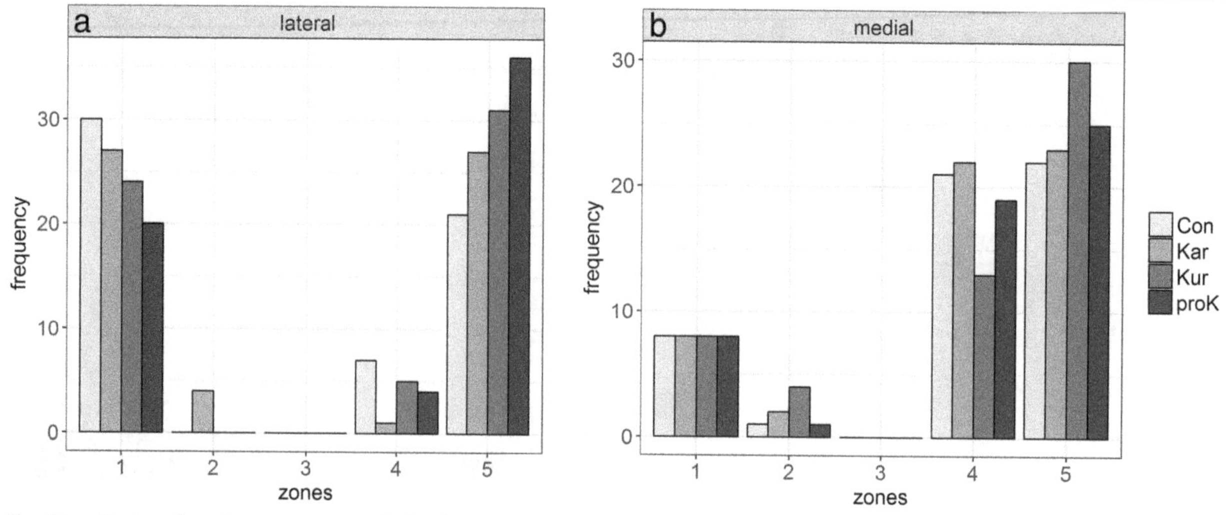

Fig. 5 Localization of maximum pressure peaks for the **a** lateral and **b** medial claw

The pressure in each zone based on the loaded area (P_{zone}) showed significant differences between concrete and rubber floorings for the lateral zones 1 and 4 and for the medial zone 4 (Fig. 7). In the loaded areas of the slope, represented by zones 2 and 3, pressure loads (14.52 N/cm^2–37.05 N/cm^2) similar to zones 1, 4 and 5 (17.19 N/cm^2–50.49 N/cm^2) occurred (Table 4).

Discussion

A physiological positioning and loading of isolated limbs is hard to achieve. The load applicator used in this study was specifically designed for static load application on isolated bovine distal limbs. The applied load of 150 kg was based on findings of Chapinal et al. [26] and Fischer [27] and was found suitable to mimic the load situation in living animals. As no specific data for cattle are available, tendon loads were adapted to tendon strains in walking ponies [28] and adjusted to a feasible order of magnitude. Previous studies suppressed the influence of most parts of the limbs' anatomical structures by fixating the flexible sections [13] or by solely using the horn capsules [29, 30] for standardized kinetic evaluation of flooring impacts on the claws. As these designs neglect physiological dynamic properties, conclusions to in vivo conditions have to be drawn carefully [29]. Therefore, the setup in the present study influenced the movement of joints and tendons as little as possible but still emphasized the normal axis of the distal phalangeal bones, which was monitored with fluoroscopy. Still, after several load applications a dorsiflexion of the pastern joint was observed in preliminary tests. Therefore this joint was stabilized with a securing strap to achieve a consistent, physiological bone axis and force transmission over all repeated measurements (Fig. 1). Despite this, the experimental setup in this study emulates the in vivo situation as closely as possible and enables a standardized examination of effects of different flooring types by exclusion of most interfering and environmental impacts such as body weight, motion, load shifting etc.

The Tekscan® Hoof™System was applied to bovine kinetic research for the first time in this study. It has frequently been used in equine research and was recommended for scientific and clinical application by Lange et al. [17]. As some authors stated a lack of accuracy in using this piezoresistive foil sensors [19, 21, 31], most equine studies were confined to relative comparisons [14, 15, 17]. However, by use of an appropriate calibration under trial conditions and expected loads, an adequate accuracy could be obtained with this sensors [19–21]. Therefore, the sensors in this study were calibrated frequently and under measurement

Table 2 Means and standard deviations of F_{side}

Zone	1	2	3	4	5
Lateral claw					
all floorings	0.40 ± 0.12^{2,3,4}	0.08 ± 0.03^{1,4,5}	0.07 ± 0.04^{1,4,5}	0.13 ± 0.05^{1,5}	0.32 ± 0.10^{2,3,4}
Medial claw					
all floorings	0.25 ± 0.16^{2,3}	0.05 ± 0.04^{1,4,5}	0.06 ± 0.08^{1,4,5}	0.35 ± 0.24^{2,3}	0.28 ± 0.14^{2,3}

Data averaged over all tested flooring types for the lateral and medial claw and all five zones. Superscript numbers indicate significant differences between the stated zones

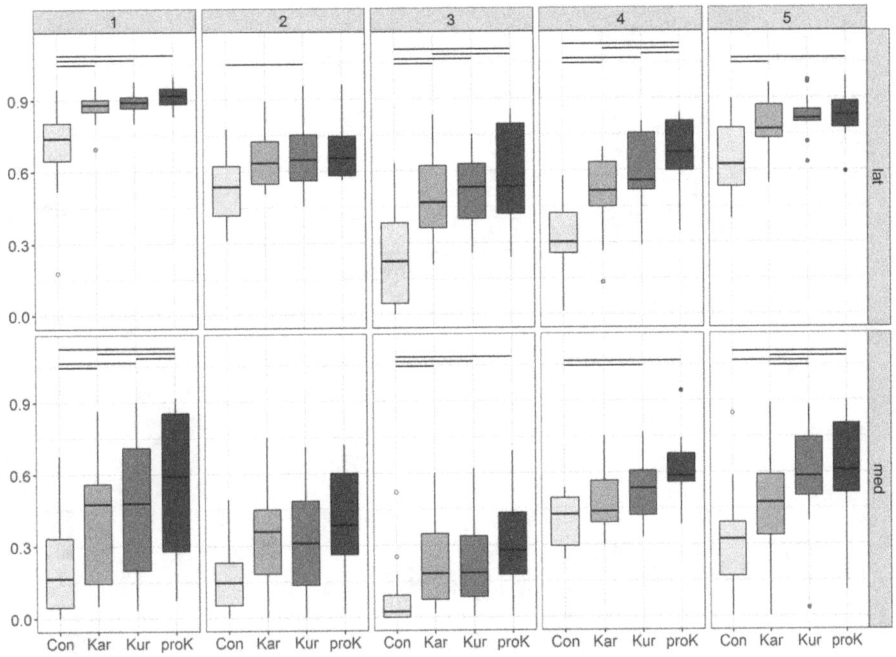

Fig. 6 Loaded area per zone relative to total zone area. Boxplots of A_{zone} for the lateral and the medial claw, all five zones and the four tested floorings (concrete (con) and the three rubber mats KARERA (Kar), KURA (Kur) and profiKURA (proK)). Horizontal lines at the top indicate significant differences between the flooring types

conditions to minimize the inaccuracy and to guarantee the same conditions for all measurements [23].

The load imbalance between the lateral and medial claw is in accordance with previous in vivo studies [32–34]. However, the lateral claw carried an even higher part of the total load in this study. This may be due to the strict perpendicular position of the limb and the lack of compensatory movement of the proximal structures. Therefore, a larger part of the force is transmitted to the longer lateral digit [35]. The reduction of imbalance on KURA and profiKURA compared to concrete shows a balancing effect of these rubber floorings on load distribution. Furthermore, the loaded area showed significant differences between the three tested rubber floorings. As all rubber mats have the same core structure and shore hardness (Table 1), the larger contact area might originate from the mats' thickness and surface structure. KARERA is thinner compared to the other tested rubber mats and has no knobs underneath, which can explain the smaller contact areas and higher pressure loads compared to KURA and profiKURA. KURA and profiKURA only differ in surface relief and a softer, thin corundum containing top layer on

Table 3 Means and standard deviations of A_{zone}

Zone	1	2	3	4	5
Lateral claw					
con	$0.69 \pm 0.21^{Kar,Kur,proK}$	0.53 ± 0.15^{Kur}	$0.24 \pm 0.21^{Kar,Kur,proK}$	$0.33 \pm 0.16^{Kar,Kur,proK}$	$0.65 \pm 0.17^{Kar,proK}$
Kar	0.87 ± 0.07^{con}	0.13 ± 0.60	$0.49 \pm 0.20^{con,proK}$	$0.50 \pm 0.17^{con,proK}$	0.78 ± 0.12^{con}
Kur	0.89 ± 0.05^{con}	0.16 ± 0.05^{con}	0.52 ± 0.16^{con}	$0.58 \pm 0.17^{con,proK}$	0.82 ± 0.09
proK	0.91 ± 0.05^{con}	0.13 ± 0.06	$0.58 \pm 0.20^{con,Kar}$	$0.66 \pm 0.17^{con,Kar,Kur}$	0.83 ± 0.11^{con}
Medial claw					
con	$0.21 \pm 0.20^{Kar,Kur,proK}$	0.16 ± 0.14	$0.09 \pm 0.16^{Kar,Kur,proK}$	$0.40 \pm 0.11^{Kur,proK}$	$0.31 \pm 0.25^{Kur,proK}$
Kar	$0.42 \pm 0.26^{con,proK}$	0.33 ± 0.22	0.22 ± 0.18^{con}	0.49 ± 0.15	$0.48 \pm 0.25^{Kur,proK}$
Kur	$0.48 \pm 0.29^{con,proK}$	0.33 ± 0.22	0.24 ± 0.22^{con}	0.54 ± 0.15^{con}	$0.58 \pm 0.23^{con,Kar}$
proK	$0.57 \pm 0.30^{con,Kar,Kur}$	0.40 ± 0.23	0.32 ± 0.23^{con}	0.61 ± 0.15^{con}	$0.63 \pm 0.22^{con,Kar}$

Relative values for the lateral and the medial claw, all five zones and the four tested floorings (concrete (con) and the three rubber mats KARERA (Kar), KURA (Kur) and profiKURA (proK)). Superscript abbreviations indicate significant differences between these floorings

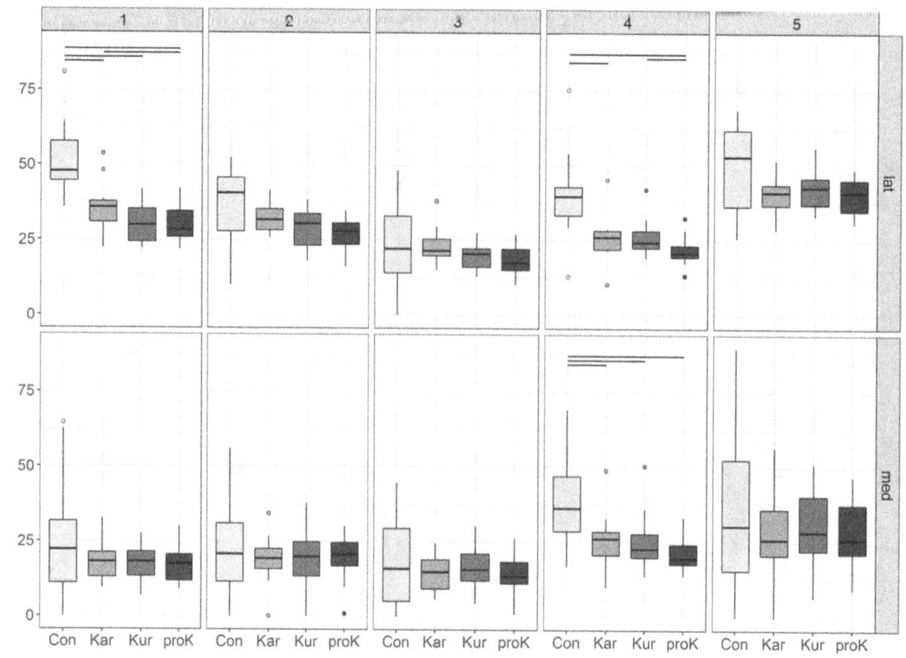

Fig. 7 Pressure in each zone, referred to the loaded area. Boxplots of P_{zone} [N/cm^2] for the lateral and the medial claw, all five zones and the four tested floorings (concrete (con) and the three rubber mats KARERA (Kar), KURA (Kur) and profiKURA (proK)). Horizontal lines at the top indicate significant differences between the flooring types

profiKURA. These differences in surface texture may provide smoother contact to the sole and obviously lead to a more regular load distribution.

The higher average and maximum pressures on concrete compared to the rubber floorings result from the smaller loaded area combined with a uniformly applied load. Especially the high pressure peaks on concrete flooring are capable of damaging the claw horn tissue, in particular in the soft bulb, which has a lower compressive strength than the wall horn [36]. As visible in Fig. 3c, these pressure peaks can be reduced considerably on the tested rubber floorings.

Additionally, the variance of pressure peaks is much higher on concrete than on the rubber floorings, which is consistent with findings of Luo et al. [37]. The high influence of the claws itself on the pressure distribution [29] may explain the generally high deviations found in the present study. Therefore it is advisable to compare the effect of different floorings and treatments rather within each claw than between entire groups of limbs.

By comparing the results of this kinetic analysis with the in vivo findings of van der Tol et al. [33] and Fischer [27], the contact area in the current study is substantially smaller. This may originate from the concrete's irregular

Table 4 Means and standard deviations of P_{zone} [N/cm^2]

Zone	1	2	3	4	5
Lateral claw					
con	51.25 ± 12.36 ^{Kar,Kur,proK}	36.51 ± 12.34	22.89 ± 14.53	40.20 ± 14.17 ^{Kar,proK}	50.11 ± 13.75
Kar	35.85 ± 8.24 $^{con, proK}$	32.33 ± 4.64	23.28 ± 5.99	26.02 ± 7.57 ^{con}	40.69 ± 6.19
Kur	30.40 ± 6.05 ^{con}	28.87 ± 6.49	19.98 ± 4.54	26.37 ± 5.95 ^{proK}	42.81 ± 7.28
proK	29.76 ± 5.78 $^{con, Kar}$	26.34 ± 5.44	18.54 ± 4.82	22.07 ± 4.91 ^{con,Kur}	40.35 ± 5.54
Medial claw					
con	25.22 ± 20.51	21.57 ± 15.50	18.17 ± 15.04	38.05 ± 14.81 ^{Kar,Kur,proK}	37.06 ± 29.35
Kar	18.55 ± 6.53	18.78 ± 8.10	14.63 ± 5.59	25.76 ± 9.04 ^{con}	27.99 ± 13.88
Kur	18.01 ± 5.95	19.34 ± 9.79	16.38 ± 7.16	25.45 ± 9.44 ^{con}	29.93 ± 12.40
proK	17.24 ± 5.93	19.35 ± 7.85	13.92 ± 6.34	21.56 ± 6.31 ^{con}	27.72 ± 10.96

Pressure values for the lateral and the medial claw, all five zones and the four tested floorings (concrete (con) and the three rubber mats KARERA (Kar), KURA (Kur) and profiKURA (proK)). Superscript abbreviations indicate significant differences between these floorings

surface compared to the smooth metal surfaces of the pressure plates used in the former surveys. The same applies to the higher maximum pressure values occurring in the present study compared to the analyzed pressure peaks of van der Tol et al. [33]. However, the average pressure was concordant with results of Telezhenko et al. [11] and van der Tol et al. [33]. Thus, the magnitude of the pressure loads determined in this study can be transferred to in vivo conditions and the effect of flooring types on the load under live cattle's claws can be estimated.

While maximum pressures occur particularly in the plantar and abaxial zones of the lateral claw, the medial claw bears maximum pressures both in the apical and plantar zones and fewer pressure peaks were found in the abaxial zone. This is largely in accordance with the findings of Carvalho et al. [25]. Although van der Tol et al. [38] used a zonal subdivision different from the present study, parallels in localization of maximum pressure peaks can be established. Still a few pressure peaks, mainly on rubber floorings, occurred in the plantar sectors of the slope, which represents the region prone to sole ulcer. This may explain the higher prevalence of sole ulcers in dairy cattle on rubber floorings found by some authors [3, 4]. This, in turn, may lead to the conclusion that the Dutch method [22], which was challenged by other authors before [25, 33], may not be appropriate for dairy cows housed on rubber flooring systems.

The relative distribution of vGRF on the different zones (F_{zone}) showed that zones 2 and 3 carried only a little proportion of the exerted load. That implies that these zones representing the slope were less burdened compared to the other zones. However, analysis of average pressure showed no general relief of these two zones. Consequently, the impacting forces might be lower but limited to a smaller area in the zones 2 and 3, which results in a pressure extent similar to the other zones.

Although an increase of loaded area between concrete and the tested rubber floorings was observed in all zones except zone 2, no associated overall decrease of mean pressure could be determined. The lack of significant differences in pressure loads between the floorings may originate from the widely varying loading patterns between the examined distal limbs. That may also be caused by the missing compensating movement of proximal structures in the used load applicator. Further, the influence of the claws itself may be a reason for the varying load distribution patterns, which has been stated by Franck et al. [12] before. Therefore, general inferences on pressure distribution over the different regions of the claws' contact area have to be drawn carefully and may require a greater number of tested limbs.

Nevertheless, the effects of the tested rubber floorings on the overall kinetic patterns of the claw (FB, A_{total}, P_{av}, P_{max}) showed a clear reduction of mechanical stress to the claws' sole in comparison to concrete. Comparable effects were recently found by Hüppler et al. [15] between concrete and soft floorings in standing and walking horses. In the present study, even differences between the tested rubber floorings in loaded area and overall pressure load could be observed. It seems, that profiKURA induces the least mechanical stress, followed by KURA and KARERA.

This investigation serves as a basis for analyzing direct kinetic influences of different flooring properties on the bovine claw in live animals. To confirm the results obtained in this study, further analysis with more specimens might be needed. Furthermore, the measuring system and the ex vivo findings have to be validated in live cows, as compensating movements of proximal structures or the location of the body's center of gravity cannot be simulated with this setting. Nevertheless, the ex vivo setup is beneficial regarding restrictions in terms of welfare reason and financial considerations and can be used to assess effects of further floorings or claw trimming methods without applying them to live animals initially.

Conclusion

In this study it was possible to detect lower pressure loads on rubber floorings compared to concrete. Furthermore, even differences between varying rubber floorings could be determined. This reduction of mechanical stress, in particular maximum pressure peaks, is one component to reduce the risk of mechanically induced claw diseases.

Acknowledgements
The authors acknowledge Enrico Reich, Nicole Röhrmann and Monique Weiß (Faculty of Veterinary Medicine of Leipzig University, Germany) for assisting with the data collection and processing.

Funding
The project was supported by funds of the Landwirtschaftliche Rentenbank (Frankfurt am Main, Germany). BO receives a scholarship from the Klaus Murmann Fellowship Program of the Foundation of German Business (SDW). Funders had neither influence on the study design and data collection, analysis and interpretation, nor in writing the manuscript. The authors also acknowledge support from the German Research Foundation (DFG) and Leipzig University within the program of Open Access Publishing.

Authors' contributions
BO made substantial contributions to conception and design, acquisition of data, analysis and interpretation of data and drafted the manuscript. SGe made substantial contributions to conception and design, acquisition of data and revising the manuscript critically for important intellectual content. SGr made substantial contributions to conception and design, acquisition of data, was involved in drafting and revising the manuscript critically for

important intellectual content. KH made substantial contributions to data processing, statistical analysis and graph preparation. JM made substantial contributions to conception and design, acquisition of data and revising the manuscript critically for important intellectual content. CM made substantial contributions to conception and design and revising the manuscript critically for important intellectual content. All authors read, revised and approved the final manuscript.

Consent for publication
Not applicable

Competing interests
The authors declare that they have no competing interests.

Author details
[1]Institute of Veterinary Anatomy, Histology and Embryology, Faculty of Veterinary Medicine, Leipzig University, Leipzig, Germany. [2]Institute of Topographic Anatomy, University of Veterinary Medicine Vienna, Vienna, Austria.

References

1. Vanegas J, Overton M, Berry SL, Sischo WM. Effect of rubber flooring on claw health in lactating dairy cows housed in free-stall barns. J Dairy Sci. 2006;89(11):4251–8.
2. Vokey FJ, Guard CL, Erb HN, Galton DM. Effects of alley and stall surfaces on indices of claw and leg health in dairy cattle housed in a free-stall barn. J Dairy Sci. 2001;84(12):2686–99.
3. Kremer PV, Nueske S, Scholz AM, Foerster M. Comparison of claw health and milk yield in dairy cows on elastic or concrete flooring. J Dairy Sci. 2007;90(10):4603–11.
4. Fjeldaas T, Sogstad AM, Osterås O. Locomotion and claw disorders in Norwegian dairy cows housed in freestalls with slatted concrete, solid concrete, or solid rubber flooring in the alleys. J Dairy Sci. 2011;94(3): 1243–55.
5. Reubold H. Entwicklung geeigneter Parameter zur Beurteilung von elastischen Laufgangauflagen in Liegeboxenlaufställen für Milchkühe [Dissertation med. vet]. Gießen: Justus-Liebig-Universität; 2008.
6. Benz B. Elastische Beläge für Betonspaltenböden in Liegeboxenlaufställen [Dissertation]. Hohenheim: Universität Hohenheim; 2002.
7. Platz S, Ahrens F, Bendel J, Meyer HHD, Erhard MH. What happens with cow behavior when replacing concrete slatted floor by rubber coating: a case study. J Dairy Sci. 2008;91(3):999–1004.
8. Bendel J. Auswirkungen von elastischen Bodenbelägen auf das Verhalten von Milchrindern im Laufstall [Dissertation med. vet]. München: Ludwig-Maximilians-Universität; 2005.
9. Bergsten C, Telezhenko E, Ventorp M. Influence of soft or hard floors before and after first calving on dairy heifer locomotion, claw and leg health. Animals (Basel). 2015;5(3):662–86.
10. Ouweltjes W, Holzhauer M, van der Tol PPJ, van der Werf JTN. Effects of two trimming methods of dairy cattle on concrete or rubber-covered slatted floors. J Dairy Sci. 2009;92(3):960–71.
11. Telezhenko E, Bergsten C, Magnusson M, Ventorp M, Nilsson C. Effect of different flooring systems on weight and pressure distribution on claws of dairy cows. J Dairy Sci. 2008;91(5):1874–84.
12. Franck A, Cocquyt G, Simoens P, De Belie N. Biomechanical properties of bovine claw horn. Biosyst Eng. 2006;93(4):459–67.
13. Nilsson C, Pålsson K-H, Ventorp M. Pressure distribution on bovine claws in slatted concrete floors. In: Proceedings of XVI CIGR world congress. Düsseldorf: VDI-Verlag; 2006.
14. Hagen J, Hüppler M, Häfner FS, Geiger SM, Mäder D. Modifying horseshoes in the mediolateral plane: effects of side wedge, wide branch, and unilateral roller shoes on the phalangeal alignment, pressure forces, and the footing pattern. J Equine Vet Sci. 2016;37:77–85.
15. Hüppler M, Hagen J, Häfner FS, Geiger SM, Mäder D. Untersuchung der auf den Huf einwirkenden Druckkräfte und deren Beeinflussung durch verschiedene Bodenbeschaffenheiten. Pferdeheilkunde. 2015;31(5):426–34.
16. Al Naem M. Untersuchungen zur Hufrehe bei Pferden mit Hilfe des HufScan®-Systems [Dissertation med. vet]. Gießen: Justus-Liebig-Universität; 2014.
17. Lange C, Kattelans A, Rohn K, Lüpke M, Brückner HP. Die kinetische Untersuchung der Fußung, der Belastung des Hufes und des Abrollvorganges an den Vordergliedmaßen von Pferden im Schritt und im Trab auf dem Laufband mit dem HoofTM-System (Tekscan®). Pferdeheilkunde. 2012;28(5):538–47.
18. Reilly PT. In-shoe force measurements and hoof balance. J Equine Vet Sci. 2010;30(9):475–8.
19. Perino W, Kawcak CE, Frisbie DD, Reiser RF, McIlwraith CW. The accuracy and precision of an equine in-shoe pressure measurement system as a tool for gait analysis. J Equine Vet Sci. 2007;27(4):161–6.
20. Hsiao H, Guan J, Weatherly M. Accuracy and precision of two in-shoe pressure measurement systems. Ergonomics. 2002;45(8):537–55.
21. Morin EL, Bryant JT, Reid SA, Whiteside RA. Calibration Issues of Tekscan Systems For Human Pressure Assessment. In: Soldier Mobility: Innovations in Load Carriage System Design and Evaluation. Neuilly-sur-Seine: NATO; 2001. p. 24–1--7.
22. Toussaint RE. Cattle footcare and claw trimming. Ipswich, Suffolk: Farming Press; 1985.
23. Brimacombe JM, Wilson DR, Hodgson AJ, Ho KCT, Anglin C. Effect of calibration method on Tekscan sensor accuracy. J Biomech Eng. 2009;131(3):34503. -1-4
24. RStudio. Integrated Development Environment for R. Boston: RStudio, Inc; 2016.
25. Carvalho V, Nääs IA, Bucklin RA, Shearer JK, Shearer L, Massafera VJR, et al. Effects of trimming on dairy cattle hoof weight bearing surfaces and pressure distributions. Braz J Vet Res Anim Sci. 2006;43(4):518–25.
26. Chapinal N, de Passilé AM, Rushen J. Weight distribution and gait in dairy cattle are affected by milking and late pregnancy. J Dairy Sci. 2009;92(2):581–8.
27. Fischer L. Analyse der Druckverteilung unter den Klauen bei Kühen der Rasse Deutsche Holsteins [Dissertation med. vet]. Hannover: Tierärztliche Hochschule Hannover; 2006.
28. Riemersma DJ, van den Bogert AJ, Schamhardt HC, Hartman W. Kinetics and kinematics of the equine hind limb: in vivo tendon strain and joint kinematics. Am J Vet Res. 1988;49(8):1344–52.
29. Franck A, De Belie N. Concrete floor–bovine claw contact pressures related to floor roughness and deformation of the claw. J Dairy Sci. 2006;89(8): 2952–64.
30. De Belie N, Rombaut E. Characterisation of claw–floor contact pressures for standing cattle and the dependency on concrete roughness. Biosyst Eng. 2003;85(3):339–46.
31. Sumiya T, Suzuki Y, Kasahara T, Ogata H. Sensing stability and dynamic response of the F-Scan in-shoe sensing system: a technical note. J Rehabil Res Dev. 1998;35(2):192.
32. Kehler W, Gerwing TH. Effects of functional claw trimming on pressure distribution under hind claws of German Holstein cows. In: Zemljic, B, editor. Proceedings of the 13th International Symposium and 5th Conference on Lameness in Ruminants. Maribor, Slovenija; 2004.
33. van der Tol PPJ, van der Beek SS, Metz J, Noordhuizen-Stassen EN, Back W, Braam CR, et al. The effect of preventive trimming on weight bearing and force balance on the claws of dairy cattle. J Dairy Sci. 2004;87(6):1732–8.
34. Gerwing TH. Beeinflussung der Lastverteilung durch Klauenpflege innerhalb der Einzelklauen und der Klauenpaare der Hintergliedmaßen von schwarzbunten Milchkühen der Rasse Deutsche Holsteins [Dissertation med. vet]. Hannover: Tierärztliche Hochschule Hannover; 2003.
35. Muggli E, Sauter-Louis C, Braun U, Nuss K. Length asymmetry of the bovine digits. Vet J. 2011;188(3):295–300.
36. van der Tol PPJ, Metz JHM, Noordhuizen-Stassen EN, Back W, Braam CR, Weijs WA. The vertical ground reaction force and the pressure distribution on the claws of dairy cows while walking on a flat substrate. J Dairy Sci. 2003;86(9):2875–83.
37. Luo ZP, Berglund LJ, An KN. Validation of F-scan pressure sensor system: a technical note. J Rehabil Res Dev. 1998;35(2):186–91.
38. van der Tol PPJ, Metz JHM, Noordhuizen-Stassen EN, Back W, Braam CR, Weijs WA. The pressure distribution under the bovine claw during square standing on a flat substrate. J Dairy Sci. 2002;85(6):1476–81.

21

A new bovine tuberculosis model for England and Wales (BoTMEW) to simulate epidemiology, surveillance and control

Colin P. D. Birch*, Ashley Goddard and Oliver Tearne

Abstract

Background: Bovine tuberculosis (bTB) is a zoonotic disease of cattle caused by *Mycobacterium bovis*, widespread in England and Wales. It has high incidence towards the South West of England and Wales, with much lower incidence to the East and North. A stochastic simulation model was developed to simulate *M. bovis* transmission among cattle, transfer by cattle movements and transmission from environmental reservoirs (often wildlife and especially badgers). It distinguishes five surveillance streams, including herd tests, pre-movement testing and slaughter surveillance. The model thereby simulates interventions in bTB surveillance and control, and generates outputs directly comparable to detailed disease records. An anonymized version of the executable model with its input data has been released. The model was fitted to cattle bTB records for 2008–2010 in a cross-sectional comparison, and its projection was compared with records from 2010 to 2016 for validation.

Results: The fitted model explained over 99% of the variation among numbers of breakdowns in four defined regions and surveillance streams in 2010. It classified 7800 (95% confidence interval c. 5500 – 14,000) holdings within high incidence regions as exposed to infectious environmental reservoirs, out of over 31,000 cattle holdings identified as potentially exposed to such sources. The model was consistent with previous estimates of low *M. bovis* transmission rate among cattle, but cattle to cattle transmission was clearly required to generate the number of cattle cases observed. When projected to 2016, the model as fitted to 2010 continued to match the distribution of bTB among counties, although it was notable that the actual distribution of bTB in 2010 was itself a close match for its distribution in 2016.

Conclusions: The close model fit demonstrated that cattle movements could generate breakdowns as observed in low incidence regions, if persistent environmental reservoirs such as wildlife maintained infection levels in the high incidence regions. The model suggests that environmental reservoirs may be a challenge for control, because, although many holdings are exposed to infection from wildlife or the environment, they are a minority of holdings. Large impacts on disease in wildlife will be required to avoid each individual transmission event to cattle.

Keywords: Bovine tuberculosis, Stochastic simulation model, Wildlife reservoir, Cattle movements, Surveillance streams, *Mycobacterium bovis*

* Correspondence: colin.birch@apha.gov.uk
Animal and Plant Health Agency, New Haw, Addlestone, Surrey KT15 3NB, UK

Background

Bovine Tuberculosis (bTB) is a worldwide bacterial disease of cattle with zoonotic potential caused by *Mycobacterium bovis*, which has increased in Great Britain from 500 to 700 cattle slaughtered for bTB control per year in the 1980s to 39,361 in 2016 [1, 2]. Currently human *M. bovis* infection is not a major problem in England and Wales, but the disease cost the English and Welsh governments £0.5 billion in the decade up to 2013 [3]. Scotland achieved Officially Tuberculosis Free Status (OTF) in September 2009, in recognition of the low incidence of bTB of native origin found in Scottish herds [4]. England and Wales also aspire to achieve OTF status before 2040, through deployment of an increased package of interventions to control all routes of transmission of disease [5, 6]. The aim is to substantially reduce bTB incidence as measured by the frequency at which herds experience 'breakdowns', in which restrictions and intense surveillance are applied following detection of bTB in one or more cattle.

Badgers were identified as a possible wildlife reservoir of infection for cattle in the early 1970s and Krebs et al. [7] concluded that there was strong evidence that badgers were a significant source of infection in areas with high bTB incidence [3, 8]. Despite the Randomized Badger Culling Trial (RBCT) of 1998–2007 confirming that badgers play a role in cattle bTB other questions have remained controversial [1]:

1. Can bTB be controlled without measures targeting badgers?
2. Can measures targeting badgers achieve net cost benefit?
3. Should culling or vaccination or both be used if badgers are targeted?

Bourne et al. [1] concluded that the answers to questions 1 and 2 were yes and no respectively. However others challenged these conclusions [9], and bTB has maintained or increased its range and incidence since 2007, while the net benefits of 5 years of proactive badger culling persisted at least 2.5 years beyond the end of the RBCT [10]. Moreover, various modelling and analytic methods have concluded that badgers or environmental sources contribute to a large proportion of herd breakdowns [11–13]. Hence culling and vaccination of badgers are potentially important tools for effective bTB control [5, 6].

The current escalation of control measures against bTB in Great Britain and the availability of multiple options confound evaluation of individual control measures. A mechanistic simulation of bTB and its management allows comparison of the progress of the epidemic with and without individual control measures [11]. However, the most prominently published model of bTB incidence among cattle herds does not explicitly represent a wildlife reservoir of infection but an abstracted environmental pool of infection, which is maintained by the presence of infectious cattle [11]. Implicitly it assumes that infection in wildlife is not self-sustaining, which is an issue of uncertainty [3]. Many hold the view that bTB in badgers is self-sustaining in parts of the UK and that transmission from wildlife to cattle must be prevented to achieve bTB eradication in England and Wales [14, 15].

In addition to Scotland, there are strong contrasts in regional bTB incidence within the rest of Great Britain. Hence contiguous counties in North and East England with low bTB incidence are designated as the 'Low Risk Area' (LRA), in contrast to the 'High Risk Area' (HRA) in the South-west and a buffer zone in between ('Edge') [6]. Explaining the persistence of bTB in England and Wales as due to wildlife reservoirs may also provide a direct process maintaining the contrast between the HRA and the LRA.

The model described here applied the hypothesis that wildlife had an important role in maintaining bTB and the contrasts between regions of England and Wales. It aimed to reproduce the distribution of bTB observed in England and Wales using a combination of relatively static environmental disease reservoirs and transfer by cattle movements. The hypothesis can be tested within the LRA, because it implies that high incidence of bTB within the HRA is maintained by sources of infection that are absent from the LRA. The test of the hypothesis was whether transfer of disease by cattle movements to regions with low bTB incidence was sufficient to match observed incidence, if disease in regions with high incidence was substantially maintained by infection from local environmental reservoirs. The model explicitly simulated detection of disease in some detail, including distinguishing between 5 surveillance streams: regular and targeted herd tests, pre-movement tests, slaughter surveillance and post breakdown tests. This was both for simulation of surveillance options and to allow model outputs to be directly compared with field records to compare epidemiological pathways as well as geographic distribution.

Results

Model development estimated a set of parameter values that defined a baseline model, as explained in the Methods and the full model description (Additional file 1). The baseline model included a distribution of probabilities that holdings would include an infectious environmental reservoir for bTB, fitting it in the process of matching the model to the observed geographic distribution of breakdowns (Fig. 1) (Additional file 1, Part 2, Section 3 "Environmental infection state"). The proportion of environmental reservoirs that were infectious in the baseline model varied widely among geographic

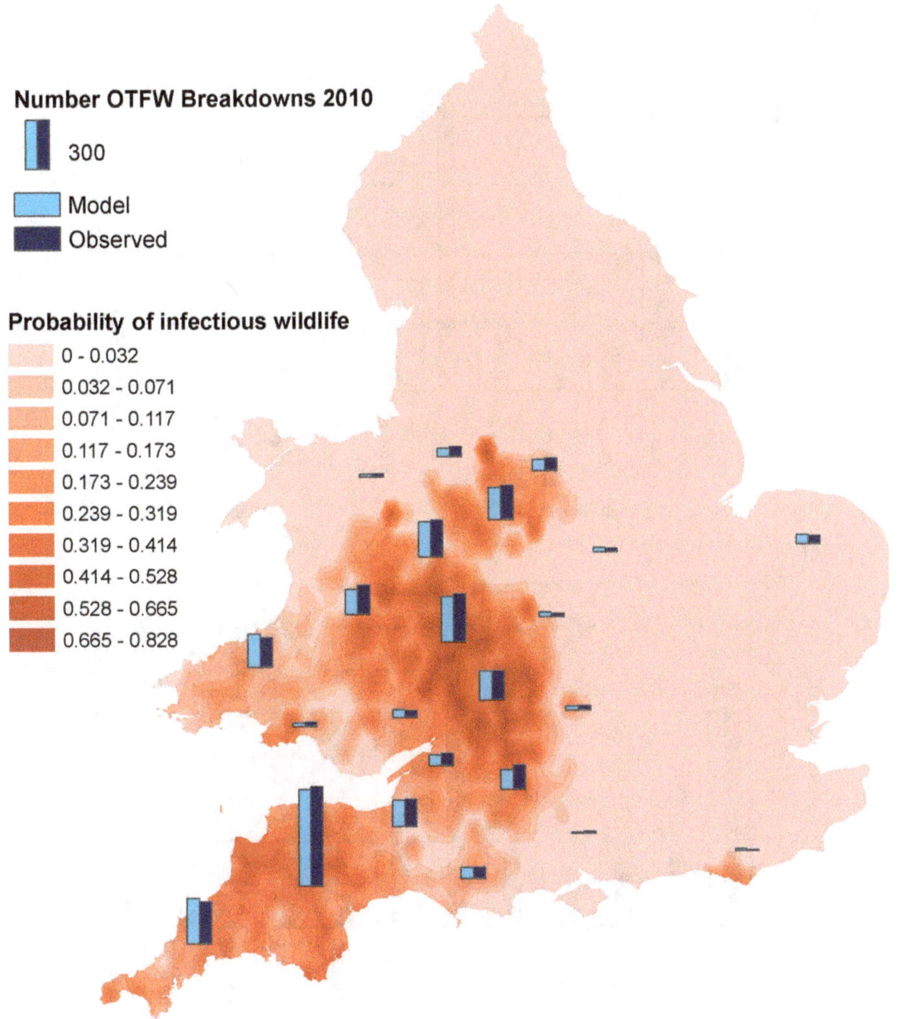

Fig. 1 Distribution of infectious environmental reservoirs after fitting the model to bTB breakdowns in 2010. Background shades in red indicate the smoothed fitted local probabilities that environmental reservoirs were infectious. Vertical bars in light and dark blue indicate the numbers of modelled and observed breakdowns in counties or contiguous groups of counties. The pair of bars farthest East indicate numbers for the whole Low Risk Area

regions. Most of the infectious environmental reservoirs were within the HRA and high incidence areas of Wales, a few were within the Edge region and none were within the LRA, as defined by the model hypothesis that bTB in the LRA was due to transfer of disease by cattle movements.

In Fig. 2 the numbers of breakdowns generated by the model are compared with observed numbers in 2010. The model outputs match very closely the distribution of breakdowns among the surveillance streams, along with the strong contrast between the HRA (over 2000 Officially Tuberculosis Free status withdrawn (OTFW) breakdowns) and the LRA (less than 60 OTFW breakdowns), with $R^2 = 0.9930$. However, the average χ^2_{24} measure of deviation (see Methods) was 101.0 (Standard deviation (SD) 22.4), whereas the 95% upper confidence limit for the difference between two χ^2_{24}statistics generated by random variation would be 51.5 (see Additional file 1, Part 2, Section 2.5), so the deviation between the baseline model fit and the observations was clearly significant, demonstrating that the model fit was not the closest possible and model assumptions and simplifications caused significant errors.

The use of a χ^2_{24} statistic to measure the fit of the model implied an assumption that the numbers of breakdowns were Poisson distributed among simulations. Indeed most SD of the numbers of breakdowns were close to the square roots of their averages, as expected from the Poisson distribution . The exceptions were the breakdowns in regular herd tests in the HRA and Wales, whose SD were about 12.5% lower than expected from the Poisson model; and breakdowns in the HRA at post-breakdown tests (SD 41% higher than

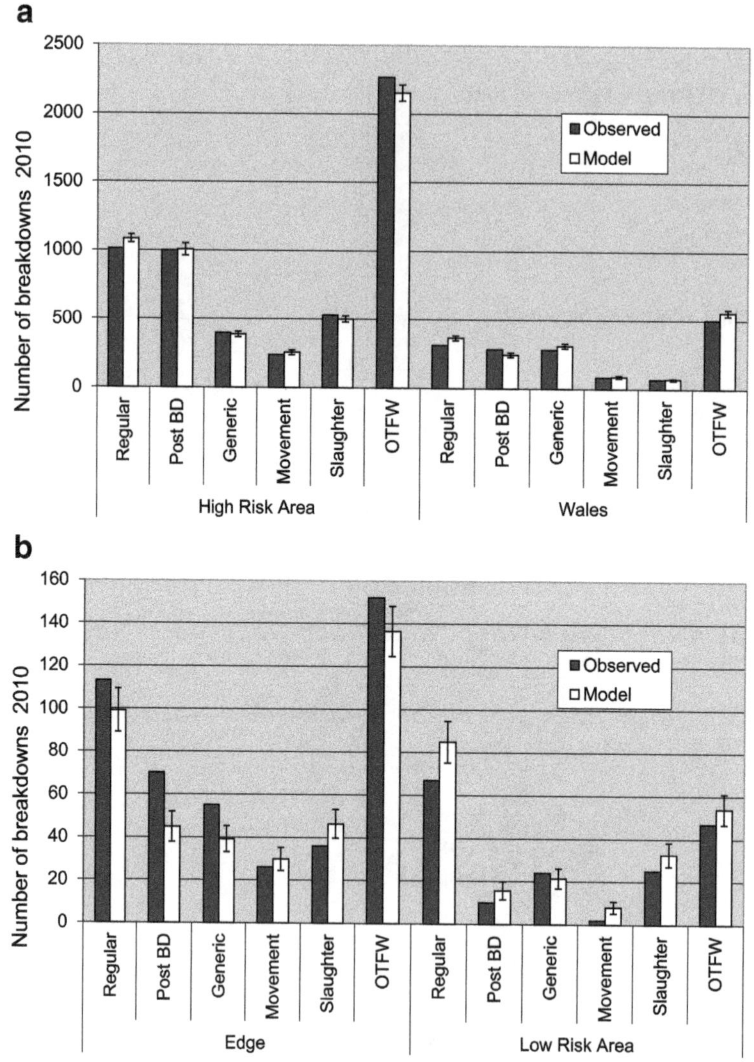

Fig. 2 Comparison of model outputs with observed numbers of breakdowns. **a** The High Risk Area and Wales; **b** The Low Risk Area and the Edge region. Error bars indicate standard deviations among 10 simulations

expected) and total OTFW breakdowns (SD 25% higher than expected).

On average the model generated 54 (95 percentile range = 41–67) OTFW breakdowns in the LRA in 2010, compared with 46 reported for that year [2]. Hence the observed number of breakdowns was within the range consistent with the model outputs. There was no evidence that any source of infection was required in the LRA beyond the transfer by livestock movements that was simulated in the model.

The outcome of a validation test is displayed for 2014 in Fig. 3 as a map comparison based on the bTB surveillance report for 2014 [16]. The model outputs for 2014 match the major features of the observed distribution of bTB, particularly the strong contrast between high incidence within the HRA and parts of Wales and much lower incidence in the LRA. A plot comparing the model with observed breakdown incidence in individual counties in 2016 demonstrated that the relationship between model predictions and observed incidence remained consistently close across the full range from 0 to 12 breakdowns per 100 herds $yr.^{-1}$ (Fig. 4a), with $R^2 = 0.8866$ (Data in Additional file 2). After square root transformation, differences between model predictions and observed incidences seemed to be more or less uniform relative to predicted values. A Cook-Weisberg test found no evidence of heteroscedasticity ($\chi^2_1 = 0.12$, $p > \chi^2_1 = 0.72$, Stata 15). However, the relationship between observed incidences in 2016 and 2010 was very similar to the relationship between observed incidence in 2016 and model predictions (compare Fig. 4b with a), with $R^2 = 0.8901$ (Data in Additional file 2). Implicitly the model achieved its match to 2016 mainly by generating little change from 2010. Comparison between observed bTB incidence and model

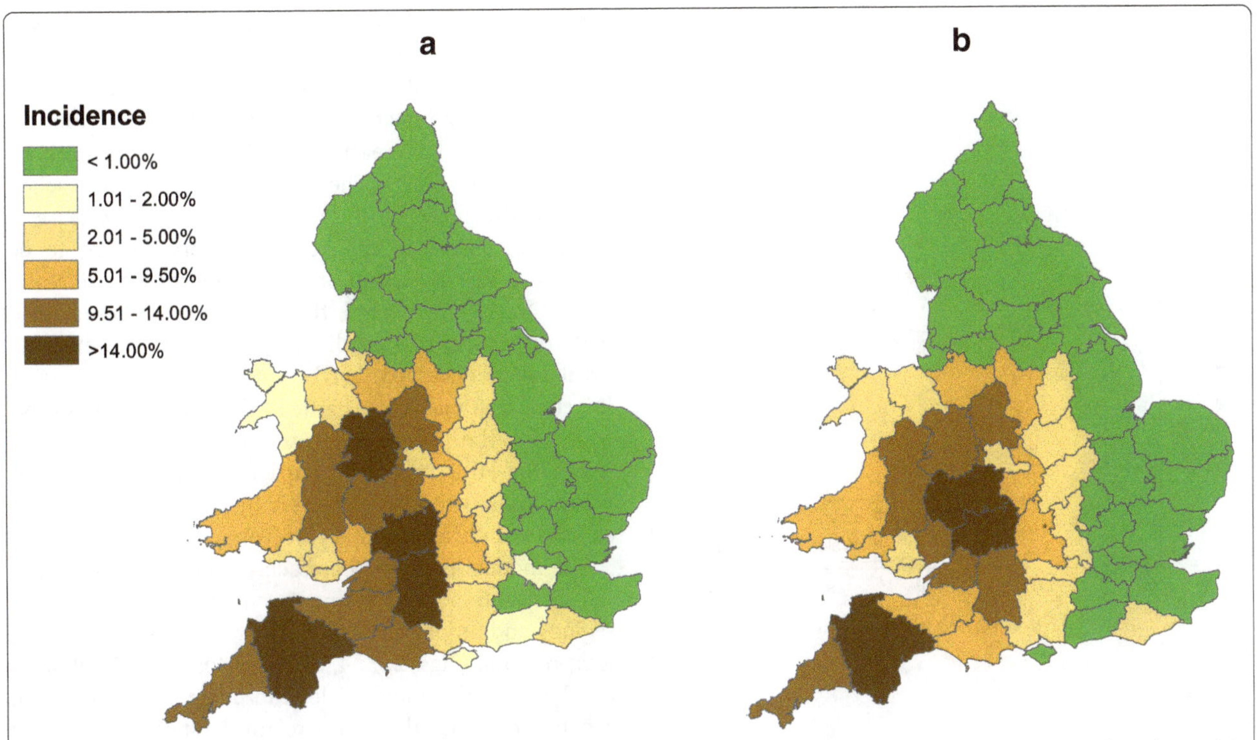

Fig. 3 Annual incidence of herd breakdowns in England and Wales during 2014. **a** As observed; **b** As the average of ten outputs from the model. In both cases incidence is calculated for herds included in the model and breakdowns at those herds

outputs from 2010 to 2015 suggested a gradual deterioration in the match, although the distinctions between counties in the HRA, Edge and LRA regions and Wales were notably maintained (Fig. 5). The closer match in 2010 appeared to be mainly due to the close fitting of incidences in the HRA counties with highest incidence, as illustrated by Fig. 1; by 2016 the relative values among the counties with high incidences had shifted in ways that were not matched by the model.

The environmental reservoirs were particularly important for this model, being the source of about 36% (SD among years 2%) of individual cattle infections in unrestricted herds during 2000–2016, while cattle to cattle transmission was the source of the remaining 64% (Data in Additional file 3). Moreover, infections from environmental reservoirs often or usually took place in previously uninfected herds, while, by definition, cattle to cattle transmission only took place in herds that were already infected. Cattle to cattle transmission must be followed by movement of infected cattle between herds to introduce infection to new herds. However, among the 6.80×10^6 cattle in the model, there were $1.95–2.04 \times 10^6$ transfers to other herds per year during 2008–2010, and $1.61–1.76 \times 10^6$ movements to slaughter (Additional file 4). Thus many or most infected cattle were detected before they left the herd where they were infected, while those that moved were almost as likely to move to slaughter as to transfer to another herd. Hence introduction of infection to previously uninfected herds was mainly due to transmission from environmental reservoirs.

The two parameters determining the environmental contribution, the number of infectious environmental reservoirs N_W (restricted to the HRA, Edge and Wales regions) and the environmental transmission parameter h, were varied, while keeping all other parameters at their baseline values, to find the potential range within which they could be consistent with observed bTB and its distribution among surveillance streams. Outputs from this sensitivity analysis suggested that the number of infectious environmental reservoirs was more constrained than the transmission rate (Fig. 6). It may be noted that the number of infectious environmental reservoirs is presented on a linear scale while the transmission rate is log-transformed. More infectious environmental reservoirs were required at lower transmission rates, but, even taking account of this trade off, the bulk of infection from environmental reservoirs should be taking place at only 5500–14,000 holdings. Note that these would be holdings with *possible* transmission of *M. bovis* from the environmental reservoir to cattle: infection of cattle from the environmental reservoir may be infrequent at many of these holdings.

Discussion

One of the benefits of a mechanistic model is that it provides a quantitative description of the system it models

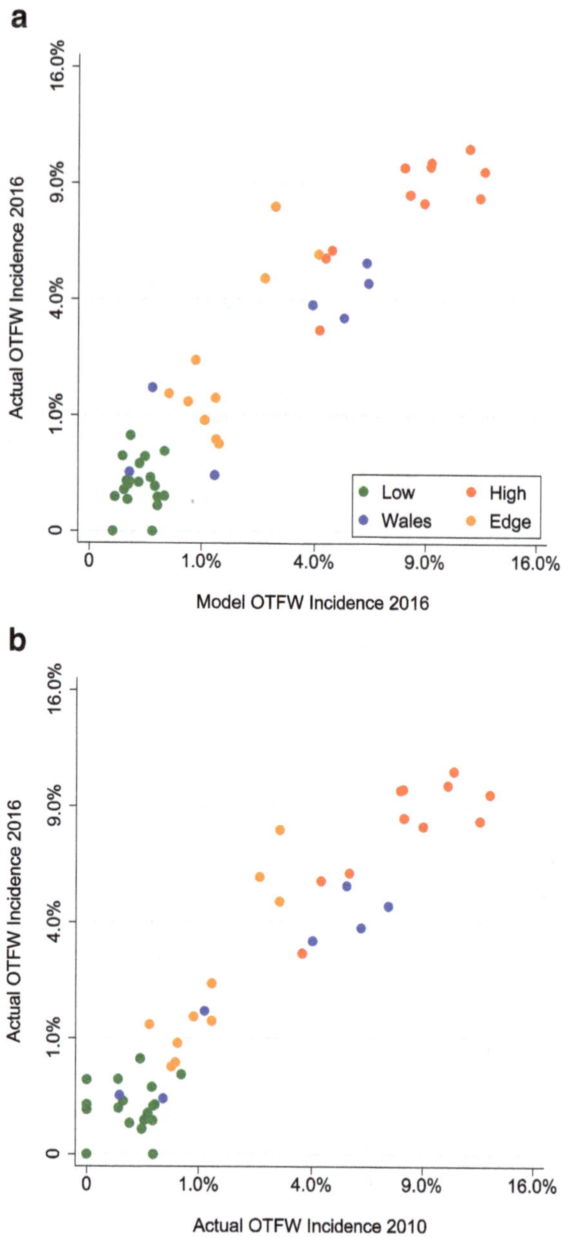

Fig. 4 Relationship between the observed and modelled incidence of breakdowns by county in 2016. **a** The observed incidence compared with the model for 2016 (average of 10 outputs); **b** The observed incidence in 2016 compared with the observed incidence in 2010

that follows current understanding, especially if its outputs are a reasonable fit to observations. Quantification builds and challenges understanding, even though the numbers are approximate and uncertain. The current model's description of the bTB epidemic in England and Wales is that introduction of *M. bovis* to previously uninfected herds is mainly driven by environmental reservoirs in areas of endemic infection, probably within a range from 5500 holdings (N_W in Table 1) exposed to environmental sources that would infect up to 1 in 20 cattle year^{-1} (*h* in Table 1) to 14,000 holdings exposed to infection of 1 in 350 cattle year^{-1} (Fig. 6). Infected cattle also transmit infection to other cattle at a rate of about 0.11 new infections per infected animal month^{-1} in a herd of 200, increasing with herd size (see Methods and Additional file 1). These conclusions can be reached because the number of cattle detected with disease is too large relative to the number of breakdowns to be generated just by infection from environmental reservoirs. On the other hand, the number of cattle detected with disease is too small to generate the number of breakdowns observed just by cattle movements and by infection that persists in cattle at the end of breakdowns. Moreover, the persistent contrast in bTB incidence between the HRA and LRA, particularly among recurrent breakdowns, strongly suggests a link with location, such as localised environmental reservoirs.

The model developed here achieved a reasonable fit to the distribution of bTB among surveillance streams and regions and among counties, which was robust up to at least 6 years, (Figs. 2, 3, 4, 5). Therefore the distribution of bTB in England and Wales can be described based on a hypothesis that there is a relatively static reservoir of infection around some herds in the relatively high incidence regions, combined with transfer of infection by cattle movements. Model deviations from observed distributions of breakdowns were significant, but those deviations could be due to the technical limitations of the model, including assumptions that parameters have been constant across large areas and several years. The model fit to observed disease in 2010 (Figs. 1, 2) was part of the process of model development, not a demonstration that the model was valid. As described in the Methods below and full model description (Additional file 1), the model includes many parameters, which can be indefinitely extended through regional and temporal variation, so the baseline set of parameter values can be assumed to be just one of an indefinite number of alternative sets. The model structure and parameterisation could also benefit from further investigation and development.

A previous bTB model estimated just seven parameters using Approximate Bayesian Computation (ABC) with sequential Monte Carlo sampling, but required significant simplification, including no regionalisation and no differentiation between surveillance streams [11]. This previous model also identified an environmental reservoir of infection as important, but assumed that it was not self-sustaining but accumulated in the presence of infected cattle. Mechanistically this is very different from the reservoirs of infection in this model, which are assumed to be self-sustaining, although reservoirs of infection can become extinct, a feature of the model that may be used more in future applications. However in practice the two models converge, because they both simulate

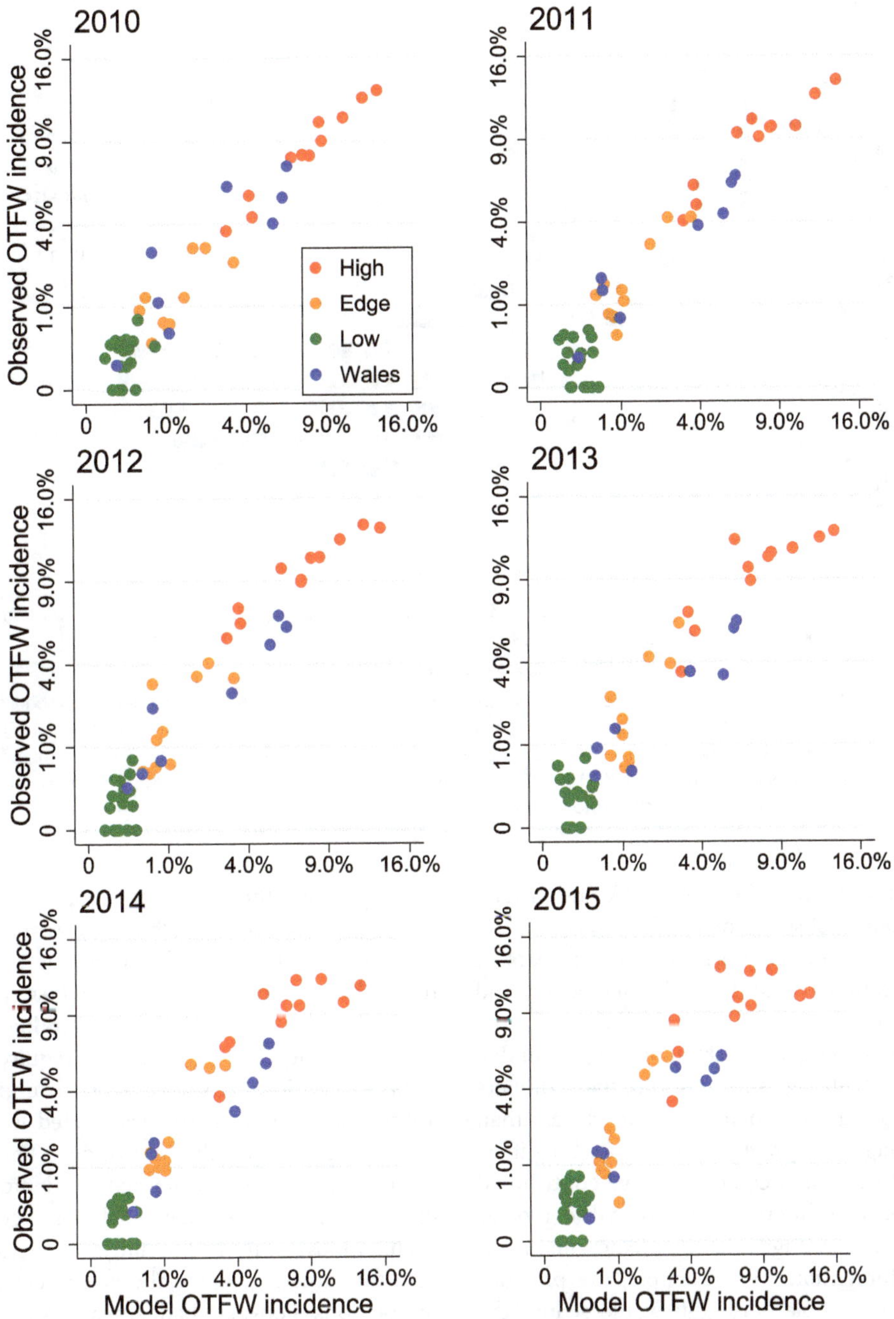

Fig. 5 Relationship between the observed and modelled incidence of breakdowns by county during 2010–2015

geographic regions in which high incidence of bTB in cattle is associated with substantial local reservoirs of infection at a minority of holdings.

Another difference between the model developed here and that of Brooks-Pollock et al. [11], which generates more differences between their predictions, is that, whereas most infected cattle are detected within this model and it has no latent period, the previous model assumed a long latent period lasting years with low detectability in cattle, which resulted in most infected cattle being slaughtered undetected [11]. However, most models of tuberculosis in cattle use relatively short periods of 1–8 weeks in which detection of infection is reduced, similar in duration to the assumption of this model [17]. The assumption of much longer periods with low detectability in cattle is contradicted by experimental evidence [18–20], although it is difficult to be certain about which features of experimental infection match natural infection. Moreover, assuming a high proportion of cattle are detected can generate distributions

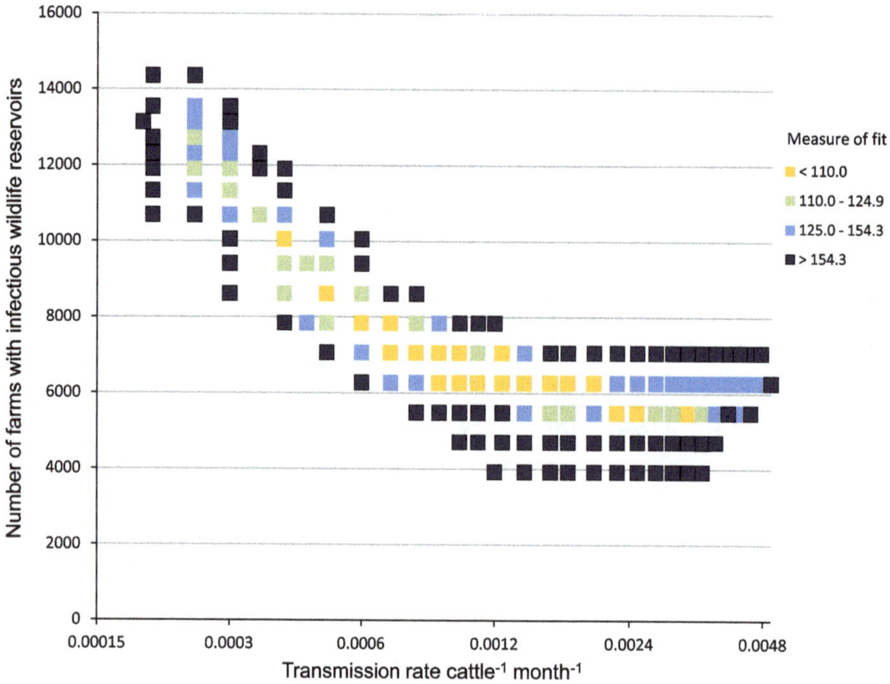

Fig. 6 Fitting parameters of transmission of *M. bovis* from environmental sources. The figure shows the dependence of model fit on the number of holdings exposed to *M. bovis* in environmental reservoirs and the rate of transmission of *M. bovis* from an environmental reservoir to local cattle. All other parameters were kept at their baseline values. Low values of the measure of fit indicate combinations of the parameter values at which the model matches observations relatively closely. Measures of fit are based on 10 simulations per point. The horizontal, transmission rate axis is plotted on a log scale but labelled with back transformed values

of breakdowns between the HRA and LRA and among counties that match observations (Figs. 2, 3, 4, 5), whereas the previous model overestimated numbers of breakdowns in counties with low bTB incidence and generated increasing numbers of breakdowns in the LRA instead of declining numbers (compare extended data Fig. 3c in [11] with Fig. 4 here, noting that numbers of breakdowns should have matched more closely than incidences), although the previous model used a different metric for fitting, which could also have contributed to this difference. The differences between the models will lead to them generating different predictions about the impact of interventions. For example, the previous model predicted that whole herd culls would reduce the number of herd breakdowns more than other interventions, whereas the current model would predict that whole herd culls would have relatively little impact.

In the current model many holdings are exposed to infectious environmental reservoirs, which are a source for many infections, but, except for a few small areas, holdings with infectious environmental reservoirs are a minority, even within regions with a high incidence of bTB (Fig. 1). According to the model fit, environment to cattle infection mainly takes place at less than 14,000 holdings, while the model set up identified over 31,000 holdings that might include infectious environmental reservoirs at the start of 2008 (Additional file 1: Table S4). Even at those holdings with infectious environmental reservoirs in the model, few or no infections from the environment to cattle take place each year. Hence regional control measures aiming to reduce infection from the environment may require substantial inputs to achieve impacts. For example, many wild animals must be culled or vaccinated for each transmission event to cattle that is avoided. Therefore targeting of control measures using statistical models, or models like this one could be beneficial. Some control measures, such as biosecurity of buildings, may also be more suitable for targeting, especially given recent evidence that interactions between badgers and cattle are rare and mainly take place at specific, identifiable locations [21]. However, improved biosecurity might reduce the cost-effectiveness of eradicating bTB from wildlife.

The benefit from quantitative, mechanistic models such as this, in comparison with broader scientific reasoning, is that quantifying impacts can have an important influence on conclusions. It was perhaps disappointing that the model's match to the observed distribution and dynamics of bTB was achieved by predicting little change, but the insight is valid and potentially useful. This model is not inert and it can generate changes in response to substantial measures such as widespread vaccination. Confidence

Table 1 Main epidemiological parameters with baseline estimates and ranges

Parameter definition	Symbol	Estimate (Range)	Source / Derivation
Initial number of undetected infected herds	Z	4400 (1850–5050)	Fitted so that simulation matched the total number of breakdowns in 2008 within 100, i.e. between 4849 and 5049.
Number of holdings in England and Wales with infectious environmental reservoirs.	N_W	7840 (SD ± 73) (5500–14,000)	Determined by the regional probabilities that individual holdings will have infectious environmental reservoirs (P_W).
Probability that individual holdings have infectious environmental reservoirs, dependent on a classification based on county and TB history.	P_W	0–0.828	The probabilities are fitted to local breakdown frequencies, see Fig. 1 and Methods.
Environment to cattle transmission rate per head of cattle at holdings with an infectious environmental reservoir.	h	0.0006 (0.0004–0.004) mth^{-1}	See Fig. 6.
Cattle to cattle transmission rate (per month) per infected animal in a herd of 200 cattle.	β	0.113 (0.10–0.14) mth^{-1}	Fitted to observed numbers of reactor animals at disclosing tests in HRA and LRA. Close to a previously published estimate [22].
Power law determining degree of density dependence of cattle-cattle transmission. (0 matches density dependence, while a value of 1 matches frequency dependence.)	q	0.5 (0.3–1.0)	Partial density dependence was demonstrated previously [22], but model outputs had low sensitivity to this parameter's precise value.
Probability of a persistent infection at the end of a breakdown in herds > 300 cattle, given that infected cattle remained after the disclosing test or there was at least 1 additional reactor.	c_p	0.6 (0.3–0.9)	Fitted to the observed number of breakdowns detected by post breakdown tests, especially in the Edge and Low Risk regions. Probability was lower in smaller herds, see Additional file 1.

'Symbol' is the symbol used in formulae. 'Estimate (Range)' indicates the baseline value, with a range indicating the potential uncertainty of the estimate, or a range dependent on location. Parameter values outside the stated ranges are likely to be associated with substantially worse model fit, or an infringement of reasonable constraints

that badgers are a source of *M. bovis* infection in cattle might suggest that reducing bTB among badgers is a way to reduce bTB incidence in cattle [9]; however, the insight that badger to cattle transmission may be unusual although very important suggests that controlling bTB in badgers may be a challenging option, even if it is essential.

Conclusions

The BoTMEW model achieved a reasonable fit to the observed numbers and distribution of BTB breakdowns in 2010 and maintained a good match to the relatively stable state of bTB since then. It demonstrated that transfer by cattle movements was sufficient to generate observed levels of bTB in the LRA, if higher incidence in the HRA was largely maintained by infection from environmental sources. To match features of bTB in England and Wales, such as the high observed numbers of breakdowns detected by post-breakdown surveillance, the model fit suggested that only a minority of herds in the HRA are exposed to environmental sources of infection. Therefore cost-effective control of bTB by reducing *M. bovis* transmission from the environment may be challenging. However, the outputs of any model depend on its structure and parameterisation, and, since the BoTMEW model has not yet been run with a wide variety of fitted parameter sets, its behaviour has not yet been fully investigated, and comparison with other models would be advisable.

Methods

The Bovine Tuberculosis Model for England and Wales (BoTMEW) is a spatially explicit, stochastic, dynamic simulation model of *M. bovis* transmission and control throughout England and Wales at the resolution of farm holdings and individual infected cattle. It focuses on the spread of *M. bovis* to individual herds by import of infected cattle, or by transmission from a local environmental reservoir (Table 1), and on the detection of infected herds by surveillance (Fig. 7, Table 2), which mainly relies on the single intradermal comparative cervical tuberculin (SICCT) skin test and detection at slaughter. The model also represents transmission among cattle within herds and the restrictions on herds detected as infected ('breakdowns'), while they are being cleared of infection. However, these within herd processes are simplified compared with specifically within herd models [17, 22]. The model is fully described and explained in a model description (Additional file 1). Here the main features of the model are described, focussing on transmission by cattle movements and from environmental reservoirs. An anonymized executable of the model with a full set of inputs has been released [23].

Introduction of *M. bovis* by cattle movements

Movement is the only process in the model by which cattle can transmit *M. bovis* to another herd. Direct transmission

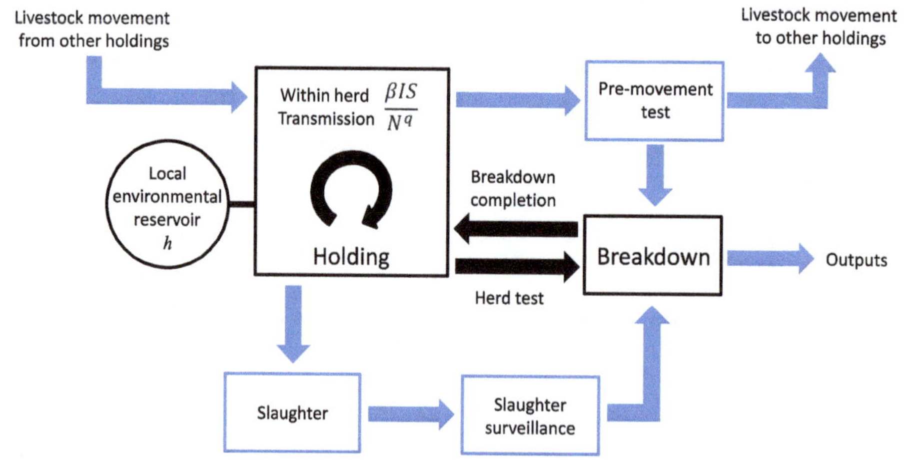

Fig. 7 Flow diagram summarising surveillance and disease transmission in the model. The relationship between the transfer and detection of TB through daily livestock movements (blue arrows) and epidemiology and detection resolved monthly at holding level (black). The pre-movement test and slaughter surveillance streams are distinguished from herd tests. Symbols in the transmission equations are: N = number of cattle in herd, I = number of infected cattle, S = number of susceptible cattle, β = cattle to cattle transmission parameter, q = power law for cattle to cattle transmission and h = environment to cattle transmission parameter

of *M. bovis* between cattle 'over the fence' is not explicitly simulated, because *M. bovis* infection rates are low and cattle to cattle contacts within herds are much more frequent than between neighbouring herds. There is no clear evidence of cattle to cattle transmission between contiguous herds that have not exchanged any animals or shared pasture [3]. In any case, actual transmission 'over the fence' is difficult to distinguish from successive infections from a shared environmental source.

Cattle movements were simulated by replaying the movements recorded in the Cattle Tracing Scheme (CTS) for England and Wales every day during the three calendar years 2008–2010. The CTS data had already been georeferenced and paired by the Data Systems Group at APHA to allow analysis of movement patterns, and series of movements via markets and other intermediate locations had been simplified to transfers between herds, or movements from herds to slaughter [24]. The quality of CTS data is more than adequate for this application: for example, missing movements or movements with errors will mostly result in incomplete pairs of movements, which constituted only 1.4% of all paired movements during 2008–2017 (Additional file 4). Further movements were omitted because CTS data excluded movements between pairs of holdings that arranged exemptions from reporting called 'CTS links', which involved about 35% of cattle holdings in 2008 but only included a small proportion of all movements [25]. CTS links no longer exist after being phased out from July 2016, a change that has not had an obvious impact on the total number of movements recorded (e.g. see Additional file 4).

The network of source and destination herds was also based on CTS data, but was static for each simulation, i.e. the list of herds and their cattle populations did not change during a simulation. The baseline set up included 6,802,196 cattle in 73,371 herds. Because the resolution of CTS is at holding level, the model treats each holding as consisting of a single cattle herd and its associated environmental reservoir. The restriction of holdings to a fixed list with only one herd at each holding meant that reference numbers of observed breakdowns for this study could be slightly fewer than in official statistics.

If a movement took place from an unrestricted herd including at least one infected animal, and the age class of an animal moved matched that of an infected animal, the infected animal had a chance of being selected for the movement. The cattle at each herd were allocated to five age classes: 0–42 days, 43 days to 15 months, > 15–30 months, > 30–60 months and > 60 months. By regulation calves up to 42 days old are too young for SICCT testing, while 30 months used to be the maximum age in the UK of cattle that could enter the human food chain due to bovine spongiform encephalopathy; the division at 15 months divides cattle up to 30 months roughly in half in time and numbers, while the division at 60 months divides the number of cattle older than 30 months roughly in half. Selection was stochastic, based on an assumption that all animals within an age class in a herd had equal probability of being selected. If the movement was to another herd, a pre-movement test might be applied to the moving animals, depending on the local control regime. If there was no reactor to tests, information about the infected animal was moved to the destination herd. If the movement was to slaughter, surveillance at slaughter was checked for every infected animal. If any were detected a breakdown was started at the source herd. Any infected animals that were

Table 2 Main surveillance parameters with current estimates

Parameter definition	Symbol	Estimate (Range)	S-T range	Source / Derivation
Maximum probability of detecting one infected animal in a herd test using the SICCT skin test	a	0.48 (0.36–0.63)	0.45–0.53	In literature [22, 33] and estimated by fitting. The number of breakdowns disclosed by slaughterhouse surveillance is sensitive to this parameter.
Maximum probability of detecting one infected animal when it is moved from a holding within an area where pre-movement testing is applied.	a_m	0.32 (0.24–0.42)	N	Detection reduced by exemptions from pre-movement testing. The recorded number of pre-movement skin tests was about ½ the number of movements from areas with pre-movement testing. Detection > 0.48/2 because infected cattle were more likely to be in movements without exemptions.
Infection duration until SICCT reaches maximum sensitivity	t_L	35 (20–200) d	N	Experimental evidence [20].
Specificity of an individual SICCT skin test.	sp_{Sk}	0.9998	N	The estimate of Goodchild et al. [34].
Proportion of age class i that is tested in a regular herd test	w_1 $w_2..w_5$	0 0.5–1.0	0 0.50–0.78	Fitted to observed numbers of regular herd tests.
Frequency of generic herd tests on infected herds	g_i	Location dependent	0.020–0.09 mth^{-1}	Fitted to observed numbers of breakdowns detected by generic herd tests.
Frequency of generic herd tests on uninfected herds	g_u	Location dependent	0.0012–0.04 mth^{-1}	Fitted to observed numbers of generic herd tests.
Maximum probability of detecting an infected animal at slaughterhouse surveillance.	a_s	0.875 (0.70–1.00)	0.557–0.875	Estimate may be high because of factors and practices absent from the model, including approved finishing units and age effects [35, 36]. Sensitivity increased during 2008–2010.
Maximum probability of confirming an infected reactor.	a_c	0.40	N	Fitted to number of OTFW (confirmed) breakdowns. This value underestimates the number of OTFW animals. (OTFW animals are unevenly distributed among breakdowns.)
Time required for lesions to reach maximum detectability.	t_{Ls}	35 d (20–200) d	N	This and other models seem to fit using a wide range of values for related parameters [22]. Reports that a high proportion of young reactors have visible lesions indicate a low value [36].

See legend for Table 1. 'S-T range', i.e. 'Spatial-temporal range' indicates the actual range of values for the parameter used for the baseline fitted model in England and Wales during 2008–2010. 'N' indicates that only a single value was used for the parameter

not detected at slaughter were recorded as an output from the model.

Transmission of *M. bovis* from environmental reservoirs

Every holding was assumed to have a local environmental reservoir, treated as being independent from environmental reservoirs in other holdings (Fig. 7). In the simulations reported here, the reservoir was assumed to be either infectious or not infectious for the years 2008–2010 of each simulation, an assumption influenced by studies at Woodchester Park, UK showing that a wildlife reservoir can be very local and maintain a stable spatial distribution for several years [26]. However, although Woodchester Park has been carefully studied for many years, it may not be typical of badger populations within the HRA. The model simulates potential changes to the infection state of environmental reservoirs during 2011–2016 (see Additional file 1).

An iterative approach was used to classify holdings in England and Wales into one of nine classes for the probability that infectious environmental reservoirs were present at the start of a simulation (see Additional file 1). It was assumed that all holdings in the LRA had zero probability of having an infectious environmental reservoir, as well as all other holdings with no bTB history and in areas with local bTB incidence < 1 reactor / 1000 cattle / year, estimated within a grid of 6.25 km^2 hexagonal cells [27].

For model fitting, all the probabilities that environmental reservoirs were infectious were increased or decreased by equal multiples, to preserve the relative distribution of infection in the environment, except the maximum probability was 1.00. Parameter settings for scenarios were summarized by reporting the number of infected environmental reservoirs, N_W. Within holdings that had infectious environmental reservoirs, every cow had the same probability of becoming infected from the environment each month, which was set by the environment transmission parameter h. Hence the actual distribution of environmental infectiousness was approximated by environmental reservoirs only having two values of infectiousness: 0 or h.

The potential ranges of the average infection rate from the environment h and the number of infected environmental reservoirs N_W were investigated by varying these two parameters from their baseline settings. The relationship between these values and measures of model fit could potentially indicate limits to their ranges. Although environmental infectiousness and the number of infected environmental reservoirs could compensate for each other's influence on total numbers of breakdowns, their relative values might be limited by the proportion of breakdowns that were disclosed by post breakdown tests, for example.

Detection of *M. bovis* infection in unrestricted herds

The model was designed to generate outputs that could be directly compared with surveillance observations, which were carefully simulated as 5 distinct surveillance streams: regular and targeted herd tests, pre-movement tests, slaughter surveillance and post breakdown tests (Fig. 7). Regular herd tests are tests carried out to meet the minimum regionally required frequency of bTB testing (currently every 4 years in the LRA and annually in the rest of England and Wales), while targeted herd tests represent all surveillance not included in the other 4 streams. The model also distinguished cattle infections and hence breakdowns as confirmed by visible lesions or bacterial culture, which is the main reason for classifying breakdowns as OTFW, or unconfirmed. Further detail is provided in the full model description (see Additional file 1).

Transmission within herds and breakdown processes

Cattle to cattle transmission was calculated once per month in herds with infected cattle, to compromise between execution speed and accuracy, while coordinating with the 2 month time intervals required for short interval testing within breakdown herds (Fig. 7). Previous models have been unable to decisively select between strongly contrasting assumptions about the development of individual infectiousness [22]. All models consider transitions between a susceptible state and an infectious state, but many models also include a latent period between infection and becoming infectious, or a period in which cattle are less reactive to surveillance tests, which may be before or after they become infectious [17, 28]. The direct evidence for a latent period and for a delay before infected cattle become reactive to surveillance tests is limited and possibly consistent with both periods being short [18], while parameters for different aspects of epidemiology may compensate for each other, e.g. a reduced transmission rate could compensate a shorter or omitted latent period [17]. Here each cow was assumed to have constant infectiousness β from the moment of infection, as assumed by the susceptible-occult-reactor (SOR) model of Conlan et al. 2012 [22]. A power law parameter, with a potential value between 0 and 1 was used to adjust the dependence on herd size of cattle-to-cattle transmission within a herd (Table 1) [22].

Several within-herd models of *M. bovis* transmission and control have been developed [17, 22, 28–30]. In contrast, the primary focus of this model was to simulate infection dynamics into and within unrestricted herds, to provide a framework for exploring the impact of control options on undetected infection. Nevertheless some representation of breakdown resolution was required, mainly because periods of restriction of herds at high risk of infection, and potential persistence of undetected infected cattle were essential to the dynamics of the system [22]. At the start of each breakdown, its duration, the numbers of confirmed, unconfirmed and false positive reactors, and whether an undetected infected cow would remain at the end of the breakdown were determined, based on the state of the holding affected (see Additional file 1). The overall dynamics of the model have low sensitivity to the duration of breakdowns, while the number of new reactors during a breakdown has no effect on any other process or output from the model.

Model initialisation and setup

The model was seeded with infected herds taken from a chronological list of herds that had reported breakdowns starting from 1st January 2008 onwards. If necessary, the number of herds seeded was adjusted until the average number of simulated breakdowns in 2008 was within 100 of the observed number, i.e. within the range 4850–5050. By 2010, the year used for model fitting, almost all these herds had either lost their infection or been detected. Herds in breakdown at the start of the simulation, or that recently had breakdowns were also identified, to minimize initial changes in model behaviour (see Additional file 1 and the full model set up [23]).

Programming and quality assurance and verification of code

The model was coded in Java Standard Edition version 7 using the Eclipse Integrated Development Environment (http://www.eclipse.org/) and the Git version control system (http://git-scm.com/). The code was written using test-driven development methods, which involve testing each model component in isolation to verify against the agreed model algorithms, as well as protecting against new components unexpectedly interacting with pre-existing code [31]. Distinct versions of the software were created as new features were added, with every code change logged and attributed to its author in the version control system.

The model is single threaded but multiple copies can be started from separate working directories on computers with multiple CPU cores. Logging and outputs data are saved as text files in CSV format in the working directory. Additional scripts were created to assist with setting up the multiple working directories for parallel runs, for scenario exploration, and to aggregate results data.

Third party open source libraries used in the model include the following.

- Mathematical functions: Colt high performance scientific computing library, developed at CERN (http://acs.lbl.gov/software/colt/), and the Apache Commons project (http://commons.apache.org/).
- Logging framework: SLF4J (http://www.slf4j.org/) and the Apache logging project (http://logging.apache.org/log4j/1.2/).
- Testing framework: JUnit (http://junit.org/) and Mockito (http://code.google.com/p/mockito/).

Model outputs

All simulated herd breakdowns during a run were recorded, including start date and duration, the surveillance stream that disclosed the breakdown, and the numbers of reactors and confirmed reactors. At the end of a simulation, data was also output for current undetected infected cattle, as well as a list of holdings with infected environmental reservoirs. In addition, summary information was output each year of the simulation for numbers of breakdowns disclosed by each surveillance stream, numbers of confirmed breakdowns, numbers of cattle infected from environmental reservoirs and from other cattle while herds were unrestricted, and total numbers of reactors, confirmed reactors and false positive reactors, including those detected during herd breakdowns. This summary information was provided for four regions: Wales, the LRA of England and two regions defined by county boundaries to approximate the HRA of England and the Edge region of England. (The regions for summary outputs were intended to be readily modified, so they were defined by counties rather than lists of individual holdings.) The hypothesis that *M. bovis* could only be introduced to herds in the LRA through cattle movements was tested by comparing model outputs with the observed number of OTFW breakdowns, because the total number of breakdowns in the LRA would be substantially dependent on the specificity of surveillance tests.

Model fitting

Data on bTB breakdowns for model development and fitting was taken from the APHA Vetnet database for 2002–2011, which has been superseded by the Sam database, APHA's herd registration and notifiable animal disease control and surveillance system, which records, for example, details of herds, TB tests, TB incidents and the details of any slaughtered (reactors, slaughterhouse cases and direct contacts) and inconclusive reactor cattle. The model was fitted by cross-sectional comparison with data for 2010 (see Additional file 3). Due to technical constraints at the time of model development, this model could not be fitted using fully computerised algorithms such as ABC. Instead parameters were estimated using a combination of external parameter estimates, applying model constraints and through iteration, comparing among batches of simulations with ranges of parameter estimates informed by previous simulation outputs. However, model behaviour was carefully investigated and is reported in the full model description (Part 2, Section 4 Sensitivity Analysis in Additional file 1).

Formalised and computerised fitting algorithms such as ABC would have the advantage that they would identify posterior distributions for fitted parameter values that would allow estimation of the uncertainty of model outputs. However, when this model was developed in 2011, it was impossible to use methods such as ABC to fit a model as complicated as this. Even now, development of a model like this using ABC or other computer fitting algorithms is dominated by the challenges of parameter estimation, fundamentally affecting the model design and purpose. Fitting also requires decisions about model structure and the prior distributions of parameters that may be subject to biological dispute. Models with different purposes and priorities can be fitted different ways to complement each other's strengths.

Each simulation included over 3000 breakdowns per year, so variation among simulations with the same set of parameter values was small relative to differences between simulations with different parameters. Hence 10 replicates were sufficient to evaluate each set of parameters. Because there was indefinite scope for improvement and there were alternative criteria for evaluating model fit, the set of parameter values used for sensitivity analysis and model exploration is referred to as the 'baseline' set of parameter values (Tables 1 and 2). For the baseline parameter set, 200 replicates were used to gain more precise estimates of its outputs and variance.

Fitting was most guided by a χ^2_{24} measure of deviation of the model from the actual number and distribution of breakdowns in 2010, which was calculated by comparing for each of the four summary regions the numbers of breakdowns from the five surveillance streams and the number of confirmed breakdowns, i.e. six times four = 24 comparisons. The χ^2_{24} measure was the sum of 24 terms of the form:

$$\chi^2 = \frac{(N_{\text{model}} - N_{\text{observed}})^2}{N_{\text{model}}} \tag{3}$$

where N_{model} is the average number of breakdowns from model runs. This measure was expected to have an approximately χ^2_{24} distribution, so it is referred to as the 'χ^2_{24} value'.

The χ^2_{24} value was used to indicate potential ranges for individual parameter values by observing the impact on the average χ^2_{24} value of varying parameters from their baseline values (Tables 1 and 2). The calculation of the χ^2_{24} value can be seen implemented in a spreadsheet in Additional file 3. Note that this metric is the same for all the surveillance streams, although it will give higher weight per breakdown to deviations from streams detecting smaller numbers of cases. Based on the conservative argument of paragraph 2.5 of Part 2 of Additional file 1, an average χ^2_{24} measure from 10 model replicates greater than 154.3 was treated as a significant deterioration of model fit at $p < 0.05$.

A χ^2 metric was chosen for evaluating model fit to allow matching to the observed contrast between large numbers of breakdowns in the HRA and parts of Wales, relative to much smaller numbers in the Edge region and the LRA. Being count data, dividing squared differences by expected values was expected to correctly weight deviations from a wide range of observed values. A previous bTB model used squared deviations as its metric, which may have reflected its emphasis on the development of bTB over time [11].

As an approximation to a large dataset, the model was almost certain to deviate significantly from observed values. Therefore R^2 was used as an index of the usefulness of the model, i.e. the portion of variation among data values after square root transformation that was explained by the model.

Model validation

Fitting a model demonstrates that it can match a set of observations of the system it models, but it does not demonstrate how stable the model behaviour is outside the range it was fitted to. When a model has many parameters there is a risk of overfitting, which may lead to the model rapidly deviating from actuality outside the range of data it was fitted to. Now a few years have elapsed since the model fit to 2010, the baseline model was projected to generate the incidence of bTB among counties in England and Wales up to 2016 (% herds starting a new breakdown during the year) to compare with data on observed incidence, which is in the public domain [2] (see Additional file 2). As mentioned above and in the full model description (Additional file 1), potential dynamic changes to the infection state of environmental reservoirs were introduced and the cattle to cattle transmission rate was increased during 2011–2016. Projections into the future using the model can only use past movement data, so it was considered appropriate to use recycled movement data for this validation exercise. A previous study reported no clear trend in cattle movement patterns during 2002–2010 [32], and the original fitting period for the model, 2008–2010, was partly chosen to follow the impacts on cattle movements from introducing pre-movement testing during 2006–2007. A recent query of CTS data has confirmed that total numbers of cattle movements in Great Britain and seasonal patterns were stable during 2008–2017, with average annual paired non birth or death cattle movements = 6.68×10^6 year^{-1} during 2008–2010 and 6.65×10^6 year^{-1} during 2011–2016 (Additional file 4).

Movements were generated for the projection by recycling the movements from 2008 to 2010, so movements in 2011–2016 repeated those recorded in 2008–2010 twice. To check that the model maintained a match to the observed distribution of bTB, modelled and observed incidences of bTB were compared visually and by regression among counties. Since the model was fitted to observed bTB in 2010, observed incidences in 2016 and 2010 were also compared with each other in an equivalent way.

Anonymised model

For submission for publication, an alternative anonymised version of the model has been released [23]. The CPH identifiers have been altered to only retain the county location of each holding, and geographical locations have been removed. The summary outputs have been confirmed to be a close match at county resolution to the full model with the same parameter sets (Section 6, Part 2 of Additional file 1).

Additional files

Additional file 1: Detailed functional description of a bovine tuberculosis model for England and Wales (BoTMEW), 41 pages. Part 1 Model specification. Part 2 Model setup and use.

Additional file 2: A spreadsheet file with outputs from a recent run of the anonymised model with observed numbers of OTFW breakdowns in 2016 as displayed in Fig. 4.

Additional file 3: A spreadsheet file with outputs from a recent run of the anonymised model with observed numbers of breakdowns in 2010 as displayed in Fig. 2.

Additional file 4: A spreadsheet providing annual numbers of movements recorded in CTS, to indicate the frequency of errors, numbers of transfers between herds during 2008–2010, and variation among years.

Abbreviations

Baseline model: The model run using a set of parameter values, including variations of some parameters between regions and times, chosen because it achieved a good fit to observed records of bTB in 2010 and matched current knowledge of bTB epidemiology.; BoTMEW: Bovine tuberculosis model for England and Wales; bTB: Bovine tuberculosis; CPH: County parish holding identifier; CTS: Cattle tracing scheme; Edge: Buffer region between the HRA and LRA; *h*: *M. bovis* transmission rate from an infectious environmental reservoir per susceptible cow; HRA: High risk area; LRA: Low risk area; MS: Mean square error; N_W: The total number of holdings with infectious environmental reservoirs present; OTFW: Officially tuberculosis free status withdrawn; P: Probability that a test statistic would be exceeded if a null model applied; R^2: The proportion of variance explained by a regression model; RBCT: Randomized Badger Culling Trial; SD: Standard deviation; SE: Standard error; χ^2: Measure of deviation with Chi-square distribution

Acknowledgements

We thank Tony Goodchild for sharing epidemiological data, analysis and discussion and Sara Downs for epidemiological advice. Paul Upton managed and provided the data used in this project, while Adam Ashton and Stuart Ashfield provided GIS advice and assisted preparation of Fig. 1. James Tiller and Paul Upton provided the data from CTS in Additional file 4. Mark Arnold discussed the model and its outputs during its development, as well as reviewing manuscript drafts. David Rae and Ian Mawhinney performed analyses during development of the model specification. We are also grateful to other staff in the Department of Epidemiological Sciences, Animal and Plant Health Agency, who extracted and prepared livestock movement data and holding data. Finally we thank Iain Simmons for preparing model runs and completing analysis for Fig. 6 during his pre-university work experience placement.

Funding

This work was supported by the Department for Environment, Food and Rural Affairs (Defra) under project codes SE3122 and SB4500.

Availability of data and materials

Software: A current executable version of BoTMEW accessible by DOI, including a complete but anonymized set of its setup files, has been released on Zenodo. The source code is not available because the project was not planned or resourced as an open source project.
Operating system: Platform independent.
Programming language: Java.
Requirements: Java 7 or later.
Licenses: Original parts of the executable and its set up data was released under an Open Government Licence (UK) v3. The software uses free to use licenses from CERN, Apache 2.0 and MIT, which are included in a file "license.txt".
The datasets used and/or analysed during the current study that are not included in this published article and its additional files are available from the corresponding author on reasonable request, subject to redaction for reasons of confidentiality.

Authors' contributions

CPDB led the project and developed the model specification, performed model fitting and simulation runs, identified model bugs during running, and prepared and wrote this document and the additional files. AG developed the model code under the supervision of OT, including translation of the model specification into the software design and architecture.

Consent for publication

Not applicable.

Competing interests

The authors completed this work as civil servants working for the Animal and Plant Health Agency, UK.

References

1. Bourne FJ, Donnelly CA, Cox DR, Gettinby G, McInerney JP, Morrison WI, et al. Bovine TB: the scientific evidence. A Science Base for a sustainable policy to control TB in cattle. An epidemiological investigation into bovine tuberculosis. London: Defra; 2007.
2. Quarterly publication of National Statistics on the incidence and prevalence of tuberculosis. (TB) in cattle in Great Britain. London: Department for Environment, Food and Rural Affairs; 2017. https://www.gov.uk/government/uploads/system/uploads/attachment_data/file/629981/bovinetb-statsnotice-quarterly-19jul17.pdf. Accessed 6 Sept 2017
3. Godfray HCJ, Donnelly CA, Kao RR, Macdonald DW, McDonald RA, Petrokofsky G, et al. A restatement of the natural science evidence base relevant to the control of bovine tuberculosis in Great Britain. Proc R Soc B. 2013;280:20131634.
4. Bessell PR, Orton R, O'Hare A, Mellor DJ, Logue D, Kao RR. Developing a framework for risk-based surveillance of tuberculosis in cattle: a case study of its application in Scotland. Epidemiol Infect. 2013;141:314–23.
5. Wales TB Eradication Programme delivery plan. Welsh government; 2017.
6. The Strategy for achieving Officially Bovine Tuberculosis Free status for England. Department for environment, food and rural affairs; 2014.

7. Krebs J, Anderson R, Clutton-Brock T, Morrison I, Young D, Donnelly C. Bovine tuberculosis in cattle and badgers. London: Ministry of Agriculture Fisheries and Food; 1997.
8. Goodchild AV, Clifton-Hadley RS. Cattle-to-cattle transmission of Mycobacterium bovis. Tuberculosis. 2001;81:23–41.
9. King D. Bovine tuberculosis in cattle and badgers. Department for Environment, Food and Rural Affairs; 2007.
10. Jenkins HE, Woodroffe R, Donnelly CA. The duration of the effects of repeated widespread badger culling on cattle tuberculosis following the cessation of culling. PLoS One. 2010;5:e9090.
11. Brooks-Pollock E, Roberts GO. Keeling MJ. A dynamic model of bovine tuberculosis spread and control in Great Britain. Nature. 2014;511:228–31.
12. Donnelly CA, Nouvellet P. The contribution of badgers to confirmed tuberculosis in cattle in high-incidence areas in England. PLOS Currents Outbreaks. 2013;1
13. Green DM, Kiss IZ, Mitchell AP, Kao RR. Estimates for local and movement-based transmission of bovine tuberculosis in British cattle. Proc R Soc B. 2008;275:1001–5.
14. Fitzgerald SD, Kaneene JB. Wildlife reservoirs of bovine tuberculosis worldwide: hosts, pathology, surveillance, and control. Vet Pathol. 2013;50:488–99.
15. Palmer MV. Mycobacterium bovis: characteristics of wildlife reservoir hosts. Transbound Emerg Dis. 2013;60:1–13.
16. Animal and Plant Health Agency. Bovine tuberculosis: infection status in cattle in GB annual surveillance report for the period January to December 2014. London, UK: Department of Environment Food and Rural Affairs; 2015.
17. Álvarez J, Bezos J, de la Cruz ML, Casal C, Romero B, Domínguez L, et al. Bovine tuberculosis: within-herd transmission models to support and direct the decision-making process. Res Vet Sci. 2014;97:S61–8.
18. McCorry T, Whelan AO, Welsh MD, McNair J, Walton E, Bryson DG, et al. Shedding of Mycobacterium bovis in the nasal mucus of cattle infected experimentally with tuberculosis by the intranasal and intratracheal routes. Vet Rec. 2005;157:613–8.
19. Khatri BL, Coad M, Clifford DJ, Hewinson RG, Whelan AO, Vordermeier HM. A natural-transmission model of bovine tuberculosis provides novel disease insights. Vet Rec. 2012;171
20. Thom ML, Hope JC, McAulay M, Villarreal-Ramos B, Coffey TJ, Stephens S, et al. The effect of tuberculin testing on the development of cell-mediated immune responses during mycobacterium bovis infection. Vet Immunol Immunopathol. 2006;114:25–36.
21. Drewe JA, O'Connor HM, Weber N, McDonald RA, Delahay RJ. Patterns of direct and indirect contact between cattle and badgers naturally infected with tuberculosis. Epidemiol Infect. 2013;141:1467–75.
22. Conlan AJK, McKinley TJ, Karolemeas K, Pollock EB, Goodchild AV, Mitchell AP, et al. Estimating the hidden burden of bovine tuberculosis in Great Britain. PLoS Comput Biol. 2012;e1002730:8.
23. Birch CPD, Goddard A, Tearne O. Bovine tuberculosis model for England and Wales (BoTMEW) including model description. 2018. https://doi.org/10.5281/zenodo.1237748
24. Mitchell A, Bourn D, Mawdsley J, Wint W, Clifton-Hadley R, Gilbert M. Characteristics of cattle movements in Britain - an analysis of records from the cattle tracing system. Anim Sci. 2005;80:265–73.
25. Orton RJ, Bessell PR, Birch CPD, O'Hare A, Kao RR. Risk of foot-and-mouth disease spread due to sole occupancy authorities and linked cattle holdings. PLoS One. 2012;7:e35089.
26. Delahay RJ, Langton S, Smith GC, Clifton-Hadley RS, Cheeseman CL. The spatio-temporal distribution of Mycobacterium bovis (bovine tuberculosis) infection in a high-density badger population. J Anim Ecol. 2000;69:428–41.
27. Brunton LA, Nicholson R, Ashton A, Alexander N, Wint W, Enticott G, et al. A novel approach to mapping and calculating the rate of spread of endemic bovine tuberculosis in England and Wales. Spat Spatiotemporal Epidemiol. 2015;13:41–50.
28. O'Hare A, Orton RJ, Bessell PR, Kao RR. Estimating epidemiological parameters for bovine tuberculosis in British cattle using a Bayesian partial-likelihood approach. Proc R Soc B. 2014;281
29. Barlow ND, Kean JM, Hickling G, Livingstone PG, Robson AB. A simulation model for the spread of bovine tuberculosis within New Zealand cattle herds. Prev Vet Med. 1997;32:57–75.
30. Fischer EAJ, van Roermund HJW, Hemerik L, van Asseldonk M, de Jong MCM. Evaluation of surveillance strategies for bovine tuberculosis (Mycobacterium bovis) using an individual based epidemiological model. Prev Vet Med. 2005;67:283–301.
31. Beck K. Test-driven development by example. Addison Wesley; 2002.
32. Vernon MC. Demographics of cattle movements in the United Kingdom. BMC Vet Res. 2011;7:31.
33. Nuñez-Garcia J, Downs SH, Parry JE, Abernethy DA, Broughan JM, Cameron AR, et al. Meta-analyses of the sensitivity and specificity of ante-mortem and post-mortem diagnostic tests for bovine tuberculosis in the UK and Ireland. Prev Vet Med. 2017;
34. Goodchild AV, Downs SH, Upton P, Wood JLN. de la Rua-Domenech R. Specificity of the comparative skin test for bovine tuberculosis in Great Britain. Vet Rec. 2015;177:258.
35. Trading options for farms under TB restrictions. *Better Returns programme*. Warwickshire: EBLEX, AHDB; 2014. https://beefandlamb.ahdb.org.uk/wp-content/uploads/2014/07/brpTB-restrictions080714a.pdf. Accessed 26/07/2018.
36. Brooks-Pollock E, Conlan AJK, Mitchell AP, Blackwell R, McKinley TJ, Wood JL. Age-dependent patterns of bovine tuberculosis in cattle. Vet Res. 2013;44:97.

22

Prevalence of bovine tuberculosis in cattle, goats, and camels of traditional livestock raising communities in Eritrea

Michael K. Ghebremariam[1,2*], A. L. Michel[3,4], J. C. M. Vernooij[5], M. Nielen[5] and V. P. M. G. Rutten[1,3]

Abstract

Background: The aim of the current study was to assess the prevalence of bovine tuberculosis (BTB) in cattle, goats, and camels, and its zoonotic potential within the traditional livestock raising communities in four regions of Eritrea. The Single Intradermal Comparative Tuberculin Test (SICTT) as indicator of *M. bovis* infection was conducted on 1077 cattle, 876 goats, and 195 camels. To elucidate possible risk factors for BTB transmission between animals and its potential zoonotic implication, questionnaire based face-to-face interviews were conducted in households of which 232 raised cattle, 128 goats, and 29 camels.

Results: The results of the SCITT were interpreted using the OIE standard (> 4 mm cut-off) for positive responses. In cattle, individual animal ($n = 1077$) and herd ($n = 413$) prevalences were 1.2% ($n = 13$) [Confidence Interval (CI) 95% CI, 1.0–1.3%] and 3.2% ($n = 13$) (95% CI, 3.0–3.4%), respectively. In goats ($n = 876$), none of the animals was positive. In camels, individual animal ($n = 195$) and herd ($n = 70$), BTB prevalences were 1.5% ($n = 3$) (95% CI,1.4–1.6%) and 2.9($n = 2$) (95% CI, 0.9–4.6%), respectively. Overall, male animals were more at risk (OR = 2.6; 95% CI:1.0–8.7) when compared to females. Sharing of water points, introduction of new animals into herds and migration of animals over large distances were common events that may contribute to intra and inter-species transmission of BTB. Consumption of raw milk, lack of BTB transmission awareness, and low levels of education were common in the farming communities.

Conclusion: The current study highlighted a low prevalence of *M. bovis* in cattle, goats and camels in extensive traditional livestock in Eritrea. Despite this, the spatial distribution of affected animals across most of the sampled regions and consumption of unpasteurized milk warrants surveillance, cautious and timely control measures for the disease.

Keywords: Bovine tuberculosis, Camels, Eritrea, Goats, Mixed crop-livestock system, Pastoral system, Single intradermal comparative tuberculin test (SICTT)

Background

Bovine tuberculosis (BTB) is a chronic bacterial disease caused by *Mycobacterium bovis* (*M. bovis*), a member of the group known as *Mycobacterium tuberculosis complex* (MTC), that has a wide host range. It predominantly affects cattle, but also other domesticated species, like goats [1–4], and camels [5, 6], as well as many wildlife species [7, 8]. In general, in traditional livestock raising systems, cattle and goats are often herded together and watering points are shared by many animal species. Such livestock husbandry and management systems can be an important risk factor for animal-to-animal, animal-to-human, human-to-animal, and human-to-human *M. bovis* transmission [9–13].

M. bovis infected animals, as indicated by SICTT, were present in the 'intensive' dairy husbandry system of the major milk producing regions in Eritrea [14, 15]. Besides, the presence of *M. bovis* was confirmed by bacterial culture and molecular diagnostic tools from bovine tissues collected at the Asmara slaughterhouse (Ghebremariam, unpublished data). However, the BTB status was never studied in the extensive traditional livestock (pastoral and mixed crop-livestock) system, which comprises the largest percentage of the livestock population (approx. > 99.9%)

* Correspondence: michaelvet@yahoo.com
[1]Department of Infectious Diseases and Immunology, Faculty of Veterinary Medicine, Utrecht University, Utrecht, The Netherlands
[2]Department of Veterinary Sciences, Hamelmalo Agricultural College, Keren, Eritrea
Full list of author information is available at the end of the article

[14]. In neighbouring Ethiopia, with similar agricultural settings, BTB was reported to be prevalent in the 'intensive' dairy cattle in different studies (11.6% and 22.1%, respectively) [16, 17], as well as in cattle in the traditional extensive livestock husbandry system (8.2% and 11%, respectively) [16–18].

In developing countries, especially in rural settings, where dwelling areas may be shared between humans and animals, humans may become infected. This may occur through the inhalation of cough sprays released by chronic coughing animals [9, 12], or/and by drinking raw milk from infected animals [1, 19, 20]. The aim of the current study was to assess the prevalence of BTB in cattle, goats, and camels, and its zoonotic potential within the traditional livestock raising communities in the four regions (Debub, Anseba, Gash Barka, and Southern Red Sea) in Eritrea that share borders with at least one of the neighbouring countries (Fig. 1).

Methods

Study population and sample size determination

We conducted a BTB prevalence study using multistage sampling in Debub, Anseba, Gash Barka, and Southern Red Sea regions of Eritrea (Fig. 1), from October to November 2013, and September to December 2014. First, we selected the regions with the target species of animals (cattle, goats and camels or at least two of the species) and those sharing common borders with neighbouring countries (Ethiopia, Sudan, Djibouti) (Fig. 1). For the second stage official lists of sub-regions, including villages that were connected by paved and dry weather roads were used to randomly select villages where two or three of the target species were present. Participation of livestock owners/herdsmen present at the testing sites was voluntary, and only animals of those who gave their consent to participate were recruited for the study. Convenience sampling was applied by selecting animals that could be caught and restrained for

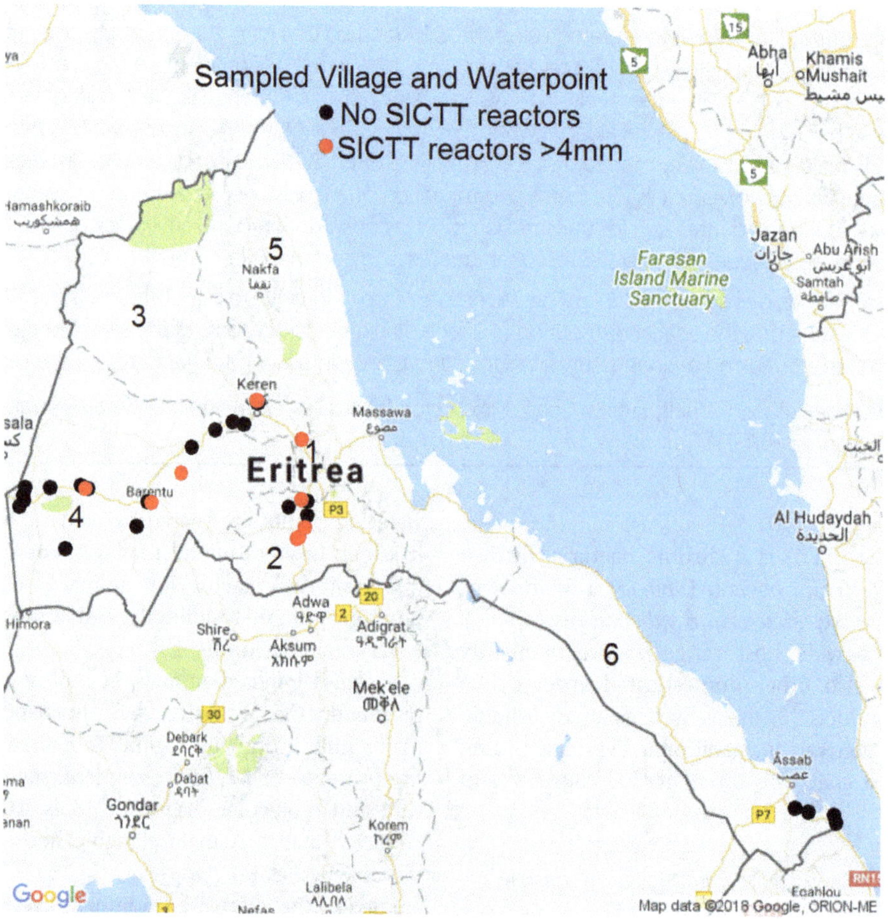

Fig. 1 Map of Eritrea depicting the study areas ($n = 31$) and water points ($n = 4$, shared by animals from 11 study areas), having skin test reactors with > 4 mm-cut-off (red dots) and those with no reactors using > 4 mm (black dots) in the selected study areas within the traditional livestock husbandry system in Eritrea. The numbers (1–6) indicated on the map show the six administrative regions of Eritrea (1 = Maekel; 2 = Debub; 3 = Anseba; 4 = Gash Barka; 5 = Northern Red Sea; 6 = Southern Red Sea) (Adapted using: Loecher and Ropkins [54]. RgoogleMaps and loa: Unleashing R Graphics Power on Map Tiles. Journal of Statistical Software 63(4), 1–18. URL http://www.jstatsoft.org/v63/i04/)

testing, and if possible, the questionnaire was concurrently completed by the owners/herdsmen. In the Debub region, as camel raising is not a common practice, hence camels were not included in the study there. Since the prevalence of BTB in the extensive livestock production system in Eritrea was unknown, required sample sizes were determined considering the livestock population size per species (data provided by the Ministry of Agriculture, not shown) of the selected sub-region and assuming 3% prevalence as found in a similar study in extensive traditional livestock farming in Ethiopia [21], and 95% level of significance, using the WIN EPISCOPE 2.0, veterinary epidemiological computer software.

Outcomes of calculations dictated a minimum of 99 animals from each species per study region to be tested to find at least one positive animal. The precision for the estimated prevalence assuming 3% prevalence and the calculated sample size is between 0.6 and 8.5%. But as animals are clustered within herds we aimed for a higher number of cattle and goats by selecting 3% of each herd per species for testing. All the target species of animals were sampled up to a herd size of 3 and at least up to a convenient higher number for larger farms.

That is, in the presence of three or less animals in a herd of a given species (cattle, goats or camels), the whole herd was subjected to testing. However, although the sample size (3% prevalence) requires 99 animals to be tested per region, it was not possible to reach that number at all times. Particularly, camels were less abundant and not represented in all herds.

The test sites were water points ($n = 4$, where animals from 11 study areas were tested), villages ($n = 21$), livestock gathering (resting) ($n = 4$ sites) and grazing areas ($n = 6$ sites) as they provide frequent opportunities for inter-and intra-species contact.

Geographic coordinates were registered at the test sites (areas) by the global positioning system (GPS) (Fig. 1). Information relating to age, sex, region, sub-regions, and study areas was recorded for all the tested cattle, goats and camels (Additional file 1). In the testing areas animals were restrained by manually handling them using 3–4 men, and were ear tagged for identification purposes.

Questionnaire survey among livestock owners

In total, 232 cattle, 128 goats, and 29 camel owners, who presented their animals for tuberculin testing, were interviewed face-to-face using a standardized questionnaire consisting of both open and closed questions, based on their consent to voluntary participation (Additional files 2, 3, 4 and 5). The questionnaire included potential risk factors (variables) for *M. bovis* transmission amongst animals, and between animals and humans: housing of animals, source of water for the animals, sharing of water points, livestock migration, presence of wild animals, farmers' BTB transmission awareness, raw milk consumption habits, and farmers' level of education, etc. The questionnaires were translated from English into two local languages (Tigre and Tigrigna) that are widely spoken in the study areas. Initially, the questionnaire was pre-tested by 10 randomly selected households in one of the villages to verify if the questions were clearly understood by the respondents. Subsequently the questionnaire was fine-tuned (ambiguous/misleading questions were corrected or omitted) and used. Administrators of the selected sub-regions, villages, including village elders, were informed of the testing program through regional administration offices to make them aware of and sensitize them to it. The respondents were informed about the confidentiality of data collected through the questionnaire and the interviews were conducted upon the consent of the respondents. It was not possible to interview all the farmers that presented their animals for testing, since at some moments, sites were too crowded to allow timely completion of the questionnaires, and the farmers were leaving the sites.

Intradermal tuberculin skin testing

The skin testing was conducted according to the OIE [22] standard. Briefly, two sites on the left side of the mid-neck in cattle, right and left neck in goats, and upper and lower neck, closer to the shoulder, in camels, were shaved 12 to15cm apart. At the indicated sites skin thickness was measured with a 'Vernier calliper' and recorded, and the sites were injected with 0.1 ml containing 2500 IU avian PPD (Prionics, Lelystad, The Netherlands) and 0.1 ml of 3000 IU bovine PPD (Prionics, Lelystad, The Netherlands) using McLintock pre-set automatic syringes. Correct injection was confirmed by palpation of a small pea-like swelling at each site of injection.

Seventy-two hours after inoculation, the skinfold thicknesses at the injection sites were re-measured by the same operator and with the calliper used before. The readings were interpreted using the standard (> 4 mm cut-off; OIE) method [22]. An animal was considered positive if the differential increase in skin thickness between the bovine and avian injection was greater than 4 mm, inconclusive when the reaction difference was 1–4 mm, and negative when the bovine reaction was less or equal to the avian reaction in the absence of any clinical signs at the injection site. For the logistic regression analysis, we considered the inclusive results as negatives. As an incentive for participation by the owners, their animals were treated for internal and external parasites with ivermectin (anthelmintics) following the reading of the results. Though we have used standard interpretation of the results throughout the manuscript, we have included severe (> 2 mm cut-off) method for comparison with other studies (Additional files 6 and 7). According to the severe method, an animal was considered positive if the bovine minus the avian reaction was greater than 2 mm.

Data analysis

Data were entered in an Excel spreadsheet and then exported to SPSS IBM version 20 for analysis and all the analyses were conducted using this statistical package. Descriptive analysis was conducted at both individual animal and herd levels. Due to repeated samples within herds a mixed effect logistic regression model would be most appropriate to account for this dependency. A considerable number of farms have only one sample which causes problems in the estimation of the farm effects (farm level is the level of the only sampled animal). Secondly the outcome is binary (pos/neg per animal) which means that with 3 animals tested in a herd only a few levels for estimation maybe observed (0, 33, 66, 100% positive). The expected prevalence is 3%, so at farm level the prevalence will be mostly 0%. Therefore, we used an ordinary logistic regression model to investigate the relationship between the potential risk factors and outcomes. First, in univariable analysis, the measure of association between each of the potential risk factors and BTB (skin test positive score as a parameter), was examined for each factor individually and evaluated for statistical significance using a Fisher's Exact test for independence; secondly, multivariable logistic regression was applied to estimate the measures of associations between the potential risk factors (species, age, and sex of the tested animals) and the outcome were tested resulting in odds ratios (OR). The outcome of all statistical analyses were individual animal and herd level binary outcomes. A herd was considered positive if it had at least one skin test positive reactor using the > 4 mm cut-off.

Results

The Single Intradermal Comparative Tuberculin Test (SICTT) was conducted in 1077 cattle, median mean herd size 3.00 (range: 1–31), from 36 study areas and 11 sub-regions, 876 goats, median herd size 4.00 (range:1–15) from 27 study areas and 9 sub-regions, and 195 camels, median herd size 6.0 (range:1–11) from 16 study areas and 7 sub-regions (Table 1). In total, 413 cattle, 243 goat, and 70 camel herds were tested (Table 2). Overall, females accounted for 71% of animals tested (n = 764) in cattle, 95.5% (n = 838) in goats, and 75% (n = 146) in camels (Table 3).

Tuberculin reactors in cattle, goats and camels at individual and herd levels

Cattle

Results of the SICTT are shown in Table 2. The overall individual animal and herd prevalences, using the standard method were 1.2% (13/1077) [Confidence Interval (CI), 95% CI, 1.1–1.3%] and 3.2% (13/413), (95% CI, 3.0–3.4%), respectively. Within the cattle rearing villages, 22% (8/36) of the herds had positive reactors (Tables 1 and 2). Whereas, using the severe method (> 2 mm cut-off), the individual and herd prevalences were 5.5% (59/1077) and 13% (54/413), respectively, and more (58%; 21/36) villages had at least one reactor cattle when compared with the standard method (22%; 8/36) (Additional files 6 and 7).

Goats

No reactor animals were detected in goats in all the tested regions using the standard method (> 4 mm cut-off). Using the severe method (> 2 mm cut-off), 2.2% (19/876) of the individual goats and 4.9% (12/243) of the herds were reactors. With this method about 30% (7/27) of villages had at least one reactor goat (Additional files 6 and 7).

Camels

In camels, the animal and herd prevalences were 1.5% (3/195) (95% CI, 1.4–1.6%) and 2.9% (2/70), (95% CI, 0.9–4.9%), respectively (Table 2), with > 4 mm cut-off. Only one village had reactors (Table 1). In contrast, using the severe method (> 2 mm cut-off), the animal (11.8%; 23/195) and herd (26.8%; 19/70) prevalences were higher and about 56% (9/16) of the villages had at least one reactor camel (Additional files 6 and 7).

Statistical analysis

In the univariable analysis male camels and male cattle were at approximately 2–3 times at higher risk to be test positive when compared to females (Table 4). Similarly, in the multivariable analysis of the risk factors, only 'sex' of the animals remained in the final reduced model, though with borderline significance as shown for the univariable results (Table 4). Overall, male animals had around 3 times higher odds to be test positive than females. The variables 'species' and 'age', though, apparent potential risk factors, were not statistically significant.

Descriptive epidemiology of the animal and human risk factors on farm level

Risk factors for animal BTB exposure

Median numbers of cattle, goats and camels owned per farmer (household) interviewed were 6 (Range:1–208), 11 (Range: 2–250), and 3 (Range: 1–47), respectively. Out of the 232 interviewed households (farmers) keeping cattle, 5.6% (n = 13) had positive animals, whereas this was the case in 6.9% (n = 2) of camel keeping households. None of the households with goats had SICTT positive animals. Out of the two households with SICTT reactor camels, one also owns dairy farm (MK Ghebremariam, personal experience). Among the interviewed households, 96.6% (n = 224) of cattle, 93.8% (n = 120) of

Table 1 Number (and herd) of skin tested cattle, goats and camels, and number (herd) of reactors at region, sub-region and study areas levels using the standard (> 4 mm cut-off) method in the selected study areas within the traditional livestock husbandry system in Eritrea. '0' = zero animals tested from zero herds. NA = not applicable

Regions	Sub-region	Study Areas	Tested cattle(herd)	Reactor cattle (herd) > 4 mm	Tested goats (herds)	Reactor goats (herd) > 4 mm	Tested camels (herds)	Reactor camels (herd) > 4 mm
Debub	Dbarwa	DRW1	61(20)	0(0)	60(13)	0(0)	NA	NA
		DRW2	44(29)	3(3)	15(6)	0(0)	NA	NA
		DRW3	38(26)	0(0)	17(7)	0(0)	NA	NA
		DRW4	48(15)	0(0)	25(6)	0(0)	NA	NA
	Total		191 (90)	3(3)	117 (32)	0(0)	NA	NA
	Mendefera	MFR1	104(35)	4(4)	38(7)	0(0)	NA	NA
		MFR22	52(23)	1(1)	16(5)	0(0)	NA	NA
		MFR3	38(15)	1(1)	0(0)	–	NA	NA
	Total		194 (73)	6(6)	54(12)	0(0)	NA	NA
Anseba	Hagaz	HAZ1	3(2)	0(0)	15(2)	0(0)	0	–
		HAZ2	12(2)	0(0)	10(1)	0(0)	1(1)	0(0)
		HAZ3	3(1)	0(0)	0	–	0	–
		HAZ4	8(2)	0(0)	0	–	0	–
		HAZ5	0	–	13(2)	0(0)	10(1)	0(0)
		HAZ6	0	–	3(1)	0(0)	0	–
		HAZ7	0	–	5(1)	0(0)	0	–
	Total		26(7)	0(0)	46(7)	0(0)	11(2)	0(0)
	Hamelmalo	HAM1	14(6)	0(0)	0	0(0)	0	–
		HAM2	1(1)	0(0)	0	0(0)	0	–
		HAM3	61(27)	0(0)	37(7)	0(0)	10(3)	3(2)
		HAM4	45(11)	0(0)	11(2)	0(0)	0	–
		HAM5	0(0)	–	0(0)	–	0	–
	Total		121(45)	0(0)	48(9)	0(0)	10(3)	3(2)
	Adi-Tekelezan	ATK1	72(35)	1(1)	39(9)	0(0)	0	–
	Total		72(35)	1(1)	39(9)	0(0)	0	–
Gash Barka	Barentu	BAR1	89(32)	1(1)	0	–	0	–
		BAR2	28(6)	0 (0)	7(2)	0(0)	0	–
		BAR3	6(4)	0(0)	49(13)	0(0	13(10)	0(0)
		BAR4	22(10)	0 (0)	72(18)	0(0)	2(2)	0 (0)
	Total		145(52)	1(1)	128(33)	0(0)	15(12)	0(0)
	Tessenei	TES1	28(12)	0 (0)	163(70)	0(0	1(1)	0 (0)

Table 1 Number (and herd) of skin tested cattle, goats and camels, and number (herd) of reactors at region, sub-region and study areas levels using the standard (> 4 mm cut-off) method in the selected study areas within the traditional livestock husbandry system in Eritrea. '0' = zero animals tested from zero herds. NA = not applicable *(Continued)*

Regions	Sub-region	Study Areas	Tested cattle(herd)	Reactor cattle (herd)	Tested goats (herds)	Reactor goats (herd)	Tested camels (herds)	Reactor camels (herd)
				> 4 mm		> 4 mm		> 4 mm
		TES2	11(4)	0 (0)	22(9)	0(0)	0	–
		TES3	0	–	0	–	71(20)	0 (0)
		TES4	30(10)	0 (0)	55(21)	0(0)	1(1)	0 (0)
		TES5	15(5)	0(0)	10(5)	0(0)	0(0)	–
	Total		84(31)	0(0)	250(105)	0(0)	73(22)	0(0)
	Hykota	HYK1	17(5)	0(0)	0	–	0	–
		HYK2	16(8)	0 (0)	0	–	1(1)	0 (0)
		HYK3	11(11)	1(1)	0	–	10(2)	0(0)
		HYK4	8(1)	0(0)	0	–	0	–
		HYK5	28(8)	0 (0)	0	–	0	–
		HYK6	14(14)	0 (0)	0	–	48 (6)	0 (0)
		HYK7	31(1)	0(0)	0	–	0	–
		HYK8	28(1)	0(0)	0	–	0	–
	Total		153(49)	1(1)	0	–	59(9)	0(0)
	Mogolo	MOG1	46(9)	1(1)	25(2)	0(0)	11(8)	0 (0)
	Total		46(9)	1(1)	25(2)	0(0)	11(8)	0(0)
	Akurdet	AKU1	25(15)	0 (0)	0	–	0	–
		AKU2	18(5)	0(0)	0	–	0	–
	Total		43(20)	0(0)	0(	–	0	–
Southern Red Sea	Debub Dankalia	DANK1	0	–	17(4)	0(0)	3(3)	0 (0)
		DANK2	0	–	10(1)	0(0)	2(2)	0 (0)
		DANK3	0	–	54(13)	0(0)	0	–
		DANK4	2(2)	0(0)	35(10))	0(0)	0	–
		DANK5	0	–	53(6)	0(0)	11(9)	0(0)
	Total		2(2)	0(0)	169(34)	0(0)	16(14)	0(0)
	Grand Total		1077(413)	13(13)	876(243)	0(0)	195(70)	3(2)

Table 2 BTB prevalence in cattle, goats and camels at individual animal and herd levels within the traditional livestock husbandry system in the study regions using standard cut-off (> 4 mm). NA = not applicable

Number and herds of cattle, goats and camels tested	Anseba	Debub	Gash Barka	Southern Red Sea	Overall
Number (%) of cattle	219 (20.3)	385 (35.7)	471 (43.7)	2 (0.2)	1077 (100)
Number (%) of goats	133 (15.2)	171 (19.5)	403 (46)	169 (19.3)	876 (100)
Number (%) of camels	21 (10.8)	NA	158 (81.0)	16 (8.2)	195 (100)
Total number (%) tested/region	373 (17.4)	556 (25.9)	1032 (48)	187 (8.7)	2148 (100)
Individual animal Prevalence (%)					
Cattle	1 (0.5)	9 (2.3)	3 (0.6)	0 (0.0)	13 (1.2)
Goats	0 (0.0)	0 (0.0)	0 (0.0)	0 (0.0)	0 (0.0)
Camels	3 (13.6)	NA	0 (0.0)	0 (0.0)	3 (1.5)
Herds of cattle, goats and camels tested					
Herds (%) of cattle	87 (21.1)	163 (39.5)	161 (38.9)	2 (0.5)	413 (100)
Herds (%) of goats	25 (10.3)	44 (18.1)	140 (57.6)	34 (14.0)	243 (100)
Herds (%) of camels	6 (8.3)	0 (0.0)	50 (72.2)	14 (19.4)	70 (100)
Herd prevalence (%)					
Cattle	1 (1.2)	9 (5.5)	3 (1.9)	0 (0.0)	13 (3.2)
Goats	0 (0.0)	0 (0.0)	0 (0.0)	0 (0.0)	0 (0.0)
Camels	2 (33.3)	NA	0 (0.0)	0 (0.0)	2 (2.9)

goat, and 100% ($n = 29$) of camel owners allowed their animals to share watering points with other animals. Besides, 18.1% ($n = 42$) of the cattle, 3.9% ($n = 5$) of the goats, and 13.8% ($n = 4$) of the camel owners reported that they bought and introduced new animals to their existing herds in the last 2–3 years (Table 5). Likewise, 22.0% ($n = 51$) of the cattle, 10.2% ($n = 13$) of the goat, and 55.2% ($n = 16$) of the camel owners reported that their animals migrate to other regions during the dry seasons (Table 5).

Of the farmers interviewed 38.4% ($n = 89$) reported that their cattle spend the nights in the open including night grazing, 52.7% ($n = 106$) in separate animal houses, 13.8% ($n = 14$) in enclosures, and 2.3% ($n = 5$) share houses with humans at nights (Table 6). They also indicated that 74.2% ($n = 95$) farmers house their goats in separate animal houses, whereas 23.4% ($n = 30$), and 2.3% ($n = 3$), keep them in 'enclosures made of thorns' and 'free roaming in compounds' at night, respectively (Table 6).

Risk factors for human BTB exposure

Out of the farmers interviewed, 24.6% ($n = 57$) of cattle, 51.7% ($n = 15$) of camel, and 38% ($n = 49$) of goat owners indicated that raw milk and milk products are consumed

Table 3 BTB prevalence as associated to sex in cattle, camel, and goats within the traditional livestock husbandry system using > 4 mm

Status	*P*-value	Female	Male
		764 (79.9)	313 (29.1)
Cattle: > 4 mm cut-off	(0.089)	Number (%)	Number (%)
Inconclusive		146 (19.1)	67 (21.4)
Negative		612 (80.1)	239 (76.4)
Positive		6 (0.8)	7 (2.2)
Camels: > 4 mm cut-off	(0.053)	146 (74.9)	49 (25.1)
Inconclusive		22 (15.1)	15 (30.6)
Negative		122 (83.0)	33 (67.4)
Positive		2 (1.4)	1 (2.0)

Table 4 Species of animals (cattle and camels) and sex as potential risk factors for the presence of BTB reactors at > 4 mm cut-off within the tested animals in the study regions within the extensive traditional livestock husbandry system analyzed by univariable and multivariable logistic regression

Species	*P*-value	OR	95% CI	
			Lower bound	upper bound
Univariable analysis				
Camel versus cattle	0.703	1.3	0.4	4.5
Sex				
Male versus female cattle	0.058	2.9	1.0	8.7
Male versus female camel	0.743	1.5	0.13	16.9
Male versus female (overall)	0.06	2.6	1.0	8.7

Table 5 Risk factors for the presence of SICTT reactors as compared between cattle, goats and camels in the extensive livestock husbandry system within the study regions in Eritrea

Variables	Cattle herds (n = 232)	Goat herds (n = 128)	Camel herds (n = 29)	Overall herds (n = 389)
	Number (%)	Number (%)	Number (%)	Number (%)
Water point sharing				
Yes	224 (96.6)	120 (93.8)	29 (100)	373 (95.9)
No	8 (3.4)	8 (6.3)	0 (0.0)	16 (4.1)
Introduction of new animals				
Yes	42 (18.1)	5 (3.9)	4 (13.8)	51 (13.1)
No	190 (81.9)	123 (96.1)	25 (86.2)	338 (86.9)
Livestock migration				
Yes	51 (22.0)	13 (10.2)	16 (55.2)	80 (20.6)
No	181 (78.0)	115 (89.8)	13 (44.8)	309 (79.4)
Source of water				
Outside farms	128 (55.2)	46 (35.9)	11 (37.9)	185 (47.6)
Inside farms	49 (21.1)	60 (46.9)	17 (58.6)	126 (32.4)
Inside and outside farms	55 (23.7)	22 (17.2)	1 (3.4)	78 (20.0)

in their households regularly (Table 7). Moreover, 4.7% (n = 11) of the cattle, and 2.3% (n = 33) of the goat owners reported the presence of respiratory diseases in their family members. Of these, one cattle herd was positive with > 4 mm cut-off and two with > 2 mm cut-off. None of goats' farms that reported the presence of tuberculosis in their families had positive animals. The questionnaires did not include this variable with reference to camels. Among the total number of farmers interviewed, 57.6% (n = 224) indicated that BTB awareness campaigns had never been held in their areas. Of the respondents, 42.9% (n = 167) had no education (cannot read and write), 37.5% (n = 146) had a low (literate; primary level education; grade 1–5), 17.5% (n = 68) medium (grade 6–8), and 2.1% (n = 8) higher levels of education (grade 9–12/college level education) (Table 7).

Discussion

This study presents the first efforts to assess the prevalence of BTB in cattle, goats and camels, and its zoonotic potential within the extensive traditional livestock husbandry system in Eritrea (Fig. 1). Focusing on cattle, our study reports low (1.2%) individual animal and 3.2% herd prevalences of BTB. Similar findings for individual animal prevalence were reported in Ethiopia, Gumi et al. [21] 5.5%, Ameni et al. [23] 1.8% and 4.7%, Ameni et al. ([24, 25], reported) 7.9% -11.6%, Tschopp et al. ([26, 27], 0.8%; 0.9%), and Mamo et al. [18] reported 11% and a herd prevalence of 44% in Ethiopia in the Afar region with similar conditions as the pastoral areas in our study. In Ghana 13.8% of individual animal prevalence was observed [28]. Although low, the prevalence in our study differed between regions, Debub showing the highest

Table 6 Housing of cattle and goats at night in Debub (in central highlands with high altitude and mild temperature), Anseba (partially in the central highlands and partially in the lowlands with hot and arid climate), Gash Barka (in the western low lands with hot and arid climate) and Southern Red Sea (in Eastern low land with hot and arid climate) regions within the extensive livestock husbandry system

Variables	Debub (n = 106)	Anseba (n = 61)	Gash Barka (n = 64)	Southern Red Sea (n = 1)	Overall (n = 232)
Cattle housing	Number (%)	Number (%)	Number (%)	Number (%)	Number (%)
Open area	9 (8.5)	32 (52.5)	48 (75.0)	0 (0.0)	89 (38.4)
Separate animal houses	95 (89.6)	8 (13.1)	3 (4.7)	0 (0.0)	106 (52.7)
Enclosures	0 (0.0)	18 (29.5)	13 (20.3)	1 (100.0)	32 (13.8)
Share houses with humans	2 (1.9)	3 (4.9)	0 (0.0)	0 (0.0)	5 (2.2)
Goat housing	n = 39	n = 17	n = 39	n = 33	n = 128
Open area	0 (0.0)	0 (0.0)	3 (7.7)	0 (0.0)	3 (2.3)
Separate animal houses	39 (100)	6 (35.3)	19 (48.7)	31 (93.9)	95 (74.2)
Enclosures	0 (0.0)	11(64.7)	17 (43.6)	2 (6.1)	30 (23.4)

Table 7 Bovine tuberculosis awareness, education levels, and raw milk consumption habit among farmers keeping cattle, goats and camels, within the traditional livestock farming system in Eritrea

Variables	Cattle owners (n = 232)	Goats owners (n = 128)	Camel owners (n = 29)	Overall (n = 389)
TB awareness	Number (%)	Number (%)	Number (%)	Number (%)
Yes	109 (47.0)	46 (35.9)	10 (34.5)	165 (42.4)
No	123 (53)	82 (64.1)	19 (65.5)	224 (57.6)
BTB animal to humans				
Yes	151 (65.1)	91 (71.1)	13 (44.8)	255 (65.6)
No	24 (10.4)	9 (7.0)	10 (34.5)	43 (11.0)
I don't know	57 (24.7)	28 (21.9)	6 (20.7)	91 (23.4)
TB human to animals				
Yes	23 (9.9)	22 (17.2)	5 (17.2)	50 (12.9)
No	79 (34.1)	37 (28.9)	15 (51.7)	131 (33.7)
I don't know	130 (56.0)	69 (53.9)	9 (31.0)	208 (53.5)
Level of education				
No education (cannot read and write)	93 (40.1)	53 (41.4)	21 (72.4)	167 (42.9)
Low (literate; grade 1–5 of formal education)	87 (37.5)	56 (43.8)	3 (10.3)	146 (37.5)
Medium (grade 6–8 of formal education)	46 (19.8)	18 (14.1)	4 (13.8)	68 (17.5)
Higher (grade 9–12 formal /college level)	6 (2.6)	1 (0.8)	1 (3.4)	8 (2.1)
Raw milk consumption				
Yes	57 (24.6)	31 (24.2)	25 (86.2)	113 (29.0)
No	175 (75.4)	97 (75.8)	4 (13.8)	276 (71.0)

prevalence, 2.3%, where also a high (7.3%) BTB prevalence in dairy cattle was recorded in our previous study [14]. Contact between the dairy cattle and cattle within the extensive system may be postulated as a potential risk factor for the transmission of BTB. This region, located in the central highland of Eritrea, where mixed crop-livestock farming is conducted, is endowed with relative mild temperature and higher precipitation, hence environmental conditions more favourable for survival of *M. bovis* as compared with the arid and semi-arid regions of Gash Barka, Southern Red Sea and Anseba (partially) [14, 29]. In the latter areas, climatic conditions, lower cattle density, housing of animals in open areas (Table 6) may explain the low prevalence of BTB in the extensive cattle production system in general. In our current study, in Anseba region, BTB prevalence was very low (0.5%) in cattle within the extensive livestock production system (Tables 1 and 2). In contrast to Debub region, in Anseba, very low (0.2%) BTB prevalence was reported in dairy cattle [14], thus, in this case, transmission of BTB from the dairy cattle to the indigenous cattle within the extensive farming may be less likely when compared to Debub region.

Although the observed BTB prevalence was low, it is noteworthy that the presence of infection was indicated in many of the study areas (Table 1 and Fig. 1). This may suggest that BTB was introduced to these areas sporadically from various sources but spread was limited. This can be explained by investigating the generally accepted drivers of BTB prevalence, i.e. breed of cattle, farming system (intensive/ extensive), housing and gathering of animals at grazing and watering points. Our current study was conducted exclusively in indigenous (zebu) cattle which are considered relatively resistant to BTB as compared to exotic breeds [13, 25, 30]. Likewise, the extensive livestock management practiced in our study areas is known to pose a far lower risk for BTB progression and transmission than the intensive dairy farming system. It can be argued that under these circumstances and in combination with the prevailing climatic conditions, the risk for BTB transmission is effectively reduced as evidenced by the current low prevalence.

Nevertheless, the existing potential for spread of BTB due to inter-species herd mixing at water points and resting areas where livestock congregate and as well as due to migration and uncontrolled livestock movement must not be underestimated, nor ignored (Table 5). On the other hand, it must be kept in mind that several host related factors like malnutrition, recent infection with *M. bovis*, co-infection with non-tuberculous mycobacteria, infestation with gastrointestinal parasites, and generalized tuberculosis [31–34] are able to decrease reactivity to the SICTT and cannot be ruled out to have influenced our study outcome.

In goats, no positive reactors were found in our current study, and this finding was in agreement with what was reported by Tschopp et al. (2011 and 2010b) [10, 27]. A similar stud in Ethiopia reported 0.5% in small ruminants, [35]. The absence of BTB positive goats in our current study (mixed crop-livestock, and pastoral systems) might be attributed to the restriction of grazing of the flocks within their villages or housing of goats separately from cattle and camels at night. In addition, separate herding may have contributed to a low contact rate between goats and the other species of animals, unlike in the neigbouring Ethiopia where there is congregation and interspecies as well as wildlife mixing in grazing areas [18]. Infestation with liver flukes (*Fasciola hepatica*) and other helminths in the relatively wet highlands and arid low land areas are commonly encountered in slaughtered goats ([36], Tsegay Ghebremeriam, senior meat inspector, MOA, personal communication, and MK Ghebremariam, personal experience) perhaps causing reduced reactivity to the SICTT as reported by several studies [31–33]. In addition, the sensitivity of SICTT in goats, as in cattle, might be compromised by co-infections with viral diseases in chronic stages such as peste des petits ruminants (PPR), Contagious Caprine Pleuropneumonia (CCPP); or sheep and goat pox (MK Ghebremariam, personal experience). In general, the possibilities for goats to become infected are less, as they are browsers and rarely graze pastures that may have been contaminated. Studies also suggest that small ruminants are only spillover hosts that cannot maintain the disease in a herd [37], unless they are in close contact with cattle with high BTB prevalence [9, 38] or managed under intensive production system [39–41]. There is no information on the status of paratuberculosis (*M. avium subspecies paratuberculosis*) in Eritrea, that is known to interfere with the skin test when present [23, 42]. Though our current study was not able to show the presence of BTB in goats, several studies in Africa and Europe with different as well as similar agricultural settings as in Eritrea showed the presence of *M. bovis* and *M. caprae* in goats [1–4, 35, 39–45] thus we need to approach the current finding cautiously since only the OIE standard was used to interpret the results. In indigenous (zebu) cattle as well as in goats, the severe (> 2 mm cut-off) method showed better sensitivity without affecting the specificity of the SICTT as compared to the standard method [21, 44–47]. The use of severe method in our study might have increased the sensitivity of the test. Results that compare the number of positive animals when both the standard (> 4 mm cut-off) and the severe (> 2 mm cut-off) methods are used are presented in additional files (Additional files 6 and 7). Might be good to emphasize the increase spatial spread in case of the severe interpretation/ implying increase risk for infection of animals and humans.

Our study has shown that camels were more at risk of being SICTT reactors as compared with cattle (Table 4), though the association was not significant as only few were positive. BTB is prevalent in dairy cattle in Anseba region, as reported by Ghebremariam et al. [14]. The overall individual animal prevalence in camels was 1.5%. This is considered low as compared with similar studies in Ethiopia and Kenya that showed 6% (29/480) and 37% (15/41) prevalences, respectively, based on standard interpretation [48, 49]. Relatively, the prevalence was higher in the Anseba region (1.5%; Table 2) when compared to other regions of Eritrea where the study was conducted. In this region mixing of camels and cattle is relatively common, and in some cases camel owners also own dairy farms and use their camels to transport animal feed to the farms that may allow camel-cattle contact. Such interactions may have contributed to the presence of more SICTT reactor camels in Anseba as compared with the other study regions. Camels in close contact with cattle were found to be more prone to *M. bovis* infection and to have more tuberculosis lesions in the abattoir than those not having contact with cattle [5, 6, 50–52]. No reactor camel was found in Gash Barka and Southern Red Sea regions. In these regions camels are herded separately from cattle, but trekked long distances where they may come into contact with other animal species en route and at water points. The low prevalence of BTB in camels in the lowlands can be understood as they are browsers, in addition to being herded separately from cattle in a region with low prevalence of BTB in cattle. However, there is no information on the status of helminths, paratuberculosis (*M. avium subspecies paratuberculosis*), other non-tuberculous mycobacteria or viral infection in camels in Eritrea, that may interfere with the skin test when present [31–34, 41, 42].

Comparatively, the overall BTB prevalence at animal level in all the tested species and the number of study areas with reactors was highest in Debub (Fig. 1). Out of the nine positive reactor animals in this region, seven were males. As mixed-crop livestock production system is practiced in this region, male cattle are mostly used either as oxen or for mating purposes, and thus kept longer in the herd than females [29].

Gash Barka is the region where approx. 60% of the livestock population is located and it is the destination for all the animals migrating from different regions of the country, especially, during the dry season. Besides, this region shares borders with Sudan and Ethiopia where uncontrolled movement of animals is possible. The low BTB prevalence in this region might be due to the arid and hot climate which is not suitable for the survival of *M. bovis* as it is readily destroyed by direct sunlight under dry condition [13], in contrast to Debub region.

Focusing on the risk factors for human BTB exposure, overall, within the traditional extensive livestock husbandry system in the selected study areas, 29% of the households consume raw, untreated milk (Table 7). Such milk consumption habit might serve as a vehicle for BTB transmission from infected animals to humans as several studies have shown the presence of *M. bovis* in camels', goats' and in cows' milk [1–4, 53]. Moreover, among the interviewed cattle owners, 2.4% share their houses with their cattle (Table 6). Sharing of the same microenvironment and dwelling between humans and animals has been identified as one of the routes of animal-to-human BTB infection or vice versa, mainly in rural areas in developing countries [12]. The presence of tuberculosis within some of the cattle rearing families and their animals warrants suspicion of the presence of *M. bovis* within the animal-human interface in this production system.

BTB in camels and goats was not considered of veterinary concern in Eritrea, thus, so far, no attempt has been made to conduct BTB testing or routine post mortem examinations in the slaughterhouses for the detection of TB-like lesions. The current study indicated the presence of BTB in cattle and camels, and its spread throughout the study regions within the extensive livestock production system in Eritrea, though at a low prevalence. It warrants future, more in-depth, studies on BTB in these livestock species.

Our study has one major limitation, i.e., the number of the farmers that completed the questionnaires were fewer than the animal herds tested. This was mainly attributed to the overcrowding of the testing sites and the hot and arid climate that forced the farmers to leave the testing sites without filling the questionnaires (even after presenting their animals for testing). As the observed level of animal prevalence is low, a multilevel analysis of the data was not estimable although such a model would give more precise estimates if feasible. Such a model also needs sufficient observations within each cluster which was not the case in our study as many very small herds are present, most of the herds consist of 1–3 animals.

Conclusion and recommendation

The current study has shown that SICTT reactors are rare in cattle and camels and were not found in goats. However, though rare, the spatial distribution of the affected animals across most of the selected regions (Fig. 1), where consumption of unpasteurized milk is common, warrants continuous surveys, cautious and timely control measures of the disease. We recommend the testing of the animals to be conducted during mild weather seasons so as to be able to conduct face-to-face interviews and complete the questionnaires that would match the number of herds tested.

Additional files

Additional file 1: Regions, sub-regions and villages included in the study, and species, breed, sex and age of each animal tested using the SICTT during the study periods

Additional file 2: Questionnaire for BTB risk factors study within the cattle raising communities in the extensive livestock husbandry system in Eritrea.

Additional file 3: Questionnaire for BTB risk factors study within the goat raising communities in the extensive livestock husbandry system in Eritrea.

Additional file 4: Questionnaire for BTB risk factors study within the camel raising communities in the extensive livestock husbandry system in Eritrea.

Additional file 5: Consolidated criteria for reporting qualitative studies (COREQ): 32-item checklist and answers.

Additional file 6: Number (and herd) of skin tested cattle, goats and camels, and number (herd) of reactors at region, sub-region and study areas levels using the standard (> 4 mm cut-off) method in the selected study areas within the traditional livestock husbandry system in Eritrea presented for comparison. '0' = zero animals tested from zero herds. NA = not applicable.

Additional file 7: BTB prevalence in cattle, goats and camels at individual animal and herd levels within the traditional livestock husbandry system in Eritrea using the standard and severe cut-offs (> 4 mm and > 2 mm) presented for comparison. NA = not applicable.

Abbreviations

BTB: Bovine tuberculosis; ERF: Eritrean Research Fund; FAO: Food and Agriculture Organization; HAC: Hamelmalo Agricultural College; *M. bovis*: *Mycobacterium bovis*; MOA: Ministry of Agriculture; NAPHL: National Animal and Plant Health Laboratory; NUFFIC: Netherlands University Foundation for International Cooperation; OIE: World Organization for Animal Health; PPD: Purified protein derivative; PPD-A: Purified protein derivative for *mycobacterium avium*; PPD-B: Purified protein derivative for *mycobacterium bovis*; SICTT: Single intradermal comparative tuberculin test

Acknowledgements

The authors thank the Ministry of Agriculture (MOA), Eritrea, for providing ear tags, anthelminthic drugs and equipment for performing the skin test. We also thank all MOA staff who have participated and provided important data for the study. We thank Dr. Uqubazghi Kifle, Dr. Habte Gaim, and Dr. Yonas Woldu for their wonderful work in organizing the field works (MOA, Heads of Animal Resources, Debub, Gash Barka, and Southern Red Sea regions, respectively). We thank Samuel Abraha (MOA, Debub region, Mendefera sub-region) and Andebrhan (MOA, Debub region, Debarwa sub-region), Mohammed (Gash Barka region, Akurdet sub-region) for their participation in the study. We thank Redie Kidane (Hamelmalo Agricultural College; HAC) for his active involvement in the field work and for translating the questionnaire from English to Tigre language. We thank Elias (MOA, Gash Barka region, Tessenei sub-region) for coordinating the fieldwork in Tessenei sub-region and for his active participation in the fieldwork. We thank Solomon (MOA, Gash Barka region, Hykota sub-region), Solomon and Mohammed (MOA, Southern Red Sea region) for their participation in the field work. We thank Tsige Kifleyesus (National Animal and Plant Health Laboratory; APHL) for organizing the automatic syringes and needles for the skin test, ice boxes and ice packs for transporting the PPDs, and for making sure that the PPDs were safely stored in the cold at NAPHL. Efrem Ghebremeskel is thanked for allowing us to use his office and the resources of NAPHL during the study period. NUFFIC is acknowledged for funding the project, and Eritrean Research Fund for partially funding this study. We thank HAC for the timely dispatch of official letters for cooperation to the regions of study and for organizing vehicles for the field works. We thank FAO representative office, Asmara, Eritrea, for providing us internet access for our research purposes.

Funding

This study was funded by NUFFIC and, partially, by Eritrean Research Fund (ERF).

Authors' contributions

MKG was involved in the design of the study, conducted the skin testing of the animals, collected and analyzed data, and did the write up of the manuscript. MAL was involved in the conceptualization of the study and write up of the manuscript. VJCM assisted in data analysis and write up of the manuscript. MN was involved in the design of the study, data interpretation and write up of the manuscript. RVPMG was involved in the conceptualization of the study and write-up of the manuscript. All authors read and approved the final manuscript.

Ethics approval and consent to participate

Participation in the study by the farmers was on a voluntary basis and those farmers who volunteered to participate have given their oral consent to include their animals in the study.

Bovine tuberculosis is one of the priority diseases for control and eradication by the government of Eritrea. Skin testing is an official method for diagnosis of BTB in the country. The skin testing in the current study was conducted with the involvement of the Ministry of Agriculture, animal health unit staff, and all the datasets included in this study were issued from animals analyzed within an official context. No deliberate killing of animals was performed for this study. For the human component, this study was submitted to and got approval by the 'Health Research proposal review and Ethical committee', unit of planning policy and HRD, Ministry of Health, Asmara, Eritrea, on 16/05/2013.

Consent for publication

Not applicable. Information that may compromise the farmers' identities are de-identified in all the files.

Competing interests

The authors declare that they have no competing interests.

Author details

[1]Department of Infectious Diseases and Immunology, Faculty of Veterinary Medicine, Utrecht University, Utrecht, The Netherlands. [2]Department of Veterinary Sciences, Hamelmalo Agricultural College, Keren, Eritrea. [3]Department of Veterinary Tropical Diseases, Bovine Tuberculosis and Brucellosis Research Programme, Faculty of Veterinary Science, University of Pretoria, Pretoria, South Africa. [4]Research Associate at the National Zoological Gardens of South Africa, Pretoria, South Africa. [5]Department of Farm Animal Health, Faculty of Veterinary Medicine, Utrecht University, Utrecht, The Netherlands.

References

1. Shanahan A, Good M, Duignan A, Curtin T, More SJ. Tuberculosis in goats on a farm in Ireland: epidemiological investigation and control. Vet Rec. 2011;168(18):485.
2. Quintas H, Reis J, Pires I, Alegria N. Tuberculosis in goats. Vet Rec. 2010; 166(14):437–8.
3. Daniel R, Evans H, Rolfe S, De la Rua-Domenech R, Crawshaw T, Higgins RJ, Schock A, Clifton-Hadley R. Outbreak of tuberculosis caused by Mycobacterium bovis in golden Guernsey goats in great Britain. Vet Rec. 2009;165(12):335.
4. Cadmus SI, Adesokan HK, Jenkins AO, Van Soolingen D. Mycobacterium bovis and M. Tuberculosis in goats, Nigeria. Emerg Infect Diseases. 2009; 15(12):2066–8.
5. Manal MY, Gobran RA. Some studies on tuberculosis in camel, Egypt. J Comp Path Clinic Path. 2008;21(4):58–74.
6. Mustafa IE. Bacterial diseases of dromedaries and Bactrian camels [anthrax, brucellosis, haemorrhagic septicaemia, plague, salmonellosis, tuberculosis, paratuberculosis, leptospirosis, clostridial infections]. Rev sci tech Off int Epiz de l'OIE. 1987;6(2):391–405.
7. Michel AL, Coetzee ML, Keet DF, Mare L, Warren R, Cooper D, Bengis RG, Kremer K, Van Helden P. Molecular epidemiology of Mycobacterium bovis isolates from free-ranging wildlife in south African game reserves. Vet Microbiol. 2009;133(4):335–43.
8. Hlokwe TM, Van Helden P, Michel AL. Evidence of increasing intra and inter-species transmission of *Mycobacterium bovis* in South Africa: are we losing the battle? Prev Vet Med. 2014;115(1):10–7.
9. Boukary AR, Thys E, Abatih E, Gamatié D, Ango I, Yenikoye A, Saegerman C. Bovine tuberculosis prevalence survey on cattle in the rural livestock system of Torodi (Niger). PLoS One. 2011;6(9):e24629.
10. Tschopp R, Bobosha K, Aseffa A, Schelling E, Habtamu M, Iwnetu R, Hailu E, Firdessa R, Hussein J, Young D, Zinsstag J. Bovine tuberculosis at a cattle-small ruminant-human interface in Meskan, Gurage region, Central Ethiopia. BMC Infect Dis. 2011;11(1):318.22.
11. Byarugaba F, Etter E, Godreuil S, Grimaud P. Pulmonary tuberculosis and Mycobacterium bovis, Uganda. Emerg Infect Diseases. 2009;15(1):124–5.
12. Amanfu W. The situation of tuberculosis and tuberculosis control in animals of economic interest. Tuberculosis. 2006;86(3):330–5.
13. Radostits OM, Gay CC, Hinchcliff KW, Constable PD. Diseases associated with mycobacterium species. In: Veterinary medicine- a text book of the diseases of cattle, sheep, pigs, goats and horses. 10th ed. London: Sounders; 2007. p. 1007–44.
14. Ghebremariam MK, Rutten VG, Vernooij JC, Uqbazghi K, Tesfaalem T, Butsuamlak T, Idris AM, Nielen M, Michel AL. Prevalence and risk factors of bovine tuberculosis in dairy cattle in Eritrea. BMC Vet Res. 2016;12(1):1.
15. Omer MK, Skjerve E, Woldehiwet Z, Holstad G. A cross sectional study of bovine tuberculosis in dairy farms in Asmara, Eritrea. Trop Anim Health Prod. 2001;33(4):295–303.
16. Fetene T, Kebede N, Alem G. Tuberculosis infection in animal and human populations in three districts of western Gojam, Ethiopia. Zoonoses Public Health. 2011;58(1):47–53.
17. Regassa A, Tassew A, Amenu K, Megersa B, Abunna F, Mekibib B, Marcotty T, Ameni G. A cross-sectional study on bovine tuberculosis in Hawassa town and its surroundings, southern Ethiopia. Trop Anim Health Prod. 2010;42(5):915–20.
18. Mamo G, Abebe F, Worku Y, Hussein N, Legesse M, Tilahun G, Medhin G, Bjune G, Ameni G. Bovine tuberculosis and its associated risk factors in pastoral and agro-pastoral cattle herds of afar region, Northeast Ethiopia. J Vet Med Anim Health. 2013;5(6):171–9.
19. Shitaye JE, Tsegaye W, Pavlik I. Bovine tuberculosis infection in animal and human populations in Ethiopia: a review. Veterinarni Medicina. 2007;52(8):317.
20. Grange JM, Daborn C, Cosivi O. HIV-related tuberculosis due to *Mycobacterium bovis*. Eur Respir J. 1994;7:1564–6.
21. Gumi B, Schelling E, Firdessa R, Aseffa A, Tschopp R, Yamuah L, Young D, Zinsstag J. Prevalence of bovine tuberculosis in pastoral cattle herds in the Oromia region, southern Ethiopia. Trop Anim Health Prod. 2011;43:1081–7.
22. OIE, 2009. http://www.oie.int/fileadmin/Home/eng/Healthstandards/tahm/2.04.07 BOVINE TB.pdf.
23. Ameni G, Tadesse K, Hailu E, Deresse Y, Medhin G, Aseffa A, Hewinson G, Vordermeier M, Berg S. Transmission of mycobacterium tuberculosis between farmers and cattle in Central Ethiopia. PLoS One. 2013;8(10):e76891.
24. Ameni G, Bonnet P, Tibbo M. A cross-sectional study of bovine tuberculosis in selected dairy farms in Ethiopia. Int J Appl Res Vet M. 2003;1(4):253–8.
25. Ameni G, Aseffa A, Engers H, Young D, Gordon S, Hewinson G, Vordermeier M. High prevalence and increased severity of pathology of bovine tuberculosis in Holsteins compared to zebu breeds under field cattle husbandry in Central Ethiopia. Clin Vaccine Immunol. 2007;14(10):1356–61.
26. Tschopp R, Schelling E, Hattendorf J, Young D, Aseffa A, Zinsstag J. Repeated cross-sectional skin testing for bovine tuberculosis in cattle kept in a traditional husbandry system in Ethiopia. Vet Rec. 2010a;167:250–6.
27. Tschopp R, Aseffa A, Schelling E, Berg S, Hailu E, Gadisa E, Habtamu M, Argaw K, Zinsstag J. Bovine tuberculosis at the wildlife-livestock-human Interface in Hamer Woreda, south Omo, southern Ethiopia. PLoS One. 2010b;5(8):e12205. https://doi.org/10.1371/journal.pone.0012205.
28. Bonsu OA, Laing E, Akanmori BD. Prevalence of tuberculosis in cattle in the Dangme-west district of Ghana, public health implications. Acta Trop. 2000;76(1):9–14.
29. Humblet MF, Boschiroli ML, Saegerman C. Classification of worldwide bovine tuberculosis risk factors in cattle: a stratified approach. Vet Res. 2009;40(5):1–24.
30. Vordermeier M, Ameni G, Berg S, Bishop R, Robertson BD, Aseffa A, Hewinson RG, Young DB. The influence of cattle breed on susceptibility to bovine tuberculosis in Ethiopia. Comp Immunol Infect Dis. 2012;35:227–32.
31. Flynn RJ, Mannion C, Golden O, Hacariz O, Mulcahy G. Experimental Fasciola hepatica infection alters responses to tests used for diagnosis of bovine tuberculosis. Infect Immun. 2007;75(3):1373–81.

32. Claridge J, Diggle P, McCann CM, Mulcahy G, Flynn R, McNair J, Strain S, Welsh M, Baylis M, Williams DJ. Fasciola hepatica is associated with the failure to detect bovine tuberculosis in dairy cattle. Nat Commun. 2012;3:853.
33. Ameni G, Medhin G. Effect of gastro-intestinal parasitosis on tuberculin test for the diagnosis of bovine tuberculosis. J Appl Anim Res. 2000;18(2):221–4.
34. De la Rua-Domenech R, Goodchild AT, Vordermeier HM, Hewinson RG, Christiansen KH, Clifton-Hadley RS. Ante mortem diagnosis of tuberculosis in cattle: a review of the tuberculin tests, γ-interferon assay and other ancillary diagnostic techniques. Res Vet Sci. 2006;81:190–210.
35. Mamo GK, Abebe F, Worku Y, Legesse M, Medhin G, Bjune G, Ameni G. Tuberculosis in goats and sheep in afar pastoral region of Ethiopia and isolation of mycobacterium tuberculosis from goat. Vet Med Int. 2012;2012: Article ID 869146, 8 pages. https://doi.org/10.1155/2012/869146.
36. Ghebremariam MK, Debesai MG, Sanjay D, Basharat AP. Hydatidosis as a major cause of liver condemnation among parasitic diseases in goats and sheep in Keren slaughterhouse, Anseba zone, Eritrea. Veterinary World. 2014;7(4):266–70.
37. Coleman JD, Cooke MM. *Mycobacterium bovis* infection in wildlife in New Zealand. Tuberculosis. 2001;81:191–202.
38. Malone FE, Wilson EC, Pollock JM, Skuce RA. Investigations into an outbreak of tuberculosis in a flock of sheep in contact with tuberculous cattle. J Vet Med B Infect Dis Vet Public Health. 2003;50:500–4.
39. Sharpe AE, Brady CP, Johnson A, Byrne W, Kenny K, Costello E. Concurrent outbreak of tuberculosis and caseous lymphadenitis in a goat herd. Vet Rec. 2010;166(19):591–2.
40. Napp S, Allepuz A, Mercader I, Nofrarías M, López-Soria S, Domingo M, Romero B, Bezos J, Pérez de Val B. Evidence of goats acting as domestic reservoirs of bovine tuberculosis. Vet Rec. 2013;172(25):663.
41. Bezos J, Álvarez J, Romero B, Aranaz A, de Juan L. Tuberculosis in goats: assessment of current in vivo cell-mediated and antibody-based diagnostic assays. Vet J. 2012;191(2):161–5.
42. Álvarez J, De Juan L, Bezos J, Romero B, Sáez JL, Gordejo FR, Briones V, Moreno MÁ, Mateos A, Domínguez L, Aranaz A. Interference of paratuberculosis with the diagnosis of tuberculosis in a goat flock with a natural mixed infection. Vet Microbiol. 2008;128(1):72–80.
43. Crawshaw T, Daniel R, Clifton-Hadley R, Clark J, Evans H, Rolfe S, de la Rua-Domenech R. TB in goats caused by Mycobacterium bovis. Vet Rec. 2008;163(4):127.
44. Tafess K, Dawo F, Sori T, Ameni G. Prevalence of caprine tuberculosis in mid-Rift Valley area of Oromia, Ethiopia. Afr J of Microbiol Res. 2011;5(12):1473–8.
45. Gumi B, Schelling E, Firdessa R, Erenso G, Biffa D, Aseffa A, Tschopp R, Yamuah L, Young D, Zinsstag J. Low prevalence of bovine tuberculosis in Somali pastoral livestock, Southeast Ethiopia. Trop Anim Health Prod. 2012; 44:1445–50. https://doi.org/10.1007/s11250-012-0085-5.
46. Ameni G, Hewlinson G, Aseffa A, Young D, Vordermeier M. Appraisal of interpretation criteria for the comparative intradermal tuberculin test for diagnosis of tuberculosis in cattle in Central Ethiopia. Clin Vaccine Immunol. 2008;15(8):1272–6.
47. Ngandolo BN, Müller B, Diguimbaye-Djaïbe C, Schiller I, Marg-Haufe B, Cagiola M, Jolley M, Surujballi O, Akakpo AJ, Oesch B, Zinsstag J. Comparative assessment of fluorescence polarization and tuberculin skin testing for the diagnosis of bovine tuberculosis in Chadian cattle. Prev Vet Med. 2009;89(1):81–9.
48. Beyi AF, Gezahegne KZ, Mussa A, Ameni G, Ali MS. Prevalence of bovine tuberculosis in dromedary camels and awareness of pastoralists about its zoonotic importance in eastern Ethiopia. J Vet Med Anim Health. 2014;6(4):109–15.
49. Paling RW, Waghela S, Macowan KJ, Heath BR. The occurrence of infectious diseases in mixed farming of domesticated wild herbivores and livestock in Kenya. Ii. Bacterial diseases 1. J Wildl Dis. 1988;24(2):308–16.
50. Pavlik I, Matlova L, Dvorska L, Shitaye JE, Parmova I. Mycobacterial infections in cattle and pigs caused by Mycobacterium avium complex members and atypical mycobacteria in the Czech Republic during 2000-2004. Vet Med (Praha). 2005;50(7):281.
51. Kinne J, Johnson B, Jahans KL, Smith NH, Ul-Haq A, Wernery U. Camel tuberculosis. A Case report. Trop Anim Health Prod. 2006;38(3):207–13.
52. Wernery U, Kinne J, Jahans KL, Vordermeier HM, Esfandiari J, Greenwald R, Johnson B, Ul-Haq A, Lyashchenko KP. Tuberculosis outbreak in a dromedary racing herd and rapid serological detection of infected camels. VetMicrobiol. 2007;122(1–2):108–15.
53. Kazwala RR, Daborn CJ, Kusiluka LJ, Jiwa SF, Sharp JM, Kambarage DM. Isolation of mycobacterium species from raw milk of pastoral cattle of the southern highlands of Tanzania. Trop Anim Health Prod. 1998;30(4):233–9.
54. Loecher M, Ropkins K. RgoogleMaps and loa: Unleashing R graphics power on map tiles. J Stat Softw. 2015;63:4.

Identification and characterization of *Staphylococcus devriesei* isolates from bovine intramammary infections in KwaZulu-Natal, South Africa

Tracy Schmidt[1,2*], Marleen M. Kock[2,3] and Marthie M. Ehlers[2,3]

Abstract

Background: Coagulase-negative staphylococci (CoNS) are among the leading bacterial causes of bovine mastitis in many dairy-producing countries. Among the challenges associated with the specific diagnosis of CoNS infections is the biochemical heterogeneity of the species in the genus and the unavailability of accurate, cost-effective and up-to-date diagnostic tests. A previous study investigating the diversity of CoNS associated with cases of bovine mastitis in South Africa, resulted in six CoNS isolates which could not be identified despite the use of a combination of different molecular assays. The identification and characterisation of the isolates was pursued further in this study.

Results: The six CoNS isolates in question were identified by sequencing multiple housekeeping genes (*dnaJ*, *hsp60*, *rpoB*, 16S rRNA) and characterized through the use of matrix-assisted laser/desorption ionization time of flight mass spectrometry (MALDI-TOF MS) and the Biolog GEN III Microplate™ bacterial identification system. Sequencing of housekeeping genes identified the isolates as *S. devriesei*. This *Staphylococcus* species was only described in 2010 and this is the first report documenting the isolation of *S. devriesei* from cases of bovine IMIs in South Africa. Analysis of mass spectra generated by the six isolates showed intra-species variation which was also observed when evaluating the metabolic profiles of the isolates using the Biolog GEN III system. Neither the MALDI-TOF MS nor the Biolog database are currently populated with data relating to *S. devriesei*, resulting in the isolates not being identified, in the case of MALDI-TOF MS analysis, or mis-identified as was observed with the Biolog GEN III system.

Conclusions: The phenotyping data collected during this investigation provides useful information concerning *Staphylococcus devriesei* which could be used to populate user system databases thereby ensuring the accurate identification of isolates in future. The availability of improved diagnostics will in turn facilitate studies to elucidate the epidemiology, pathogenicity and true prevalence of this species in dairy herds.

Keywords: *Staphylococcus devriesei*, Bovine mastitis, CoNS, MALDI-TOF MS, Biolog

* Correspondence: Tracy.Schmidt@kzndard.gov.za
[1]Allerton Provincial Veterinary Laboratory, KwaZulu-Natal Department of Agriculture and Rural Development, Private Bag X2, Cascades 3202, South Africa
[2]Department of Medical Microbiology, University of Pretoria, Pretoria 0001, South Africa
Full list of author information is available at the end of the article

Background

Coagulase-negative staphylococci (CoNS) are a common cause of bovine intramammary infections (IMIs) or mastitis, a disease condition which causes significant financial losses to the dairy industry worldwide [1, 2]. The importance of CoNS as mastitis pathogens has been disputed as infections caused by this group of bacteria are usually mild and remain subclinical [3]. However, this perception is changing as CoNS have emerged as one of the leading bacterial causes of IMIs worldwide, including South Africa [4–6]. Research emanating primarily from European countries, has provided evidence to suggest that some CoNS, for example *S. chromogenes*, are more specifically adapted to the bovine udder and are capable of causing more serious infections than some of the other species [7]. The importance of differentiating and identifying CoNS to species-level has become increasingly important but is currently limited by the availability of appropriate, cost-effective diagnostic tests [8].

The genus *Staphylococcus* currently comprises of 49 validly described species and 22 subspecies [9]. At least 24 different CoNS have been reported from bovine mastitis cases [2]. In order to phenotypically differentiate CoNS, a large number of biochemical tests need to be carried out which is both time consuming and costly [10, 11]. Manual and automated bacterial identification systems are commercially available to facilitate the identification of *Staphylococcus* species, however, many of these systems have been found to produce inaccurate results when testing veterinary isolates [11–13]. The reason for this is attributed to the limited number of veterinary strains evaluated and incorporated into the reference databases [14]. Variability in the expression of phenotypic characteristics and the occurrence of phenotypic variants may also lead to the misidentification of *Staphylococcus* species or subspecies [15–17]. Advances in the molecular field have seen the development and application of DNA-based methods for the differentiation and identification of staphylococcal species [10, 14, 18]. Since these methods do not rely upon gene expression they are not subject to the same pitfalls as phenotyping methods [14]. In general, DNA-based methods have been shown to have a higher discriminatory power and show greater reproducibility and typeability than conventional phenotyping methods [14]. The inclusion of low-quality DNA sequence data in databases may however present challenges [15, 19].

Amplification and sequencing of variable regions in the 16S rDNA gene is one of the most commonly applied approaches used to differentiate bacterial species [17, 20]. Some reports have suggested that partial 16S rDNA sequencing of staphylococcal isolates may not always be discriminatory enough at the subspecies level [10, 16, 20]. Other housekeeping gene targets that have been evaluated for their suitability in discriminating between *Staphylococcus* species and subspecies include: chaperone dnaJ (*dnaJ*) [21]; elongation factor Tu (*tuf*) [15], the 60 kDa heat shock protein (*hsp*60) [22, 23], glyceraldehyde-3-phosphate (*gap*) [24], RNA polymerase B (*rpo*B) [16, 25] and superoxide dismutase (*sod*A) [26].

An alternative, protein-based approach to species and subspecies-level bacterial identification which is gaining traction is matrix-assisted laser desorption/ionization time of flight spectroscopy (MALDI-TOF MS) [11]. This method examines the pattern of proteins directly from intact bacterial cells, generating a unique mass-spectral fingerprint for isolates [27]. Comparison with a spectral database facilitates the identification of isolates based on their cellular protein 'fingerprints' [28]. The ability of MALDI-TOF MS to correctly identify CoNS has been evaluated in both medical and veterinary laboratories and the technology has been found to provide accurate and highly reproducible results [29]. Furthermore, MALDI-TOF MS analysis is quick, technically easy to perform and cost-effective compared with many molecular identification techniques [30].

Our research group recently conducted an extensive study to characterize CoNS implicated in cases of bovine IMIs recovered from dairy herds across the province of KwaZulu-Natal (KZN), South Africa [31]. A combination of techniques including multiplex-PCR (M-PCR), MALDI-TOF MS and sequencing of the *tuf* gene were used to identify the CoNS isolates to species-level. Despite the combinatorial approach used, five CoNS isolates could not be identified. The purpose of this investigation was to characterize the isolates further and investigate the taxonomic standing of the isolates within the *Staphylococcus* genus. A sixth isolate, which showed only a 97% similarity to *S. haemolyticus* when comparing the *tuf* gene sequence with sequences in GenBank, was included with the five isolates.

Results

Phenotypic characterization of isolates

A summary of the preliminary phenotyping and molecular results of the six CoNS isolates is presented in Table 1. After 48 h of incubation, a small zone of clear hemolysis was evident around colonies of all the test isolates. Some variation in colony appearance was observed between the isolates. Colonies of isolates AN22, AN25, BN7 and FN2 were grey-yellow in color, whilst colonies of AN3 and AN19 were yellow-orange in appearance.

The phenotypic characterization of the test isolates using the Biolog GEN III MicroPlate™ identification system (Biolog, USA) was carried out in triplicate for each test isolate. The phenotyping results are summarized in Table 2. All isolates were uniform in their ability to utilize: L-alanine, acetoacetic acid, citric acid, dextrin,

Table 1 Preliminary characterization results for the coagulase-negative *Staphylococcus* isolates recovered from bovine intramammary infections

Isolate reference (farm of origin)	Phenotyping assays		Multiplex PCR 1				Species-level identification of CoNS isolates			Antimicrobial susceptibility testing											
	Coagulase production	Pastorex StaphPlus test	16S rRNA genus-specific target	*S. aureus nuc* gene	*mecA* gene	*luk* S/F gene	Bovine M-PCR[a]	MALDI-TOF MS (score)[b]	*tuf* gene sequencing	Penicillin (10 IU)	Ampicillin (10 µg)	Amoxicillin/ Clavulanate (20 µg / 10 µg)	Cefoxitin (30 µg)	Cephalothin (30 µg)	Tetracycline (30 µg)	Trimethoprim/ Sulfamethoxazole (1.25 µg / 23.75 µg)	Moxifloxacin (30 µg)	Clindamycin (2 µg)	Erythromycin (15 µg)	Gentamicin (10 µg)	Streptomycin (10 µg)
AN3 (commercial farm A)	(–)	(–)	(+)	(–)	(–)	(–)	(–)	No reliable identification	c	R	R	S	S	S	S	S	S	S	S	S	S
AN19 (commercial farm A)	(–)	(–)	(+)	(–)	(–)	(–)	(–)	No reliable identification	c	S	S	S	S	S	S	S	S	S	S	S	S
AN22 (commercial farm A)	(–)	(–)	(+)	(–)	(–)	(–)	(–)	No reliable identification	c	S	S	S	S	S	S	S	S	S	S	S	S
AN25 (commercial farm A)	(–)	(–)	(+)	(–)	(–)	(–)	(–)	No reliable identification	c	R	R	S	S	S	S	S	S	S	S	S	S
BN7 (commercial farm B)	(–)	(–)	(+)	(–)	(–)	(–)	(–)	*S. lugdunensis* (1.845 and 1.727)	c	R	R	S	S	S	S	S	S	S	S	S	S
FN2 (commercial farm F)	(–)	(–)	(+)	(–)	(–)	(–)	(–)	No reliable identification	d	S	S	S	S	S	S	S	S	S	S	S	S

[a]The bovine M-PCR assay comprised of primers specific for: *S. chromogenes*, *S. epidermidis*, *S. haemolyticus*, *S. simulans*, *S. warneri* and *S. xylosus*
[b]MALDI-TOF MS interpretation: 2.00 to 2.299 = secure genus identification, probably species identification; 1.70 to 1.99 = probable genus identification; < 1.7 = not a reliable identification
[c]Amplicon sequence showed 99–100% similarity with data in GenBank (EU571021)
[d]Amplicon showed 97% similarity to *S. haemolyticus*; (+) = positive; (–) = negative; S = sensitive; R = resistant

Table 2 Metabolic fingerprints for six *Staphylococcus devriesei* isolates obtained using the Biolog GENIII Microplate™ identification system

Assay		AN3	AN19	AN22	AN25	BN7	FN2
Carbon sources	Dextrin	+	+	+	+	+	+
	D-Maltose	+	+	+	+	+	+
	D-Trehalose	+	+	+	+	+	+
	D-Cellobiose	b	b	b	b	b	b
	Gentiobiose	+	+	b	b	b	+
	Sucrose	+	+	+	+	+	+
	D-Turanose	+	+	+	+	+	+
	Stachyose	b	b	b	b	(−)	b
	D-Raffinose	(−)	(−)	b	b	(−)	b
	α-D-Lactose	b	+	+	+	+	+
	D-Melibiose	b	b	b	b	b	b
	ß-Methyl-D-Glucoside	+	b	(−)	(−)	b	+
	D-Salicin	b	(−)	(−)	(−)	(−)	(−)
	N-Acetyl-D-Glucosamine	+	b	b	+	+	+
	N-Acetyl-β-D-Mannosamine	+	b	b	+	b	+
	N-Acetyl-D-Galactosamine	(−)	(−)	(−)	(−)	(−)	(−)
	N-Acetyl-Neuraminic Acid	+	+	+	+	+	+
	α-D-Glucose	+	+	+	+	+	+
	D-Mannose	b	b	b	b	b	+
	D-Fructose	+	+	+	+	+	+
	D-Galactose	+	+	+	+	+	+
	3-Methyl Glucose	b	b	b	b	b	b
	D-Fucose	b	b	b	b	b	b
	L-Fucose	b	b	b	b	b	b
	L-Rhamnose	b	b	b	b	b	b
	Inosine	b	+	[illegible]	+	+	+
	D-Sorbitol	b	b	b	b	b	b
	D-Mannitol	+	b	+	+	+	+
	D-Arabitol	+	+	b	b	+	+
	myo-Inositol	(−)	(−)	(−)	b	(−)	b
	Glycerol	+	+	+	+	+	+
	D-Glucose-6-PO4	b	b	b	b	b	b
	D-Fructose-6-PO4	b	b	b	b	b	b
	D-Aspartic acid	b	b	(−)	b	(−)	b
	D-Serine	(−)	(−)	(−)	(−)	(−)	b
	Gelatin	b	b	(−)	(−)	(−)	b
	Glycyl-L-Proline	(−)	+	b	(−)	b	b
	L-Alanine	+	+	+	+	+	+
	L-Arginine	+	+	(−)	(−)	(−)	+
	L-Aspartic Acid	b	b	(−)	b	(−)	+
	L-Glutamic Acid	+	+	b	b	[illegible]	+
	L-Histidine	(−)	(−)	(−)	(−)	(−)	(−)
	L-Pyroglutamic Acid	(−)	(−)	b	b	b	+

Table 2 Metabolic fingerprints for six *Staphylococcus devriesei* isolates obtained using the Biolog GENIII Microplate™ identification system *(Continued)*

Assay		AN3	AN19	AN22	AN25	BN7	FN2
	L-Serine	+	+	(−)	+	+	+
	Pectin	+	+	+	+	+	+
	D-Galacturonic Acid	b	b	b	b	b	b
	L-Galactonic Acid Lactone	b	b	b	b	b	b
	D-Gluconic Acid	+	+	+	+	+	+
	D-Glucuronic Acid	b	b	b	b	b	b
	Glucuronamide	b	b	b	b	b	b
	Mucic Acid	(−)	(−)	(−)	b	b	b
	Quinic Acid	(−)	(−)	(−)	(−)	(−)	b
	D-Saccharic Acid	b	b	(−)	b	b	b
	p-Hydroxy-Phenylacetic Acid	(−)	(−)	(−)	(−)	(−)	(−)
	Methyl Pyruvate	+	+	+	+	+	+
	D-Lactic Acid Methyl Ester	b	b	b	b	b	b
	L-Lactic Acid	+	+	+	+	+	+
	Citric Acid	+	+	+	+	+	+
	α-Keto-Glutaric Acid	(−)	(−)	b	b	b	b
	D-Malic Acid	(−)	(−)	(−)	(−)	(−)	(−)
	L-Malic Acid	b	b	+	+	+	+
	Bromo-Succinic Acid	(−)	(−)	b	(−)	(−)	b
	Tween 40	b	+	b	b	b	+
	γ-Amino-Butyric Acid	(−)	(−)	(−)	(−)	(−)	(−)
	α-Hydroxy-Butyric Acid	b	+	+	b	+	+
	β-Hydroxy-D,L-Butyric Acid	(−)	(−)	(−)	(−)	(−)	(−)
	α-Keto-Butyric Acid	+	b	+	b	b	b
	Acetoacetic Acid	+	+	+	+	+	+
	Propionic Acid	+	b	b	+	b	+
	Acetic Acid	+	+	b	+	b	+
	Formic Acid	+	(−)	+	+	+	+
Chemical sensitivity assays	pH 6	+	+	+	+	+	+
	pH 5	b	b	(−)	(−)	(−)	b
	1% NaCl	+	+	+	+	+	+
	4% NaCl	+	+	+	+	+	+
	8% NaCl	+	+	+	+	+	+
	1% Sodium Lactate	+	+	+	+	+	+
	Fusidic acid	(−)	(−)	(−)	(−)	(−)	(−)
	D-Serine	+	+	+	b	+	+
	Troleandomycin	(−)	(−)	(−)	(−)	(−)	(−)
	Rifamycin SV	(−)	(−)	(−)	(−)	(−)	(−)
	Minocycline	(−)	(−)	(−)	(−)	(−)	(−)
	Lincomycin	(−)	(−)	(−)	(−)	(−)	(−)
	Guanidine HCl	(−)	(−)	(−)	(−)	(−)	(−)
	Niaproof 4	(−)	(−)	(−)	(−)	(−)	(−)
	Vancomycin	(−)	(−)	(−)	(−)	(−)	(−)

Table 2 Metabolic fingerprints for six *Staphylococcus devriesei* isolates obtained using the Biolog GENIII Microplate™ identification system *(Continued)*

Assay		AN3	AN19	AN22	AN25	BN7	FN2
	Tetrazolium Violet	(−)	(−)	(−)	(−)	(−)	(−)
	Tetrazolium Blue	(−)	(−)	(−)	(−)	(−)	(−)
	Nalidixic Acid	+	+	+	+	+	+
	Lithium Chloride	+	+	+	+	+	+
	Potassium Tellurite	+	+	+	+	+	+
	Aztreonam	+	+	+	+	+	+
	Sodium Butyrate	+	+	+	+	+	+
	Sodium Bromate	b	b	(−)	b	b	b

+ = Positive; (−) = Negative; b = Borderline

D-fructose, D-galactose, D-gluconic acid, α-D-glucose, glycerol, L-lactic acid, D-maltose, methyl pyruvate, N-Acetyl-D-neuraminic acid, pectin, sucrose, D-trehalose and D-turanose.

The Biolog GEN III database identified isolates AN3, AN19, AN22 and BN7 as *S. haemolyticus*, with similarity indices (SIM) of 0.612 (0.529–0.720), 0.622 (0.524–0.674), 0.589 (0.582–0.596) and 0.66 (0.635–0.689). No identification was assigned to isolate AN25 after 48 h incubation. The closest match to AN25 was *S. haemolyticus* but the SIM value, 0.221 (0.171–0.267) was lower than the default cut-off of 0.5; precluding the assignment of an identification. Isolate FN2 was identified as *S. haemolyticus* (SIM = 0.525) on the first test whilst the subsequent two tests identified the isolate as *S. aureus* subsp. *aureus* (SIM = 0.646 and 0.682).

Sequence analysis of housekeeping genes

Amplification and sequencing of the 16S rDNA gene was carried out in duplicate for each of the test isolates. Isolates AN19, AN22 and AN25 showed 99.8 to 100% homology with sequences deposited for *S. devriesei* in the European Bioinformatics Institute database (http://www.ebi.ac.uk). Both attempts at sequencing a partial fragment of the 16S rRNA gene of AN3, BN7, FN2 resulted in low quality sequences. No further attempts were made to analyze this data.

The *rpo*B gene sequences obtained for the six isolates under evaluation showed 100% homology with *S. devriesei* sequences in the database. Similarly, partial *hsp*60 sequences for five of the isolates (AN3, AN22, AN19, AN25, BN7) showed 100% homology with sequences deposited for *S. devriesei* whilst the *hsp*60 sequence for isolate FN2 showed 99.4% homology with *S. devriesei* sequences. Partial *dnaJ* sequences for the test isolates showed between 99.4 and 100% similarity with *S. devriesei* sequences. The species-level cut-off values previously reported for the 16S rRNA, *dnaJ*, *hsp*60, *rpo*B and *tuf* genes are 98.7, 97, 97, 94 and 97% respectively [8].

Matrix-assisted laser desorption/ionization time of flight mass spectrometry analysis

The minimal spanning tree generated from the protein spectral fingerprints of the six CoNS isolates (Fig. 1) showed intra-species variation with isolates AN22 and AN25 clustering apart from the other four isolates. When the spectra were visualized in conjunction with the spectra of all *Staphylococcus* species contained in the MALDI biotyper database (not shown) the test isolates grouped together in a cluster comprising of *S. aureus*, *S. haemolyticus*, *S. hominis*, *S. hyicus* and *S. simulans*.

Discussion

The sequencing of multiple housekeeping genes unequivocally identified the six isolates as *S. devriesei*. This *Staphylococcus* species was described 8 years ago by Belgian researchers following the characterization of several isolates recovered from bovine teat apices as well as a case of bovine subclinical mastitis [8]. Following the original description of the species only sporadic reports can be found in the literature relating the isolation of *S. devriesei*. Most of the reports document the isolation of this coagulase-negative *Staphylococcus* species from bovine teat skin, milk or milk by-products [19, 32–34]. A single report documenting the isolation of *S. devriesei* from a foot ulcer of a diabetic patient was found [35]. To the best of our knowledge this is the first report documenting the occurrence of *S. devriesei* in South Africa.

The six *S. devriesei* isolates were part of a collection of 102 CoNS isolates recovered from bovine IMIs in a previous study [31] *Staphylococcus chromogenes* was the predominant species identified in the study comprising almost 80% of the CoNS isolates. The *S. devriesei* isolates represent the second most common *Staphylococcus* species identified in the study with a prevalence of 5.9% (6/102). The *S. devriesei* isolates were recovered from lactating cows in three different herds. In order to assess the true prevalence and significance of this species as a causative agent of bovine IMIs, further investigations will need to be undertaken.

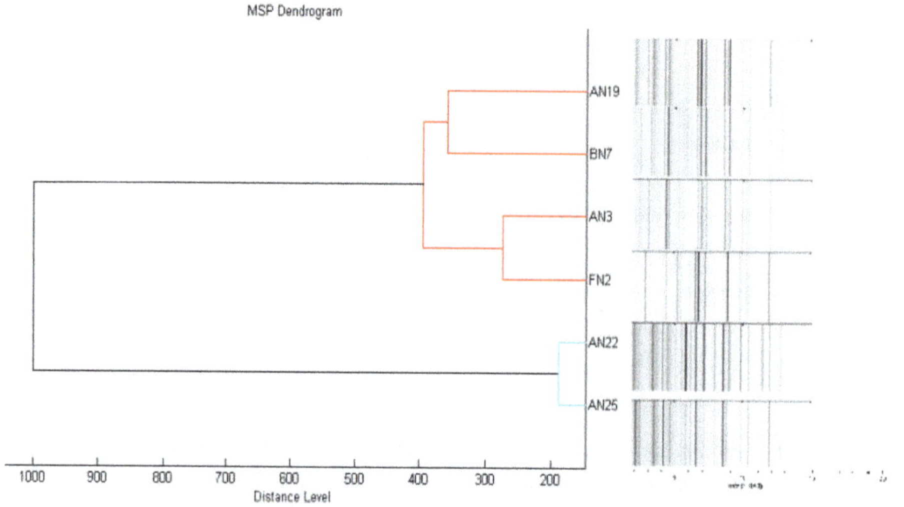

Fig. 1 Minimal spanning tree generated by MALDI biotyper 3.0 software showing the six *Staphylococcus devriesei* isolates under investigation. The tree was created by the standard MALDI biotyper MSP creation method where distance values are relative and are always normalized to a maximum value of 1.00

Preliminary phenotyping and molecular approaches used to identify the six CoNS, including a M-PCR assay, MALDI-TOF MS evaluation and *tuf* gene sequencing, failed to identify the isolates. Retrospectively, closer evaluation of the MALDI-TOF MS biotyper database (version 3.0) shows that there are no reference spectra for *S. devriesei*; accounting for the failure to achieve a successful identification. The current reference database comprises of only 38 *Staphylococcus* species with a number of the more recently described species not being represented including: *S. agnetis* (described in 2012), *S. argensis* (2015), *S. argenteus* (2015), *S. jettensis* (2013), *S. massiliensis* (2010), *S. microti* (2010), *S. petrasii* (2013), *S. rostri* (2010), *S. schweitzeri* (2015) and *S. stepanovicii* (2012) [9]. The accuracy of MALDI-TOF MS analysis, and essentially any commercially available phenotyping system, is wholly dependent upon the quality of the reference database used to match protein or phenotyping profiles. It is therefore imperative that regular updates are made to system databases to keep abreast with the description of new species as well as any changes in taxonomy.

Whilst MALDI-TOF MS was used in the initial study to identify the CoNS isolates, the technology was subsequently utilised to assess the relatedness of the six isolates as well as generate mass spectra which could be used to build up the local instrument database. Whilst all six isolates clustered together, sub-clusters were noted indicating intra-species variation with respect to cellular protein content. Unfortunately, with so few isolates available for evaluation it is difficult to make any inferences regarding the sub-clusters as no direct correlation can be seen between the clusters and the farm of origin or the biochemical profiles of the isolates.

To date, all reports documenting the isolation of *S. devriesei* have utilised genotyping methods to identify the bacterial isolates. The original description of the species by Supré and co-workers [8] provides the only information currently available concerning the phenotypic and biochemical diversity of this species. The six isolates characterized in this study conformed to the description provided by Supré and co-workers [8] in that, colonies of all isolates were surrounded by a zone of complete hemolysis. Furthermore, as noted by Supré and co-workers [8], variation with respect to the color of colonies was observed with some isolates presenting grey-yellow colonies whilst other isolates formed colonies which were yellow-orange in color.

The six *S. devriesei* isolates identified in this study were characterized using the Biolog GEN III MicroPlate™ (Biolog, USA) identification system to provide metabolic fingerprints for the field isolates and contribute to existing knowledge available concerning this species. In agreement with the *S. devriesei* type strain, KS-SP 60^T (=LMG 25332^T = CCUG 58238^T), all six of the field isolates in this study had the ability to utilize D-fructose, D-galactose, α-D-glucose, glycerol, maltose, sucrose, D-trehalose and turanose. Five of the six isolates evaluated also utilized D-mannitol and α-D-lactose with the sixth isolate giving a borderline test reaction. The inability to utilize L-histidine, D-malic acid, p-hydroxy-phenylacetic acid, β-hydroxy-D,L-butyric acid, D-salicin and D-serine were consistent among the test isolates in this study. However, apart from D-salicin, no comparisons can be made with the report by Supré and co-workers [8], as the other substrates listed were not reported. Supré and co-workers [8] utilized the API50CH, API Staph ID 32 (bioMérieux, France) and the Staph Zym

(Rosco, Denmark) phenotyping systems to characterize the original *S. devriesei* isolates. The profile of substrates incorporated into the aforementioned phenotyping systems are not the same as those included in the Biolog GEN III test system.

The Biolog GEN III database (version 2.3.1.404) currently comprises of data for only 35 of the 50 described *Staphylococcus* species. None of the *Staphylococcus* species described in the last decade, including *S. devriesei*, have been incorporated into the database so far. Consequently the assignment of a correct bacterial identification by the Biolog system was not expected for the six field isolates under evaluation. It is interesting to note however, that based upon the outcome of the carbon source utilization and chemical sensitivity assays, the Biolog database did, in fact, assign identifications to five of the six isolates. Most notably, four of the isolates were identified as *S. haemolyticus*. Phylogenetic studies [8, 10] have shown that *S. devriesei* is most closely related to *S. haemolyticus* and *S. hominis*. Consequently, the mis-identification of the isolates as *S. haemolyticus* is understandable. The similarity index values, which represent the level of agreement with the assigned identification, were low (0.589 to 0.66) indicating multiple mismatches with the profiles of the type strains in the database. However, since the SIM values were above the threshold of 0.5 used by the system, a positive identification was made. As was discussed for the MALDI-TOF MS system, the accuracy of any commercial bacterial identification system is underpinned by the quality of the database supporting the system. Failure to maintain current records, will negatively impact the accuracy and reliability of the system output. Diagnosticians and researchers should be cognisant of these shortcomings to avoid misidentification of bacterial isolates.

Conclusions

To the best of our knowledge this is the first report documenting the isolation and analysis of *S. devriesei* from bovine IMIs in South Africa. Further sampling is required to ascertain the true prevalence and relative importance of *S. devriesei* as a mastitis causing pathogen in local dairy herds. The characterization studies carried out in this investigation using both MALDI-TOF MS and the Biolog GEN III Microplate™ system, will contribute to the limited knowledge currently available concerning this bacterial species.

Methods

Bacterial isolates

Six CoNS isolates, cultured from sub-clinical cases of bovine mastitis from three commercial dairy farms in KZN, were isolated in a previous study [31]. Preliminary bacteriological testing of the isolates, including the examination of isolates for coagulase production, the presence of clumping factor, as well as antimicrobial susceptibility testing of the isolates is described elsewhere [31].

Phenotypic characterization of isolates

Isolates were streaked onto Columbia blood agar (Oxoid, England) supplemented with 5% sheep blood and the plates were incubated (Nuaire, USA) aerobically at 35 ± 1 °C. Bacterial colony morphology was documented after 24 and 48 h incubation. All six isolates were tested in triplicate using the Biolog GEN III MicroPlate™ identification system (Biolog, USA). Suspensions of the test isolates were prepared in inoculating fluid A (Biolog, USA) and the turbidity adjusted to between 90 and 98% transmittance using a turbidometer (Biolog, USA). One hundred microliters of the cell suspension was inoculated into each well of a GEN III microplate. This was repeated for each test suspension. The plates were incubated (Labotec, SA) at 33 °C. Microplates were read using the Biolog MicroStation™ after 24 and 48 h, and results were captured and analyzed by the system software (GEN III database, version 2.3.1.404).

PCR amplification and sequencing of housekeeping genes

Genomic DNA was prepared from overnight cultures using the QIAGEN® DNeasy blood and tissue kit (Germany) according to the manufacturer's instructions. Partial fragments of the 16S rRNA gene and three other housekeeping genes, *dnaJ*, *hsp60* and *rpoB* were amplified as previously described [8, 16, 21, 36]. Details of the primers and PCR amplification conditions are summarized in Table 3. Separate PCR reactions mixtures were prepared for the amplification of each gene target. Each reaction mixture comprised of: 1X PCR buffer (Promega, USA); 4 mM $MgCl_2$, (Promega, USA), 0.2 mM dNTPs (Promega, USA), 0.25 mM of each primer (Inqaba Biotechnical Industries, South Africa) and 1 U HotStart DNA polymerase (Promega, USA). Five microliters DNA template was added to each reaction mixture before amplification was carried out in a MJ Mini thermocycler (Bio-Rad, USA). Thermocycling conditions for all PCR assays are shown in Table 3.

The resulting PCR amplicons were electrophoresed in a 1.5% agarose gel (SeaKem® LE, Lonza, USA) with 1X Tris-borate-EDTA buffer (Melford, UK) at 100 V for 60 min. A 100 bp molecular weight marker (Thermo Scientific, USA) was included as a reference standard on all gels. Following staining in a 0.5 μg/mL ethidium bromide solution (10 mg/mL) (Bio-Rad, USA) gels were visualized under ultraviolet light (UVIPro, Uvitec, UK) and documented. The PCR amplicons were submitted to Inqaba Biotechnical Industries (South Africa) for sequencing. Trace data files were uploaded into CLC Genomics Workbench, version 6.0 (CLC, USA) and

Table 3 Amplification of the 16S rRNA, *dnaJ*, *hsp60* and *rpoB* genes of the coagulase-negative *Staphylococcus* isolates

Gene target	Primer name	Oligonucleotide sequence[a] (5′ – 3′)	PCR amplification conditions	Size of amplicon (bp)	Reference
16S rRNA	16F27 16R1522	AGAGTTTGATCCTGGCTCAG AAGGAGGTGATCCAGCCGCA	95 °C for 5 min, 30 cycles of [95 °C for 30 s, 58 °C for 60 s, 72 °C for 90 s] and a final extension step at 72 °C for 10 min	1500	[8]
dnaJ	SA-(F) SA-(R)	GCCAAAAGAGACTATTATGA ATTGYTTACCYGTTTGTGTACC	94 °C for 3 min, 5 cycles of [94 °C for 30 s, 45 °C for 30 s, 72 °C for 60 s], 30 cycles of [94 °C for 30 s, 50 °C for 30 s, 72 °C for 60 s], 72 °C for 3 min	920	[21]
hsp60	hsp60-F hsp60-R	CAIIIGCIGGIGAYGGIACIACIAC YKIYKITCICCRAAICCIGGIGCYTT	94 °C for 2 min, [94 °C for 60 s, TD[b] for 2 min, 72 °C for 5 min], 20 cycles of [94 °C for 60 s, 35 °C for 2 min, 72 °C for 5 min], 72 °C for 10 min	600	[36]
rpoB	1418f 3554r	CAATTCATGGACCAAGC CCGTCCCAAGTCATGAAAC	94 °C for 5 min; 30 cycles of [94 °C for 45 s, 52 °C for 60 s, 72 °C for 90 s]; 72 °C for 10 min	899	[16]

[a]All primers were synthesized by Inqaba Biotechnical Industries (South Africa)
[b]TD = Touchdown PCR from 45 °C to 35 °C with temperature decrease of 0.5 °C per cycle; *I* Inosine, *K* Keto (G or T), *Y* pyrimidine (C or T)

consensus sequences were generated. Consensus sequences were analyzed using the European Bioinformatics Institute website (http://www.ebi.ac.uk).

Matrix-assisted laser desorption/ionization time of flight mass spectrometry

Overnight bacterial cultures (< 24 h) were submitted to the Department of Microbiology and Plant Pathology (University of Pretoria) for MALDI-TOF MS analysis (Bruker Daltonik GmbH, Germany). An ethanol/formic acid extraction was used to prepare isolates before ten spots were prepared on a steel target plate (MSP 96 polished-steel target) (Bruker Daltonik GmbH, Germany). Following air-drying, the spots were overlaid with 1 μL α-cyano-4-hydroxy-cinnamic acid (HCCA) solution (Bruker HCCA matrix portioned, Art. #255344). Three spectral readings were captured from each spot before being evaluated manually and analyzed using the MALDI biotyper software (version 3.0). Two minimal spanning trees were generated to (i) depict the relationship between the six CoNS under investigation (Fig. 1) and (ii) evaluate the relatedness of the test isolates to other *Staphylococcus* isolates in the MALDI biotyper reference library version 3.0 (Bruker Daltonik GmbH, Germany) based on spectral patterns (not shown).

Abbreviations

CoNS: Coagulase-negative staphylococci; IMI: Intramammary infection; MALDI-TOF MS: Matrix-assisted laser/desorption ionization time of flight mass spectrometry; M-PCR: Multiplex polymerase chain reaction; SIM: Similarity indices

Acknowledgements

The authors would like to thank Mr. Michael Relihan and Ms. Archana Nunkumar for facilitating and assisting with the testing of isolates using the Biolog system. Ms. Amelita Lombard is also acknowledged for arranging and assisting with the MALDI-TOF MS work.

Funding

The research presented herein is funded, in part, by the University of Pretoria, National Health Laboratory Services, RESCOM, the National Research Foundation (NRF) Research Technology Fund and the KZN Department of Agriculture and Rural Development. The MALDI-TOF MS work is supported in part by the National Research Foundation (NRF) of South Africa (Grant specific unique reference number, UID 74426).

Authors' contributions

All authors contributed to the design of this study. TS carried out all laboratory analyses. TS, MMK and MME analyzed data and prepared the manuscript. All authors have read and approved the final manuscript.

Consent for publication

Not applicable.

Competing interests

The authors declare that they have no competing interests.

Author details

[1]Allerton Provincial Veterinary Laboratory, KwaZulu-Natal Department of Agriculture and Rural Development, Private Bag X2, Cascades 3202, South Africa. [2]Department of Medical Microbiology, University of Pretoria, Pretoria 0001, South Africa. [3]Tshwane Academic Division, National Health Laboratory Service, Pretoria 0001, South Africa.

References

1. Schukken YH, González RS, Tikofsky LL, Schulte HF, Santisteban CG, Welcome FL, Bennett GJ, Zurakowski MJ, Zadoks RN. CNS mastitis: nothing to worry about? Vet Microbiol. 2009;134:9–14.
2. Vanderhaeghen W, Piepers S, Leroy F, Van Coillie E, Haesebrouck F, De Vliegher S. Identification, typing, ecology and epidemiology of coagulase negative staphylococci associated with ruminants. Vet J. 2015;203:44–51.
3. Vanderhaeghen W, Piepers S, Leroy F, Van Coillie E, Haesebrouck F, De Vliegher S. Effect, persistence, and virulence of coagulase-negative *Staphylococcus* species associated with ruminant udder health. J Dairy Sci. 2014;97:5275–93.
4. Petzer I-M, Karzis J, Watermeyer JC, van der Schans TJ, van Reenen R. Trends in udder health and emerging mastitiogenic pathogens in south African dairy herds. J S Afr Vet. 2009;80:17–22.
5. Pyörälä S, Taponen S. Coagulase-negative staphylococci – emerging mastitis pathogens. Vet Microbiol. 2009;134:3–8.
6. De Vliegher S, Fox LK, Piepers S, McDougall S, Barkema HW. Mastitis in dairy heifers: nature of the disease, potential impact, prevention, and control. J Dairy Sci. 2012;95:1025–40.
7. De Visscher A, Supré K, Haesebrouck F, Zadoks RN, Piessens V, Van Coillie E, Piepers S, De Vliegher S. Further evidence for the existence of environmental and host-associated species of coagulase-negative staphylococci in dairy cattle. Vet Microbiol. 2014;172:466–74.
8. Supré K, Haesebrouck F, Zadoks RN, Vaneechoutte M, Piepers S, De Vliegher S. Some coagulase-negative *Staphylococcus* species affect udder health more than others. J Dairy Sci. 2010;94:2329–40.

9. Euzéby JP. List of prokaryotic names with standing in nomenclature – Genus *Staphylococcus*. 2015. Available from: http://www.bacterio.net/staphylococcus.html. [Accessed 14 Mar 2018].
10. Lamers RP, Muthukrishnan G, Castoe TA, Tafur S, Cole AM, Parkinson CL. Phylogenetic relationships among *Staphylococcus* species and refinement of cluster groups based on multilocus data. BMC Evol Biol. 2012;12:171.
11. Becker K, Heilmann C, Peters G. Coagulase-negative staphylococci. Clin Microbiol Rev 20014. 27:870–926.
12. Bes M, Guérin-Faublée V, Meugnier H, Etienne J, Freney J. Improvement of the identification of staphylococci isolated from bovine mammary infections using molecular methods. Vet Microbiol. 2000;71:287–94.
13. Sampimon OC, Zadoks RN, De Vliegher S, Supré K, Haesebrouck F, Barkema HW, Sol J, Lam TJGM. Performance of API staph ID 32 and staph-Zym for identification of coagulase-negative staphylococci isolated from bovine milk samples. Vet Microbiol. 2009;136:300–5.
14. Zadoks RN, Watts JL. Species identification of coagulase-negative staphylococci: genotyping is superior to phenotyping. Vet Microbiol. 2009; 134:20–8.
15. Heikens E, Fleer A, Paauw A, Florijn A, Fluit AC. Comparison of genotypic and phenotypic methods for species-level identification of clinical isolates of coagulase-negative staphylococci. J Clin Microbiol. 2005;43:2286–90.
16. Mellmann A, Becker K, von Eiff C, Keckevoet U, Schumann P, Harmsen D. Sequencing and staphylococci identification. Emerg Infect Dis. 2006;12:333–6.
17. Wragg P, Randall L, Whatmore AM. Comparison of biolog GEN III MicroStation semi-automated bacterial identification system with matrix-assisted laser desorption ionization-time of flight mass spectrometry and 16S ribosomal RNA gene sequencing for the identification of bacteria of veterinary interest. J Microbiol Methods. 2014;105:16–21.
18. Dubois D, Leyssene D, Chacornac JP, Kostrzewa M, Schmit PO, Talon R, Bonnet R, Delmas J. Identification of a variety of *Staphylococcus* species by matrix-assisted laser desorption ionization – time of flight mass spectrometry. J Clin Microbiol. 2010;48:941–5.
19. Lange CC, Brito MAVP, Reis DRL, Machado MA, Guimarães AS, Azevedo ALS, et al. Species-level identification of staphylococci isolated from bovine mastitis in Brazil using partial 16S rRNA sequencing. Vet Microbiol. 2015;176:382–8.
20. Ghebremedhin B, Layer F, König W, König B. Genetic classification and distinguishing of *Staphylococcus* species based on different partial *gap*, 16S rRNA, *hsp*60, *rpo*B, *sod*A, and *tuf* gene sequences. J Clin Microbiol. 2008;46: 1019–25.
21. Shah MM, Iihara H, Noda M, Song SX, Nhung PH, Ohkusu K, et al. *dnaJ* gene sequence-based assay for species identification and phylogenetic grouping in the genus *Staphylococcus*. Int J Syst Evol Microbiol. 2007;57:25–30.
22. Kwok AY, Su SC, Reynolds RP, Bay SJ, Av-Gay Y, Dovichi NJ, et al. Species identification and phylogenetic relationships based on partial HSP60 gene sequences within the genus *Staphylococcus*. Int J Syst Evol Microbiol. 1999; 49:1181–92.
23. Kwok AYC, Chow AW. Phylogenetic study of *Staphylococcus* and *Macrococcus* species based on partial *hsp*60 gene sequences. Int J Syst Evol Microbiol. 2003;53:87–92.
24. Park JY, Fox LK, Seo KS, McGuire MA, Park YH, Rurangirwa FR, et al. Comparison of phenotypic and genotypic methods for the species identification of coagulase-negative staphylococcal isolates from bovine intramammary infections. Vet Microbiol. 2011;147:142–8.
25. Drancourt M, Raoult D. *rpo*B gene sequence-based identification of *Staphylococcus* species. J Clin Microbiol. 2002;40:1333–8.
26. Poyart C, Quesne G, Boumaila G, Trieu-Cuot P. Rapid and accurate species-level identification of coagulase-negative staphylococci by using the *sod*A gene as a target. J Clin Microbiol. 2001;39:4296–301.
27. Wieser A, Schneider L, Jung J, Schubert S. MALDI-TOF MS in microbiological diagnostics – identification of microorganisms and beyond (mini review). App Microbiol Biotech. 2012;93:965–74.
28. Carbonnelle E, Mesquita C, Bille E, Day N, Dauphin B, Beretti J-L, et al. MALDI-TOF mass spectrometry tools for bacterial identification in clinical microbiology laboratory. Clin Biochem. 2011;44:104–9.
29. Barreiro JR, Ferreira CR, Sanvido GB, Kostrzewa M, Maier T, Wegemann B, et al. Identification of subclinical cow mastitis pathogens in milk by matrix-assisted laser desorption/ionization time-of-flight mass spectrometry. J Dairy Sci. 2010;93:5661–7.
30. Kadlec K, Wendlandt S, Feßler AT, Schwarz S. Methods for the detection of antimicrobial resistance and the characterization of *Staphylococcus aureus* isolates from food-producing animals and food of animal origin. In: Chen C-Y, Yan X, Jackson CR, editors. Antimicrobial resistance and food safety. London: Elsevier; 2015. p. 207–32.
31. Schmidt T, Kock MM, Ehlers MM. Diversity and antimicrobial susceptibility profiling of staphylococci isolated from bovine mastitis cases and close human contacts. J Dairy Sci. 2015;98:6256–569.
32. Verdier-Metz I, Gagne G, Bornes S, Monsallier F, Veisseire P, et al. Cow teat skin, a potential source of diverse microbial populations for cheese production. Appl Environ Microbiol. 2012;78:326–33.
33. Srednik ME, Grieben MA, Bentancor A, Gentilini ER. Molecular identification of coagulase-negative staphylococci isolated from bovine mastitis and detection of β-lactam resistance. J Infect Dev Ctries. 2015;9:1022–7.
34. De Visscher A, Piepers S, Haesebrouck F, De Vliegher S. Teat apex colonization with coagulase-negative *Staphylococcus* species before parturition: distribution and species-specific risk factors. J Dairy Sci. 2016;99: 1427–39.
35. Smith K, Collier A, Townsend EM, O'Donnell LE, Bal AM, Butcher J, et al. One step closer to understanding the role of bacteria in diabetic foot ulcers: characterising the microbiome of ulcers. BMC Microbiol. 2016;16:54.
36. Zadoks RN, Schukken YH, Wiedmann M. Multilocus sequence typing of *Streptococcus uberis* provides sensitive and epidemiologically relevant subtype information and reveals positive selection in the virulence gene *pau*A. J Clin Microbiol. 2005;43:2407–17.

PERMISSIONS

All chapters in this book were first published in VR, by BioMed Central; hereby published with permission under the Creative Commons Attribution License or equivalent. Every chapter published in this book has been scrutinized by our experts. Their significance has been extensively debated. The topics covered herein carry significant findings which will fuel the growth of the discipline. They may even be implemented as practical applications or may be referred to as a beginning point for another development.

The contributors of this book come from diverse backgrounds, making this book a truly international effort. This book will bring forth new frontiers with its revolutionizing research information and detailed analysis of the nascent developments around the world.

We would like to thank all the contributing authors for lending their expertise to make the book truly unique. They have played a crucial role in the development of this book. Without their invaluable contributions this book wouldn't have been possible. They have made vital efforts to compile up to date information on the varied aspects of this subject to make this book a valuable addition to the collection of many professionals and students.

This book was conceptualized with the vision of imparting up-to-date information and advanced data in this field. To ensure the same, a matchless editorial board was set up. Every individual on the board went through rigorous rounds of assessment to prove their worth. After which they invested a large part of their time researching and compiling the most relevant data for our readers.

The editorial board has been involved in producing this book since its inception. They have spent rigorous hours researching and exploring the diverse topics which have resulted in the successful publishing of this book. They have passed on their knowledge of decades through this book. To expedite this challenging task, the publisher supported the team at every step. A small team of assistant editors was also appointed to further simplify the editing procedure and attain best results for the readers.

Apart from the editorial board, the designing team has also invested a significant amount of their time in understanding the subject and creating the most relevant covers. They scrutinized every image to scout for the most suitable representation of the subject and create an appropriate cover for the book.

The publishing team has been an ardent support to the editorial, designing and production team. Their endless efforts to recruit the best for this project, has resulted in the accomplishment of this book. They are a veteran in the field of academics and their pool of knowledge is as vast as their experience in printing. Their expertise and guidance has proved useful at every step. Their uncompromising quality standards have made this book an exceptional effort. Their encouragement from time to time has been an inspiration for everyone.

The publisher and the editorial board hope that this book will prove to be a valuable piece of knowledge for researchers, students, practitioners and scholars across the globe.

LIST OF CONTRIBUTORS

Víctor Rodríguez-Prieto, Deborah Kukielka, Belén Rivera-Arroyo, Ana Isabel de las Heras and José Manuel Sánchez-Vizcaíno
VISAVET, Veterinary School, Complutense University of Madrid, Puerta de Hierro s/n, 28040, Madrid, Spain

Beatriz Martínez- López
VISAVET, Veterinary School, Complutense University of Madrid, Puerta de Hierro s/n, 28040, Madrid, Spain
SaBio (Sanidad y Biotecnología) research group, National Wildlife Research Institute IREC (CSIC - Universidad de Castilla-La Mancha), Ronda de Toledo s/n, 13071, Ciudad Real, Spain

Joaquín Vicente
SaBio (Sanidad y Biotecnología) research group, National Wildlife Research Institute IREC (CSIC - Universidad de Castilla-La Mancha), Ronda de Toledo s/n, 13071, Ciudad Real, Spain

Chiara Nogarol, Luigi Bertolotti, Barbara Colitti, Bryan Iotti, Giovanni Perona and Margherita Profiti
Department of Veterinary Science, University of Turin, Turin, Italy

Sergio Rosati
Department of Veterinary Science, University of Turin, Turin, Italy Largo Paolo Braccini, 2, 10095 Grugliasco, Italy

Nicola Decaro, Maria Stella Lucente, Gabriella Elia and Canio Buonavoglia
Department of Veterinary Medicine, University of Bari, Bari, Italy

Stefano Petrini
Istituto Zooprofilattico Sperimentale dell'Umbria e delle Marche, Perugia, Italy

Martin Faldyna
Veterinary Research Institute, Hudcova 296/70, Brno 621 00, Czech Republic

Nada Zemankova, Katarina Chlebova, Jan Matiasovic, Jana Prodelalova and Jan Gebauer
Veterinary Research Institute, Hudcova 296/70, Brno 621 00, Czech Republic
Faculty of Science, Masaryk University, Brno, Czech Republic

Jan B. W. J. Cornelissen, Freddy M. de Bree, Fimme J. van der Wal, Engbert A. Kooi, Miriam G. J. Koene, Alex Bossers, Bregtje Smid, Adriaan F. Antonis and Henk J. Wisselink
Department of Infection Biology, Central Veterinary Institute of Wageningen UR, AB, Lelystad, The Netherlands

J. Berman, D. Francoz, J. Dubuc and S. Buczinski
Département des sciences cliniques, Faculté de médecine vétérinaire, Université de Montréal, Saint-Hyacinthe, QC, J2S 7C6, Canada

Evy Goossens, Stefanie Verherstraeten, Leen Timbermont, Freddy Haesebrouck, Richard Ducatelle and Filip Van Immerseel
Department of Pathology, Bacteriology and Avian Diseases, Faculty of Veterinary Medicine, Ghent University, Salisburylaan 133, B-9820 Merelbeke, Belgium

Piet R. Deprez, Bonnie R. Valgaeren and Bart Pardon
Department of Internal Medicine and Clinical Biology of Large Animals, Faculty of Veterinary Medicine, Ghent University, Salisburylaan 133, B-9820 Merelbeke, Belgium

Stijn Schauvliege and Diego Rodrigo-Mocholí
Department of Surgery and Anesthesia of Domestic Animals, Faculty of Veterinary Medicine, Ghent University, Salisburylaan 133, B-9820 Merelbeke, Belgium

Leandro Tana-Hernández and Katherine Navarrete-Arroyo
Grupo de Investigación en Sanidad Animal y Humana (GISAH), Carrera Ingeniería en Biotecnología, Departamento de Ciencias de la Vida y la Agricultura, Universidad de las Fuerzas Armadas ESPE, Sangolquí, Ecuador

Armando Reyna-Bello
Grupo de Investigación en Sanidad Animal y Humana (GISAH), Carrera Ingeniería en Biotecnología, Departamento de Ciencias de la Vida y la Agricultura, Universidad de las Fuerzas Armadas ESPE, Sangolquí, Ecuador
Investigador-Prometeo, Departamento de Ciencias de la Vida, Carrera Ingeniería en Biotecnología, Universidad de las Fuerzas Armadas ESPE, Sangolquí, Ecuador

María Augusta Chávez-Larrea
Grupo de Investigación en Sanidad Animal y Humana (GISAH), Carrera Ingeniería en Biotecnología, Departamento de Ciencias de la Vida y la Agricultura, Universidad de las Fuerzas Armadas ESPE, Sangolquí, Ecuador
Carrera de Ingeniería en Biotecnología, Departamento de Ciencias de la Vida y la Agricultura, Universidad de las Fuerzas Armadas - ESPE, Av. General Rumiñahui s/n, Sangolquí, Ecuador

Jorge Ron-Román
Grupo de Investigación en Sanidad Animal y Humana (GISAH), Carrera Ingeniería Agropecuaria, Departamento de Ciencias de la Vida y la Agricultura, Universidad de las Fuerzas Armadas ESPE, Sangolquí, Ecuador

W. Bahir, O. Omar and G. Ziay
Central Veterinary and Diagnostic Research Laboratories, Kabul, Afghanistan

R. S. Rosales, M. Hlusek and A. M. Whatmore
Animal and Plant Health Agency, Woodham Lane, Addlestone, Surrey KT15

W. Schauwers
NB, UK. 3Landell Mills, Trowbridge, Wilts BA14 8HE, UK

R. A. J. Nicholas
Consultant, The Oaks, Nutshell Lane, Farnham, Surrey GU9 0HG, UK

Peili Hou, Guimin Zhao, Chengqiang He, Hongmei Wang and Hongbin He
Key Laboratory of Animal Resistant Biology of Shandong, Ruminant Disease Research Center, College of Life Sciences, Shandong Normal University, No. 88 East Wenhua Road, Jinan City, Shandong Province, China

Rahmeto Abebe, Mesele Abera, Bekele Megersa and Kassahun Asmare
School of Veterinary Medicine, Hawassa University, Hawassa, Ethiopia

Hagere Hatiya
Animal and Fisheries Resources Development Bureau of Southern Nations, Nationalities and Peoples' Regional State, Hawassa, Ethiopia

Ziwei Li, Qiumei You, Philip Mead, Margaret Quinton and Niel A. Karrow
Department of Animal Biosciences, University of Guelph, Guelph N1G 2W1, ON, Canada

Faisury Ossa
Lallemand Inc., Montréal H4P 2R2, QC, Canada

Veslemøy Sunniva Oma, Thea Klem and Maria Stokstad
Life Sciences, Dep, 0033 Oslo, Norway

Madeleine Tråvén and Stefan Alenius
Department of Clinical Sciences, Swedish University of Agricultural Sciences, 75007 Uppsala, Sweden

Britt Gjerset
Norwegian Veterinary Institute, Sentrum, 0106 Oslo, Norway

Mette Myrmel
Department of Food Safety and Infection Biology, Norwegian University of Life Sciences, Dep, 0033 Oslo, Norway

V. Kaiser, L. Nebel, R. G. Zanoni and M. Schweizer
Institute of Virology and Immunology, Federal Food Safety and Veterinary Office (FSVO) and Vetsuisse Faculty, University of Bern, Laenggass-Strasse 122, POB, CH-3001 Bern, Switzerland

G. Schüpbach-Regula
Veterinary Public Health Institute, Vetsuisse Faculty, University of Bern, Schwarzenburgstrasse 155, CH-3097 Liebefeld, Switzerland

Li Mao, Wenliang Li, Leilei Yang, Wenwen Zhang, Fei Hao and Jieyuan Jiang
Institute of Veterinary Medicine, Jiangsu Academy of Agricultural Sciences; Key Laboratory of Veterinary Biological Engineering and Technology, Ministry of Agriculture, National Center for Engineering Research of Veterinary Bio-products, Nanjing 210014, China

Yong Wei and Jianhui Wang
Suining Animal Husbandry and Veterinary station, Suining 221200, China

Yonglong Ding, Yinhua Sun, Suping Cheng and Qiusheng Wang
Hai'an Animal Husbandry and Veterinary Station, Hai'an 226600, China

Henk J. Wisselink, Jan B. W. J. Cornelissen, Fimme J. van der Wal, Engbert A. Kooi, Miriam G. Koene, Alex Bossers, Bregtje Smid, Freddy M. de Bree and Adriaan F. G. Antonis
Wageningen Bioveterinary Research, AB, Lelystad, The Netherlands

Frederike Reichmann, Annette Pfitzner and Guenter Rademacher
Clinic for Ruminants with Ambulatory and Herd Health Services at the Centre for Clinical Veterinary Medicine, LMU Munich, Sonnenstrasse 16, 85764 Oberschleissheim, Germany

Carola M. Sauter-Louis
Clinic for Ruminants with Ambulatory and Herd Health Services at the Centre for Clinical Veterinary Medicine, LMU Munich, Sonnenstrasse 16, 85764 Oberschleissheim, Germany
Institute of Epidemiology, Friedrich-Loeffler-Institute, Suedufer 10, 17493 Greifswald, Isle of Riems, Germany

Elke Schwedinger and Klaus Cussler
Paul Ehrlich-Institut, Paul-Ehrlich-Straße 51-59, 63225 Langen, Germany

Gerard M. Murray
Marine, Doonally, Sligo, Ireland

Simon J. More and Tracy A. Clegg
Centre for Veterinary Epidemiology and Risk Analysis, UCD School of Veterinary Medicine, University College Dublin, Belfield Dublin 4, Ireland

Bernadette Earley and Dayle Johnston
Animal and Bioscience Research Department, Animal & Grassland Research and Innovation Centre, Teagasc, Grange, Dunsany, Co. Meath, Ireland

Máire C. McElroy and Rónan G. O'Neill
Central Veterinary Research Laboratory, Department of Agriculture, Food and Marine, Backweston, Celbridge, Co. Kildare, Ireland

John Gilmore and Mikhail Nosov
Farmlab Diagnostics, Emlagh Lodge, Elphin, Co. Roscommon, Ireland

Thomas J. Inzana
Department of Biomedical Sciences & Pathobiology, Virginia-Maryland College of Veterinary Medicine, Virginia Tech, Blacksburg, VA, USA

Joseph P. Cassidy
School of Veterinary Medicine, University College Dublin, Belfield Dublin 4, Ireland

Paulina Jawor and Tadeusz Stefaniak
Department of Immunology, Pathophysiology and Veterinary Preventive Medicine, Wroclaw University of Environmental and Life Sciences, 31 C.K. Norwida, 50-375 Wrocław, Poland

John F. Mee
Animal and Bioscience Research Department, Teagasc, Moorepark Research Centre, Fermoy, Co. Cork, Ireland

Yao-Chi Chung, Li-Ting Cheng, Jia-Yu Zhang, Yue-Jyun Wu and Chun-Yen Chu
Graduate Institute of Animal Vaccine Technology, College of Veterinary Medicine, National Pingtung University of Science and Technology, 1, Shuehfu Road, Neipu, Pingtung 91201, Taiwan

Shyh-Shyan Liu
Department of Veterinary Medicine, College of Veterinary Medicine, National Pingtung University of Science and Technology, 1, Shuehfu Road, Neipu, Pingtung 91201, Taiwan

Benjamin Oehme, S. Grund, K. Hainke, J. Munzel and C. K. W. Mülling
Institute of Veterinary Anatomy, Histology and Embryology, Faculty of Veterinary Medicine, Leipzig University, Leipzig, Germany

S. M. Geiger
Institute of Veterinary Anatomy, Histology and Embryology, Faculty of Veterinary Medicine, Leipzig University, Leipzig, Germany
Institute of Topographic Anatomy, University of Veterinary Medicine Vienna, Vienna, Austria

Colin P. D. Birch, Ashley Goddard and Oliver Tearne
Animal and Plant Health Agency, New Haw, Addlestone, Surrey KT15 3NB, UK

Michael K. Ghebremariam
Department of Infectious Diseases and Immunology, Faculty of Veterinary Medicine, Utrecht University, Utrecht, The Netherlands
Department of Veterinary Sciences, Hamelmalo Agricultural College, Keren, Eritrea

V. P. M. G. Rutten
Department of Infectious Diseases and Immunology, Faculty of Veterinary Medicine, Utrecht University, Utrecht, The Netherlands
Department of Veterinary Tropical Diseases, Bovine Tuberculosis and Brucellosis Research Programme, Faculty of Veterinary Science, University of Pretoria, Pretoria, South Africa

A. L. Michel
Department of Veterinary Tropical Diseases, Bovine Tuberculosis and Brucellosis Research Programme, Faculty of Veterinary Science, University of Pretoria, Pretoria, South Africa
Research Associate at the National Zoological Gardens of South Africa, Pretoria, South Africa

J. C. M. Vernooij
Department of Farm Animal Health, Faculty of Veterinary Medicine, Utrecht University, Utrecht, The Netherlands

Tracy Schmidt
Allerton Provincial Veterinary Laboratory, KwaZulu-Natal Department of Agriculture and Rural Development, Cascades 3202, South Africa
Department of Medical Microbiology, University of Pretoria, Pretoria 0001, South Africa

Marleen M. Kock and Marthie M. Ehlers
Department of Medical Microbiology, University of Pretoria, Pretoria 0001, South Africa
Tshwane Academic Division, National Health Laboratory Service, Pretoria 0001, South Africa

Index

www.ingramcontent.com/pod-product-compliance
Lightning Source LLC
LaVergne TN
LVHW080714210926
848485LV00003B/463